GEOGRAPHICAL INFORMATION SYSTEM
Concepts and Business Oportunities

Also Available

- — Census Mapping Survey for International Geographical Union Commission on Population Geography/ **P. Nag** *(Ed.)*
- — Digital Remote Sensing/ **P. Nag** and **M. Kudrat**
- — Geography and Environment (In 3 Volumes)/ **P. Nag, V.K. Kumra** and **Jagdish Singh** *(Eds.)*
- — Geography of India/ **P. Nag** and **Smita Sengupta**
- — Introduction to Geographical Information Systems/ **P. Nag** and **Smita Sengupta**
- — Population, Settlement and Development in Zambia/ **P. Nag**
- — Thematic Cartography and Remote Sensing/ **P. Nag**
- — Tourism and Trekking in Nainital Region/ **S.K. Singh** and **P. Nag**

About the Authors

Prithvish Nag, (*b.* 1951), M.Sc. (Gold Medalist), Ph.D., is the Director of the National Atlas and Thematic Mapping Organisation (NATMO) of the Ministry of Science and Technology, Government of India. He has been the President of the Indian National Cartographic Association, Indian Association of Special Libraries and Information Centre, Institute of Indian Geographers, Indian Institute of Geomorphologists, Professor S.P. Chatterjee Memorial Foundation and Indian Geographical Foundation. At the international level, he was the Chairman of the International Cartographic Association (ICA), Commission on Population Cartography, UN/PCGIAP Committee on Institutional Strengthening for Asia and the Pacific, International Steering Committee on Global Mapping (ISCGM) Working Group on Data Standardization; and Full Member of the International Geographical Union (IGU) Commission on Population Geography, and Member, Editorial Board, International Journal on Geographical Information System.

Dr. Nag was the Visiting Member of the Staff of the School of Oriental and African Studies, London. He is an author and editor of over 35 books and 110 research papers. His well known books are *Geography of India* and *Digital Remote Sensing*.

Dr. Nag was UN Consultant in Cartography in the Sultanate of Oman and the Surveyor General of India.

Smita Sengupta, (*b.* 1956), M.A., Ph.D., had a varied experience of working with universities, institute, government and the GIS industry. At present she is working as GIS Consultant. Dr. Sengupta is a recipient of the UGC and CSIR fellowships. She has brought out books on *Residential Patterns of Suburbs, Geography of India* and *Environment, Population and Development*, apart from over 35 research papers in well known journals.

GEOGRAPHICAL INFORMATION SYSTEM

Concepts and Business Oportunities

Prithvish Nag

and

Smita Sengupta

Concept Publishing Company, New Delhi-110059

Cataloging in Publication Data—DK
Courtesy: D.K. Agencies (P) Ltd. <docinfo@dkagencies.com>

Nag, P. (Prithvish), 1951-
Geographical information system : concepts and businessopportunities /
Prithvish Nag and Smita Sengupta.
p. cm.
In Indian context.
Includes bibliographical references (p.)
Includes index.
ISBN 8180694356

1. Geographic information systems—India. 2. Geographic information systems—Economic aspects—India. I. Sengupta, Smita, 1956-, joint author. II. Title.

DDC 910.285 22

ISBN 81-8069-435-6 (HB)
ISBN 81-8069-436-4 (PB)

First Published 2007

Published and Printed by
Ashok Kumar Mittal
Concept Publishing Company
A/15-16, Commercial Block, Mohan Garden
New Delhi-110059 (INDIA)

Phones : 25351460, 25351794
Fax : 091-11-25357103
Email : publishing@conceptpub.com

Dedicated to
Jack Dangermond
Who Promoted
GIS World Over

Preface

After publishing the book on *Digital Remote Sensing*, my publishers were keen on bringing out another book, and this time it was on the Geographical Information System. They were pursing me to do so since 1998. But due to my hectic activities and involvement in several scientific and professional activities, I was not able to concentrate on this volume. Nevertheless, I managed to plan the whole scheme and was successful in convincing my co-author to work with me again. Earlier, we brought out books on *the Geography of India* and *Environment, Population and Development*. But the problems were of different type. Even with fast means of communications available now, it was not always possible to interact easily. I was based in Kolkata and my co-author in Mumbai. Later, my movement to Dehra Dun did not make our life easy. I returned to Kolkata with a determination to complete the pending work. Hence this book.

In the beginning of eighties I came to know about the power of computers and start thinking of its application, particularly in thematic mapping where information from a variety of sources has to be processed for a meaningful topic. Initially my exposure to the digital technologies in the Department of Geography, University of Durham, United Kingdom in 1982 gave a big boost. Professor John I. Clarke and Professor Michael Blakemore were extremely helpful in this regard. Following this, I wrote a paper for the Auto-Carto London 1984 which was accepted but I could not go there for presentation. Nevertheless, it attracted the attention of Professor J.T. Coppock of the University of Edinburgh. He not only approached me to get this paper published in the then newly brought out *International Journal on Geographical Information System* but also to accept the Membership of its Editorial Board. My paper entitled *A proposal for the development of a geographical information system for India* was published in the first volume of this journal. A few years later I joined the Indian Space Research Organisation as Scientist/Engineer. There I was involved in several committees to assess the GIS requirements and to analyze GIS software for various activities of ISRO. I soon returned to the National Atlas and Thematic Mapping Organisation or NATMO and was entrusted with similar activities.

Meanwhile, my co-author, Dr Smita Sengupta, after completing her Ph.D., was on a look out of a thrust area which is exciting. I suggested her to explore GIS related topic for post-doctoral work. Her proposal was accepted by the Council for Scientific and Industrial Research, Government of India. She continued her interest in this subject even after shifting to Indian Institute of Technology, Mumbai and other GIS related industries in the same city. Due to her exposure of the potentials of GIS in industry, we were able to contribute on business opportunities and applications in the realm of GIS.

I was greatly benefited by the experience I gained in carrying out a project of the Technology, Information, Forecasting and Assessment Council (TIFAC) on Government-Industry Co-operation in Cartography in 1997. With this survey I could get an insight of the status of the GIS related activities in the government institutions, and the industry and how the combination works in the country. The support of Shri Y.S. Rajan, then Executive Director of TIFAC was extremely beneficial. Around the same time, I had close connection with several professional bodies related to geography and cartography. These associations gave me an opportunity to know the status of GIS related activities in the universities, institutions and collages. Recently my association with Survey of India and the National Spatial Data Infrastructure allowed me to look into the issue entirely from a different angle, i.e. availability of basic data for the development of GIS. I am sure that the new National Map Policy and acceptance of the NSDI by the government will give a boost to GIS activity in the country.

This book is our combined effort to promote GIS technology in India. The language has been made simple as possible in order to make the subject easily understandable. Some may find it useful as a text book, however, the intension was never to do so. We did our best to get Indian examples of GIS applications in business. We went through the company profiles, their presentations in different conferences and web sites. But we were only partially successful. Firstly, the Indian companies hardly take up local projects; and if they do so, it is difficult to get the details. Hence, a lot of examples have been taken from abroad. Perhaps this will help in exploring the potential of GIS in different businesses. For this purpose, a world wide search was launched through the web, which was time consuming and cumbersome as well. By the time we could finalize the manuscript, many web sites were withdrawn. However, we have included the case studies here. Nevertheless, after going through the book, one could gather that the scope of GIS is mind boggling. Another objective of this book is also to promote entrepreneurship in the field of GIS, not only of the projects in the country but also of abroad. With our success in Business Process Out-sourcing (BPO), which basically deals with non-spatial data, there is no reason why India can become hub for GIS BPO as well. The man-power, technology and infrastructure are getting ready to take up such a challenge as well. We recently brought out a book focusing on this subject. This book is a modified version prepared to meet the requirements of the students' community of the country.

This book could not have been brought out without the support of our friends and colleagues in the government, industries and institutes. My association with the Department of Science & Technology, National Atlas & Thematic Mapping Organisation, Survey of India, different institutions under the Department of Space and ISRO including NNRMS, Indian National Cartographic Association; and my coauthor's association with IIT and industry is worth mentioning. We are extremely grateful to Shri A.K. Malik, Dr A.K. Dasgupta, Dr. G.C. Debnath, Shri A.J. Kurrian, Dr. R. Siva Kumar, Dr. P.S. Acharya, Dr. P. Venketachalam, Brig Girish Kumar, Dr. B. Nagarajan, Shri H.B.Madhwal, Dr. Devi Dutt Chuniyal, Dr. M.Kudrat, Dr. N.C. Gautham, Dr. K.V. Sundarum, Shri Sanjay Kumar, Shri Ayon Tarafdar and Dr. A. Perumal.

Our special thanks are to Dr. T. Ramasami, Secretary, Professor V.S. Ramamurthy, former Secretary of the Department of Science and Technology for their constant encouragement. Dr. K. Kasturirangan, Hon'ble Member of Parliament and former Chairman, ISRO has been constantly in touch with our scientific and academic pursuits. We are grateful to him as well.

Shri Ashok Kumar Mittal of the Concept Publishing Company, New Delhi has always supported us in our endeavors and this is no exception. Unfortunately we could not bring out this book during the life time of his father, Shri Naurang Rai. However, we would like to dedicate this book to him.

Kolkata **Prithvish Nag**
19 October 2006

Content

List of Figures

List of Tables

1

Introduction

Tradition of map-making is as old as human civilization. Geographical knowledge traces its historical background from very beginning of civilization. Knowledge of maps as a science grew with time and religious thinkers also started delving into its mysteries. In the ancient literature, there are descriptions of the territories or *Bharatavarsha* or *Jambu-Dvipa* or different regions of India, named as *janapadas, khanda,* and *dvipas*. The descriptions are fairly correct and have relevance even today, particularly to understand India as a concept. There is an ancient Indian concept of the spherical world with the *Jambu-dvipa* (*i.e.* India) as the central part and Mount Meru (Kailas) as the centre of the world. In *Vedas, Puranas,* Buddhist and Jain literature, there are references about time, longitudes, latitudes, rhumb lines and the like. In *Puranas* (500 BC to 700 AD), particularly *Brahman Purana* the concept of map scale, and generalization of geographical features were realized. Preservation of collection of maps in the form of atlases has been mentioned in *Padma Purana.*

Man's first conception of the earth's figure was a plane. Greek philosopher thought it a sphere. Later, Sir Issac Newton showed it must be spheroid. However, development of Science of Surveying and Mapping of earth's surface in an orderly manner is ascribed to Eratosthenes (276-198 BC) who based the compilation of his comprehensive map of the world upon his famous determination of the circumference of the earth. Claudius Ptolemy, a mathematician, astronomer and geographer produced a map of India in the 2nd century AD which was subsequently made use of and commended in 18th century. The golden age of Indian renaissance started when in 5th century, genius Aryabhat who calculated the earth circumference to be 25,080 miles. The next contributors to Indian geography were Chinese and Arab travellers who left the interesting detailed accounts of their journey.

In Kautilya's *Arthasastra* (4th century BC), the significance of village boundaries and methods for land measurements have been described. The accounts given by him about the *Siva Purana* and Panini's *sutras* indicate the importance given to geodesy, cadastral surveys, plans and sketch maps for army movements.

References to regions and places with the help of astronomical calculations were included in traditional Indian system. Ujjain was considered the oldest prime meridian much before Ptolemy. The main source of traditional Indian cartography is *Sata Patha Brahmana* which gives account of the concept of earth in the 5th and 6th centuries. Aryabhat's astronomical contributions are well known, particularly in calculating the diameter of the earth, its axial rotation, position of planets in space, and planetary orbits in 5th century AD. The land surveying manuals were part of *Sulba Sutras* (science of mensuration). In Vedic texts, apart from the description of physical features like mountains, rivers and plants, there are *sutras* for trigonometry, co-ordinate and solid geometry, analytical conics, calculus (differential and integral), equation of straight lines and even Pythagoras theorem. Some of the above *sutras* of Vedic mathematics have relevance in modern surveying and mapping as well. It is said that *trikonamiti* is what we understand in the form of 'triangulation' today. This forms the basis of modern map-making, including geodesy.

The rich Arab traditions and their knowledge of stars and related information, the skills of Portuguese cartographers and experiences from Egypt, China and Japan (particularly calligraphy and artistic designing), Mesopotamia (Iraq), and other colonial powers contributed to the Indian cartographic traditions as well. Navigational charts, route charts for land and sea were commonly used in this part of the world. Compass lines and related navigational details were included in such charts. Nevertheless, the Indian knowledge of sea, oceans, other neighbouring countries and coasts, navigation skills etc., was excellent. *Markandeya* and *Bhagavata Puranas* have references to sea-borne trade and commerce. In the *Rigveda* and *Satapatha Brahmana,* there are references to navigation in *purva* and *paschim samundra* (Bay of Bengal and Arabian Sea). Raghuvamsa records the presence of a strong naval force. The Cholas, Oriyas, Marathas and Gujaratis used their own navigational charts. Indians had direct port to port established routes in the Indian Ocean Realm. Trade and commercial links with Oman including Muscat, the East African coast including Zanzibar, Sofala coast and different islands and territories in south and Far East Asia are well known. When the Portuguese appeared in the Indian Ocean, they were surprised to see that the Indian pilots were using compass and charts. These charts were good enough to navigate through the ocean. Such charts or maps included relevant information, such as islands, reefs, shoals, ports, cities, pole star, latitude co-ordinates, depth of water, lighthouse, and distances expressed in *zams*. But, with the emergence of colonial powers in the Indian Ocean, this tradition got affected.

The age of great exploration started with the quest for uncorrupted truth. The major exploration got underway at the end of 15th century with Columbus crossing the Atlantic in 1492, Vasco da Gama circumnavigated Africa in 1497 and the Magellan's expeditions circumnavigated the world between 1519-1522. The Dutchman Snell (1591-1626) carried out the first accurate triangulation and made the rigorous study of refraction. The French Clergyman Picard, in 1670 made the measurement of size of earth. His result of 6,275 km for the radius of earth was the first improvement on Eratosthenes determination of the size of the earth. Newton's theory, that the earth should be oblate because of the centrifugal force, caused by the spin was validated by the measurements two meridian arcs, one at the equator and the other closer to pole measured by two survey expeditions organized by French Academy of Sciences.

The dissemination of geographical knowledge was limited till 15th century before the invention of printing technique due to the labour and skill involved in duplicating graphic data. Once, however, lithographic techniques were discovered map making got a stimulus, particularly in Europe where adventurous men were out to conquer new worlds for commerce. Father Monserrate, a Jesuit Missionary brought out a map after a visit to Akbar's court. This was a first of those maps which are based on measured routes and astronomical observations.

During the early Pathan period, Sher Shah Suri was instrumental in preparing revenue maps in northern India. Later, Raja Todar Mal, a Minister in Emperor Akbar's court, made valuable contributions in preparing revenue maps for a larger part of the country. Sher Shah Suri and Todar Mal's revenue map, based on regular land survey systems were well known in the medieval period and continued to be in practice during the mid-eighteenth century. These are available information of the surveys instituted of Akbar during 16th century; measurements being made by a hempen rope which was replaced by a "Jarib" of bamboos joined by iron rings. The other noted cartographers of the time also published their vision of maps of India, but in 1752 the French geographer, Jean-Baptiste Bougigon d'Anville published map of India and put Indian geographical knowledge on a scientific footing. In the south, Rajaraja, the Chola king (985-1001 AD), carried out revenue assessment survey of his territory. Further, plans, sketches, religious maps, concept maps, route maps, charts, *parikrama* maps and the like were very popular. The layouts of palaces, forts, cities, tombs and gardens were based on topography, directions and astronomical calculations. Sun light, position of moon, wind directions, seasonal variations, calculations of geographical co-ordinates (as mentioned in *Ain-i-Akbari*), etc., were included in the plans which cannot be done without proper knowledge of the geographical space. Attempts were made from very early times to establish the geographical locations of important places. Raja Jai Singh, Sawai of Jaipur of pre-British period established astronomical observatories at Jaipur, Delhi, Mathura, Ujjain and Varanasi and determined geographical positions in India. Even today we can find these traditional materials in old cities like Allahabad, Mathura and Varanasi. A lot of materials have made their way to the libraries and collections in the West.

The pioneering work done by French expeditions showed that terrestrial measurements (angles and distances) are the main tools for the task of relative positioning. The technique of triangulation, astronomical determination of positions and azimuths as well as levelling were started in mid-seventeenth century. In the early days of the influx of foreigners in India, the geographical knowledge of southern parts of the country was greatly increased during the period of wars of political supremacy but northern parts of the country (Bengal, Uttar Pradesh etc.) was sketchy.

During the middle of 18th century, latitude could be easily determined by observing the meridian altitude of the sun or star. For longitude determination one had to wait for favourable phenomenon such as an eclipse of sun, moon or satellites in clear sky with similar observations to be made at some known places. In 1787, astronomical observations for latitude and longitude at various places were made covering India. The proper astronomical centres were established in due course of time using sextants, chronometer and telescopes. The results of these instruments were imperfect because of their accuracy level and faulty mathematical tables. It is interesting to note that because of difficulty in determining longitude early maps of India do not indicate longitude with reference to Greenwich meridian but are referred to arbitrary meridian passing through Madras or Calcutta.

Most of the early maps were local surveys carried out by cursory method. In 1776, comprehensive instructions for preparing the maps on the scale two mile to an inch formulated: in which distances were measured by parambulator rather with chains and bearing to the conspicuous hills. Short base lines laid out and measured and distant points to be fixed by triangulation.

Development of the Indian Cartographic Empire

The history of Survey of India is a part of the saga of nation building. It was perhaps never realized that the foundation of modern India is synonymous with the activities of this department. But its contribution has never been emphasized—not even by the *Survey* itself. How scientific foundation or development initiatives could have been possible without the anticipatory actions taken by this department. The process of understanding of the priorities in development and defence has been the forerunner in all respect.

All such developments were of course not without price and efforts. The scientific measurement of the country had several ramifications. Surveyors had to traverse from regions to regions, waiting for an opportune time free from natural, man-made and logistic problems in order to continue their task. Hazards, diseases, snakebites, dacoity, hostile local people, battles and the like came in the way of their mission. Surveyors penetrated the jungles, climbed mountains, crossed rivers and fixed poles, stations and control points through out the length and width of the country. There was no respite whether it may be on the slopes of Western Ghats, or swampy areas of Sundarbans, or ponds and tanks, ox-bow lakes and meandering rivers of Bengal, Madurai or Ganga basin. The deserts, mighty Himalayas, *Rann*, river Chambal or Gandak, *terai* or *dooars* were also not spared. The survey has been associated with the waves of Arabian Sea or Bay of Bengal, dust storms of Rajasthan, cyclones of Eastern Coast, cold waves of the north and the widespread monsoons and heat all over the country.

It was against this price, determination and missionary zeal that the foundation of the country was laid. The knowledge assets built over years and generated with whatever technology then available proved to be valuable. The best part of the whole process is that a new information set which may be based on latest technology like aerial photography or global positioning system take benefits of the earlier generated information. No piece of information is wasted and has relevance even after decades. How such an empire of knowledge has been built up? It is not only a technological feat that is most important but it is an emblem of pursuance of one of the longest scientific experiment carried out in the world, i.e. mapping the nation against all odds. The legacy and traditions continue. It is a matter of great interest to know how such a superstructure of information has been built ? How this great scientific experiment is carried out for so long. What were the compulsions and apparent benefits ? Why the colonial rulers and later the independent government with limited resources continued their interest in this expensive exercise. These are some of the questions that will keep on bothering the people interested in the history of science of this country.

The Survey of India traces its birth starting from Major James Rennell appointed as Surveyor General by the East India Company on the 1st January 1767. In those days, the urgent need was for a picture of the country showing general course of main rivers and the situation of principle towns. This task was taken up with speed and produced serviceable maps of the area of Bengal and Bihar in less than twelve years, an area of over 1,50,000 sq. miles, producing a continuous and uniform set of maps. These maps, however, could lay no claim to accuracy of details but were sufficient to meet the needs of the time. Rennell also produced a map "Map of Hindoostan" in 1783 after relinquishing the post of Surveyor General (Fig. 1.1).

Fig. 1.1: Map of Hindoostan by Rennell

The progress of topographical surveys in Madras and Bombay presidencies was more or less independent and uncoordinated with the work of Bengal presidency till the beginning of 19th century. This was not a satisfactory state of affairs and retarded the progress considerably. It was only in 1787 that an accurate survey of the sea coast from Madras to southernmost extremity of the peninsula was taken up by running a 300 mile line of triangles along the coast with a view not only to ascertain the actual line of the sea coast but a complete survey of the peninsula of India. This survey was the first Indian survey to have base on triangulation.

Towards the close of 18th century, primitive theodolites had been brought in use. The angles and bearing were measured with theodolite and pocket compass for direction of the road etc. The technique of plane tabling was first used in 1792. Plane tabling survey was subsequently developed into an art and has been extensively used down to present time for topographical surveys in all types of terrain. Even now, though the modern techniques of surveying are coming up, yet this simple technique is widely used in various surveys such as large scale mapping, engineering survey, cadastral survey etc.

Beginning of Scientific Surveys

The period of piecemeal surveys was over by end of 18th century. A new era began with the William Lambton and George Everest which signalled the consolidation of coordinated efforts. The foundation of a truly scientific Survey of India was laid which was the beginning of a stupendous work and occupied the lifetime score of noble and devoted surveyors. A network of primary triangles was established by the trigonometrical surveys. It was a magnificent scheme, timely conceived and brilliantly executed. Although, technique of triangulation, astronomical determination of positioning and azimuths as well as levelling were started in mid-seventeenth century but the scientific procedures started only by the end of 18th century when a project for the measurements of an arc of the meridian through a network of trigonometrical surveys covering the Indian peninsula was formulated.

The actual work of the great trigonometrical survey was commenced on 10th April 1802 by the measurement of a base line near Madras, (now Chennai). Madras base line was established using steel chain which consists of 40 links of 2½ feet each, measuring in the whole 100 feet. The baseline was measured with the aid of coffers (long boxes) as it was required for the triangulation of the "Great Arc" where utmost possible accuracy was aimed at. From this base line a series of triangles was carried up to the Mysore and the second base was measured near Bangalore in 1804. To start with primary reference station of origin was the Astronomical Observatory at Madras. Having connected the two sides of the Peninsula, Lambton devoted much of his labours to the measurement of an arc of meridian. The series measured for the purpose is known as the *"Great Arc Series"*. In addition to the measurements of Great Arc Series, webs of triangle extended for establishing the positioning of main cities. This idea of web of triangles was replaced due to cost effectiveness by an All India Grid composed of cris-crossing 'chains' or 'bars' of triangle centred on the Great Arc. The holes on the grid could be filled later by cheaper and less rigorous topographical surveys. This idea gave birth the name "grid-iron" (Fig. 1.2).

The grid-iron layout consisted generally of an outer frame of two extreme meridional and two longitudinal series closing at each junction on a measured base line. To bring the Great Arc across the plains, masonry tower stations were built which were about 50 feet high. The first essentials of every observation station, whether on hill top or tower, or otherwise was the stability of the instrument and immovability of the mark over which instrument and signals were centred. Since these marks were established after putting hard work, therefore marks were handed over to the civil authorities when all connections had been completed. In 1866, it was ordered that all stations of Great Trigonometrical Survey should be placed under the official protection of district magistrates and visited periodically. This practice is still in vogue till the present time for all primary Great Trigonometric (GT) stations and benchmarks.

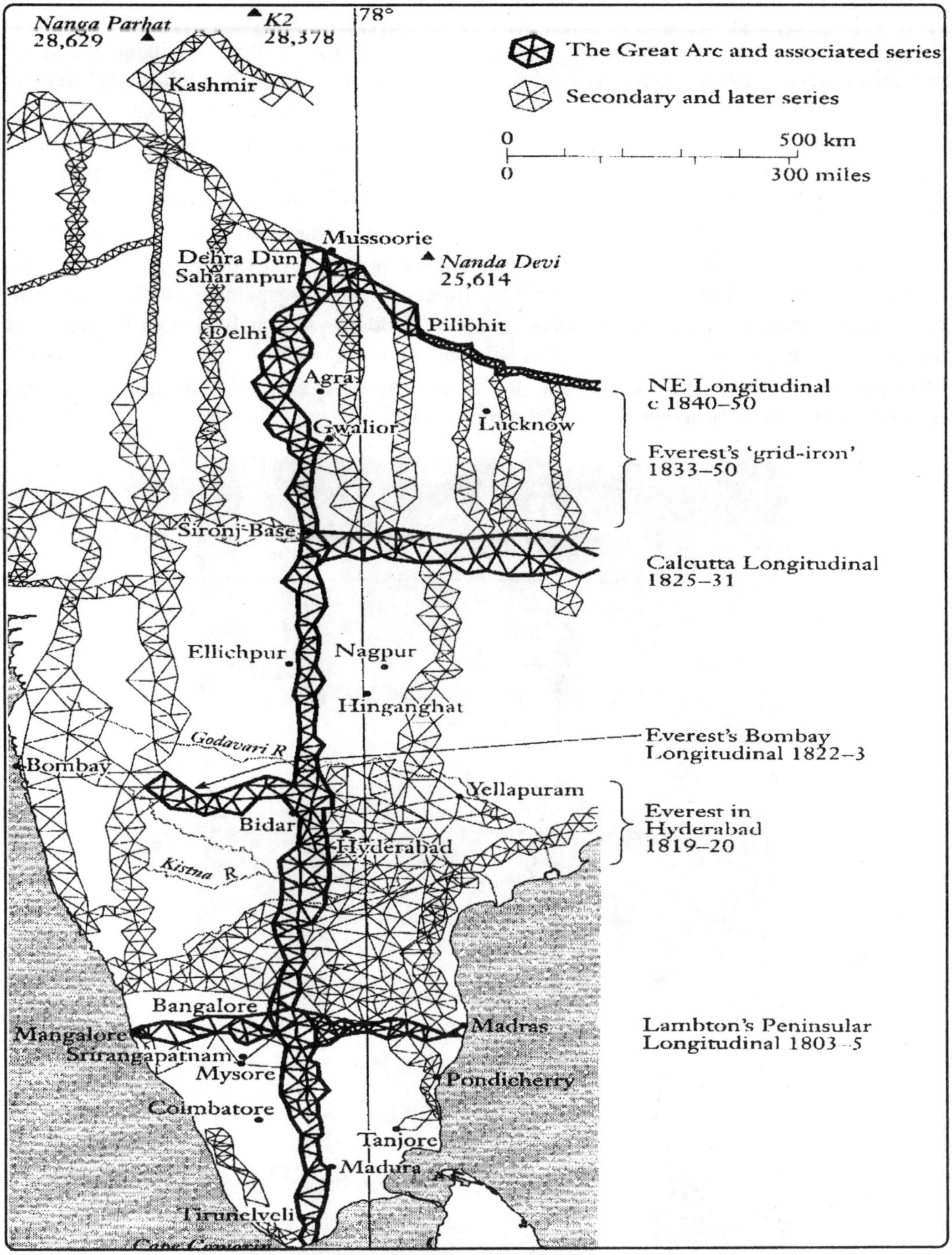

Fig. 1.2: The Great Arc

Lambton's main instrument referred as the Great Theodolite which was a marvel workmanship of those days (Fig. 1.3). The horizontal circle was 36 inches in diameter and the vertical circle 18 inches; each was read by two microscopes. This theodolite was used by Lambton and his assistant and then by Everest and others till 1866. Various other theodolites were used for observations for meridional series, viz. 36 inch theodolite, which was built up from Lambton great theodolite, 34 inch theodolite, 24 inch, 18 inch and 15 inch theodolites. For laying out the series and running secondary and minor triangulation, small theodolites were used i.e. 14", 12" and 7". For baseline measurements, compensation bars and other baseline apparatus were used. The compensation bars remained, however, the only means available for measuring the base line of the main triangulation framework—Vizagapatnam and Cape Comorin base were measured during 1862 to 1869. In 1856 'standard yard' arrived from England and the following year a special room at Dehra Dun was set aside where subsidiary standard could be laid off or compared by microscope as and when marking of staves for levelling operations required. Standard spirit levels were used during those days.

Fig. 1.3: The great theodolite

Triangulation and levelling computation were made to a regular routine, adapting rules and formulae to the requirement of the department. Computation forms were lithographed at Calcutta under the direction of the Chief Computer. One of the greatest contributions made to the Great Trigonometrical Survey was the preparation and publication of a set of tables to be used with departmental formulae and the computation forms by Babu Radhanath Sickdhar, Chief Computer. The first edition entitled *Tables to facilitate the computation of a Trigonometrical Survey and the Projection of Maps* was published in 1851. Auxiliary tables to facilitate the calculations of the survey department were published in 1868. The first official lists of geographical coordinates were published in 1842.

For the dispersal of triangular error, method followed by Everest was tested against new method devised by Gauss. Radhanath first tried two simple figures and obtained results closely agreeing with old method. He tried this on complicated figures that occur in the trigonometrical survey. The results were highly satisfactory, showing that the greatest discrepancy between Gauss's and Everest's methods would not exceed 0.14". The final distribution of errors and reduction of results of the triangulation of the Great Trigonometrical Survey was carried out under the direction of General James Walker whilst he was Superintendent of the Trigonometrical Survey. For this purpose the whole triangulation was divided into five figures—North-West, North-East, South-West, South-East quadrilaterals and the southern trigon.

Lambton and Everest did not go deeply into the subject of heights above sea level. Lambton first connected to the sea at Madras in 1802, but for his great Central Arc he preferred the connection made at Cape Comorin in 1809. From this he brought up his height by vertical angles from station to station. Surveyors in India had no professional interest in the measurement of the vertical rise and fall of tides along the coast except to find the level of the sea from which to calculate their land heights. Lambton followed the practice of his time in calculating his heights from low water. It was only in 1837, when the mean between high and low tide observations for at least for half a month were proved to agree closely from one place to another. Self-registering tide gauges had been invented during 1830-33. Self-registering tide gauge was established at Colaba observatory, Bombay in 1842. Tide gauges at various other places were also connected through spirit level during 1851 to 1860. Indian levelling was started upon correct and scientific lines in 1858. Again it was General Walker who was the founder of Indian levelling for the care and the thought he bestowed upon the subject before he initiated the fieldwork.

Waugh took over in 1843 from Everest and during seventeen years of his administration, triangulation series was advanced eastward to over Ganga valley to Calcutta and extendedup to Assam. Regular observations to the Himalayan peaks were taken. Waugh was responsible for the discovery of the highest mountain in the world, 29002 feet above the sea level, which he recommended should be named after George Everest who had built up the triangulation system by which the discovery was alone made possible (Fig. 1.4).

Regular astronomical observations for azimuth and meridian were continued along both meridional and longitudinal chains of triangle as check against accumulation of errors in direction. Both Lambton and Everest had been well aware that their observations were influenced by visible mountain masses and variation of density. Various other mathematicians and geodesists worked on this subject and attracted wide attention suggesting the value of pendulum observations for the determination of variations of gravity and introduced the theory of compensation or isostacy.

Fig. 1.4: Picture of Mount Everest (Photographed by P. Nag).

Development of Great Arc

The measurement of Great Arc from Cape Comorin to Himalayas was completed by 1843. The 'grid iron' system consists of meridional chains of triangle tied together at upper and lower end by longitudinal chains. Ambitious scheme of triangulation commenced with the Great Arc Series, having Dehra Dun base at the north and Sironj base in central India as the southern end. From Dehra Dun base NW Himalayan series was extended while from Sironj Base the Calcutta longitudinal series was extended up to Karachi base. In 1887 observation of a series from NW Himalayan known as Kashmir series was initiated. The height of stations averaged 17,000 ft. in this series. The second highest peak, next to Mount Everest was found during this triangulation which was 28,290 ft. high and was named K2.

Indian triangulation was first adjusted in 1880. In 1937-38, Bomford attempted to readjust the triangulation after incorporating the new triangulation series observed, Laplace stations to arrest azimuths errors and new baseline measured between 1930 and 1934 using invar wires. He carried out preliminary figural adjustments for the distribution of discrepancies in scale and azimuth in a rational way. The triangulation adjustment carried out in 1937-38 suffered weakness due to ill fitting nature of the ellipsoid, measurement of base lengths by 10 feet bar suspected to have undergone changes in length by 3 ppm, lack of Laplace observations, non-application of various geodetic corrections like deflection, skew normal, geodesics, polar motion etc.

The adjustment of 1937-38 was not considered comprehensive and explicit. Instead of resolving simultaneous normal equations, formed after incorporation of new data, graphical technique of adjustment was employed. Adjustment of 1880 has remained the basis of all India triangulation and mapping except for a constant change of—(2′27″.18) in longitude. This change was essential as the longitude of Madras Observatory was revised in 1905.

The mapping activities was scientifically executed and based on triangulation series adjusted in 1880. Most of the maps were on the scale half inch or larger. Mapping activities opened up new possibilities with the introduction of aerial photographs early years of the war. The science of photogrammetry had not developed but the immense value of aerial photos impressed the surveyors. The air survey techniques were, however, not much used in the department up to 1939 except for Training for Military Survey purpose and for survey of inaccessible areas. At the time of independence (1947) about 60 per cent of the country was covered by primary scale of mapping on 1 inch to 1 mile. This primary map scale 1 inch to a mile was adopted in Survey of India in 1905. It was also decided that these maps may be produced in colours and show the relief by rigorous contourings. The requirement of printed maps increased during 1939. The main printing organization at Calcutta could not meet the mapping requirement. Therefore, the present site at Dehra Dun was selected and the printing machines were installed and by middle of 1943, the printing group was in operation with three large high speed machines.

In post-independence era, there was immense increase in the developmental activities which is continued till today. The Survey of India was the only premier organization in surveying and mapping which could take up the survey work for developmental scheme, due to which normal topographic survey was kept to a secondary place.

During the First and Second Five Year Plans, the department faced with demands for survey which required for more survey potential than it had. During the First Five Year Plan (1951-56), 70 per cent of the department's potential employed on developmental surveys and in Second Five Year Plan about 60 per cent potential could be directed for this purpose. The department had diverted work load with the switching over to metric system in 1956, the basic map scale having been changed to 1:50,000 from 1 inch to 1 mile. It was during second five year plan, when advancement in various techniques of mapping were available and there was pressing demands for maps for the development of the country. Various techniques in the field of photogrammetry, cartography, printing and training were adopted in Survey of India.

Geodetic and Geophysical Surveys

The Great Trignometrical Survey triangulation network in India and adjacent country was started in 1802 and by about 1880 sufficient triangulation series had been observed. The responsibility of establishing geodetic and geophysical control in India lies with Geodetic and Research Branch of Survey of India. With the introduction of modern instruments in improving the precision and accuracy of all geodetic and geophysical survey, the activities of this branch have been diversified and today it engaged in number of research and developmental programmes. Theodolites used in the past were replaced by glass arc theodolites in the late forties of twentieth century which read directly to an accuracy of 0.2 second of arc. Geodetic bases of length 10-12 km were established during the period 1831-82 have been measured with India 10 foot bars, while the modern base lines were measured with invar wires. Astronomical azimuths have also been measured at a few stations to keep check on the orientation of the triangulation series. The triangulation network of India for providing first order control by chain of primary and secondary series every 4° (about 400 km) apart was completed by the year 1956. The work for densifications to bring control at 2° (about 200 km) apart and upgradation of the existing series classified as secondary has been taken up by the Survey of India. There has been tremendous growth in the field of space geodesy in the last couple of decades. TRANSIT system adopted in Survey of India in seventies to establish geodetic control. A more rigorous system called NAVSTAR GPS (Navigation Satellite with Time and Ranging Global Positioning System) or simply GPS, a satellite based radio navigation system, providing 3-dimensional position, navigation and time information to suitably equipped users all over the globe at any instant of time, was introduced in the year 1980. GPS technique was adopted in Survey of India around the year 1990 and enormous work has been since been done in production mode. Electronic distance measuring device have been introduced since 1962 and number of bases have been added to the principal Indian triangulation. After incorporation measured base lines, lap lace observations etc. in the Indian Triangulation net, simultaneous adjustment of the primary and secondary series has been taken up to redefine the Indian Geodetic Datum.

Various geodetic and geophysical modern instrumentations are being put in use for geodetic, gravity, geomagnetic observations. Glass Arc theodolites, astrolab, chronograph, crystal clocks for geodetic work, Frost, Worden, Lacoste Romberg and Automated (CG-3M) gravimeters for gravity measurements and QHM (Quartz Horizontal Magnetometer), ZBM (Zero Balance Magnetometer), Proton Magnetometer for geomagnetic observations are in use for geodetic and geophysical surveys. The appearance of the computers is extensively used for better and more homogenous adjustments and data processing.

Topographical Mapping

After establishing the ground control points either by conventional ground method or by space based techniques, the subsequent steps are: survey ground details and verify, annotate and introduce names of rivers, roads, villages, towns etc. and depict topographic features by contour and other suitable symbols. These steps, in the past, were carried out through the ground techniques. With the emergence of aerial photography, Survey of India introduced the technique of surveying aerial photographs. The development of photogrammetry in India has been rapid only since 1954 when the first stereo-plotting machine a Wild Autograph A7, was processed and installed, though multiplexes were purchased as early as 1950s. The photogrammetric Techniques available became the standard procedures in quite a few units of the department. The introduction of photogrammetric techniques has expedited many developments in the country.

In Survey of India, the process of map making with aerial photograph starts with planning of aerial photograph. The Ground Contour Points (GCPs) at selected locations are connected with existing geodetic control by field survey and their positions are captured. With the photographs and minimal ground contour and their position identified, the stage is set for undertaking Independent Model Photogrammetric Triangulation and subsequent computation provides requisite control points in each model. The model properly oriented is then used for generating cartographic data which are verified on ground before preparing the final map.

By early fifties of twentieth century precision analogue photogrammetric instruments were used for topographical mapping. Analytical plotters captured the data in digital form, which had added advantage for acquiring directly digital map. The SPOT, LANDSAT and IRS imagery for topographical mapping on small scale are being used in the department. Digitally generated data are being used for creating cartographic data base in various scale, viz. 1:250,000 and 1:25,000. In addition to this data base for district planning map series on scale 1:250,000 is being carried out. Digital Terrain Model (DTM) being generated for various developmental activities in the country.

Survey of India as a part of its activities to keep abreast with latest technology in the science of map making, made digital map making programme in early eighties with a view to adopt digital technology in the country. Survey of India adopted Computer Asserted Cartography (CAC) system, the "AUTOMAP" in 1981. In this system the geographical and other data are stored in digital form and processed on demand to draw the required map output by using computer operated drafting table. Survey of India has taken a giant leap forward in 1986 when integrated digital map production systems were installed at 3 locations, i.e. Modern Cartographic Centre (MCC), Dehra Dun, Digital Mapping Centre (DMC), Dehra Dun and Digital Mapping Centre (DMC), Hyderabad. Having overcome the initial difficulties of new technology induction in early nineties, the department aimed to achieve flexibility in satisfying users demands by establishing structured Nations Cartographic Data Base (NCDB) and Geographic Information Systems (GIS).

Recently, a major initiative has been taken in Survey of India. Development and business potentials have been included in the activities of the organisation. As a result, reorganization has taken place in 2003. Directorates for each state, known as Geospatial Data Centres (GDCs), have been created which are equipped with digital and GPS technologies. DMCs and MCC have been converted to new directorates, namely National Geospatial Data Centre and GIS and Remote Sensing. A new Business and Publicity Directorate has been opened.

With the completion of 1:50,000 Topographical Survey of whole country by 1982 and mapping by 1985, India has joined a select group of countries which have completed map cover on the national scale. The department now has an ambitions programme of covering the whole of India by maps on 1:25,000 scale. The country covers 394 maps on scale 1:250,000 and 5,106 sheets on scale 1:50,000. In 2003 all the 1:50,000 scale maps were digitized. With the new National Map Policy, these maps are now being converted to WGS84 datum and UTM map projection. In the process, a seamless map data is being generated (Appendix I).

Responsibilities and Challenges

Survey of India is the premier national survey and mapping organization in the country. In its assigned role as a national principal mapping agency, SOI bears a special responsibility to ensure that the country's domain is explored and mapped suitably to provide base maps for expeditious and integrated development of the nation. In the process, SOI has been producing all purpose topographical and various other series maps required by defence, general administration, internal security, developmental needs, irrigation, watershed management, resource management and various types of engineering projects. It is also responsible for establishing precise planimetric control, heights above MSL, gravity, geomagnetic and tidal prediction as prerequisite for mapping activities and other scientific applications. The department is committed to provide technical expertise to other countries in the field of geodesy, surveying, cartography and survey education.

With the introduction of digital technology in the department, three digital centres are generating Digital Topographical Database for entire country for use in various planning processes and creating GIS. Its specialized Directorates such as Geodetic and Research Branch, Research and Development Directorate and Survey Training Institute have been further strengthened to meet the growing requirement of user community. The department has also been immensely contributing to a number of multi-institutional scientific programmes related to the field of geophysics, remote sensing, glaciology, study of seismicity and seismotectonics, scientific expedition to Antarctica and digital data transfer.

India is fast moving into information and knowledge society. Emphasis is increasingly placed on IT driven transparent e-governance. The nation has been generating voluminous field/digital data, i.e. information through systematic topographical surveys, geological surveys, soil surveys, cadastral survey etc. Access and availability of such information to the citizen, society, private enterprise and government are of immense importance. As a part of this vision NSDI is being involved through partnership approach among various data generating agencies to facilitate integration, easy access and networking of databases. NSDI has been conceived as a national system that synergistically combines the resources and infrastructure of various players with the power of IT and enabling information support for decision making in government, industry, academia and other organisation besides serving the public needs. NSDI strategy and action plans have been formulated through multi-institutional approach involving Department of Science and Technology (DST), Survey of India (SoI), Indian Space Research Organisation (ISRO), Geological Survey of India (GSI) and National Atlas and Thematic Mapping Organisation (NATMO) etc.

Atlas and Thematic Mapping

At the time of independence it was felt that different types of maps are required for development, planning and reconstruction of the newly formed country. These maps should focus on development issues in accordance with the priorities of the country as envisaged by the Five Years Plans. The progress of cartography in India for its development and reconstruction process was then initiated. The earlier approach for capture, control and colonization was given reorientation. From Table 1.1, the planning priorities in different Five Year Plans may be gathered. Initially, attention was more on the development of agriculture and related activities that were necessary following the Second World War and the partition of the country. In fact, a better understanding of the newly born country was required with all its facets. Hence, the preparation of the national atlas was felt necessary. Apart from maps on the physical aspects, the Hindi version or the first *National Atlas of India* included plates on soils, forests, land use, irrigation, crops, livestock, power and industries. There were on social aspects as well, such as on population and health (Massey, 2002). The Second Five Year Plan was meant to give a shape to the dreams of the first Prime Minister of the country, Pandit Jawaharlal Nehru. Iron and steel, heavy machinery and industry was given more priority. This required power and appraisal of the available resources in different parts of the country. Considering the size of the country and the varieties, it became imperative to prepare the enlarged version of the *National Atlas of India* (Fig. 1.5).

During the Third Plan, planning and development at regional and state level got impetus. This lead to the preparation of atlases at sub-national level. One of the most important one was *The Planning Atlas of the Damodar Valley Region*. Following this, several regional atlases came into existence. These atlases followed the state framework however, the focus was not always the same. As a result, we have statistical, administrative, resources, planning or simply

Fig. 1.5: National Atlas of India: Physiography

state atlases. If we compare the contents, we will find there is a great degree of commonality though they have come out under different arrangements and funding. This process is not yet complete. Earlier, some agricultural atlases at state levels were prepared. In this plan period, the preparation of the *National Atlas of India* consisting of 300 plates was in full swing. It consists of various aspects of the country, which were required for any planning strategy at national and regional levels. This period also witnessed the diversification of the thematic mapping activities and applied geography.

Table 1.1: Planning Priorities and Cartographic Initiatives

Plan	*Priority sectors*	*Objectives*	*Cartographic initiatives*
First Plan (1951-56)	Agriculture, irrigation and power projects	To correct disequilibrium in economy caused by Second World War and partition of the country	Hindi edition of the *National Atlas of India* and state atlases.
Second Plan (1956-61)	Iron and steel industry, heavy machinery and industry	To achieve a socialistic pattern of society	*National Atlas of India;* State and regional resources and planning atlases.
Third Plan (1961-66)	Agriculture and industry	To achieve self-sustaining growth	DVC Atlas, Census atlases, DCHs.
Fourth Plan (1969-74)	Employment and education, especially for the poor	To combine equity with growth	Forest, Agricultural, Irrigation and Tourist Atlases.
Fifth Plan (1974-79)	Infrastructure relating to irrigation, power and transport	To go by self-reliance and control inflation	Land use, Geomorphological, mapping, urban studies.
Sixth Plan (1980-85)	Technological upgradation of agriculture, industry and energy	To work towards satisfaction of basic needs and removal of poverty	Land use and landform maps of Development blocks and districts. Modernization in cartography. Water Resource Development Atlas.
Seventh Plan (1985-90)	Productive employment	To work towards satisfaction of basic needs and removal of poverty	*Indian Ocean Atlas, Students' Reference Atlas and Socio-Economic Atlas.* Digital mapping system, wasteland maps. RS based mapping.
Eighth Plan (1992-97)	Human development-employment, fertility control, education and health	To upgrade efficiency through restructured economic management system and redefined role of government	District planning maps. Digital topographical maps. Forest maps by FSI. Block maps. Environmental sensitive maps.
Ninth Plan (1999-04)	Growth with social justice and equity	Priority to agriculture and rural development for productive employment and eradication of poverty; providing minimum basic services; environmental sustainability; promotion and developing people's participatory institutions like Panchyati Raj, co-operatives and self-help groups; and building self reliance.	Panchyat maps. Block and taluk profiles. Large scale environmental maps for formulation of polices. Digital maps for urban management. Cartographic base for decision support system.

Source: Based on Nag (1997).

The focus of the Fourth Plan was on employment and education. Though several new mapping activities were initiated but they did not have any direct bearing on this focal issue. But a close scrutiny will indicate that our contribution was to prepare a larger base that would ultimately help in generating employment and education. NATMO took up the preparation of the *Tourist Atlas of India* and several resource atlases on water, forest and agriculture. On behalf of the Irrigation Commission of India, Irrigation Atlas was also take up during this period.

The Next Plan, i.e. Fifth Plan (1974-79) will be considered to be a watershed in the history of thematic cartography in the country. The present name of NATMO came into existence in order to include thematic mapping in addition to atlas cartography. Apart from the preparation of the National Atlas, following two objectives were included in the manifesto:

(a) Research studies on environmental aspects and their impact on social and economic development; and
(b) Preparation and compilation of land use and land capability maps on 1: 1,000,000 scale from data collected by the relevant agencies as a pilot scheme to begin with.

The above modifications paved the way for taking up research work and large scale mapping, such as landuse mapping, geomorphological survey and mapping and urban studies. Obviously training was required to meet the new challenges. Hence training programmes were strengthened. At the instances of different bodies, project work on Dandakaranya, Sundarban, maps for land use planning of CMDA area were taken up.

The Sixth Plan (1980-85), the focus was on technological upgradation of agriculture, industry and energy. This objective could only be met by providing a sound spatial base in the form of maps. To meet the requirements at larger scales, landuse and landform maps were prepared for selected districts and development blocks. This was also the time when we start getting the satellite imagery, particularly from Landsat. Hence, satellite imagery and air photographs were widely used for such large-scale mapping. On the other hand, water was considered important for agricultural development. Due to this reason, initiative was taken to prepare the Water Resource Development Atlas as well as to revise the Irrigation Atlas of India.

The focus of the Seventh Plan was on more socio-economic aspect, i.e. on productive employment. Several proposals were suggested to the Department of Science and Technology and through then to the Planning Commission. Finally, three projects were accepted which were basically to strengthen the organisation infrastructure. These are:

Establishment of Thematic Atlases Division with the preparation of the following atlases:

(i) Establishment of Research, Development and Training Division
(ii) Establishment of Digital mapping and Geographical Information System

The Eighth Plan was having a focus on human development-employment, fertility control, education and health. Obviously the attention was shifting to micro level. Government took a lot of initiatives for planning and development at the local level by providing employment, authority of decision-making and infrastructure development. The 73rd and 74th Amendment

a multi-dimensional approach. Obviously, we have to wake up to meet this challenge in thematic cartography. A lot of research and development work has to be put in to make thematic maps more meaningful (Fig. 1.7).

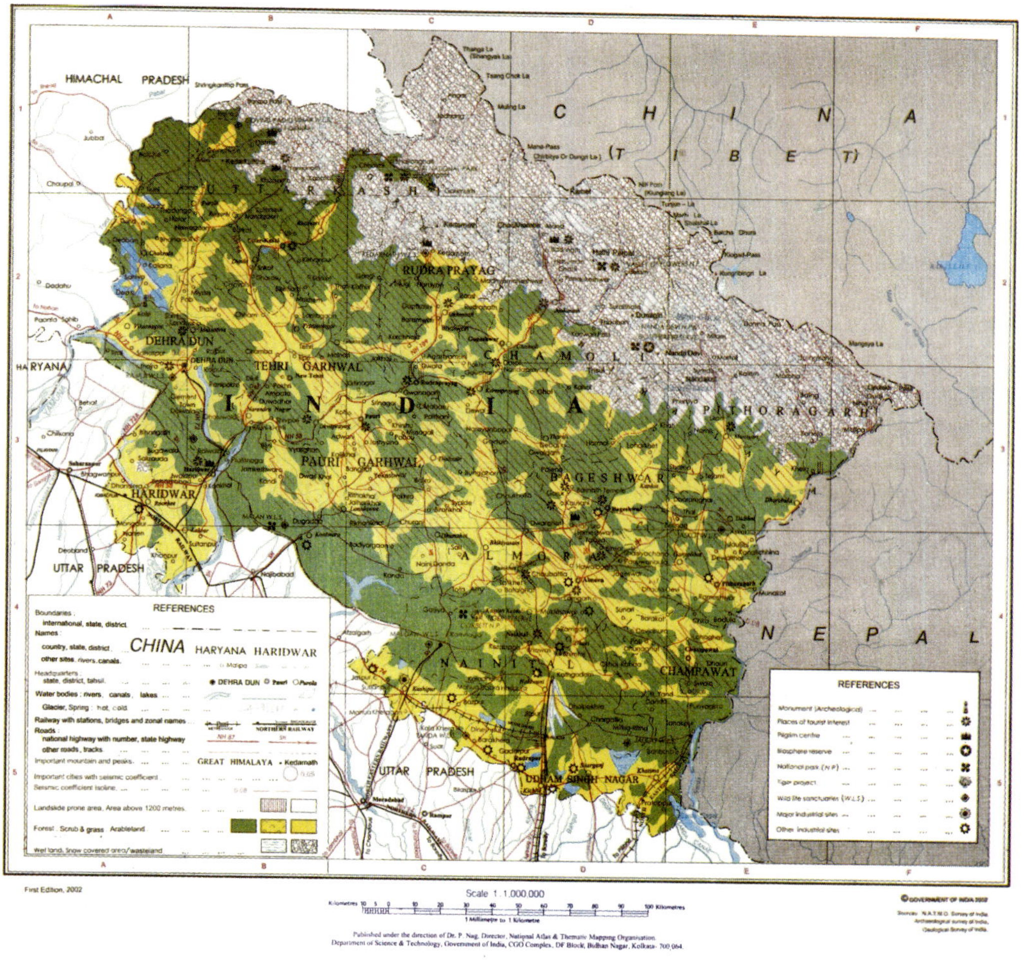

Fig. 1.7: Environmentally sensitive Zone of Uttaranchal (*Source*: NATMO)

In collaboration with the Central Pollution Control Board (CPCB), Mapping of Environmentally Sensitive Zones and Industrial Sites of several states have been carried out. NATMO will soon bring out a publication entitled *India: Environmental Status* which will contain these maps. The digital data of these are also very much in demand. Again, in collaboration with the Central Pollution Control Board, NATMO has brought out an atlas on environment. It consists of maps on physical environment, sensitive areas, pollution sources, environmental quality and pollution control. This atlas is at the national level.

Ninth Plan Priorities

The Ninth Plan is yet to get a shape. But from the documents available, some ideas about its objectives can be gathered. The focus is on *"Growth with Social Justice and Equity"*. The specific objectives of this plan arising from the greater reliance on market forces and the imperative policy and action are as follows:

1. Priority to agriculture and rural development with a view to generating adequate productive employment and eradication of poverty.
2. Accelerating the growth rate of the economy with stable prices.
3. Ensuring for all, particularly, the vulnerable sections of society.
4. Providing the basic minimum services of safe drinking water, primary health care facilities, universal primary education, shelter, and connectivity to all in a time bound manner.
5. Containing the growth rate of population.
6. Ensuring environmental sustainability of the development process through social mobilization and participation of people at all levels.
7. Empowerment of women and socially disadvantaged groups such as scheduled castes and scheduled tribes and other backward classes and minorities as agents of socio-economic change and development.
8. Promoting and developing people's participatory institutions like Panchyati Raj institutions, co-operatives and self-help groups.
9. Strengthening efforts to build self-reliance.

At least some of the above objectives are meant for (a) infrastructure development at the grassroots level with the help of Panchyati Raj and other institutions, (b) identification of priority areas for socio-economic change and development, and (c) mapping of environment-related issues. Regarding the former issue, the block development maps would be useful (Fig 1.8).

Further development of cartography or cartography for development in the country will depend on providing detail maps for *Panchayats*. This requires maps at 1:10,000 scale. Fortunately, IRS panchromatic data at 1:12,500 scale is available which can be used for this purpose. With the availability of Cartosat-1 2.5m data, mapping at this scale will be strengthened. Cartosat-2 is also due to be launched soon having better than 1m data. This will be beneficial for the cartographic activities and GIS-related activities in the country.

The development of thematic cartography is a feature of post-independence period. Several government institutions and departments have made valuable contributions in this regard. The office of the Registrar General of India produced a series of publications containing maps, exclusively or partially. The Census atlases of the country and states, administrative atlases and maps contained in the district census handbooks proved to be very handy for all sorts of users though they were not very high quality cartography work. The other survey institutions were also not lacking behind. Geological Survey of India, All India Soil and Land Use Survey, National Bureau of Soil Science and Land Use Planning, Forest Survey of India did produce good theme specific maps.

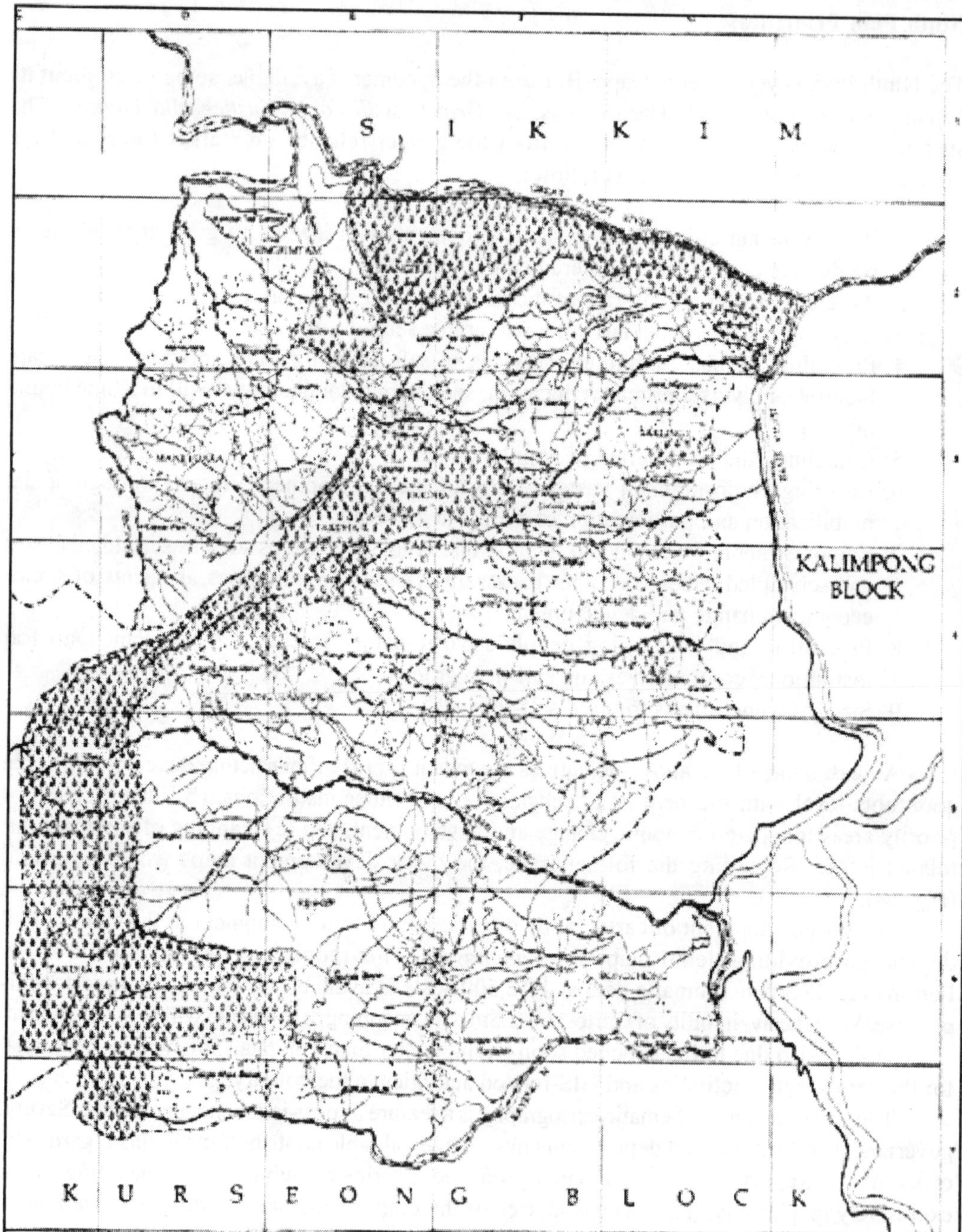

Fig. 1.8: Darjiling District Block Development Block Map

Department of Space

The Department of Space and the Indian Space Research Organisation (ISRO) were established with several objectives basically focusing on ways and means to utilize space for development. One of the major activities is on remote sensing. Some of the major institutions came over the national scene much before the establishment of the National Natural Resource Management System (NNRMS) in 1984. These institutions are the Indian Institute of Remote Sensing (earlier Indian Photo-interpretation Institute), National Remote Sensing Agency and Space Applications Centre—all under the umbrella of the Department of Space. These institutions are engaged in space borne data based cartography. They have best equipped technologically as well. Under the aegis of NNRMS, five Regional Remote Sensing Service Centres were established and several state and university centres were supported. But the real boost was following the successful of the IRS series of satellites, starting from IRS-1A in 1988. There are now five orbiting Indian satellites providing remote sensing data, the best in the world among the civilian satellites. As mentioned earlier, another IRS series, known as Cartosat, has been in action.

Other Government Initiatives

The other departments under the Union Government also have different activities related to cartography. The Department of Science and Technology (DST), apart from having two major mapping institutions (Survey of India and NATMO), has a major cartographic activity under its programme on Natural Resources Data Management System (NRDMS). Apart from developing a GIS software known as Gram++, it sponsors projects related to cartography on the management of natural resources. Similarly, the National Informatic Centre under the Planning Commission has countrywide set up for data collection and transmission. They are also involved in the GIS activities.

The state governments are now having state councils for science and technology. Through these councils' projects are granted which are exclusively or partly related to cartography. In some states, there are institutions funded by the state governments which are engaged in mapping activities. Further, the Directorates of Land Records in each state are also involved in map preparations. Some state governments are trying to modernize their set up. Under the state governments, there are planning authorities, municipalities and *panchayats*. Some of them do have very rich collections of maps meant for infrastructural development. State Ministries of Urban Development and Ministries of Rural Development and *panchayats* are supporting cartographic activities. With the 73rd and 74th Amendments of the Constitution, infastructural development at the grassroots level has been given high priority where maps are unavoidable.

The universities do have departments or centres on geography, geology, remote sensing, civil engineering or forestry which are involved in the preparation of maps. But they are basically meant for training and development of expertise. However, some centres under the university set up do take up major cartographic jobs, like the Institute of Remote Sensing in Anna University, Chennai or Visservaria Technological University of Karnataka or Jawaharlal Nehru Technological University, Hyderabad. Nevertheless, there are no university level courses specifically for the discipline of cartography. Though there are special papers on this subject in geography, but not a single university awards a degree in cartography. However, some universities are awarding M.Tech. degrees in GIS and Remote Sensing which do contain several aspects of cartography as well.

GIS Activities in Industries

The scenario on the other side, i.e. industry, is entirely different. Before 70s the cartographic activities were limited to some publishing houses engaged in producing school atlases and some wall maps. Only a very few publishing houses were serious in this matter. On the other hand, there were land surveyors who carried out specific surveys for layout plans of housing colonies, buildings, plantations such as tea estates, area development and the like. The maps produced by them very close to cadastral maps which not necessarily be linked to the geographical co-ordinate system. The surveyors were either passed from Industrial Training Institutes spread all over the country or were ex-Survey of India personnel.

The real boost in the cartographic activities in the Industry in India came with the introduction of the computers. The systems capable for digital mapping, image processing and GIS were having multi-national backing. The principals were located in the US or in Europe or even in Australia and Japan. The selling of the hardware and software lead them to explore in the service sectors. This sector also got benefit of the cheap-trained manpower in India. As a result, bulk jobs from abroad started coming to India, however, these very basically conversion of hard copy maps to vector data. In the major cities in the southern India like Hyderabad and Bangalore, a good number of industries are engaged in this activity. They do employ a good number of skilled persons. With the advent of internet and web technologies, the transfer of digital data has become easy. The time difference between India and the Western World is an additional advantage. However, these digital conversion centres are looking towards the west for their survival and steady flow of projects. The Indian jobs or projects are optional in their business strategy, which is very unfortunate. This is due to bureaucracy, delay in accepting tenders and payments, complicated procedures and non-availability of existing information. Their interaction with the government is negligible but they look forward to the government for jobs. The industry has clearly indicated that they are well equipped to take up projects of national importance, but the occasions are rare. Due to the delays in granting of jobs, the infrastructure cost becomes exorbitant. The cost of space, equipment and trained manpower is very high and cannot be sustained by the industry for a long time if they do not diversify their activities. This has resulted in to two major trends: (a) the small scale industry may not survive in long run, and (b) bigger players are taking up GIS projects as their side business along with supplying hardware, software or exporting services through wire.

The healthy sign on the industrial front is the Industry-Industry co-operation. With the liberalization of the economy, the multi-nationals and NRI investments have multiplied. They require maps or information in GIS format for the Indian operations, which includes retail-selling strategies. In such situations, the accuracy and reliability is not high on agenda. Further, in case of telecommunication and infrastructural development the government is heavily dependent on the industry. They require maps, preferably in digital form, or GIS data as well. Hence there is growing co-operation within the industry.

The third major force, which is likely to emerge, is the Non-Governmental Organisations (NGOs). Almost in all major and medium size cities, several NGOs have emerged who have some activities in the field of GIS. In fact their outfits can orient them selves to take up medium size GIS projects. But, as for the industry, they suffer from non-availability of existing data. Perhaps, the formation of the National Spatial Data Infrastructure or NSDI will be useful to them. But for them it is slightly easier to obtain government jobs. Due to this reason, some of the industries have a NGO outfit as well. Retired government officials, university teachers and unemployed youths are actively engaged in GIS related activities under the banner of NGOs. Further, there is a series of professional bodies who have been registered as NGOs but engaged in the promotion of such activities, such as the Indian National Cartographic Association (INCA), Institute of Surveyors, Indian Society of Geomatics, and the Indian Society of Remote Sensing. They hold conferences, workshops and training courses every year.

The GIS scenario in the country regarding the government and the industry is quite confusing. Several surveys reveal that there is definitely a scope for co-operation provided certain barriers are removed. There is strength on both sides, which has to be exploited for taking up jobs of national and international importance. The growth of the country can no longer be staggered for desirable co-operation in this specialized field. The weaknesses have to be identified and measures are to be taken for removing the bottlenecks.

Conclusion

If we glance through the planning, environmental or business priorities and the cartographic or GIS initiatives, we will find both the activities are not always compatible. Sometimes cartographic or GIS activities are late in translating their action in tune with the government or industry requirements. This is not only due to the lack of visualization, but the procedure for preparing the maps is lengthy and it takes time to bring out the product. Fortunately, digital mode, in the form of GIS, is changing this scenario. In lot of cases, the cartographic initiatives have got a boost. Nevertheless, such a healthy situation has not always been beneficial because by the time the digital data could actually be used, the information contained is already old.

Another trend, which is very apparent, is that the concept of single theme maps is becoming outdated considering the application of maps in planning and decision-making. The example is the series of maps being prepared on Environmental Sensitive Zones and Industrial Sites. These maps contain selective information from a variety of thematic maps. Hence, the situation is ripe to collect and collate spatial information where queries can be made for decision-making. In other words, GIS has to become an essential part of the development or business initiatives.

Since the planning and retail business priorities are becoming more area specific, the spatial unit of planning and development or even operational unit is becoming smaller. Hence, from national and state atlases, and later district maps, the priorities are now at block and panchyat level mapping. Similarly, in urban areas, digital maps for planning, development, management and taxation are being preferred. This makes a case for a variety of GIS applications in the day to day life. In a country like India, which is so diversified, the scope of GIS applications is still under explored.

Lastly, maps are not being always being prepared as a solution or a blue print for planning. The digital technologies have provided other options as well. It can provide digital base and software options to play around with the spatial information and work out initiatives required for given set of priorities. Decision support system with the help of GIS can look into the planning options specific to the area concerned. Obviously, such demands and trends make a case for the promotion of GIS-related activities in the country, not only for government-sponsored activities but also for business promotion. India is poised for a developed country from fifteen years from now. Both, government and business, have to be better organized for the quantum jump based on the better organized and managed information.

REFERENCES

Audet, Richard and Gail Ludwig (2000), *GIS in Schools,* ESRI Press, Redlands.

Annonymous (1999), Highlights of the Ninth Five Year Plan, *Yojana,* May, pp. 29-49.

Arunachalam, B. (1997), Cartography of the Chola administrative hierarchy, *Indian Cartogarpher*, pp. 3-8.

Chandrasekhara, C.S. and S.P. Chatterjee (1972), Planning Regions of India, in S.P. Chatterjee and S.P. Das Gupta, *op. cit.*

Dastidar, Sachi G. and Shefali S. Dastidar (1991), *Regional Disparities and Regional Development—Planning of West Bengal,* Firma KLM Pvt. Ltd., Calcutta.

Dutt, A. K. (1984), The National Atlas of India, *Geographical Review*, Vol. 74, pp 94-100.

Dutt, G. K. (1979), Thematic mapping in India—contribution of the National Atlas Organisation, *Landscape System*, Vol. II, pp 6-12.

Dutt, G. K. and P. Nag (1985), *Life and Work of Professor S. P. Chatterjee*, NATMO, Occasional Monograph-1, Calcutta.

Dutt, G. K. and P. Nag (1987), *NATMO Scenario by 2001 AD*, NATMO Occasional Monograph-4, Calcutta.

Gogate, Prasad P. and B. Arunachalam (1997), Two eighteenth century Maratha maps of Administration, *Indian Cartographer*, pp. 30-34.

Gupta (eds), *Regional Geography and Cartography*, Selected Papers Vol. 4, 21st International Geographical Congress Publication, Calcutta, pp. 49-67.

Krishan, Gopal (1997), From centralised planning to liberal economy: The development experience of India since independence, in Abha Lakshimi Singh (ed), *Land Resource Management*, B.R. Publishing Corporation, Delhi, pp. 93-105.

MacEachren, Alan (1994), *Some Truth with Maps: A Primer on Symbolization and Design*, Association of American Geographers.

Massey, Jack (2002), A tribute to social GIS in India, *Geospatial Today*, Vol. 1, issue 4, pp. 75-77.

Monmonier, Mark (1993), *Mapping It Out*, University of Chicago Press.

Monmonier, Mark (1996), *How to Lie with Maps*, University of Chicago Press.

Muehreke, Philip and Julina Muehke (1992), *Map Use—Analysis—Interpretation*, J.P. Publication.

Nag, P. (1997), Cartography and administration, *Indian Cartographer*, pp. 299-305.

Nag, P. and M. Kudrat (1998), *Digital Remote Sensing*, Concept Publishing Company, New Delhi.

Nag, P. (1992), Role of NATMO in promoting Indian Geography, in R.L. Singh and Rana P. B. Singh (eds). *The Root of Indian Geography*, NGSI Publications, Varanasi, pp. 301-10.

Nag, P. (1992), Professor S. P. Chatterjee (1903-1989): A Life Sketch, in Singh and Singh (eds), *op. cit.*, pp. 5-8.

Nag, P., A. K. Kundu and D. Saha (1992), Changing concept of national atlas in P. Nag (ed), *Thematic Cartography and Remote Sensing*, Concept, New Delhi, pp 39-49.

Nag, P. (1997), NATMO: His Dream and our action, *Manchitran*, No. 2, pp. 1-5.

Nag, P. (2003), *The Great Arc,* Department of Science and Technology, New Delhi.

Nath, V. (1972), Geography and development planning in India, in S.P. Chatterjee and S.P. Das Gupta (eds), *Regional Geography and Cartography*, Selected Papers Vol. 4, 21st International Geographical Congress Publication, Calcutta, pp. 148-61.

Rao, K. Narshima, V. Bhanuja Kumari, N.S. Sekhar, V.S.S.S.R. Murthy and S.V.L. Bhavani (1997), Relevance of cartography in planning process, *Indian Cartographer*, pp. 18-26.

TIFAC (1999), *Report on the Government-Industry Co-operation in Cartography*, Technology Information, Forecasting and Assessment Council, Department of Science and Technology, New Delhi.

2

Basic Concepts

Geographic Information System or GIS is so versatile that it is difficult either to define or to imagine or even to visualize it. By the end of the 1970s the term GIS had emerged in recognition both of common technological requirements and of the opportunity to build the system that could potentially satisfy all of various applications. Nevertheless, GIS is information about geography, i.e. information tied to some specific set of locations on the earth surface, including the zones immediately adjacent to the surface and subsurface, ocean and atmosphere. 'Spatial' is often used synonymously with or even in preference to 'geographical' in the context, although in principle it might be taken to include information that is tied to frames other than the earth surface, such as the human body (as in medical imaging) or a building as in architectural drawings (Faust *et al.,* 1991). Goodchild (1992) identifies two distinct primitive types of geographical information: (a) field information and (b) discrete information.

Field Information

Field information in which geography is conceived as a set of spatially continuous functions, each having a unique value everywhere, and information about discrete entities, where the world is conceived as populated by geometric objects that either and otherwise empty space and are characterize by attributes, such that any point in space may lie in any number of discrete entities. The field or object dichotomy underlies many areas of GIS including its data models, data qualities, and analysis and modelling.

Discrete Information

Modern GIS has enhanced the utility of the map by replacing it with a large number of coverage that can be automatically analyzed and their themes combined to give meaningful answer for decision-makers (Demers, 1999). Today there is an ever increasing recognition of the need to perform large scale mapping and map analysis operation for wide variety of traditionally manual task. Some of the examples are:

- Foresters are interested to keep an updated inventory of their timber resources. GIS is an efficient management tool for their day-to-day operations.
- Fire departments have a need for GIS to enhance their routing capabilities to ensure rapid response in emergencies.

- Military or defence forces could use GIS to determine appropriate battle places and to organize troops movements.
- Local governments may employ GIS to develop growth and development plans and to modify zoning regulation to account for increasing population pressures.
- Businessmen are in use of GIS to market products and even to develop mailing list based on selected spatial criteria.
- Real estate companies are beginning to use GIS to isolate available housing on the basis of customer criteria such as proximity to school, type of neighbourhood or access to highway.
- Police departments are currently using GIS to compile information to characterize the movements and operational settings of suspected serial killers
- Academic disciplines such as geography, biology, geology, landscape architecture and wild life management now have the capability to employ the technology to develop and test hypothesis concerning pattern of natural phenomenon on the earth.

The potential users of GIS are limitless and the types and number of users are growing at a logarithmic pace. More examples have been given in the following chapters.

The modern GIS has the utility of the maps by replacing it with a large number of mapped 'coverages', each with an interrelated theme. These coverages can be automatically analyzed and the themes are combined to give meaningful answers for decision-makers. The GIS is gradually changing the way we do think with maps, the way we think about geographic Information, even the way in which the geographic data is collected and compiled. The task, which was impossible with the traditional maps, is now common place. To answer such question requires access to the geographical information, which is characterized by its multidimensional nature (x, y, z coordinates and time), its large volume and high processing cost. Even to answer apparently simple questions needs that data from several sources must be integrated into consistent form and be analyzed.

In the late twentieth century, the awareness of the importance for data on topography and specific themes of the earth surface, such as natural resources, has increased greatly. Aerial photography and satellite based remote sensing data ease the means of mapping large areas with a greater degree of accuracy. The same technology also gave the resource scientist enormous advantage for reconnaissance and semi detailed mapping. The resulting thematic maps were the sources of useful information for resource exploitation and management.

GIS has been defined by its objective, as in Cowen's definition of GIS as spatial decision support system (1988). In the context of remote sensing, GIS tends to be reserved for system that integrate remotely sensed data with other types or processed data that have already been cleaned and transformed. Similarly, an atmospheric scientist or oceanographer will tend to associate GIS with system used more for multidisciplinary work and policy study, and will use other software environments for modelling and analyses within the confines of his or her own discipline. In short, because GIS implies generalized software environment which is exclusive to geographical information implying that there is a tendency for it to be most strongly associated with multidisciplinary, integrative work and applications in more narrowly defined environments.

As mentioned earlier, the term GIS is difficult to define and represents the integration of many subject areas. It is becoming hybridized and modified in order to conform to intellectual, cultural, economic and political objectives. Geographic Information Systems have three major components. These are as follows:

- Computer hardware
- Sets of application software
- Proper organizational context including skilled persons

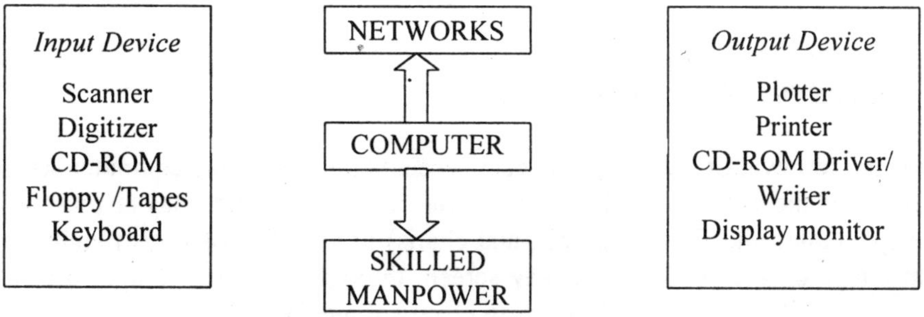

Fig. 2.1: Components of GIS

The general hardware components of GIS are presented in the above Fig. 2.1. The computer has a hard disk for storing data and operational software package (programmes), but extra storage can be provided via network or by digital tape, optical CD-Rom and other devices. A digitizer and or scanner is linked with a personal or desktop computer or work station in order to convert maps into digital form so that they can be used by the specific software. A plotter or printer or any other kind of display device or monitor is used to present the results of the data processing. Inter—local and global electronic networks using special data lines with optical fibers or over ordinary telephone lines by using a device is known as modem which provide computer communication. The user controls the computer and the peripherals (a general term for plotter, printer, digitizer and other apparatus linked to computer), via the computer screen and keyboard, added by a mouse.

Software for GIS

The software for a GIS may be categorized into five functional groups:

- Data inputting and verification or checking
- Data storing and data base management
- Data transformation and projection
- Data processing, modelling and analysis
- Interactions with users

Data input and verification

Data input covers all the aspects of capturing spatial data from existing maps, field observations, and sensors (aerial photography, satellite, lazar mapping and recording instruments) and covering them to a standard digital forms. Many tools are available including the interactive computer screen, mouse, and digitizer or scanner, word processors and spread sheet programmes, and devices necessary for reading data already written on a media (CD-ROM, magnetic tape etc.). Once the data have entered, there is a need to check them for errors—possible inaccuracies, omission, and other problems. This is to be done prior to the linkage of the spatial and attribute data.

Data Storage and Data Management

This means the retrieval of data from the data base and display on the computer screen. This also includes how the results of analysis are reported to the users. Data may be presented in the form of maps, tables and figures (graphs and charts) in a variety of ways ranging from the ephemeral image on the computer display screen, through hard copy output drawn on the printer, plotter or information recorded on an electrical media (CD-Rom, data etc.) in digital form.

Data Transformation

Data transformation embraces two types of operations, namely: (i) transformation is needed to remove errors from the data or to bring them up-to-date or to match them to other data sets; and (ii) the large array of analytical methods that may be applied to the data in order to solve the problems in question. Transformation can operate on spatial, topological and non-spatial aspects of the data, either separate or in combination. A few examples are given below:

- Those associated with changing of scale
- Those fitting data to new projection systems
- Logical retrieval of data
- Computation of area, perimeters, length and so on.

Many of these transformations are of general nature that one should expect to find them in every kind of GIS softwares in one form or another. Other kinds of manipulation may be extremely application specific and their incorporation into a particular GIS may be only to satisfy the particular users of that system. The kind of transformations methods available, their optimum use and misuse, the way in which sets of simple transformations may be combined in order to achieve certain types of geographical or spatial modeling are the major themes of GIS. Certain spatial transformations are necessary to ensure the integrity of a database.

Data Processing, Modelling and Analysis

It is long felt that the requirements of users to retrieve and transform data are unlimited. Therefore, most systems provide a range of interfaces by which the user can interact with the system. The simplest one is menu-driven commands that can be selected by simply pointing, checking and clicking with the mouse and this is an efficient way of performing complex function for ordinary users. An alternative one is for typing the simple commands with the help of command language. This is very much similar to word processors. Many commands in GIS are written in straightforward, modular language so that the users can understand exactly what is to be done. There are options for the users to create reproducible sets of commands that carry out well-defined operations linking several modular commands together. For this there is no need that the users must be very skilful and expert in programming.

Of course, menu driven GIS may not be ultimate solution for all the GIS problems. In many situations users may need to write some programme to meet their specific needs. Now a days, some GIS are equipped with so called macro languages—simplified and formal programming languages that may be used to link many basic applications together. The latest development, particularly for simulation modelling with GIS is to provide the user with a high level compliable programming language in which the models can efficiently be written. In other situations, the GIS is used to assemble the data for a complex model that is programmed outside the system using standard computer language (Burrough and McDonnell, 1998). After the modelling, the results of the models can be transferred back to GIS for display and query.

Interactions with Users

Geographic information comprises data about the surface, subsurface, and atmosphere of the earth, interpretations and explanations applied to those data, and an organizational framework for understanding the information. Commonly, geographic information is considered that which comes from maps, but the concept embraces any type of information, which may be positioned, on or relative to the earth. It may be acquired through direct measurement, remote sensors, *in situ* observations; defined through survey or legal description; interpreted from data analysis or simulation. Geographic information obtains meaning from both spatial and non-spatial aspects of the data, i.e. "where" and "what".

The availability of computerized geographic data has led to profound changes in how we use and understand information about the earth (Fig. 2.2). Map-making in its traditional form has evolved into systems for enabling user-directed views of selected information. Cartographic principles are essential to successful automation of geographic information—the concepts of abstraction, classification, delineation, and interpretation is equally applicable to a computerized system. Other map-making concepts such as projection and symbology have taken new forms as coordinate algorithms and database attributes, respectively. Although earlier systems for computerizing geographic data focused on faithful automation of the components of map sheets, such as a geometric shape with a label, newer approaches consider the underlying problem of effectively modelling real-world conditions, such as a bounded area associated with its description.

Other realms of geographic information have also been transformed by the increased prevalence of automation. Remotely sensed data and direct measurements are frequently acquired as digital data straightaway, and traditional aerial photography is being exploited using new scanning technologies. Even information not traditionally considered as purely 'geographic', such as reports about a location or videotape of an event, are often incorporated into databases of geographic information.

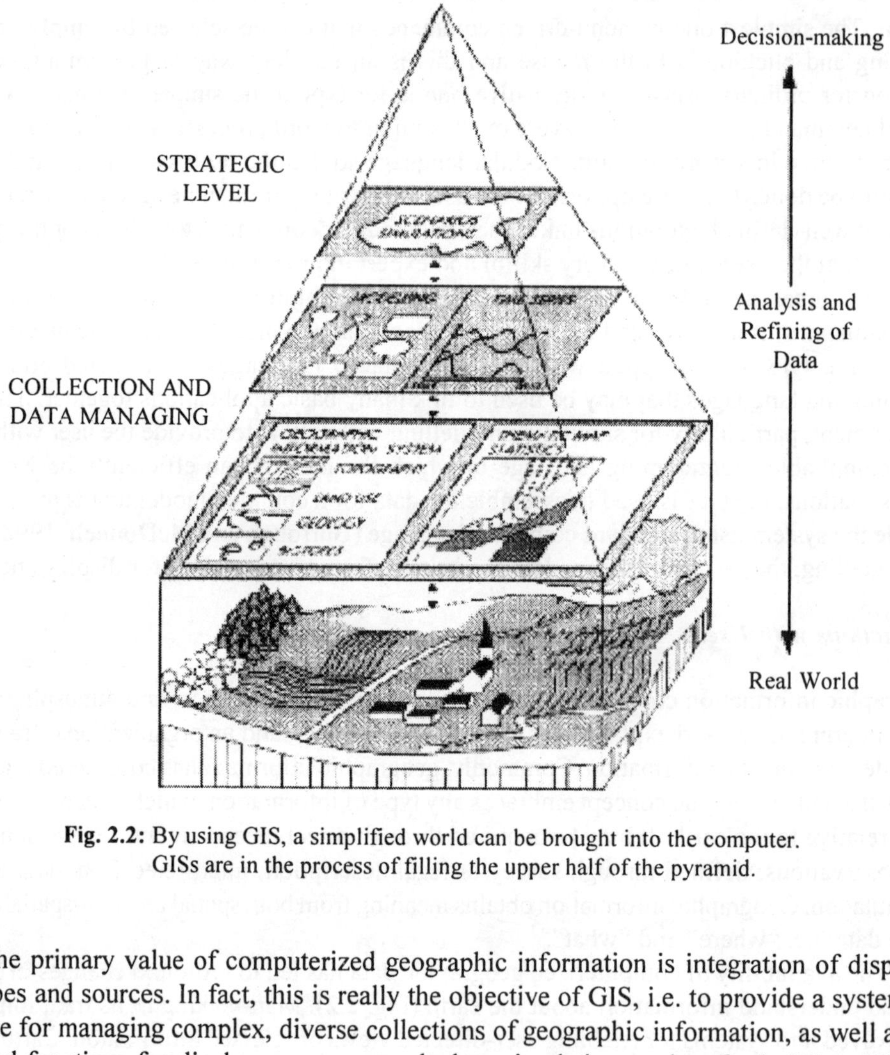

Fig. 2.2: By using GIS, a simplified world can be brought into the computer. GISs are in the process of filling the upper half of the pyramid.

The primary value of computerized geographic information is integration of disparate data types and sources. In fact, this is really the objective of GIS, i.e. to provide a systematic structure for managing complex, diverse collections of geographic information, as well as the tools and functions for display, query, map algebra, simulation, and so forth. What GIS has wrought is a new way of *thinking spatially*. Spatial analysis goes beyond simple retrieval, and allows us to explore spatial relationships and process (Fig. 2.3).

System Components

GIS have five primary components, namely, (a) *People,* computer staff, experts from various science disciplines, GIS operators, GIS experts, applications developers; (b) *Data,* which may be of type *spatial*, *temporal*, or *attribute*; (c) *Engines* that perform various data storage, retrieval, analysis, reporting, and communication functions; (d) *Interfaces* such as GUIs having widgets based on toolboxes such as X-Windows or MOTIF, Microsoft Windows; and (e) *Hardware,* including workstations and networks, disk and tape storage, digitizers, plotters, and communications devices (Fig. 2.4).

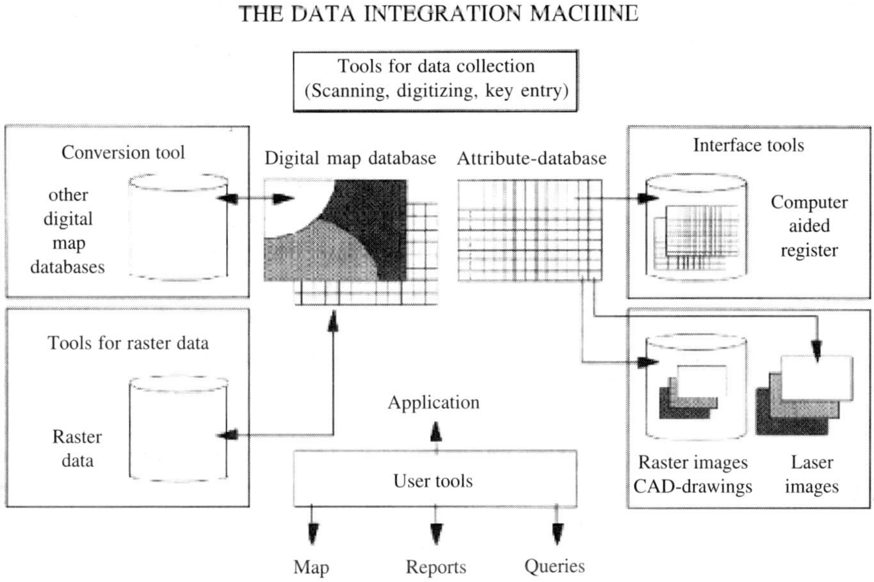

Fig. 2.3: GIS data integration machine

Fig. 2.4: Elements of GIS (*Source*: ISRI)

People

Whether a GIS is operating in a university environment or in a business or government organization, it is not in a vacuum. Inside the system there are interactions with the people responsible for day-to-day operations of the GIS, with those who operate the system, and with those responsible for project management and oversight (Fig. 2.5).

Fig. 2.5: Qualified and trained people are success key of every GIS (*Courtesy*: RSI, Hyderabad).

The system users—those who will use the GIS to solve spatial problems—are most often people who are well trained in GIS, perhaps in a specific GIS. Digitizing, error checking, editing, analysis of the raw data, and output of the final solutions to queries are the primary tasks of the system users. Whether trained in university courses or on the job, the system users will require additional training because of the constant changes both in software and in the demands of the system for new analytical techniques. In many GIS operations, this group is likely to be the most transitory, with new employees replacing those who have moved on to different positions.

System operators are responsible for the day-to-day operations of the system, more often performing tasks that allow the system users to function efficiently. Their work involves troubleshooting when programmes hang up or the complexity of the analysis requires additional insight. They are also responsible, in most cases, for training the users. In many cases system operators have experience as system users. They also know enough about hardware and software configurations to be able to make adjustments for upgrades. In addition to acting as system administrators, they frequently act as database administrators, keeping track of the security and integrity of the database to prevent possible loss or corruption of data.

GIS suppliers are responsible for providing software support and updates of the software as new and improved methods are put into the system. They may work as concert with hardware development companies to provide bundled packages of software and hardware. In many cases,

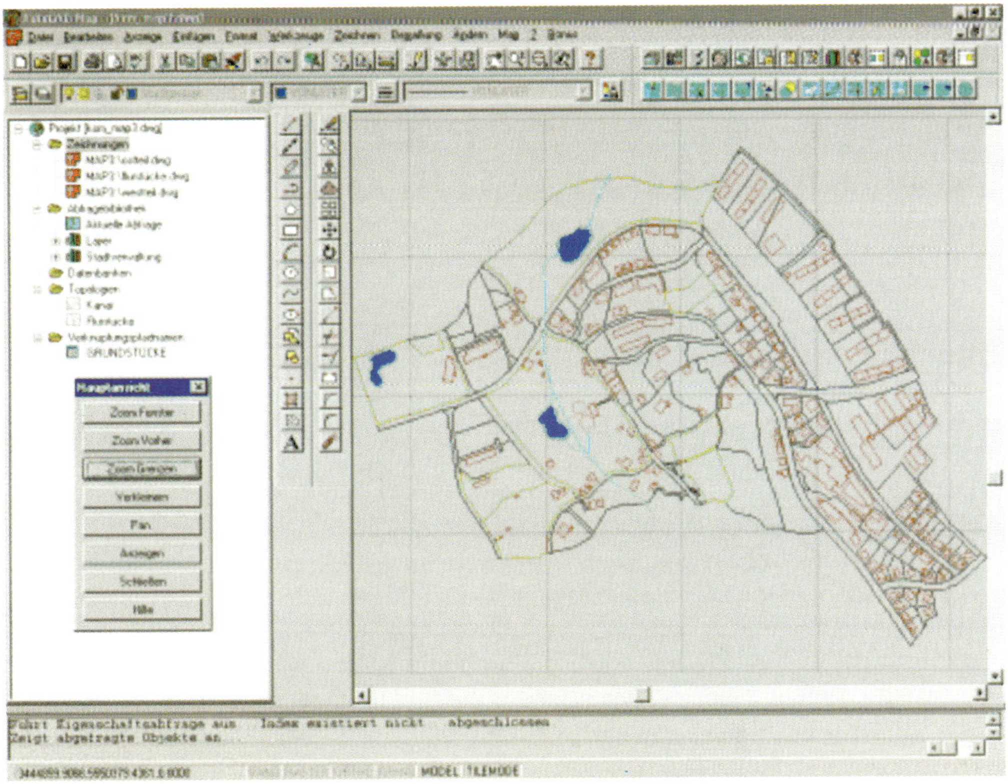

Fig. 2.6: Map production in AutoCAD Map R3

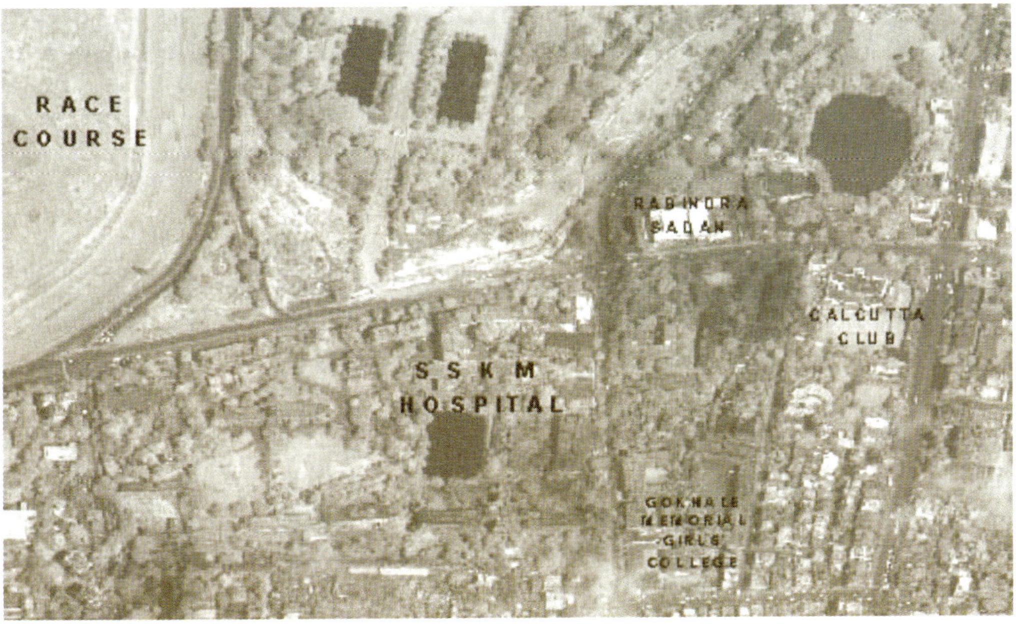

Fig. 2.7: Satellite image of the Part of Kolkata

training is provided for GIS users, either on site or at the supplier's facility, through contacts with the system administrator.

The data supplier could be either private or industry or public. The private company or industry may provide internally generated data or data obtained from public agencies modified to better fit needs expressed by the user community. Public agencies, primarily federal government agencies, provide data for large portions of the country. In many cases these data are designed for use by the agencies themselves, but the data are available to external users as well. Some public agencies offer data at cost, or in some cases at no cost at all, provided the use is for non-profit GIS operations.

Application developers are generally trained programmers who will provide user interface to reduce the reliance on specialized GIS professionals to perform common tasks. In many cases the programming is done in macro languages provided by the GIS supplier to support applications development without the need for interfacing with traditional computer languages. While in the past, most application developers were internal personnel who had years of experience with the analytical capabilities of the GIS software, today there are many small consulting firms specializing in the development of applications for third-party clients.

GIS systems analysts are group of people specialize in the study of systems design. Most often system analysts are part of a team of professionals responsible for determining the goals and objectives of the GIS system within an organisation, fine tuning the system so that it provides the right analytical techniques, and assuring the successful integration of the system in to organizational framework. Commonly the systems analysts are part of a larger consulting firm specializing in the implementation of GIS for third-party users, but in some cases they are employees of the GIS suppliers.

Data

The every GIS system should be able to absorb data in a variety of formats, not just in the native format of the particular GIS. For example, an outline map of India may be available as an AutoCAD DXF format file. The GIS should at a minimum be capable of absorbing the DXF file without further modification (Fig. 2.6). Similarly, attributes may already be stored in DBF format, and should be absordable either directly or through the generic ASCII format. Support for data formats is important to a GIS when data are to be brought in from outside (e.g. public-domain data from the internet). Ideally, the GIS software should be able to read common data formats for both raster (DEN, GIFF, TIFF, JPEG, EPS) and vector (TIGER, HPGL, DXF, Postscript, DLG). Some GIS packages have import functions only into a single data structure, usually either an entity-by-entity structure or a topological structure (for details *vide* Chapter 3: Data Structure).

For three-dimensional data, these systems usually support only the triangular irregular network. Others support only raster structures based on the grid, including the quad tree, and either convert all data into this structure or just ignore it. Some GIS packages continue to support only data in a proprietary format, available only at cost from the software vendor. A rather critical GIS function is the ability to convert between raster and vector data, an absolutely essential feature for the integration of multiple data sources such as global positioning system or GPS data and satellite images (Fig. 2.7).

Of increasing interest in recent years has been the development of GIS functions that support data in standard exchange formats. At the national and international levels, several data transfer standards have now been developed, such as the Spatial Data Transfer Standard (SDTS) and DIGEST (for details see Chapter 3). Much of the power of GIS software comes from the ability to manage not just map data but also attribute data. Every GIS is built around the software capabilities of a database management system (DBMS), a suite of software capable of storing, retrieving selectively, and reorganizing attribute information. The database manager allows us to think that all the data are available, that the data are structured in a simple flat-file format, and that they constitute a single entity. In fact, the database manager may have partitioned the data between files and memory locations and may have structured it in anyone of several formats and physical data models.

GIS Functions and Capabilities

A GIS is often defined not for what it is but for what it can do. This functional definition of GIS is very revealing about its use, because it shows us the set of capabilities that a GIS is expected to have. The functional capabilities can be grouped in the six categories. These are capabilities for data capture, data storage, data management, data retrieval, data analysis and data display. These six functions must always be present for the software to qualify as a GIS. For detail analysis of the GIS software features offered by different vendors, Appendix VI may be seen.

Data Capture

Getting the map into the computer is a critical first step in GIS and a main task of data capture functions. Before a map can be digitized it needs to be prepared. If the package supports scanning, the map needs to be clean, fold-free, free of handwritten annotation and marks, and on a stable base. GIS software could have support for digitizing and sophisticated editing systems for detecting and eliminating digitizing errors. Level of this support depends on software vendor. One of the essential data capturing functions is editing of the maps after they have been captured. For a vector data set, at a minimum we should be able to delete and reenter a point or a line. For raster data set, we should be able to modify the grid by selecting subsets, changing the grid spacing, or changing a specific erroneous grid value. Other functions typical of an editor are node snapping, in which points which are close to each other that should indeed be the same point, such as the endpoints of a line segment, are automatically placed into the graphic database with the identical coordinates; dissolve, when duplicate boundaries or unnecessary lines are eliminated automatically or manually; and mosaicing or zipping, in which adjacent map sheets scanned or digitized separately are merged into a seamless database without the unnecessary discontinuities caused by the lack of edge matching of the paper maps.

Another important editing ability is the ability to deal with map generalization (Fig. 2.8). Many digitizing modules of GIS systems, and certainly scanning, generate far more points than are necessary for the use of the GIS. This extra detail can complicate data reformatting and display, slow the analysis process, and lead to memory problems on the computer. Many

GIS packages allow the user to select how much detail to retain in a feature. Most will retain points that have a minimum separation, and snap together all points within a fuzzy tolerance. For point data sets, most GISs will eliminate or average duplicate points with the same coordinates. Some will allow line generalization, using any one of many algorithms that reduce the number of points in a line (Joao, 1998). Area features can be eliminated if they become too small, or can be grouped together, a process many GISs call clumping.

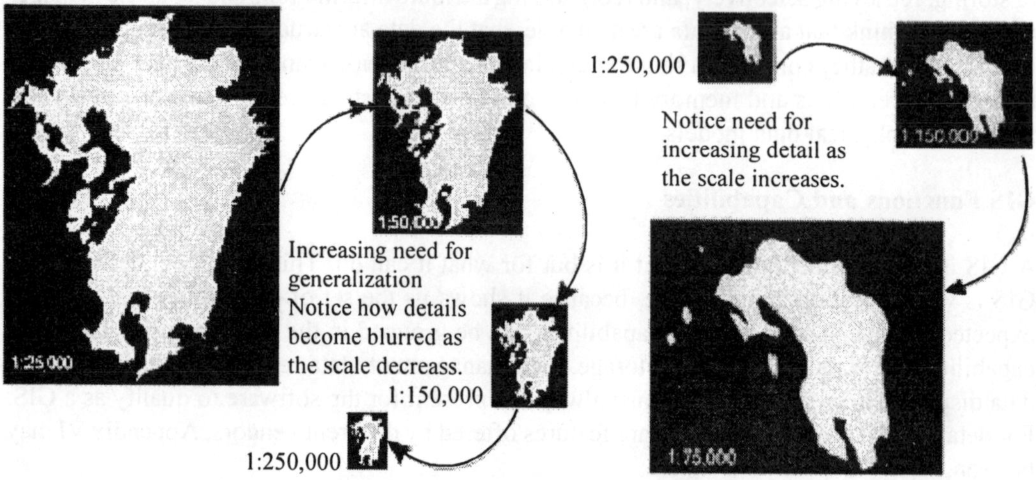

Fig. 2.8: Effect of generalization

Topological data based GIS can check automatically to make sure that the lines meet at nodes and that the entire map area is covered by polygons without gaps or overlaps. Beyond simply checking, they allow automatic cleaning of topology, snapping nodes, eliminating duplicate lines, closing polygons and eliminating slivers. The GIS user should be careful when using automatic cleaning, since the tolerances may eliminate important small features or move the features around in geographic space without accountability. Some GISs have ability for direct GPS data conversion, with survey-type data from COGO (coordinate geometry) systems, or with remotely sensed imagery.

Essential to *geocoding* capabilities, since GIS allows maps from many sources to be brought into a common reference frame and overlaid, is the *geocoding* software's ability to move between coordinate systems and map projections. Most GIS packages accomplish these operations using affine transformations. Affine operations are related to plane geometry; they manipulate the coordinates themselves by scaling the axes, rotating the map, and moving the coordinate system's origin. In some cases, when no good control is available, maps must be statistically registered together, especially when one layer is a map and one an image or photograph. The statistical method known as *rubber sheeting* or *warping* is used for this and is a function inside many GIS packages.

In order to be useful, a GIS must provide tools above and beyond the editor to check the characteristics of the database. Checking the attributes is the responsibility of the database manager. The database system should enforce the restrictions on the GIS that are specified during the data definition phase of database construction and stored in the data dictionary. Most of this checking is done at data-entry time. It checks to determine that values fall within the correct type and range.

Data Storage

Data storage within a GIS has historically been an issue of space: (a) usually how much disk space the system requires and access, or (b) how flexible a GIS is in terms of making the data available for use. The massive reductions in the cost of disk storage, new high-density storage media, and the integration of compression methods into common operating systems have made the former less critical. Although disk storage is less critical than in the past, it can still be a constraint. GIS software on a microcomputer can occupy tens of megabytes even without data, and on a workstation perhaps hundreds of megabytes. As data become higher resolution, as more raster layers are used, and as finer and finer detail becomes available, many GIS data sets can easily move into the gigabyte range size.

Database Management System

Database management system (DBMS) allows data entry, data editing and supports tabular or other list types of output, sometimes independent of the GIS. Many operations on data are very important from a mapping perspective. For example, very often maps captured from different sheets must be merged together, or sometimes a mask must be placed over the data to exclude features entirely from the GIS. Even more complex, sometimes line features such as latitude/ longitude grid, a river, or a political boundary must be sectioned up or have points added as new features or layers are introduced. This feature, called dynamic segmentation, can be done automatically by the GIS.

Data Retrieval, Analysis and Display

Another major area of GIS functionality is that of data retrieval. GIS supports the retrieval of features by both their attributes and their spatial characteristics. Geographic analysis requires a close association between the spatial elements and their attribute data. Previously these issues have been held and managed separately within automated cartography packages and databases respectively. GIS provides the technology to perform more sophisticated analysis which makes use of the links between the two. Queries to a GIS can thus be graphics driven or data driven. Graphics driven queries involve spatial based searches for objects and retrieval of the associated attribute data, or point and query functions to select displayed features and retrieve the associated information. Data driven queries involve the use of data values selectively to display the matching spatial features or the use of attribute values to determine shading pattern or colour coding of the relevant spatial elements. Operations to retrieve, summarize, selectively display and analyze both the alphanumeric and graphical data are usually embedded within a GIS. Examples of the types of spatial analysis usually available within a GIS include (Fig. 2.9).

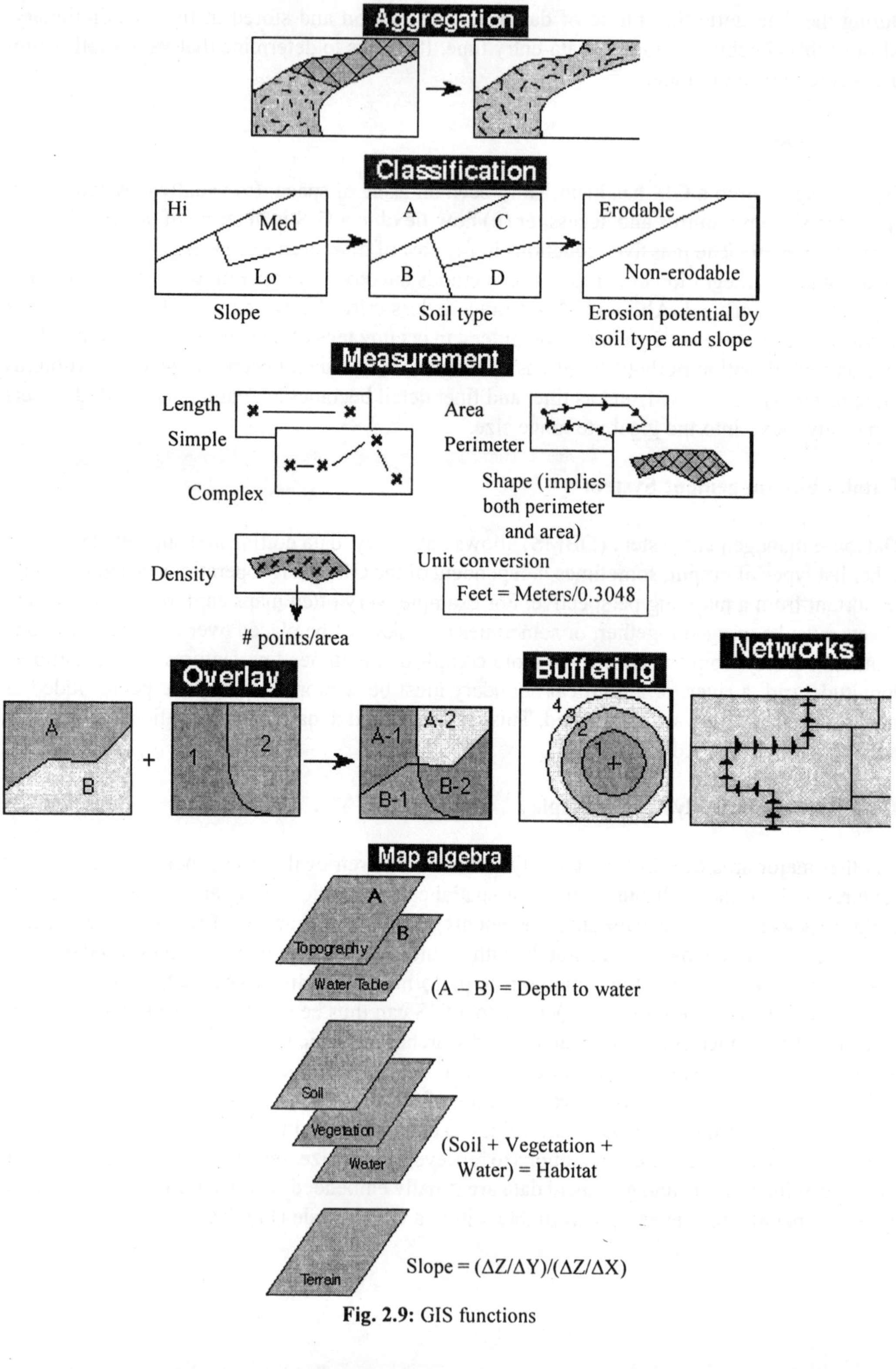

Fig. 2.9: GIS functions

Choropleth Mapping

Area features are classified according to their attributes. A legend acts as a look-up table with each range of attribute values being associated with a particular colour or shading pattern, e.g. shading enumeration districts according to population density.

Buffer Generation

Boundaries are created around points, lines or areas at an equal distance in all directions. The result being circular, corridor or polygon shaped features, respectively, which represent areas at set distances from the original object, e.g. creation of buffers around an airport to calculate zones of different noise intensity.

Polygon Overlay

Area features on one data layer are overlaid onto those of other data layers in order to calculate areas which have a certain combination of attributes or lack certain values, e.g. overlaying soil, drainage and slope data layers will provide information on the land most suitable for crop production.

Contouring

The interpolation of lines of the same elevation from spot height data, e.g. the calculation of height contours from which to generate slope and elevation data for land surfaces or any other statistical surfaces (Fig. 2.10).

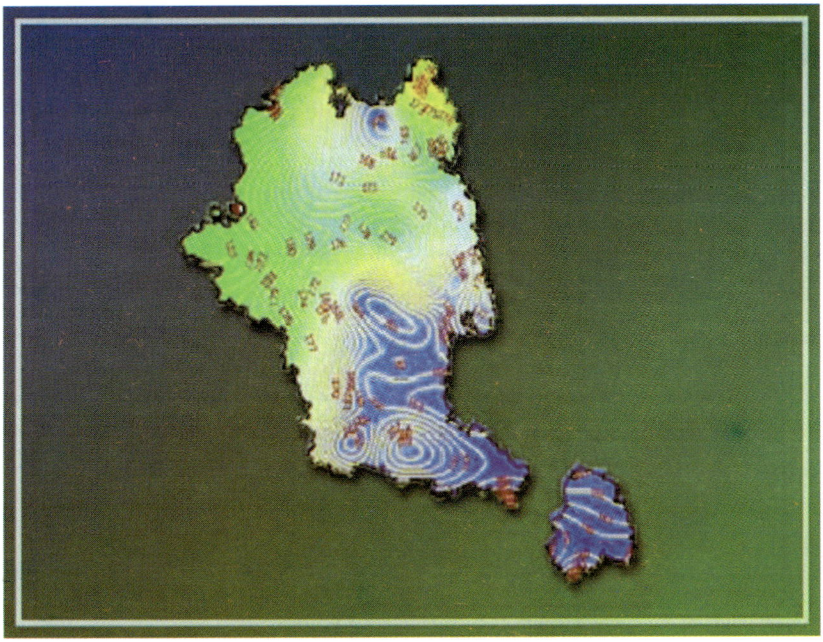

Fig. 2.10: Contours (*Source*: NRDMS)

Terrain Analysis

The creation of three dimensional views or digital terrain models (DTM) from height data, e.g. a three-dimensional view of a landscape can be used to assess the visual impact of an afforestation programme or for watershed planning (Fig. 2.11).

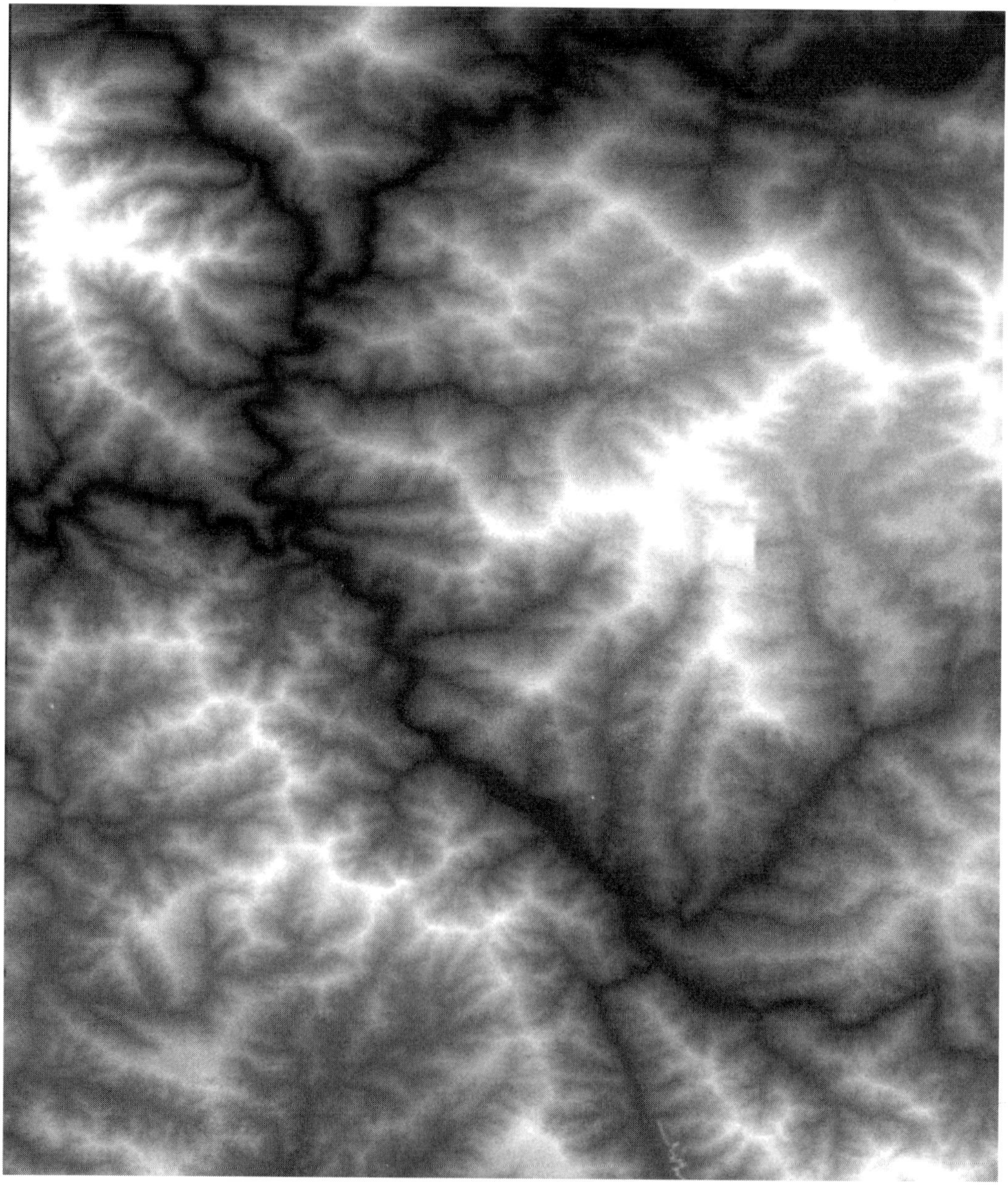

Fig. 2.11: Digital terrain Model Created from Contours (*Source*: Survey of India)

Network Analysis

Tracing through a network of connected line features in order to simulate flows of traffic or water. Often used to calculate shortest or quickest paths between two points in a network, e.g. use of a network to calculate optimum routes for emergency vehicles (Fig. 2.12).

Area and Length Calculations

The calculation of the area of polygon features or the length of linear features. Often these simple calculations can be constrained to include only those features with attributes satisfying certain selection criteria, e.g. calculation of the area of parkland within a city which needs to be maintained or calculation of the length of pipeline within a sewage network which needs replacing. Examples of how the attribute information may be manipulated within a GIS include the retrieval of data values for selected features, the computation of attribute means and the calculation of statistical summaries and samples.

Output from a GIS can be in the form of an interactive computer display of the graphic and/or attribute data or as traditional paper maps, tables, graphs and reports. Automated cartography techniques are employed to produce the graphical output from GIS. Commonly used output devices include pen plotters, inkjet printers, electrostatic plotters, laser printers and screen copiers. Export to other systems such as statistical packages, word processors and multi-media will usually require the GIS to be able to convert internal data into other standard formats.

Interfaces

GIS was heavily influenced by the types of operating systems in use. Early operating systems were quite unsophisticated but were used with GIS nevertheless. Among these were IBM mainframe operating systems, MSDOS by Microsoft and DEC VMS.

In the microcomputer environment, the GUI-based operating systems include Windows and Windows NT, now updated as Windows 2000XP, Apple Macintosh GUI and IBM OS/2. These operating systems added two critical elements to the microcomputer's capabilities, multitasking and device independence, meaning that plotters and printers could be taken out and assigned to the operating system instead of the GIS package, in somewhat the way that printing and screen fonts are handled centrally, rather than duplicated in every Windows package.

One system that had encompassed these capabilities since its inception, and that swept the workstation environment, was UNIX. It has been the dominant workstation environment for two reasons: first, because it has complete integrated network support, and second, because several full GUIs exist for UNIX in the public domain, the most important being the X-Windows system.

X-Windows implementations of most leading GUIs exist, including OpenLook and the Open Software Foundation's MOTIF interface. In many UNIX systems, the user can switch the GUI to suit particular needs or applications. A full GUI programming toolkit, including such tools as Xt, Xview, Tcl/Tk and the X-Windows libraries Xlib are part of the X-Windows release.

As a final benefit, several versions of UNIX and all of the GUI systems run extremely efficiently on microcomputers, including shareware UNIX releases such as Linux, not only outperforming the Windows type GUIs, but being available free or as shareware on the Internet or CD-ROM. A key element here has been the Free Software Foundation's releases, including GNU (GNU is NOT UNIX) versions of virtually every key element of UNIX.

Thus two main avenues for GISs have evolved as far as operating systems are concerned. On the microcomputer platform a lingering set of DOS applications is rapidly being rewritten for the updated versions of Microsoft's Windows. In this GIS environment, the number of systems installed, the mobility of laptop and subnotebook computing, and low cost of software have been major strengths. On the workstation platform, UNIX and X-Windows, often with MOTIF as the GUI, reign supreme. This work environment has led to high end applications, large data sets, networking, depth software, and high quality graphics. Both are healthy and prospering workplaces for GIS.

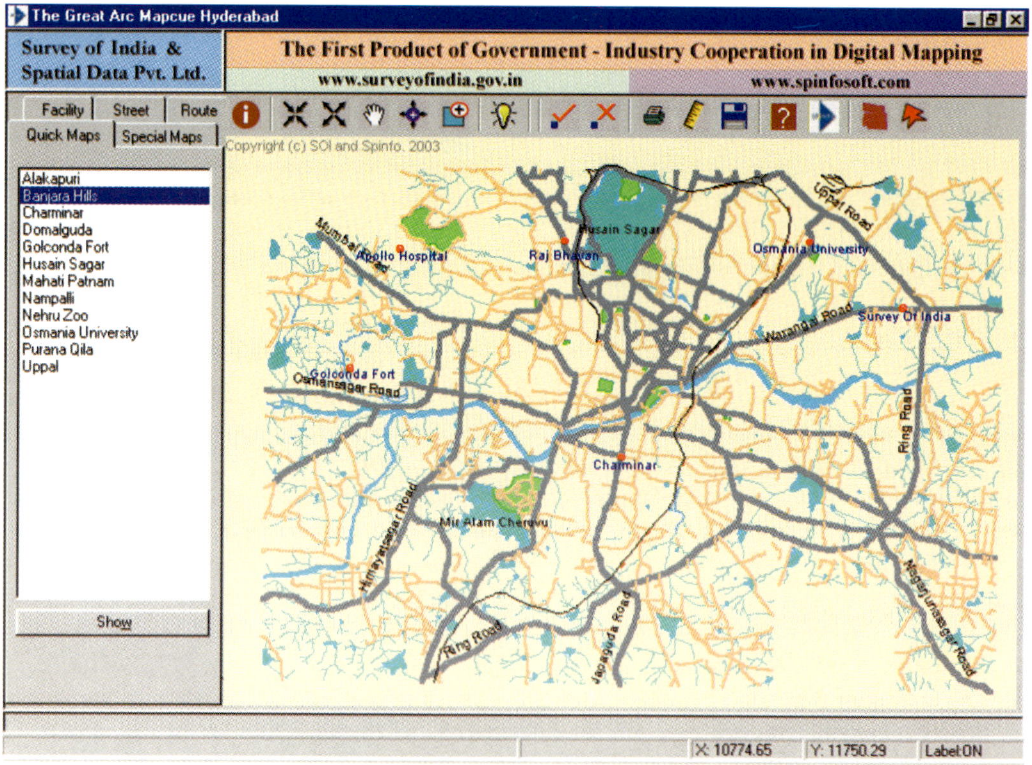

Fig. 2.12: Network of roads in Hyderabad (*Source*: Survey of India and Spinfo, Bangalore)

Summary

In any set up, big or small, the geographical information system or GIS will be having different combinations of the above hardware. As mentioned earlier, the choice of hardware is software dependent as well. The prices of hardware and even software are falling. Further, in case of the latter, the licencing policy also determines the cost. In addition, newer items are coming to the market everyday. Hence, the decision to have hardware, software and networking is a complex, technically and financially. In the following chapters we will find several examples of such combinations procured by different agencies to bring out GIS based business solutions.

REFERENCES

Burrough, P.A. (1992), Development of intelligent geographical information system, *International Journal of Geographical Information System*, Vol. 6, No. 1, 1-12.

Burrough, P.A. and McDonnell, R.A. (1998), *Principles of Geographical Information Systems*. Oxford University Press Inc. New York.

Cowen, D.J. (1988), GIS versus CAD versus DBMS: What are the differences ?, *Photogrammetric Engineering and Remote Sensing*, Vol. 54, pp. 1551-5.

Demers, M.N. (1999), *Fundamentals of Geographic Information Systems*, 2nd Ed., John Wiley and Sons. Inc.

ESRI (u.d.), *ESRI Map Book*, ESRI Press, Vol. 19, Redlands.

Faust, N.L., W.H. Anderson and J.L. Star (1991), Geographic information systems and remote sensing future computing environment, *Photogrammetric Engineering and Remote Sensing*, Vol. 57, Issue 6, pp. 655-68.

Goodchild, M.F. (1992), Geographical information science, *International Journal of Geographical Information System*, Vol. 6, No.1, pp. 31-45.

Joao, Elsa Maria (1998), *Causes and Consequences of Map Generalisation*, Taylor and Francis.

Maguire, D.J.,Goodchild, M.F. and Rhind, D., (1991), *Geographic Information System*, Longman, London.

Reddy, M. Anji (2001), *Remote Sensing and Geographical Information System*, 2nd. Ed. BS Publication, Hyderabad.

Sengupta, Smita (1992), Geographical information system, in P. Nag (ed), *Thematic Mapping and Remote Sensing*, Concept Publishing Company, New Delhi, pp. 69-78.

3

Data Structure

Geographic database management systems are more complex than database management systems used for banking, library searches, airline bookings and medical records. Three general features of the data within a GIS must be maintained are: (a) information on the position of the feature being stored, (b) topological information on the spatial relationships of the features (topology is the way in which geographic features are connected and provides a mechanism to identify the positional relationships among features), and (3) attributes of the feature (Burrough, 1986). The spatial component of geographic data describes the location of a feature and the possible topological relationships among features. The attribute component describes various attributes of the feature. There are five kinds of data to be represented in a GIS (Figs. 3.1 to 3.5).

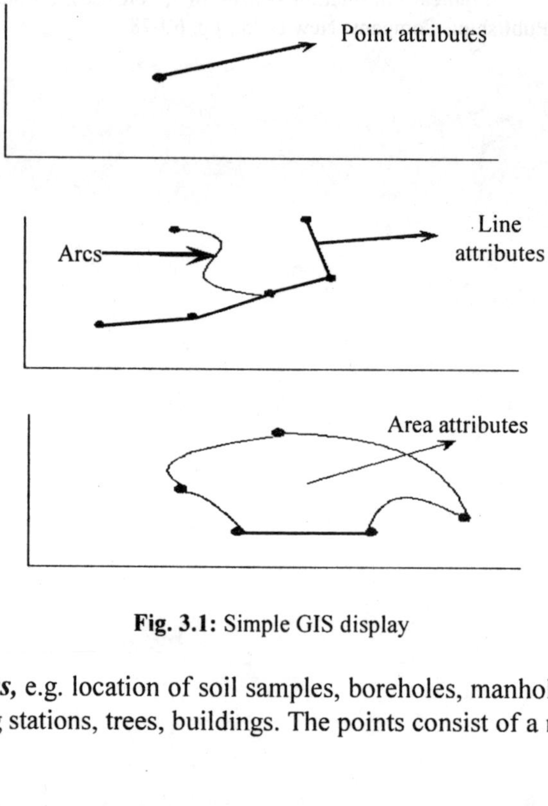

Fig. 3.1: Simple GIS display

*1. **Point features,*** e.g. location of soil samples, boreholes, manholes, rain gauges, burst water mains, pumping stations, trees, buildings. The points consist of a number of nodes with no thickness and are often referred to as zero dimensional. One method to store a point feature in a GIS is as a table in the data base management system.

Point ID	*X coordinate*	*Y coordinate*	*Pointer to attribute data*
1	30123.6	19782.4	Point 1
2	30167.3	19745.7	Point 2
3	34952.2	19648.1	Point 3

Where the pointer to attribute data is a link into another data base table when other data about that point is kept, for example that it represents an access chamber to a sewer system and that it has properties such as date of construction, condition, size, material etc. This is a link into a full data base management system so further relations are permitted from this point.

2. Linear features, e.g. roads (on small scale maps), rivers, pipelines, power lines, elevation contours. The nodes are linked with arcs, each with a number of vertices (the simple arc is a straight line). Between vertices the arc is usually considered a straight line but curved links are possible. Line data can either be non-branching lines, or tree or network structures. In a network there is more than one route between two nodes. This data has one dimension, i.e. it does not have thickness and care must be taken in the definition of the system that a loop is not confused with a polygon. A simple structure for a line feature or network is:

Line reference		
Attribute Pointer		
arc 1		
arc 2		
...		
etc.		
Arc reference	X coordinate	Y coordinate
Node 1	30123.6	19782.4
Node 2	30167.3	19745.7
Vertex 3 etc.	30952.2	19648.1

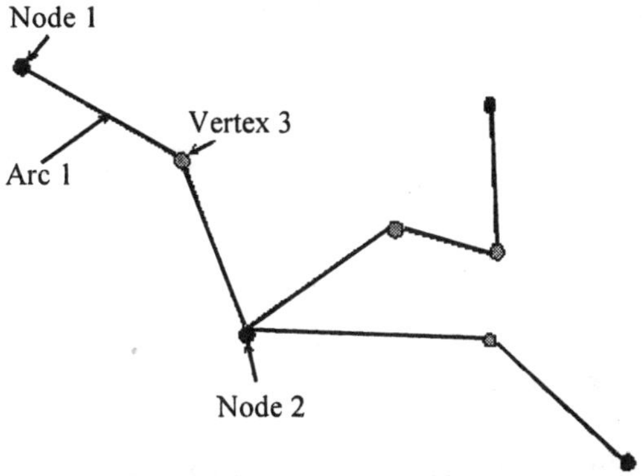

Fig. 3.2: Branching Network of 3 Arcs

3. Areas or polygons with common properties, e.g. pressure zones, catchments, contributing areas, soil association mapping units, climate zones, administrative district areas, buildings and other land cover. The polygon consists of a number of arcs or linear features that form a closed loop without crossing over one another. The arcs are usually straight between vertices but may be curved.

Polygon Reference	*Attribute Pointer*
3	Point 1
X coordinate	*Y coordinate*
30123.6	19782.4
30167.3	19745.7
30952.2	19648.1
30123.6	19782.4

Simple Polygon Structure

The simple polygon representation shown above where a quadrangle is represented, as used in CAD (or DXF format), is of little use in GIS. The three major problems with simple polygons are:

(a) The boundary between two polygons needs to be stored twice. There is always a possibility that the nodes for each boundary polygon are in slightly different positions resulting in artificial gaps between polygons, or slivers where an area is assigned to two or more polygons (Fig. 3.3). These problems are particularly acute with manually digitizing.

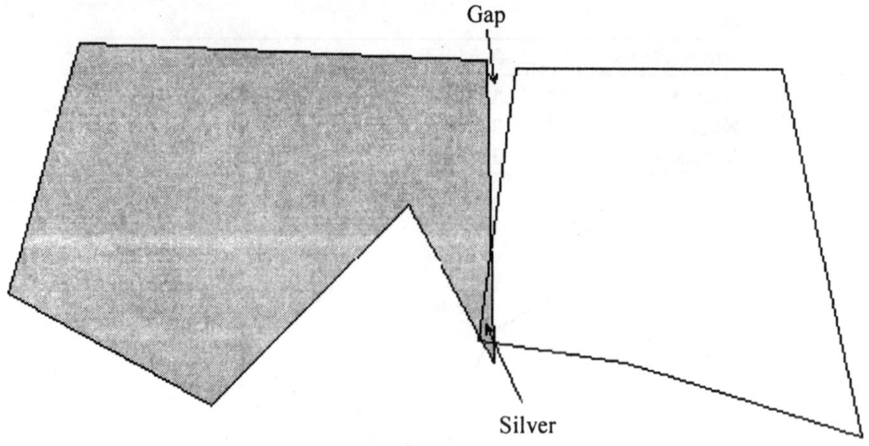

Fig. 3.3: Gaps and silvers between adjacent polygons

(b) When manually digitizing it is possible to accidentally cross over from one polygon to another creating a totally false polygon or to pass from one node to another in incorrect order giving rise to a weird polygon (Fig. 3.4).

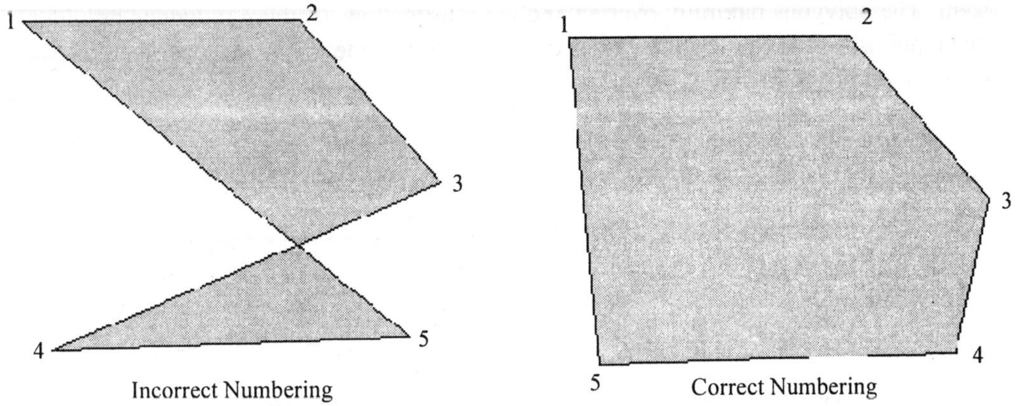

Fig. 3.4: Weird polygons

(c) Complex geographical objects are difficult to represent, for example islands or disjointed polygons (Fig. 3.5). If we consider an example from urban drainage where a garden area is completely surrounded by car park, for example, at a prestige office complex, then if we calculate the area of simple polygons the area of the car park will include, erroneously, the area of the gardens and grossly overestimate the impermeable area.

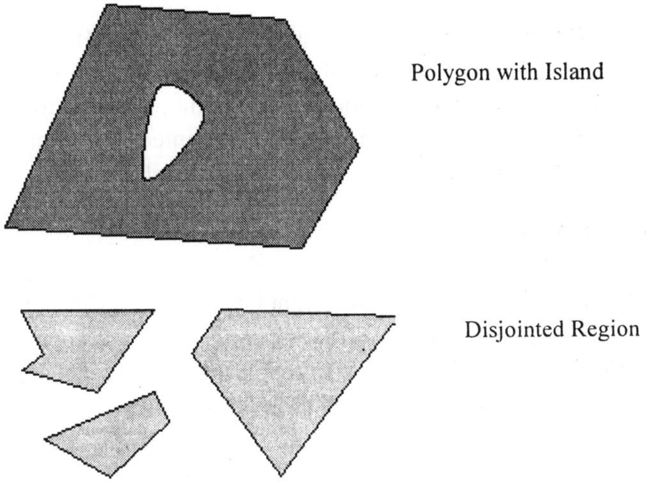

Fig. 3.5: Polygons with Islands and disjointed regions

In GIS, therefore, area data is represented as topological structure in one of a number of ways. The ArcInfo method of storing this information is shown in Fig. 3.6. A separate list is used to hold information about islands and disjointed structures. Different themes can be represented on the same coverage and there is no requirement that polygons do not overlap. For example, a single coverage may contain polygons representing land cover, whereas another

other polygons may contain the contributing areas to inlet nodes of a storm water drainage system. The polygons naturally overlap and the intersections of these polygons provides one of the main uses of GIS and is known as overlay to reflect the graphical process of overlaying one theme upon another (Fig. 3.6).

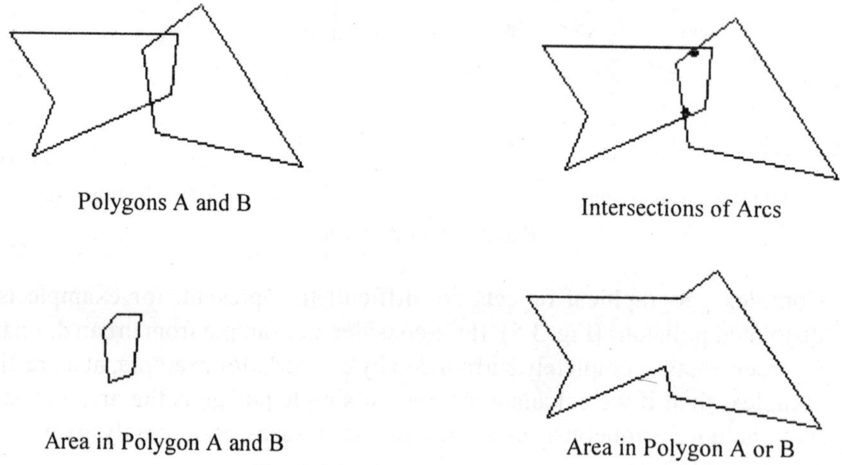

Fig. 3.6: Intersections of polygons

4. Actual or potential surfaces, e.g. ground elevation, variation of mean annual temperature, spatial distributions of rainfall, population densities.

5. Temporal elements, e.g. changes in land use over time, changes to a pipe network, rainfall records or stream flow records. These are not well represented in current GIS technology, but newer object oriented GIS should make this more readily available

Representation of Spatial Data

Spatial data are represented in GISs in two very different ways. Fig. 3.7 shows the two different ways that a stream could be represented in a GIS. Spatial data can be represented as either rasters or vectors. In raster format, a grid is used to represent the study area. The location of features in the study area is depicted by the values in the cells overlaying the feature. Vector data represent geographic features by coordinates of points, lines, and polygons. Points represent small features such as wells, towers, or nest locations. Linear features such as roads and streams are represented by lines. Areas such as cities, forests, wetlands, and soil units are represented by polygons. Polygons are bounded on all sides by a series of straight line segments.

Raster Data

Raster data are stored in the computer as a matrix. The cells are referenced by lines and elements (Fig. 3.8). In the simplest form, each line is a computer record. Each record will contain the values for all elements in the line. Any cell not containing a feature would have the value of "0". In the simplest raster system, the value stored for each cell is the attribute component of the geographic data. In Fig. 3.8(a) cells with value of "1" are forests, cells with value of "2" are croplands, and cells with value of "3" are rangelands. In more sophisticated raster systems, the cell value is a label that will link to records as an attribute file. In the above example, cells labeled as "1" could have many attributes, such as species composition, age of forest stand, and estimated volume of marketable timber.

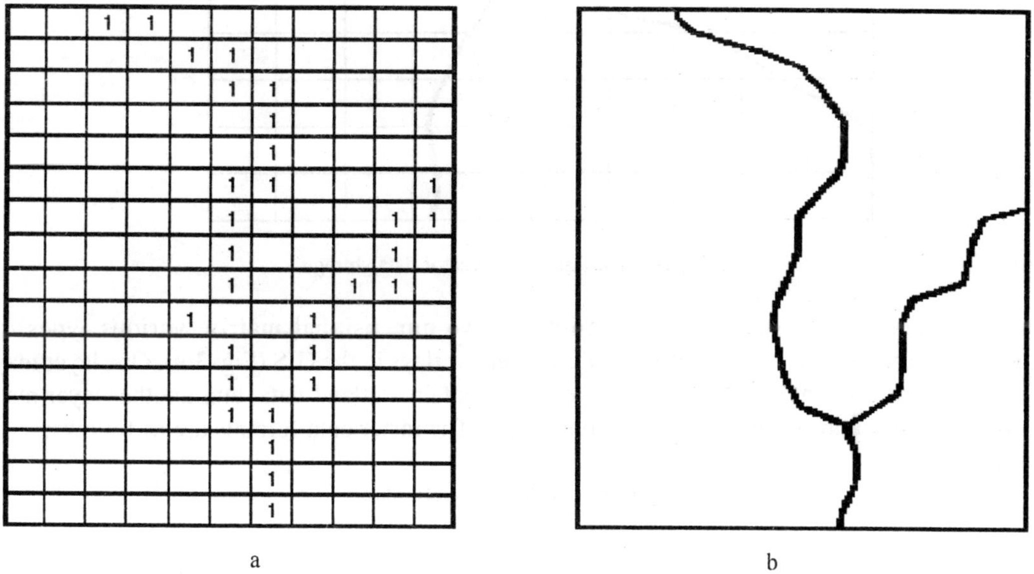

Fig. 3.7: Raster and vector data format

Note: A stream can be represented in a GIS either by a raster (a) or vector (b) format.

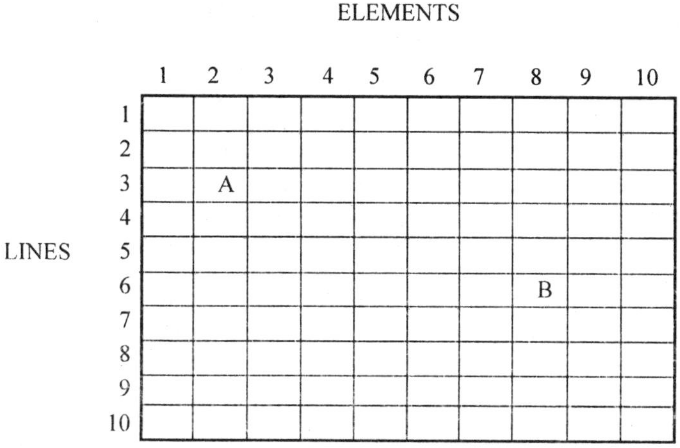

Fig. 3.8 (a): Raster data are stored in computers as a matrix. Each cell is referenced by its line and element number. The example shown is for a small file with 10 lines and 10 elements. Cell A is located at line 3, elements 2. Cell B is located at line 6, element 8.

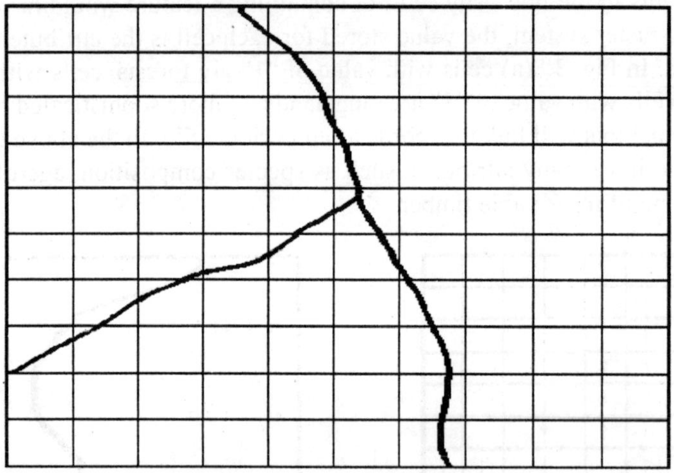

Fig. 3.8 (b): Raster and vector data storage

Because the raster system is strictly a two-dimensional matrix, various types of geographical data are stored as different layers or overlays in the GIS (Fig. 3.9). One layer may contain land use/land cover, another layer may contain wetland data, and another layer may contain information on the transportation system. The raster data formats are as follows:

Table 3. 1: Raster data formats

- Raster ADRG—ARC digitized raster graphic
- ADRI—ARC interchange raster image file
- BIL—Band interleaved by line
- BIP—Band interleaved by pixel
- BMP—Bitmap file
- BSQ—Band sequential
- CADRG—Compressed ADRG
- CEOS—Radarsat image file
- CIB—Controlled image base file
- CGM—Computer graphics metafile
- DEM—Digital elevation model
- DIB—Device independent bitmap
- DOQ—Digital orthophoto quad
- DTED—Digital terrain elevation data
- ECW—ERMapper compressed image file
- EPS—Encapsulated postscript
- GeoTIFF—TIFF image with georeferencing
- GIF—Graphic interchange format
- GIS—ERDAS Imagine file
- IMG—ERDAS Imagine native format
- JFIF—JPEG file interchange format
- JPEG—Joint photographic experts group format
- LAN—ERDAS Imagine 7.x file
- MrSID—MrSID compressed file
- NITF—National imagery transmission format
- PCX—Bitmap graphics file
- PICT—Apple's graphics file
- PNG—Portable network graphics
- PSD—Adobe Photoshop image file
- RLC—Run length encoding compressed format
- RPF—Raster product format
- SDTS—Spatial data transfer standard
- SUN/RAS—Sun raster file
- TGA—Targa graphics file
- TIFF—Tagged image file format

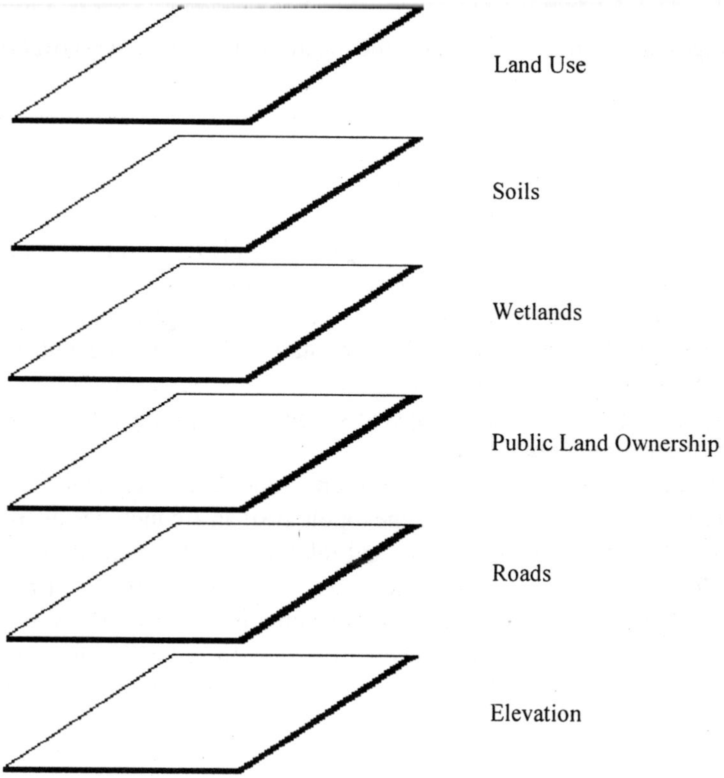

Fig. 3.9: GIS data layers

The user of a raster system must determine the size of the cells to be used. This size is referred to as spatial resolution. The cell size can vary tremendously depending upon the size of the study area and the objectives for the GIS. Cell size as large as 20 ha for state or regional planning may be adequate. For a wildlife management area, a cell size of 0.05 ha or smaller might be required, depending upon the application of the GIS and the size of the wildlife management area. Storage requirements increase drastically as the cell size is reduced. Reducing the cell size by one-half will increase the data storage requirements by a factor of four. Conversely, as cell size increases, the precision of the representation of the land feature is reduced. Choosing the appropriate cell size for a particular GIS application is a compromise between cost of data storage and computer time and reliability of the representation of the land feature.

Vector Data

Vector data provide for high precision in representing the location of features. Aronoff (1989) described how vector data can be used to define the location of a point, a line, and an area. A point is represented by a simple pair of coordinates. The line is represented by an ordered list of pairs of coordinates. The area is represented as a polygon with ordered pairs of coordinates that close the polygon (the first and last pair being the same).

The coordinates can be any arbitrary units but usually are stored as UTM, state plane, or latitude and longitude coordinates. The first vector system used simple techniques to store the X and Y coordinates for polygons. In this simple system the coordinates for the common boundary between two areas were stored twice, once for the first area and again for the adjacent area. These duplicate storage techniques, simplified computations and plotting but wasted storage space and, more importantly, provided no information as to adjacency or connectivity of geographic features (topology). Most vector systems now use topological models for representing the location of areas (Aronoff, 1989).

Topology deals with the spatial relationships between connecting or adjacent features (e.g. arcs, nodes, polygons and points). For example, the topology of an arc includes its from—and to-nodes, and its left and right polygons. Topological relationships are built from simple elements into complex elements: points (simplest element), arc (sets of connected points), areas (set of connected arcs) and routes (sets of sections, which are arcs or portions of arcs). Redundant data (coordinates) are eliminated because an arc may represent a line feature, part of the boundary of an area feature, or both. Topology is useful in GIS because many spatial modelling operations do not require coordinates, only topological information. For example, to find an optimal path between two points require a list of the arcs that connect to each other and the cost to traverse each arc in each direction. Coordinates are only needed for drawing the path after it is calculated.

In topological models (Fig. 3.10), a polygon is defined by a series of arcs. Arcs begin and end at nodes, which occur wherever two or more arcs meet. Each arc is defined by a series of coordinates, starting with the coordinates for the beginning node and ending with the coordinates for the ending node. Topological relationships are stored in three tables. The polygon topology table describes the arcs that bound each polygon, the node topology table describes the arcs that end at each of the nodes, and the arc topology table describes which end points (nodes) occur on each arc and which polygons are to the left and right of each arc. These three topology tables provide the tools required to efficiently determine the positional relationships of one feature to other features. A coordinate table defining the coordinates for each arc also is used in topological models. In addition to these topological databases, the attributes for the features are stored in an attribute database.

The vector data formats are as follows:

Table 3.2: Vector data formats

- AOI—ERDAS Imagine area of interest
- ARC—ARC export file
- ARCE—ArcInfo export format
- ARCG—ArcInfo generate format
- BNA—Atlas GIS file
- CDF—Cartographic data format
- CGM—Computer graphics metafile
- DAT—MapInfo native data format
- DFAD—Digitized feature analysis data
- DGN—Microstation design file
- DLG—Digital line graph file
- DWG—AutoCAD drawing file
- DXF—Drawing exchange format
- E00—ArcInfo export format
- EPS—Encapsulated postscript file
- GDF—Geographic data file
- GRID—ArcInfo GRID files
- IGDS—Interactive graphic design software export format
- IGES—Initial graphics exchange standard format
- MGE—Intergraph file format
- MIF/MID—MapInfo interchange format
- OVR—ERDAS Imagine annotation layer
- PICT—Apple's graphics file
- SHP—ESRI shapefile
- SIF—Standard interchange format
- SDTS—Spatial data transfer standard
- TAB—MapInfo format
- TIGER—Topologically integrated geographic encoding and referencing/line census file
- VPF—Vector product format
- VFR—Vector relational format
- WMF—Windows metafile format

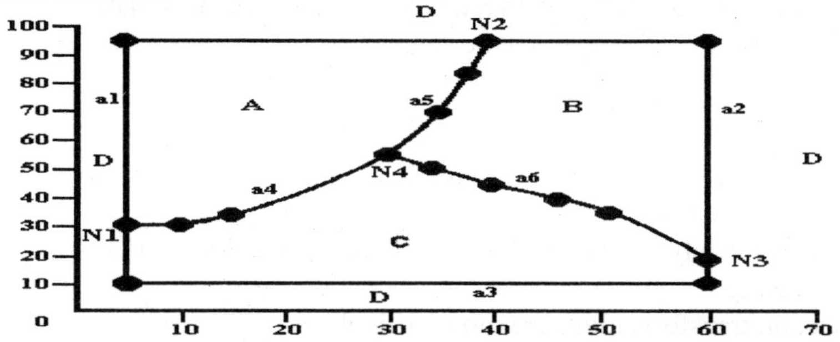

Polygon Topology	
Polygon	**Arcs**
A	a1, a4, a5
B	a2, a5, a6
C	a3, a4, a6
D	outside study area

Node Topology	
Node	**Arcs**
N1	a1, a3, a4
N2	a1, a2, a5
N3	a2, a3, a6

Arc Topology				
Arc	**Start Node**	**End Node**	**Left Polygon**	**Right Polygon**
a1	N1	N2	D	A
a2	N2	N3	D	B
a3	N3	N1	D	C
a4	N1	N4	A	C
a5	N4	N2	A	B
a6	N3	N4	C	B

Fig. 3.10: Typological model

Raster *versus* Vector Systems

Early GISs were either raster or vector systems. Various advantages and disadvantages of vector and raster data systems are mentioned below (Burrough, 1986, Aronoff, 1989). Both approaches are equally valid ways of representing spatial data. The advantages and disadvantages of raster and vector systems have been heavily debated. Most modern GISs handle both raster and vector data but usually are designed primarily for one data type. The complete integration of raster and vector data capabilities will be common in GISs of the future (Faust *et al.*, 1991). These new GISs will quickly and efficiently convert among rasters, vectors, and other data structures most appropriate for the application being performed. GISs must function equally well with both raster and vector data. Comparison advantages and disadvantages of vector and raster methods as revised from Nag and Kudrat, Burrough (1986) and Aronoff (1989).

Raster Method

Advantages

- Data structure is simple.
- The method is compatible with remotely sensed or scanned data.
- Procedures for spatial analysis are simple.

Disadvantages

- Greater disk storage is often required.
- Topological relationships are difficult to represent.
- Unless extremely small cell sizes are used, the graphic output is often aesthetically less pleasing.
- Projection transformations are more difficult.

Vector Method

Advantages

- Compact data structure requires less disk storage.
- Topological relationships are readily maintained.
- Graphic output is aesthetically more pleasing and more closely approximates hand-drawn maps.

Disadvantages

- Data structures are complex.
- Overlaying multiple vector maps is often time consuming.
- Output graphics may take hours to draw on plotters.
- Some spatial analysis procedures are difficult.
- Software and hardware for vector systems are often more expensive.
- The method is not as compatible with remote sensing data.

Database Planning and Design

The primary purpose of this phase of the GIS development process is to specify "how" the GIS will perform the required applications. Database planning and design involves defining how graphics will be symbolized (i.e., colour, weight, size, symbols, etc.), how graphics files will be structured, how non-graphic attribute files will be structured, how file directories will be organized, how files will be named, how the project area will be subdivided geographically, how GIS products will be presented (e.g., map sheet layouts, report formats, etc.), and what management and security restrictions will be imposed on file access. This is done by completing the following activities:

- Select a source (document, map, digital file, etc.) for each entity and attribute
- Set up the actual database design (logical/physical design)
- Define the procedures for converting data from source media to the database
- Define procedures for managing and maintaining the database

The database planning and design activity is conducted concurrently with the pilot study and/or benchmark activities. Clearly, actual procedures and the physical database design cannot be completed before specific GIS hardware and software has been selected while at the same time GIS hardware and software selection cannot be finalized until the selected GIS can be shown to adequately perform the required functions on the data. Thus, these two activities (design and testing) need to be conducted concurrently and iteratively.

In many cases, neither database design matters nor hardware and software selection are unconstrained activities. First, the overall environment within which the GIS will exist must be evaluated. If there exist "legacy" systems (either data, hardware or software) with which the new GIS must be compatible, then design choices may be limited. Both GIS hardware and software configurations and database organizations that are not compatible with the existing conditions should be eliminated from further consideration. Secondly, other constraints from an organizational perspective must be evaluated. It may, for example, be preferable to select a specific GIS or database structure because other agencies with whom data will be shared have adopted a particular system. Finally, assuming that the intended GIS (whether it will be large or small) will be part of a corporate or shared database, the respective roles of each participant need to be evaluated. Clearly, greater flexibility of choice will exist for major players in a shared database (e.g., county, city, or regional unit of government) than for smaller players (town, village, or special purpose GIS applications). This does not mean that the latter must always go with the majority, but simply that the shared GIS environment must be realistically evaluated. In fact, one way for the smaller participants in a shared GIS to ensure their needs are considered, is to fully document their needs and resources using procedures recommended in these guidelines.

Finally, with the completion of both the database planning and design and the pilot study and benchmark activities, sufficient detailed data volume estimates and GIS performance information will be known to calculate reliable cost estimates and prepare production schedules. This becomes the final feasibility check before major resources are committed to data conversion and GIS acquisition.

What is Already Known about the GIS Requirement

Prior phases of the GIS development process should have produced the following information which is needed at this time:

- A complete list of data, properly defined and checked for validity and consistency (from the master data list, data model and metadata entries).
- A list of potential data sources (maps, aerial photos, tabular files, digital files, etc.) catalogued and evaluated for accuracy and completeness (from the available data survey). This inventory would also include all legacy data files, either within the agency or elsewhere, which must be maintained as part of the overall shared database.
- The list of functional capabilities required of the GIS (from needs assessment).

Selecting Sources for the GIS Database

This activity involves matching each entity and its attributes to a source (map, document, photo, digital file). The information available for this task is as follows:

- List of *entities and attributes* from the conceptual design phase

Table 3.3: Master data list

Master Data List Entity	*Attributes*	*Spatial Object*
Street_segment	name, address_range	Line
Street_intersection	street_names	Line
Parcel	section_block_lot#, owner_name, owner_address, sites_address, area, depth, front_footage, assessed_value, last_sale_date, last_sale_price, size (owner_name, owner_address, assessed_value (as of previous January 1st))	Polygon
Building	building_ID, date_built, building_material, building_assessed_value	Footprint
Occupancy	occupant_name, occupant_address,occupancy_type_code	None
Street_segment	name, type, width, length, pavement_type	Polygon
Street_intersection	length, width, traffic_flow_conditions, intersecting_streets	Polygon
Water_main	type, size, material, installation_date	Line
Valve	type, installation_date	Node
Hydrant	type, installation_date, pressure, last_pressure_test_date	Node
Service	name, address, type, invalid_indicator	None
Soil	soil_code, area	Polygon
Wetland	wetland_code, area	Polygon
Floodplain	flood_code, area	Polygon
Traffic_zone	zone_ID#, area	Polygon
Census_tract	tract#, population	Polygon
Water_District	name, ID_number	Polygon
Zoning	zoning_code, area	Polygon

- The list of *surveyed data sources from the Available Data Survey* and their recorded characteristics in the metadata tables *Source Documents, Entities Contained in Source, and Attributes by Entity.*

Table 3.4: Source documents

Source Document Name:	Smarts Map
Source ID #:	1
Source Organization:	Kolkata Municipal Corporation (KMC)
Type of Document:	Map
Number of Sheets (map, photo, etc.):	100
Source Material:	Paper maps
Projection Name:	UTM
Coordinate System:	State Plane
Date Created:	5-Oct-1910
Last Updated:	8-Nov-95
Control Accuracy Map:	National Map Accuracy Standard
Scale:	Variable; 1:600 or 1" = 50 feet
Availability:	Current
Reviewed By:	NATMO
Review Date:	18 August 1997
Spatial Extent:	Original Calcutta Municipal Corporation area
File Format:	N/A
Comments:	Covers about 60% of today KMC area

Table 3.5: Entities contained in source

Source ID #:	1
Entity Name:	Parcel
Spatial Entity:	Polygon
Estimate Volume Spatial Entity:	100 per map sheet
Symbol:	None
Accuracy Description Spatial Entity:	Earlier Map Accuracy Standard
Reviewed By:	NATMO
Review Date:	02-Jan-94
Scrub Needed:	Yes
Comments:	

Table 3.6: Attributes by entity

Source ID #:	1
Entity Name:	Parcel
Attribute Name:	Property Number
Attribute Description:	Borough, Ward, Street and plot Number
Code Set Name:	N/A
Accuracy Description Attribute:	N/A
Reviewed By:	NATMO
Review Date:	08-August-97
Comments:	

If there is a choice between sources, that is, two or more sources are available for a particular entity attribute, then criteria for deciding between them will be needed. In general, these criteria will be:

- Accuracy of resulting data
- Cost of conversion from source to database
- Availability of the source for conversion
- Availability of a continuing flow of data for database maintenance.

Occasionally, alternative sources may result in different representations in the database, such as a vector representation *versus* a scanned image. In this situation, the ability of each representation to satisfy the requirements of the GIS applications will need to be evaluated. Once a source has been selected, the metadata tables that record source data information need to be completed as appropriate. These are:

- Data object information
- Attribute information
- Spatial object information
- Source document information

To complete the accuracy information, the accuracy expected from the conversion process will need to be determined. This accuracy target will also be used later in the database construction phase by the quality control procedures. The metadata tables that need to be completed at this time are shown below:

Table 3.7: Data Object Information

Data Object Name	Parcel
Type:	Simple
Data Object Description:	Land ownership parcel
Spatial Object Type:	Polygon
Comments:	

Table 3.8: Attribute Information

Data Object Name:	Parcel
Data Attribute Name:	Property Number
Attribute Description:	Borough, Ward, Street and Plot Number
Attribute File Name:	Parcel.PAT
Codeset Name/Description:	N/A
Measurement Units:	N/A
Accuracy Description:	N/A
Comments:	

Table 3.9: Spatial Object Information

Data Object Name:	Parcel
Spatial Object Type:	Polygon
Place Name:	Amherst
Projection Name/Description:	UTM
HCS Name:	State Plane Coordinate System
HCS Datum:	NAD83
HCS X-offset:	1000000
HCS Y-offset:	800000
HCS Xmin:	25
HCS Xmax:	83
HCS Ymin:	42
HCS Ymax:	98
HCS Units:	Feet
HCS Accuracy Description:	National Map Accuracy Standard
VCS Name:	
VCS Datum:	
VCS Zmin:	0
VCS Zmax:	0
VCS Units:	
VCS Accuracy Description:	
Comments:	

Table 3.10: Source Document Information

Data Object Name:	Parcel
Spatial Object Type:	Polygon
Source Document Name:	Parcel Map
Type:	Map
Scale:	Variable: 1" = 50 feet To 1" = 200 feet
Date Document Created:	17-Nov-89
Date Last Updated:	05-Oct-94
Date Digitized/Scanned:	24-Apr-95
Digitizing/Scanning Method Description:	Manual digitized with Wild B8
Accuracy Description:	90% of all tested points within 2 feet
Comments:	

For some of the above tables, information will be available for only some of the entries. The remaining entries will be completed later as the database is implemented. The examples shown are from the metadata portion of the GIS design software package that accompanies these guidelines. This package is a Microsoft Access programme that runs "stand-alone" (user does not need a copy of Microsoft Access) on a regular PC. Where the same information is needed for multiple tables, this information is only entered once. The information is then automatically transferred to the other tables where it is needed.

Logical/Physical Design of the GIS Database

This activity involves converting the conceptual design to the logical and physical design of the GIS database (hereafter referred to as the physical design). The GIS software to be used dictates most of the physical database design. The structure and format of the data in a GIS, like ARC/INFO, Intergraph, MapInfo, System 9 etc. have already been determined by each vendor respectively. If one separates the conceptual entity and its attributes from the corresponding spatial entity and its geometric representation, it can be seen that the physical database design for the spatial entity has been completely defined by the vendor and the GIS designer does not need to do anything more for this part of the data. The attributes of the entities may, however, be held in a relational database management system linked to the GIS. If this is the case, the GIS analyst needs to design the relational tables for the attribute information. It will not always be the case where one entity from the E-R diagram translates into a single layer. More complex representations will be needed. Generally this will involve two or more entities forming a single layer with, possibly, several relational database tables.

The water main segments, the valves and the fire hydrants have been placed together in one layer as line segments, and two sets of nodes. However, each entity has its own relational table to record its respective attributes. The relationship is maintained by unique keys for each instance of each entity. Every entity shown on the E-R diagram must be translated to either a GIS layer, a relational table(s), or both, as indicated by the information to be included. In addition, every relationship of the type "relationship represented in database" (single line hexagon on the E-R diagram) must be implemented through the primary and secondary keys in the tables for the entities represented.

The entity "parcel" may "contain" the entity "building." The table for each entity would have its own primary key (ID#), however, the table for building must also have a secondary key (parcel ID#) to maintain the relationship in the database. The completed physical database design must account for all entities and their attributes, the spatial object with topology and coordinates as needed, and all relationships to be contained in the database. The remaining items on the E-R diagram, the two types of spatial relationships, must be accounted for in the list of functional capabilities, that is, the implied spatial operations must be possible in the chosen GIS software.

Procedures for Building the GIS Database

Developing a GIS database is frequently thought of as simply replicating a map in a computer. As can be inferred by the nature and detail of the activities recommended upto this point in these guidelines, building a GIS database involves much more than replicating a map. While substantial portions of the GIS database will come from map source documents, many other sources may also be used, such as aerial photos, tabular files, other digital data, etc. Also, the map representation is only part of the GIS database. In addition to the map representation and relational tables, a GIS can hold scanned images (drawings, plans, photos), references to other objects, names and places, and derived views from the data. The collection of data from diverse sources and its organization into a useful database requires development of procedures to cover the following major activities:

- ***Getting the Data*** which may include acquiring existing data from both internal and external sources, evaluating and checking the source materials for completeness and quality, and/or creating new data by planning and conducting aerial or field surveys. Contemporary GIS projects attempt to rely on existing, rather than new, data due to the high cost of original data collection. However, existing data (maps and other forms) were usually created for some other purpose and thus have constraints for use in a GIS. This places much greater importance on evaluating and checking the suitability of source data for use in a GIS.

- ***Fixing any problems in the data source***, often focused only on map source documents, this activity has been called *map scrubbing*. Depending on the technology to be used to convert the map graphic image into its digital form, the source documents will have to meet certain standards. Some conversion processes require the map to be almost perfect which other processes attempt to automate all needed *fixes* to the map. What is required here is for the GIS analyst to specify, in detail, a procedure capable of converting the map documents into an acceptable digital file while accounting for all known problems in the map documents. This procedure should be tested in the pilot project and modified as needed.
- ***Converting to digital data***, the physical process of digitizing or scanning to produce digital files in the required format. The major decision here is whether or not to use an outside data conversion contractor or to do the conversion within the organization. In either case, specifications describing the nature of the digital files should be prepared. In addition to including the physical database design, specifications should describe the following:

  —Accuracy requirements (completeness required, positional accuracy for spatial objects, allowable classification error rates for attributes).
  —Quality control procedures that will be conducted to measure accuracy.
  —Partitioning of the area covered by the GIS into working units (map sheets) and how these will be organized in the resulting database (including edge matching requirements).
  —Document and digital file flow control, including logging procedures, naming conventions, and version control.

- ***Change control***, most map series are not static but are updated on a periodic basis. Once a portion of the map has been sent to digitizing (or whatever process is used), a procedure must be in place to capture any updates to the map and enter these into the digital files.
- ***Building the GIS Database,*** once digitizing has been completed, the sponsoring organization has a set of digital files, not an organized database. The system integration process (a subsequent guideline document) must take all the digital files and set-up the ultimate GIS database in a form that will be efficient for the users.

Procedures for Managing and Maintaining the Database

Because the physical world is constantly changing, the GIS database must be updated to reflect these changes. Once again, the credibility of the GIS database is at stake if the data is not current. Usually, the effort required to maintain the database is as much as, or more than that required creating it. This ongoing maintenance work is usually assigned to in-house personnel as opposed to a contractor. The entire process should be planned well in advance. Once again, the equipment and personnel must be ready to take over the maintenance of the database when the data conversion effort and database building processes are complete.

Database maintenance requires two supporting efforts: (a) ongoing user training and (b) user support. Ongoing user training is needed to replace departing users with newly trained personnel. This will enable the data maintenance to be carried out on a continuous and timely basis. It is also important to offer advanced training to existing users to provide them with the opportunity to improve their skills and to make better use of the system. GIS is a complicated technology, making operating problems inevitable. User support will help users solve these problems quickly. It will also customize the GIS software to enable them to execute processing tasks more quickly and more efficiently. User support is usually provided by in-house or contract programmers. It requires knowledge of the operating system and macro programming language as well as troubleshooting common command and file problems.

GIS Data Sharing Cooperatives

The establishment of data sharing cooperatives within the public sector is a cost-effective means of database development and maintenance which is encouraged. Cooperative multi-participant database projects allow for data exchange, and the opportunity to create new means for developing, maintaining, and accessing information. The sharing of data in the public sector, especially between government agencies and offices which are funded by the same financial resources, should be expected. It does not make fiscal sense for public funds to be utilized in the development of two GIS databases of the same geographic area for two different agencies. Benefits of data sharing thus would include: the development of a much larger database, for far less cost; the development of more efficient interaction between public agencies; and through the utilization of a single, seamless database the availability of more accurate information, since all agencies would share the same, up-to-date information (see Chapter 17: Spatial Data Infrastructure).

The goal of a data sharing strategy is to maximize the utility of data while minimizing the cost to the organization. It is important that all parties involved have clear and realistic expectations as well as common objectives to make the data sharing work. Under any circumstance, however, database management and maintenance will require us to redefine our relationships with those we routinely exchange data with, whether they are within an organization or part of a multi-participant effort including outside agencies. Work flow and information flow must be reviewed and changed, if necessary. Procedures and practices for the timely exchange and updating of data must be put in place and data quality standards adhered to, whether it is hard copy data which must be converted for inclusion or digital files which might be available for importing to our system. Systematic collection and integration of new and/or updated data must be employed in order to safeguard the initial investment, maintain the integrity of the database and assure system reliability to meet function needs.

Database Structure and Database Management

Basics of Computer File/Database Structures

The essential features of any data storage system are that they should allow data to be accesses and cross referenced quickly. There are several ways of achieving this, some of which are more efficient than others. Unfortunately, there seems to be no single 'best' method that can be used for all situations. A goal of GIS is to represent and store the graphic entities of mapped information along with relevant attributes in such a way to make all the data easily retrievable and scope for manipulation. This is done by taking advantages of the ways computers handle data in logical fashion through file and database structures. A brief overview of the ways computers can handle data is offered here.

Simple list—The most basic file structure is a *simple list*. In this file structure, there really is no absolute ordering of the data. The data occur in the file is essentially the same way in which they were entered. Simple lists may start out in logical fashion but whenever modifications are made, they rapidly get out of order because new data are appended to the end of the list. It is very easy to add data to this system but retrieving is data is inefficient.

Ordered sequential files can be thought of as a rolodex in which we keep everything in alpha-numeric order. As new data are added, the file is restructured (sorted) to maintain that order. Very often ordered sequential files are accessed by binary search procedure.

Indexed file structures provide pointers to more efficiently search data. The most efficient system is to develop an index that is based on a commonly searched attribute in the database. In many applications, particularly in geographical information, the individual items (pixels, points, lines or areas) will not have any key attribute such as an identification number or a name, but will also carry information about the associated attributes. Very often it is the information of the associated attributes that is required. With *indexed files* original data file can be speeded up in two ways. If the data items in the files themselves provide the main order of the file, then the filers are known as *direct files*. The location of items in the main file may also be specified according to topic, which is given in second file, known as *inverted file*.

Types of Database Systems

A database is a collection of files that contain interrelated information, and a Database Management System (DBMS) is a program that lets the user add, delete and modify records in the database. All GIS databases include links between graphic (map) features and tabular data that provides descriptive information for the graphic objects. If it were not for this link, a GIS would not exist but would be similar to computer aided design or computer mapping systems. The tabular data provides the *information* in GIS. These days most GIS's provide links to a variety of databases allowing users a choice between a hosts of commonly used systems. Therefore, the GIS can be a combination of two distinct systems generated individually, but used for a common purpose. Database systems come in three main varieties:

- Hierarchical database systems,
- Network database systems, and
- Relational database systems.

Of these three types of database systems, the hierarchical is the fastest but the most restrictive for the user. This system is appropriate if you can naturally organize the information in the database into the same structure as a hierarchy chart. The network system is more flexible than the hierarchical system but more difficult for a user than the relational database system. The relational system is rapidly becoming the most popular of the three. It is the most flexible and easiest to use. But this flexibility comes at the cost of low speed compared to other database systems.

Hierarchical Database Systems

A hierarchical database system is one in which the DBMS supports a hierarchical structure of records organized in files at various logical levels with connections between the levels. A record at a particular level contains data common to a set of records at the next lower level. There are no connections between records of the same level. Each record contains a field defined as the key field, which organize the hierarchy.

Records of the same type are collected in files known as elements. Several types of elements may reside at the same logical level. Starting from the highest-order element, the hierarchy permits a set of elements to be accessed at the next lower level. Any one element may retain connections to only one set of lower level elements and itself may be a member of only one such set. Construction of a hierarchical structure begins with a main object at its top. The main object has a range of characteristics which can be collected at the various levels of hierarchy (Fig. 3.11).

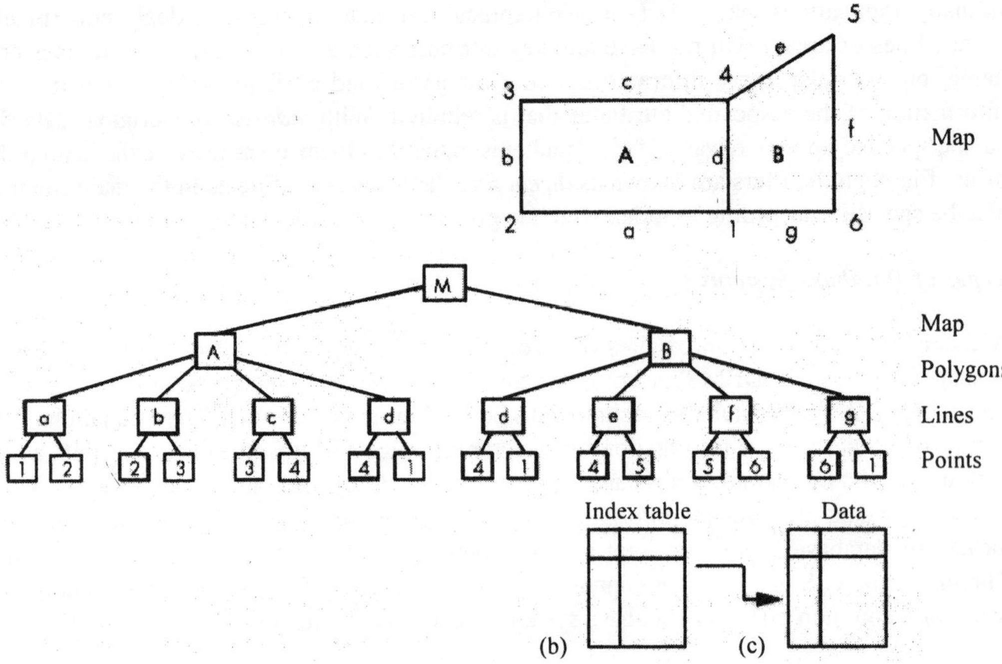

Fig. 3.11: Map data stored in a hierarchical database system.
Note: The double storage of lines and points (Bernhardsen, 1992)

Geographic data which is organized hierarchically fit into this model. Geographical data may be stored in a hierarchical structure in a manner that reflects the real world. An example is when the levels of a hierarchical model correspond to the levels of real-world administration, for example in a country, county, township or town/city. A good example of spatial data that can fit into this model is the data of the Census of India in which information for one enumeration area has no lateral relation to another enumeration area, but is related to the higher levels in the hierarchy to city, *tahsil* or *taluk*, district and the state. In the hierarchical model an owner record (file) can have many members, but the member records can only be owned by one owner record.

Hierarchical database systems are easily expanded and updated. However, they require large index files, must be frequently maintained and are susceptible to multiple entries. Searches are rapid, but search routines are fixed and constrained by the structures. The elements or the structure are related only through one-to-many connections. This constraint imposes the presupposition that all queries are known in advance and accounted for in structuring and entering data. This constraint is not always natural or suitable in GIS applications. As a result, hierarchical database systems are usually restricted to storing digital map data in GIS.

Network Database Systems

A network database system is one in which the DBMS supports a network type of organisation. Each element, or collection of like records, has connections to several different level elements. Interconnections are made in the hierarchical organisation, and a characteristic may be associated with two main objects. The resultant network structure more closely represents the complex interrelationships which often exist between real-world geographical objects. The elements of the structure may be related through one-many, many-one and many-many connections. The purpose of the network structure is to improve the flexibility and reduce the multiple entries of the hierarchical structure. Points and lines are entered only once. Searches need not pass through all levels, but can take shortcuts. However, the volume of index data is greater than that of the hierarchical structure (Fig. 3.12).

A network structure permits rapid connection between data which physically are stored in different disk sectors. Maintaining data stored in a network structure is complex, so although a network structure is better suited to geographical data than is a hierarchical structure, it is infrequently used in GIS applications.

Comparison between Hierarchical and Networks Data Structure

A record type represents each entity type in the E-R diagram. A 1:N relationship is represented as hierarchy type where the record type corresponding to the one side of the relationship is the parent (a 1:1 relationship is a special case of the 1:N relationship). A week entity can be represented as a separate record type. This record type becomes a dependent in a hierarchy where the record type, corresponding to the strong entity in identifying relationship, is the parent. An M:N relationship is not easy but it requires introducing duplicate record types or using multiple hierarchies and introducing virtual records.

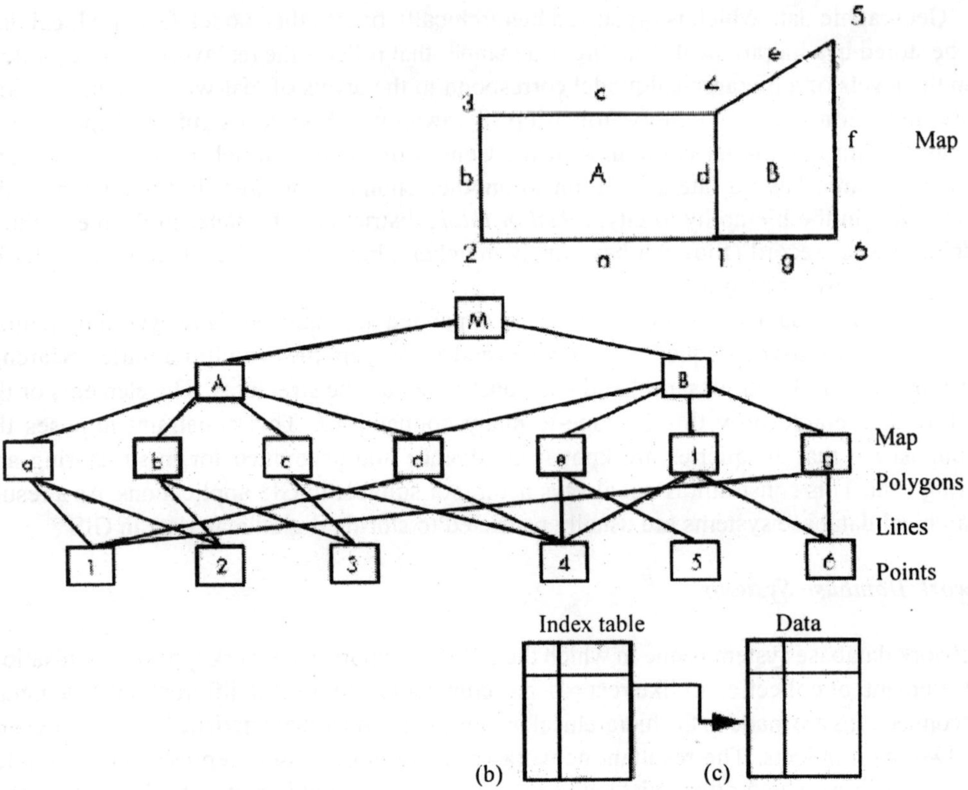

Fig. 3.12: Storage structure in a network-based database system.
(a) shows a map section; (b) shows polygons stored in the network;
(c) is a pointer table which specifies where the map data
(d) are in fact stored (Bernhardsen, 1992)

Each entity type in E-R diagram is represented by record type and each instance of the entity is represented by record occurrence. A week entity can be represented as a separate record type. In the case of networking data model the identifying relationship is represented as a set type wherein the record type corresponding to the week entity type forms the member and the record type corresponding to the strong entity is owner. A 1:1 or 1:N relationship is represented as a set type. An M:N relationship requires introducing an intermediate record type. This record type is a common member in two set type, one of which is owned by each record types involved in M:N relationship.

Merits and Demerits

The major limitation of hierarchical system cannot handle many to many relationship and GIS data bases have, in addition to one to one and many to one relationship, many to many relationships in which a single entry may have many attributes, and each attribute is linked explicitly to many entities. However, network system is the improvement over hierarchical structure for GIS. Networking system can handle many to many relationships. In network system each piece of data may be associated with an explicit computer structure called a pointer which directs it to all of the other pieces of data to which it relates. Unlike hierarchical data structure it is not restricted to the branching tree structure, each individual piece of data can be linked directly anywhere in the database without any parent-child relationships. The major drawbacks is that in very complex GIS database, the number of pointers can get quite large, coming to comprise a substantial portion of storage space. In addition, although linkages between data elements are more flexible, they must still be explicitly defined with the use of pointers. The numerous possible linkages may become an extremely tangled web, often resulting in confusion and missed and incorrect linkages. This limitation may be frustrating for the novice GIS users. The disadvantages of the network data model are: the query language is procedural and requires the user to navigate through the database by specifying sets, owners, and members. This in turn means that the user has to be cognizant of the data structure of the database.

Relational Database Systems

This has become the most popular form of a database model since it has the most flexible approach to linkages between individual databases. There is no hard structure to a relational model and the form of the model is determined by the needs and concepts of the user. In a relational database each record (file) has a set of attributes and a set of rows (entries) that form the table. Separate tables are linked, or related, to one another through a common attribute. Any attribute in the table can be used as the key link between that table and another table. Links can be formed in any direction and between any nodes (table). This database model allows maximum flexibility since ownership is not defined as it is in the hierarchical and network models (Fig. 3.13).

Building I.D.	Property	Owner	Year	Type
589	44 110			
610	44 50			
955	44 99			

Property I.D.	Owner	Area	Address
44 99			
44 50			
44 110	Nils Nilsen	6,51	9999 Toppen

Fig. 3.13: A relations database.

Note: Each field in a table can be the key to locating data in another table (Bernhardsen, 1992)

There are no pointers in the data tables. Internal index tables are used to direct inter-table communications, which means that the system must open at least one index table for each connection between data tables. This can result in relatively large databases and slow access. Moreover, as there are no pointers, searches for records or fields must be sequential through the tables. The result is that relational database systems are inherently slower than hierarchical or network database systems.

A field may be located quickly in a given table, and the data accessed may be manipulated. The simple structures of relational database systems have permitted the development of standard query languages, one of which is Standard Query Language (SQL). Relational algebra may be performed using two classes of storage and retrieval operations. The set of operations include union, intersection, difference and product. The relational operations include selection (accessing rows), projection (accessing columns), joining and dividing. Joining creates a new table from data retrieved from various tables. The new table need not physically be stored in the database.

Relational DBMS are now the type used most frequently in GIS applications, primarily because of their simple, flexible structures. Another reason is that they support the complex relationships common among real-world geographical objects. Multiple entries are less frequent than in hierarchical or network database systems, although search times tend to be longer. This is particularly noticeable in specific operations on digital map data, for instance when forming polygons and with automatic text placement (Fig. 3.14).

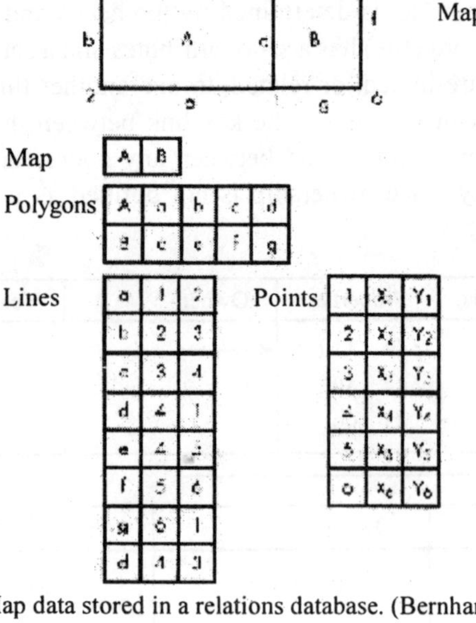

Fig. 3.14: Map data stored in a relations database. (Bernhardsen, 1992)

What is SQL?

SQL is an abbreviation for Structured Query Language. It is a tool for organizing, managing, and retrieving data stored in a database. It works with a specific type of database model, a relational database.

History of SQL

SQL is a language used for accessing relational databases. As with relational databases, the origin of SQL dates back to 1970 when Dr. Codd proposed a relational model for databases. Codd's article triggered research on relational databases. IBM started a research project called System/R in 1974. The goal of the project was to prove the workability of the relational concept and to provide some experience in actually implementing a relational DBMS. In 1978, System/R was distributed to a number of IBM customers for evaluation. With the successful implementation of System/R, IBM proved that relational databases were feasible, and could be the basis for commercial products. SQL was born during the implementation of System/R. In 1979, Relational Software Inc. (now Oracle Inc.) came to market with a relational Database Management System (DBMS) based on SQL. IBM worked hard to produce commercial relational products. In 1982, IBM introduced its first commercial product SQL/DS, and in 1983, DB2, DB2's SQL became the *de facto* standard database language.

In 1986, the American National Standards Institute and the International Standards Organization (ANSI/ISO) published a standard for SQL, and gave SQL official status as a standard. The standard is largely based on DB2's SQL. The standard was revised and expanded in 1989, and is called SQL89 (SQL1) standard. The ANSI committee continued its work on the standardization of SQL, and in 1992 a new more rigorous standard SQL92 (SQL2) was published. After some extensions to SQL2, ANSI published standard SQL99 (SQL3) in 1999.

The ANSI has published three standards over the years: SQL89 (SQL1), SQL92 (SQL2), and SQL99 (SQL3). Today most relational database products are designed to meet the SQL92 standard. However, none of them implement all features of the SQL2 standard and many of the advanced features of SQL92 have yet to be implemented. SQL3 addressed more advanced issues in SQL, but no vendor has implemented the standard at the present time. The majority of the language has not changed through these updates. All of the core functions of SQL, such as retrieving, adding, and modifying data, are the same. The SQL statements given in this tutorial are the core part of SQL. They are common to most available relational products today.

Object Oriented Database Systems

Hierarchical, network and relational database systems are essentially intended for administrative tasks. Consequently, they are not particularly well suited to representing a conceptual data model of geographical reality. All these systems are record oriented, which is to say that the data they contain are filed record by record. For instance, in a relational database, all the elements which together comprise a particular map object may reside in several records in various tables. Furthermore, geometrical data and attribute data are often physically separated, each residing in a separate database. Topological storage of vector data is object oriented, but the relevant database structure remains a problem. Object-Oriented Database Systems (OODS) attempt to overcome these difficulties by representing more faithfully the real world, which comprises homogeneous, through complex, objects with varying internal and external relations. There are several different object-oriented technologies, all sharing some of the same concepts and philosophical principles, but the direct benefits of OODS come from the following ideas:

(a) feature classes (a *shema*) in which each feature type has several attributes;
(b) active data-features that have their own behavior ("methods") stored in the repository not in application programmes; and
(c) inheritance between the classes of data (attributes) and behaviour (methods).

An active object database allows geographic feature data to be grouped into classes that represent similar entities. This technology approach allows that spatial data could be stored in complex, richly structured *shema*—complex features or complex feature classes. Some other important features of object-oriented approach are:

Integrity control—active data can check their own integrity, the validity of their relationships with other features, topological constraints, etc.;

Version control including virtual check-in and check-out of repository segments—spatial object repositories make it easy to provide novel versioning mechanism so each user can see a virtual copy of all data while editing only a part of them;

Advanced topology support—geometric topology results from the sharing of geometry—poles placed at cable intersections, rivers forming parts of administrative boundaries. Spatial object repositories can identify spatial coincidence and dynamically model them in the database. This ability exploits two important spatial object database features: the ability to hold all types of data and the ability to effectively manage a vast web of connections among objects;

Rich representation—display methods allow objects to display themselves, providing important information about the object (a valve/switch can indicate whether it's open or closed). Display method is centralized as part of the feature, and appearance of objects can be made consistent across a utility—local authority, conforming to standards;

Multiple presentation from a master dataset—spatial objects can be associated with several geometries, and display methods can use any of the geometries. The same objects can appear differently in different contexts.

As the first commercial GISs built upon the object-oriented technology were sold more than ten years ago (in 1991) OO GISs are not widely used (U.S. Geological Survey, U.K. Ordnance Survey, Land Information New Zealand, Czech Land Survey Office, some other national mapping agencies, marine and air chart producers, military spatial data users and automated mapping/facilities management spatial resource planning utilities) but initial success is proof that OO GIS is a future requirement, not an option.

REFERENCES

Aronoff, Stan (1989), *Geographic Information System: A Management Perspective,* WDL Publication, Ottawa.

Bernhardsen, T. (1992), *Geographic Information Systems,* VIAKIT IT, Norway.

Burrough, Pater and R. McDonnell (1998), *Principles of Geographical Information Systems,* Oxford University Press, Oxford.

Burrough, P.A. (1986), *Principles of Geographic Information Systems for Land Resources Assessment,* Oxford University Press, Oxford.

Faust, N.L., W.H. Anderson and J.L. Star (1991), Geographic information systems and remote sensing future computing environment, *Photogrammetric Engineering and Remote Sensing,* Vol. 57, Issue 6, pp. 655-68.

Nag, P. and M. Kudrat (1998), *Digital Remote Sensing,* Concept Publishing Company, New Delhi.

4

Software and Hardware Options

In last few years a huge volume of spatial information has been available, stored and managed in the form of Geographical Information System (GIS) for various applications which has opened a new vista for business opportunities. International Geo-sphere Biosphere research, digital earth, digital regions and several global and national projects related to digital information have given a boost to a new type of applications. These applications or projects are based on geo-spatial data and knowledge about such data-bases. The management of such data, information or knowledge depends on efficient and user-friendly GIS. The GIS technology now permits us to collect basic common sense knowledge linked with geo-spatial data, and also allows this knowledge to be closely associated with such data. The association of knowledge with spatial data makes the option of having feature-oriented objects in the GIS environment. The new applications look towards the database which can produce intelligent levels. Such a combination of data and knowledge has opened several types of applications not only in the developed world but also in the developing world. Hence, there has been a greater urgency to understand the spatial or geospatial data infrastructures and their various facets.

We live in information age which has become essential to solve problems of national development. Further, spatial information has proved to be an important element underpinning decision-making for many disciplines because it deals with location. In the sophisticated digital environment, it has expanded to include databases, satellite positioning, communication networks including internet, web technology and wireless applications. It draws experiences from the geographic information systems, computer science, land administration, geography, surveying and mapping, legal and public administration disciplines. GIS is so versatile that it is more difficult to define than to imagine or visualize. By the end of the 1970s the term GIS had emerged in recognition both of common technological requirements and of the opportunity to built the system that could potentially satisfy all of the various applications. GIS is information about geography, i.e. information tied to some specific set of location on the earth surface (including the zones immediately adjacent to the surface and subsurface, ocean and atmosphere.

The potential users of GIS are limitless and the types and number of users are growing at a logarithmic pace. Further, the modern GIS has the utility of the maps by replacing it with a large number of mapped coverages, each with an interrelated theme. These coverages can be automatically analysed and these themes combined to give meaningful answers for decision-makers. The GIS is gradually changing the way we do think with maps, the way we think about geographic information (GI), even the way in which geographic data collected and compiled. The task, which was impossible with the traditional maps, is now common place. To answer such question requires access to the geographical information, which is characterized by its multidimensional nature (x, y, z coordinates and time), its large volume and high processing cost. Even to answer apparently simple questions, the data from several sources must be integrated into consistent form and be analyzed.

In the late twentieth century, an awareness of importance for data on topography and specific themes of the earth surface, such as natural resources, have accelerated greatly. Aerial photography and satellite-based remote sensing data ease the means of mapping large area with great accuracy. The same technology also gave the resource scientist enormous advantage for reconnaissance and semi-detailed mapping. The resulting thematic maps were the sources of useful information for resource exploitation and management.

GIS has been defined by its objective, as in Cowen's definition of GIS as spatial decision support system (Cowen, 1988). In the context of remote sensing GIS tends to be reserved for system that integrate remotely sensed data with other types or processed data that have already been cleaned and transformed. Similarly, an atmospheric scientist or oceanographer will tend to associate GIS with system used more for multi-disciplinary work and policy study, and will use other software environments for modelling and analyses within the confines of his or her own discipline. In short, because GIS implies generalized software environment that is exclusively to geographical information, i.e. a tendency for it to be most strongly associated with multi-disciplinary, integrative work and applications, in more narrowly defined environments less general solutions may be adequate.

The term GIS is difficult to define and represents the integration of many subject areas. The term itself is becoming complex and modified to conform to intellectual, cultural, economic and political objectives. The terminology has in fact become extremely elastic resulting in an increasingly confusing jargon due to new definition that constantly creeps into both scientific and popular literature.

A preliminary definition for consideration might be that of David Rhind (1988), who defines GIS as "a computer based system for collecting, checking, integrating and analysing information related to surface of the earth." This definition has some highly worthwhile elements that should be examined. First, it indicates that it deal with the surface of the earth. Although this is not an absolute requirement, the vast majority of GIS applications do deal with the portion of the earth. Moreover, the statement that GIS is used to collect, check, integrate and analyse information enumerates a large number of the necessary groups of operations for any GIS.

Components of Geographical Information Systems

GIS refers to an information technology capable of collecting, storing, analysing and displaying both spatial (geographically-referenced) and non-spatial data. Components of a GIS include both computer hardware and software, spatial data features (raster grid cells, or points, lines, and polygons) and associated tabular attribute data. Utilization of GIS technology occurs in a wide range of applications, from Land Information Systems (LIS), Automated Mapping and Facilities Management (AM/FM), Census and Legislative Redistricting and "transportation" GIS (GIS-T), to natural resource management, environmental modelling and Spatial Decision Support Systems (SDSS). Consequently, each area of specialization has unique needs for data and computing. This standard focuses on general criteria for those aspects of GIS technology specifically associated with computing architecture (e.g., hardware and software). In addition, guidelines are provided for two other GIS standard categories: spatial data transfer format and metadata content. GISs have three major components. These are as follows:

—Soft ware
—Hard ware
—Proper organizational including skilled persons

The computer has a hard disk for storing data and operational software package (programmes), but extra storage can be provided via network or by digital tap, optical CD-ROM and other devices. A digitizer and or scanner are used to convert maps into digital form so that they can be used by the specific software. A plotter or printer or any other kind of display device or monitor is used to present the results of the data processing. Local and global electronic networks using special data lines with optical fibers or over ordinary telephone lines by using a device known as modem provide computer communication. The user controls the computer and the peripherals (a general term for plotter, printer, digitizer and other apparatus linked to computer), via the computer screen and keyboard, added by a mouse. Now a variety of mouse is available.

GIS SOFTWARE

Selection of a GIS software is highly dependent on a number of factors, including the intended use of the chosen application, software and maintenance cost, software "learning curve" and the current technical expertise of agency personnel. The software standards identified are representative of two broad GIS application categories (i.e., professional and desktop). GIS application software should be capable of running under a Microsoft Windows (Windows-95, Windows-98, Windows-NT, Windows-2000XP) or X Windows (e.g., UNIX) environment. While it is recommended that software incorporate both vector and raster data models (Chapter 3), priority should be given to vector data model capabilities. GIS software incorporating a relational database must conform to the Structured Query Language (SQL) standard described in the database management systems (DBMS) for multiuser applications standard. GIS software not using a relational database should be able to export and import SQL data.

Beyond the differing data model specifications described above, GISs may be categorized, compared and contrasted based on their functional capabilities, ranging from basic mapping packages to desktop and professional GIS software applications. Typically, mapping packages are not true GISs, possessing a flat file structure data model with no topology or spatial analysis capabilities; such applications will not be addressed in this standard. Becoming more powerful every year, desktop GIS software applications most often run within a Windows environment, and make use of a relational database management structure. In addition to map creation, desktop GIS functionality typically includes address geocode matching capabilities, proximity analysis (e.g., *what is near?*) and buffering, as well as polygon overlay utilities. Desktop GIS packages are gaining in popularity particularly due to their point-and-click graphical user interfaces and relative ease of use. Professional GIS applications are exemplified by software running under either a UNIX or Windows-NT operating system (OS), incorporating both vector, raster and, potentially, TIN data models, as well as some level of CADD integration capabilities. The best choice for database development and management, professional GISs are capable of performing complex spatial operations from feature overlays to nearest neighbor analysis. Such systems also feature programming capabilities for user-customization of the system.

Two levels of software application standards have been identified based on the *professional* and *desktop* functionality levels described above. For the professional level, the ESRI ARC/INFO suite of products is the standard. At the desktop level, the standard consists of two vendor packages: the MapInfo product suite and ESRI's ArcView GIS and associated extensions. In addition to contrasts in functionality, professional and desktop applications may also be differentiated based on cost and *ease-of-use/learning curve*. Typically, professional level GISs require a higher software investment (up to five times more) than a desktop application. Cost of supporting hardware platforms also tends to be greater when operating on a UNIX-based platform, but is negligible under Windows-NT, etc. Designed as much for the end-user as the analyst or the programmer, desktop packages are generally menu-driven, containing a number of generic query, display and mapping utilities which allow the user to quickly access, manipulate and utilize spatial digital data without extensive specialized training and/or experience. Certified training is available throughout the region for each of the product suites identified below. Nevertheless, the choice of software is generally user oriented. There are standard packages now available off the shelf. The components of a GIS software are more-so-ever standardized. The detail criteria are as follows:

- — *Data entry*: digitizing, scanning, automated data capture, interface with existing digital files, key board entry.
- — *Analysis*: map overlay, proximity analysis, mathematical analysis, enclosure, buffer, measurements, attribute analysis and interpolation.
- — *Surface modelling*: 3-D surfaces, slope analysis, and draping.
- — *Network modelling*: route selection, allocate demand to network and facility sitting.
- — *Manipulation*: image merging, projection, clip, window, update, generalize and aggregate.
- — *Query*: spatial and attribute.
- — *Output*: tabular list, map display, digital outputs.

There are a number of key components to data and its management.

- **Database Support Environment**
  - —Data extractions
  - —Data Integrity
  - —Data Version Control
  - —Storage Management
  - —Schema
  - —Data Dictionary
  - —Security/authorization
- **Database Management**
  - —GIS Data components

Layers
Attributes

—GIS Data modeling
Satisfying queries
Dangers of data modelling
Make data model flexible for future projects

—Develop Data Standards and Procedures
Conventional data standards
Framework Layers
Conventional QA/QC standards
Distribution standards
Backup Standards

—Organization
Hierarchical (tree structure with data dependency defined)
Pros and Cons
Fast access
Large storage
Not dynamic
Limited sharing
Network (multiple hierarchical, multiple data paths)
Pros and Cons
Increased flexibility
Difficult to update
Limited sharing
Better than hierarchical
Relational (organize data into tables of rows and columns)
Pros and Cons:
Flexible
Standard Query Language (SQL)
Performance is system dependent
Widely used

—Linkages
Combining information contained in existing production systems and files with your GIS application
2 Methods:
Copied files: exporting the data to a file and importing that file into your GIS. This is the most common method and is time consuming when matching data structure of source application to GIS.
Direct link: importing data into the GIS from a live information application. This is least common and very difficult if the data structure of the source application does not match with existing GIS.

—Integration, Ownership, and Sharing
On going data management policy. Should be an organization's IT/IS Policy.
Ownership/stewardship—who's responsible for data maintenance and accuracy?

Sharing—define procedures for distribution and maintenance (updates, corrections, etc).

Cost effective

- **Geospatial Database Conversion and Creation**

—Quality, Quality, Quality

Data is only as good as its source.

Can be costly

Greater the accuracy, higher the cost.

"Measure twice, cut once"—More expensive if data is developed twice.

Costs are coming down (DOQQ base maps)

$$cost = area \times density \times features \times process$$

What level of accuracy to get?

Follow Federal Standards?

Base quality on needs of users?

Base quality on needs of future too!

—Data Conversion and Creation

Define data conversion as taking what exists and making it useful to GIS

Paper Maps to digital map

Tabular data joined to geospatial data.

Expense of Conversion/Creation:

Quantity of conversions

Quality of Conversions

Availability of organizational resources:

Staff, Hardware, Software, Expertise

Conversion/Creation Methods:

Manual Digitizing

Land Surveys

Data Purchases

In order to bring some order to all of this complexity, this chapter summarizes some of the available software products according to their intended application area. Four categories of software are identified: (a) desktop mapping; (b) data viewers; (c) components, and (d) internet mapping. First, the general characteristics and application area of each product is briefly described. Second, a table with a list of vendors and products listed according to each application area is presented. Details related to GIS vendors, applications and capabilities have been included in the book (Appendices III, IV and VI).

Desktop Mapping

Desktop mapping is the traditional software application that is thought of when a product is labeled a GIS. A full featured GIS includes built-in ability to:

- display and query spatial data and non-spatial attribute data,
- perform spatial topological operators such as clip, union, and join,

- alter the coordinate system or projection of spatial data,
- create high-quality cartographic output,
- convert between spatial data formats available from federal and commercial sources,
- framework for implementing customizations either through a proprietary or third generation programming language.

Data Viewer

Data viewers are a more recent development among the increasingly specialized products offered by vendors. Strictly speaking, a data viewer application allows you to browse spatial data without the ability to perform spatial topological operations. Typically, these programs support a limited number of data formats and require that all data to be displayed is in a uniform coordinate system or projection. Functionality relating to map cartography may be present in such applications, but is generally limited in comparison to a GIS.

GIS Software Components

Like data viewers, deployment of GIS component software is also a relatively recent trend in the field. Developments in software engineering during the last 10 years, such as object-oriented programming languages and the availability of dynamically linked libraries, have enabled the increased reuse and improved modularity of code written for commercial software. Customer demand for increasingly specialized applications coupled with the establishment of several software component specifications has enabled vendors to re-sell some of the technology used to rebuild their own software as component technology. Components are particularly well suited for building standalone applications which require GIS functionality, but will not be operated by GIS personnel.

Software Component Overview

The GIS software components shall fully support and exploit the capabilities of the proposed hardware platform and shall provide full functionality for entry, editing, maintenance, analysis, display, and hard copy output of both graphics and tabular data on a continuous and interactive basis. For purposes of this procurement, software component capabilities have been grouped into the functional categories of:

- Database structure
- User interface
- Data entry
- Data editing/maintenance
- Data query and analysis
- Data display/output
- Application development
- Operating system requirements

Data Editing and Maintenance

The vendor shall describe the tools and capabilities of the proposed system to modify and manipulate spatial and attribute data in the GIS for the following categories:

- Interactive Graphic Editing
- Attribute Editing
- File Copying
- Deletion of Features
- Edit Controls
- Rubber Sheeting
- Coordinate Registration and Transformations
- Quality Control/Error Detection
- Merging, Extraction, Edge Matching of Data
- Data Transactions, including the capabilities of the proposed system to translate data into and out of the following formats:

  —GFIS to Proposed System Format (specify how attribute data is addressed)
  —AutoCad DXF (specify how attribute data is addressed)
  —AutoCad CWG (specify how attribute data is addressed)
  —Intergraph IGDS (specify how attribute data is addressed)
  —USGS DLG and DEM
  —TIGER Line Files
  —ArcInfo Export Files
  —Exchange data with KVS Computer Assisted Mass Appraisal (CAMA) System

Data Query and Analysis

The proposed software shall support the following data query and analysis capabilities:

- Graphic Data Query
- Area/Perimeter/Distance Calculation
- Attribute Data Query
- Spatial Aggregation
- Buffer Analysis
- Address Matching
- Polygon Overlay Analysis
- Linear Network Analysis
- Area Districting and Zoning

Data Display and Output

The data display and output tool capabilities that the proposed software shall support including the following:

- Graphic Display

- Tabular Display
- Raster Image Display/Production
- Vector Map Overlay
- Hard Copy Map Production
- Hard Copy Report Production
- Map Plot/Display Relationship with Scale
- Graph/Chart Production
- Interactive Map Composition

Application Development

The vendor shall propose one of more software components that, in a well integrated manner, provide the following capabilities and features:

- Menu Design and Custom Application Development
- Programming Features
- Supporting High-Level (4 GL) Programming
- Subroutine Libraries

Basic Operating System Requirements

The operating system component of the software shall be the primary operating system of the proposed hardware platform and shall provide all of the traditional features of current operating systems as described below:

- Multi-user Support
- Multi-tasking, Multi-threading Support
- Security Management
- File Management
- Memory Management
- Database Backups
- Error Monitoring/Disaster Recover
- System Diagnostics
- Anti-viral Protection
- Electronic Mail (E-Mail)

Network Management Functions

The proposed system shall provide capabilities for monitoring and managing all data and devices on the GIS network as one unified system and support the following capabilities:

- Multi-user Database Access and Maintenance
- Monitoring of Network Activity
- Network Problem Diagnostics
- Print and Plot Management

GIS Database Structure

Database Model

A GIS database model defines the nature and usage of spatial (geographic) data with a database. The proposed software shall support a spatial data model that is capable of creating, managing, and manipulating data sets, defined on the basis of spatial coordinates and associated attribute data sets.

Feature Types: The data model shall support multiple feature types including point, node, line, polygon, and text features.

Data Storage: Features shall be stored as double precision x and y coordinates.

Data Types: The data model shall support multiple graphic and nongraphic data types.

Database Organization: Vendors shall describe strategies for organizing data into logical groups on the basis of data themes, and shall describe the capabilities of the data model for supporting simple and complex feature types.

The geographic data model shall support the creation and maintenance of topological data. Topology shall be created through execution of a software function to structure graphic data sets. Vendors shall describe the ability of the proposed data model to support logical polygons, networks, and user-defined topological structures. Software capabilities that support large-scale engineering and design activities should be outlined as well as specific engineering functions and appropriate modules. It shall support for storage of raster map images (e.g., scanned bluelines, orthophotos) and for raster scanned documents, and the geographic data model shall support the creation and storage of a continuous geographic database. A Relational Database Management System (RDBMS) will be able to maintain a minimum of 30,000 records of parcel ownership information in a single table and shall provide functionality for updating database content, queries, and production of reports. The recommended RDBMS must be either a part of the GIS software or have a direct access capability. Some of the major functions of standard GIS packages are shown in Table 4.1.

Table 4.1: Popular GIS Software and their functions

Vendor Name	*Desktop Mapping*	*Data Viewer*	*Component*	*Internet Mapping*
Autodesk	AutoCAD Map	Map Guide Viewer	N/A	Map Guide
Bentley	MicroStation	N/A	MicroStation/J	Model Server Imager
Blue Marble Geographics	N/A	Geographic Explorer	GeoObjects GeoView	no out-of-box solution see BeyondGeo.com
Environmental Systems Research Institute, Inc	ArcInfo ArcView	ArcExplorer	MapObjects ArcObjects	ArcIMS
Intergraph	MGE (Modular GIS Environment)	GeoMedia Professional	GeoMedia	GeoMedia WebMap
MapInfo, Inc	MapInfo Professional	MapInfo ProViewer	MapX	MapXtreme
Object FX	N/A	N/A	SpatialFX	N/A

N/A: Not Available in Category

The details of the capabilities of industry standard software are mentioned in Appendix VI. The list of free GIS software is enclosed in Appendix V; and the list of GIS vendors and related software developers is resting in Appendix III.

Selection of Software

Initially it is required to evaluate the software independently of hardware. The software will be selected based on the functionality it offers. The hardware selection is based on the selected GIS software and the operating system strategy in the organization uses. It is needed to test the hardware and software together making sure it works uniformly without any hindrances. The nature of hardware and software technology is that it changes. In recent years it has been changing very quickly. Do not let this stop the efforts. It is easy to get intimidated. The important thing to remember is to get a product that has been proven in the market place and continues to have a clear development path. Avoid technology that is outdated or is on the bleeding edge and has not been proven. Software is evaluated on functionality and performance.

Functionality

What is important here is the ability of the software to do the things you need it to do in a straightforward manner. As an example, if the intended users are relatively new to using computers, the software has to have an easy-to-use Graphical User Interface (GUI). If the organisation needs to develop specific applications, the software should have a programming language that allows the software to be modified or customized.

Standards

Standards are a way of making sure that there is a common denominator that all systems can use. This can be in the form of data formats that can import and export data into the system, guidelines used for developing software, supporting industry developed standards that allow different applications to share data. Standards are generally developed by a neutral trade organization or in some cases are defined by the market. There is a group that has formed for the GIS industry called *Open GIS.* This organisation is developing standards for developers to use as they engineer software. *Open GIS* is made up of representatives from the software developer companies.

Performance

The performance of the software is dependent on two factors: (a) how it is engineered and (b) the speed of the hardware it is running on. GIS software is complex and will use a large amount of the system resources (memory, disk, etc.). The more complex the software, the more resources it will need. Performance will be impacted if you have a minimally configured computer. Look for the developer's software specifications to see what configuration is needed to run the software. This will give the minimum requirements. It has to be followed up by getting the recommended specifications from the developer or a user group. These recommendations will give a more accurate idea of the type of configuration one will need.

Expandability

The software needs of today will change over time. More than likely the system will need to expand. Is the software being evaluated able to provide networking capabilities? Will it share data with other applications? Will it grow as the organisation's GIS grows? Evaluate software based on the ability to grow with the organisation. This may mean that there are complimentary products that can be used in conjunction with the package one is evaluating today or the developer has clearly defined plans for added functionality. Discussions with other users to see if the developer has a good track record for providing these enhancements would be beneficial.

Licensing

GIS software is not purchased, it is licensed. There is normally a one-time licence fee with an on-going maintenance fee that provides us with the most current versions of the software as they are released. In large systems this will be spelled out in a licensing agreement with a corresponding maintenance agreement. For desktop software a shrink wrap license is used with subsequent releases being offered to existing users through a discounted upgrade. The maintenance fees and upgrade costs generally run between 15 to 30 per cent of the initial licence fee. The terms in most software packages spell out how the software can and cannot be used. The terms of the licence should be reviewed by an attorney before signing up. This can save hassles later as one is developing and using the system.

HARDWARE

Computer hardware used to support GIS is a highly variable part of the overall system. Users will customize their hardware environment to best meet their own individual needs. In all cases, however, a fully functional GIS must contain hardware to support data input, output, storage, retrieval, display and analysis. During its infancy, GIS processing was supported only on large mainframe computers. Today, a variety of platforms support GIS processing, ranging from large mainframe computers to mini-computers to scientific work stations to personal computers. In many cases, hardware used to support other applications (e.g., payroll or accounting or digital image processing) is also used for GIS. Hardware configurations for GIS span a tremendous range in terms of start-up and acquisition costs. All systems carry associated costs for maintenance and support.

When discussing hardware, there are terms and concepts that need to be understood. However, GIS software selection drives the hardware requirements. Therefore, before launching a full scale evaluation of hardware, the GIS software selection should be made. The GIS hardware standard establishes consistent configurations for GIS-based computing platforms and associated peripherals. PC computing hardware should be Intel-compliant. Intel platforms utilized for GIS should conform to the desktop system with minimum specification of (a) minimum 32MB RAM, (b) minimum 17" monitor size with 1280×1024 by 256 colour with 4MB VRAM, and (c) 101-key Keyboard.

In addition to PC hardware platforms, GIS technology has been widely implemented on workstation computers utilizing a UNIX operating system. Because so many brands (e.g., non-compatible variations) of UNIX exist, an ITCC UNIX standard has not been established. However, in considering purchase of such a workstation, one should consider six major vendors, including (in alphabetical order): DEC, Hewlett-Packard, IBM, Intergraph, Silicon Graphics, and Sun Microsystems. Other hardware specifications to be taken into consideration include RAM requirements (min. 32MB), monitor size and resolution, and hard disk requirements (min. 1 GB in shared (e.g., networked) disk-space environment; 2 GB for stand-alone workstation).

Deciding what hardware to use for the GIS is a difficult yet important task. It will make up the foundation on which one will build his system. There is no clear-cut formula to use to make the selection process easier. When discussing hardware, there are terms and concepts that one need to understand. Hardware has following basic components (Fig. 4.1):

- Operating System
- Processor
- Disk
- Memory
- GIS Hardware Used in Digital Data Conversion
- Communications

Fig. 4.1: Hardware required for GIS

Operating Systems

An operating system is the software that runs the computer hardware. It is this programme that tells the computer what to do and how to do it. One may already be familiar with some of the operating systems that are on the market such as Microsoft's Windows product or various brands of the UNIX operating system.

It is an important to have an Operating System (OS) plan within the organisation. The plan should take into account the departments that will be using the computer system, the type of network being used (or being planned), what operating systems are currently being used, how large the database is and what kind of technical support skills one have access to (in-house or contractor). The GIS will need to fit into the OS plan. This will be important as to add other departments onto the system.

Pen plotters can handle most plotting jobs, but they do not produce good results in area shading such as in the production of cholorpleth maps. *Raster plotters,* on the other hand, are excellent in producing shading results. *Raster plotters* usually cost more than vector plotters, but are substantially more versatile and have better capabilities (Fig. 4.4).

Fig. 4.3: Scanner

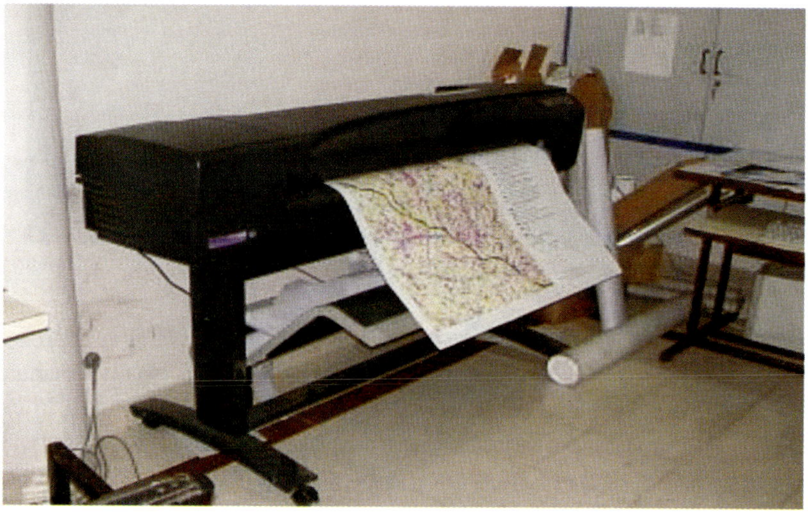

Fig. 4.4: Plotter

Other output devices for the creation of hardcopies of GIS data include: screen copy devices, used for copying screen contents onto paper without having to produce a plot file; computer FAX (facsimilie) transmissions, often used in communications between conversion contractors and clients, produce small letter-size plots, and the fax transmission files (as raster images) can be saved and viewed later; printers are used to output tabular data which is derived from the GIS, and if configured correctly, can produce small letter size plots (Fig. 4.5).

Fig. 4.5: Printer

Communications

The trend in most systems today is to link up users throughout the organization on a network. This is an area in the computer industry that is advancing very rapidly. It is recommended that the organisation retain a competent consultant who works with networks to give the detailed and current information. In simple terms, a network is a connection between computers that allows information to be passed around from computer to computer. In a typical organisation, this is a Local Area Network (LAN). In order to connect a computer to the network it will need a network card for the wiring to plug into and network software to allow the computer to transmit and receive signals over the net (Fig. 4.6). Of course, the physical network (wiring) is also needed.

A small network within a department is inexpensive and can allow the users to share network resources such as printers and database servers. The network can provide services like e-mail, disk sharing and using common hardware facilities. It can also be the entryway into larger networks that go outside the building or campus your organisation is located on.

This is called a Wide Area Network (WAN). A WAN requires a more structured network architecture. It does give users access to more resources (Fig. 4.7).

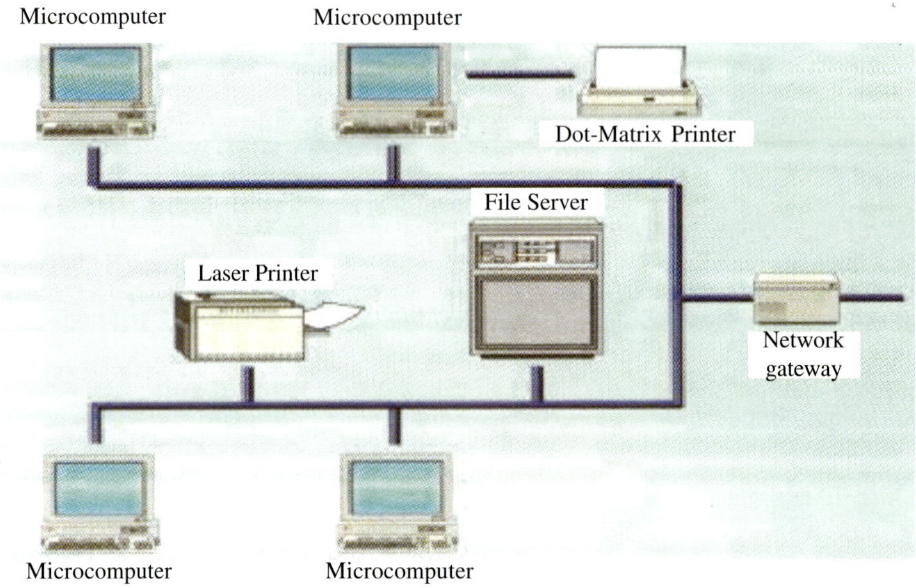

Fig. 4.6: Local Area Network—Intranet

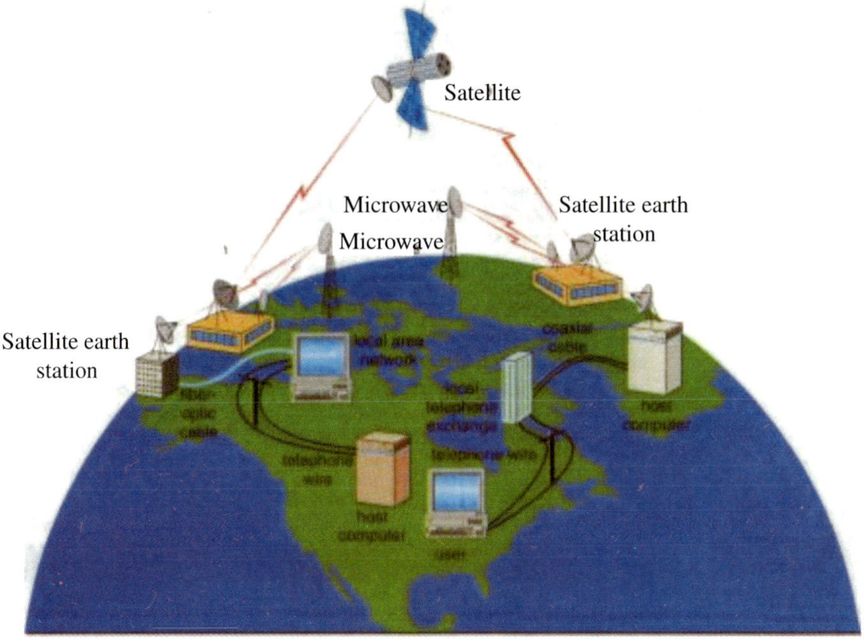

Fig. 4.7: Wide Area Network (WAN) Internet

Another important point to consider is developing access to the internet. This specialized network is growing rapidly and provides an incredible amount of resources for a user. The internet is an area to share ideas in a GIS forum, download data for use in the system, get technical support for a problem, get the latest information on a product from a vendors' home page or develop one of your own. The amount of information is overwhelming and too diverse to list here. The point is that one should seriously be considering getting a connection to the internet. When considering the network requirements, this factor must be considered. Further, internet based GIS has opened new areas for applications. Networks should have features to support storage of data which is accessible by users on the network by specifying particular files, collections of features or attributes, and geographic areas; access security at different levels of the GIS database; ability to support and connect query workstations on the network and to allow database queries directly from workstations on the network; and ability to allow network-wide access to plotters and printers. Further, the network should also be able to perform the following functions:

- Access to data on remote nodes by reference to the node, disk, directory, and file
- Access to programs on remote nodes by similar reference
- Assignment of logical names or aliases for programs or data locations on remote nodes
- Control of peripheral devices from any node on the network
- Passing of mail messages across nodes
- Programme-to-programme communications across nodes
- Monitoring of traffic and errors on the network

The proposal shall include all cabling and devices required to implement all data communication connections, utilizing existing facilities. The proposed system must operate at a minimum raw data speed of 10 megabits per second.

Benchmarking a System

Benchmarking a GIS can be a very involved process. The level of effort needed for the benchmark should be proportional to the size and complexity of the overall system being developed. A benchmark is the process of testing various combinations of hardware and software and evaluating their functionality and performance. The benchmark is usually part of a tender process and is only done with a limited number of selected vendors (i.e., those that have responded and been short listed). Each combination is tested under similar conditions using a predefined data set that is indicative of the database. This dataset should be used with all of the hardware and software configurations selected for evaluation. When completed, an organisation will have results that can used to objectively evaluate the systems.

Installation

When putting a benchmark together there is strength in numbers. A committee may be appointed for this purpose. The committee will take the burden off of a single person and give the process more objectivity. The members of the committee should be drawn from all the interested departments and agencies within the organisation. A working group of about 8-10 committee members is reasonable. The committee should develop the criteria that will be used to evaluate the systems. These criteria will form the basis of the benchmark. A series of tasks that each vendor will need to complete during the benchmark is to be worked out. The tasks should be measurable (i.e., time, ease of use, can the function be done). Also prepare a form that each of the committee members will use to rate the tasks performed in the benchmark.

In the benchmark one will not only be to rating various aspects of the system, but also going to be rating the vendor. One has to be sure to include some measurement for teamwork, communication, and technical skills of the vendor. It might be useful to work with a consultant that has experience setting up benchmarks or to get advice (and examples of documentation) from another local government who has recently completed a benchmark. Well in advance of the scheduled benchmarks, information that outlines the tasks the vendor will need to perform is to be sent out and any rules they will need to follow (how much time for set up, time given to perform various tasks, how many people can be present for the benchmark, etc.).

Vendor Support

The selected vendor will become an extended team member for the GIS group. The vendor will be a good source of support and information. All vendors provide some type of technical support. Ask current users how it has worked for them. If there have been problems in the past, do existing users see improvement? The GIS industry has been growing very fast over the last few years, there are bound to be some growing pains. What one should be looking for is a vendor who listens to the need and makes improvements based on the user input.

Spatial Data Transfer Formats

The standardization of spatial data transfer formats is intended to provide efficient interchange of data between users. In reality, the methods selected for such interchange is often dependent on limitations imposed by hardware and software resident at both the source and target sites. Currently a number of efforts are ongoing to refine existing standards or develop alternatives, including the Open GIS Consortium or OGC. Spatial Data Transfer Standard (SDTS) components include a conceptual model, specifications for a quality report, transfer module specifications and definitions of spatial features and attributes. Most major GIS software vendors now support conversion utilities to and from the SDTS format. In addition, the U.S. Geological Survey has developed a set of public-domain software tools designed to support, among other tasks, the encoding and decoding of logically compliant SDTS data into and out of the SDTS's ISO 8211-FIPS 123 general information interchange format. Additional information on data transfer standards may be obtained by accessing the following sites on the World Wide Web:

SDTS:ftp://sdts.cr.usgs.gov/pub/sdts/standard/latest_draft/pdf
OGIS:http://www.opengis.org/techno/interop/index.htm

Mctadata Content

Documentation provides information to users about the development, content and format of digital spatial data. Known as metadata, this information serves to describe the content, quality, condition, and other characteristics of various datasets. The GIS standard for documentation for digital spatial data shall be the Content Standard for Digital Geospatial Metadata, developed by the Federal Geographic Data Committee. Regarding India, the National Spatial Data Infrastructure or NSDI has formulated the metadata standards (Appendix II). The standard specifies the information content of metadata for a set of digital spatial data, including a common set of terminology and definitions for documentation related to these metadata. The standard does not specify how this information is to be organized or transferred within and between computer systems, or how the information is to be transmitted or communicated to users. Additional information on the FGDC metadata content standard may be obtained by accessing the following site: http://www.fgdc.gov/metadata.

Sample Hardware Specifications

Specifications for a system configuration to support GIS development and operational applications follow the system configuration consists of various devices that will be networked together to support data capture, storage, processing and display in both digital and hard copy forms, including:

- mapping/analysis workstations (2)
- colour laser printer (1)
- black and white laser printer (1)
- cartridge tape drive (1)
- colour raster plotter (1)

The proposal shall include technical and functional capabilities of the devices offered to meet these specifications. Provision of the following information should be included for each device:

- manufacturer
- year of production
- model number and version
- capabilities/configuration of each device in comparison to the device specifications
- documentation provided with the device (i.e., manuals: CD or hardcopy)
- features of warranty and guarantee included in the purchase price
- post purchase user support, duration and minimum response time

GIS Workstations

The mapping or analysis workstations will support a widerange of GIS activities, including database development, database quality control, user application development, database maintenance and all GIS applications supported by fully functional GIS software such as cartographic production, geographic database queries, and advanced geographic analysis using both spatial and attribute information. One of the GIS workstations must support high capacity data storage, and multi-user GIS processing, and should perform all GIS operations and applications within acceptable user response times. General specifications for workstations are as follows:

- Mass storage may be configured within the workstations' cabinetry and/or as external drives
- High resolution (1280 × 1024 or greater) color monitor with at least 19" minimum diagonal screen dimension
- Keyboard and a pointing device such as a mouse
- Network-ready and should be capable to connect a local area network (LAN) and supporting a minimum transmission speed of 10 megabits per second (mbps).
- Multi-user, multi-tasking OS supporting logical security measures such as user name/ password validation, and user access privileges.
- Virtual memory operations, either through a dedicated hardware controllers(s) or through software OS functions.

Specific Details of Workstations

- 32-bit processor supporting both 64-bit address and data buses, minimum of 75 MHz clock speed and may have multiple CPUs on board.
- 128 MB (megabytes) of main memory and shall support 32 MB memory modules and be expandable to 256 MB.
- mass storage disk drive(s) for direct access of data and software functions with a minimum of 3 GB of mass storage.
- CD-ROM drive with one 1.44 MB floppy drive.
- Should support multi-user/multi-tasking operations.
- Options for upgrading the speed and performance.

Small-Format Colour Printer

One colour printer will be used for the production of colour hard copy graphic plots and non-graphic report generation. The colour laser printer should meet the specifications or equivalent described below:

- 300 dots per inch (dpi) resolution
- 100 sheet paper tray
- 4 MB memory onboard with capacity for memory upgrades
- Support for letter, legal size, and 11" × 17" paper sizes
- Built-in postscript compatibility
- Serial and parallel interface

Large-Format Colour Raster Plotter

A colour raster plotter shall be included for the production of high quality, large format cartographic products. This device must provide a high-volume colour plotting capacity. The plotter shall support 36" × 42" plots and produce colour plots at a minimum resolution of 300 dots per inch. The plotter shall be compatible with the proposed LAN hardware and communications protocols and must be accessible by all workstations on the LAN. A sample hard copy output from the proposed plotter shall be included in the proposal. Additionally, the plotter should meet the specifications described below:

- Capable of supporting true colour plotting
- 8 MB memory onboard with capacity for memory upgrades
- Support for all paper sizes, A through E size
- Built-in postscript compatibility
- Serial and parallel interface

Conclusion

A proper combination of software, hardware and communication technology has become a core issue for GIS development in any set up. One has to be very careful in working out the details which is not easy. Fast changing technology, increasing volume of data to be handled, new features, licensing protocols, falling prices and options of integrations with other technologies have made such decision more complex. As mentioned earlier, a consortium approach to decision making may reduce the pitfalls.

REFERENCES

Burrough, P.A. (1992), Development of intelligent geographical information system, *International Journal of Geographical Information System,* Vol. 6 No. 1, 1-12.

Burrough, P.A. and McDonnell, R.A (1998), *Principles of Geographical Information Systems.* Oxford University Press Inc., New York.

Cadoux-Hudson, J. and Heywood, I. (eds), (1991) *Geographic Information System,* Taylor and Francis, London.

Carstensen, Jr., Laurence, W. and Campbell, James, B. (1991), Desktop scanning for cartographic digitization and spatial analysis, *Photogrammetric Engineering and Remote Sensing,* Vol. 57, No.11, pp. 1437-46.

Cowen, D.J. (1988), GIS versus CAD versus DBMS: What are the differences ? *Photogrammetric Engineering and Remote Sensing,* Vol. 54, pp. 1551-55.

Frank, Andrew U., Egenhofer, Max J. and Kuhn, Werner (1992), A perspective on GIS Technology in the nineties, *Photogrammetric Engineering and Remote Sensing,* Vol. 57, No.11, pp. 1431-36.

Goodchild, M.F. (1991), Geographical information science, *International Journal on Geographical Information System,* Vol. 6, No. 1, pp. 31-45.

Goodchild, B.F. (1992), Geographic data model, *Computer and Geosciences,* Vol. 18, pp. 401-08.

Green, N.P., Finch, S. and Wiggins, J. (1986), The state of art in geographic information system, *Area,* Vol. 17, No. 4, pp. 295-301.

Groff, James R. and Paul N. Weinberg. (1999), *SQL, the Complete Reference.* Berkeley, California: Osborne/McGraw-Hill.

Knapp, W.G.M. van der (1992), The vector to raster conversion: (mis)use in geographical information systems, *International Journal on Geographical Information Systems,* Vol. 6, No. 2, pp.159-70.

Maguire, D.J., Goodchild, M.F. and Rhind D., (1991), *Geographic Information System,* Longman, London.
Mark, David, M. and Gould, Michael, D. (1991), Interacting with Geographic information: A commentary, *Photogrammetric Engineering and Remote Sensing,* Vol. 57, No.11, pp. 1427-30.
Mounsey, H.M. and Tomlinson, R.F., (1988), *Building Data Bases for Global Science,* Taylor and Francis, London.
Nag, P. (1987), A proposed geographical information system for India, *International Journal on Geographical Information System,* Vol. l, No. 2, pp. 181-87.
Rhind, D.W. (1988), A GIS Research Agenda, *International Journal of Geographical Information System,* Vol. 2, p. 23-28.
Worrall, Les (1990)(ed), *Geographic Information System: Development and Applications*, Taylor and Francis, London.

5

GIS in State and Local Governments

Over the years, there is growing need to decentralise the process and practice of development, planning and administration. This is to ensure that the enough attention is given to the day to day problems of the people. Further, the administration is becoming more and more professional and bank on spatial information which should be quickly accessible and readily available. Hence, the application of GIS in the administration of state and local governments has become an essential component. Further there is a renewed emphasis on finding out solution of problems locally. Planning at regional level for sorting out local issues is an outdated concept. In India, the institutional changes have been made by bringing 73rd and 74th Amendments in the Constitution for empowering the states governments and *panchyats* or local self-governments in rural areas consisting of a group of villages. Further, eleventh and twelfth schedules of the Constitution specify a set of twenty-one and thirty-four sectors for the *panchyats* and the *Nagarpalikas* (urban municipalities) respectively where the local bodies are expected to prepare and implement the development plans.

GIS technology helps in streamline business practices while adhering to complex political and regulatory requirements. ESRI is the world leader in GIS software, which means you have access to an unparalleled wealth of data, third party software, and services to meet your needs and budget. In each case there are different combinations of approaches and technologies for finding out solutions. GIS may be at the centre of such solutions, but its link with other technologies is a mater of great interest. The variation in the solutions is due to the varieties of issues involved in each project.

Tourism

Bulgaria is promoting its scenic delights, with a number of local authorities using 3-D visualisations to whet the appetite of tourists. The examples shown here were generated by Sofia-based ET Troeva Consult as part of a feasibility study for a comprehensive development plan in the town of Chepelare and its surrounding tourist zone in the south of the country. The study analysed natural and man-made aspects of the terrain, including relief, water resources, climate, soils and vegetation, protected sites, transportation and many other economic and social factors.

The 3D visualisations were developed in ArcView 3-D Analyst and Spatial Analyst extensions and employed during public presentations and at business forums organised by the municipality. In this capacity, they facilitated public discussion among those who had limited knowledge of planning tools. With the ideas behind the proposals fully understood, they were adopted by the local planning commission. (www.geoplace.com/ge/2001/)

Advertising Billboard in the Slope

In one recent assignment undertaken by Malthas Tegnestue of Aalborg, Denmark, a town council wished to parcel out an area of sloping terrain for development but from whose higher point scenic views of an inlet could be enjoyed. To preserve these views, parcel placement, building elevations and maximum heights all had to be written into the plan document. To ease the task, 3-D GIS analysis was employed using ArcView 3D Analyst software. With this viewshed visibility analysis, it is possible to see, from any given viewpoint, which buildings 'cast shadows' and whose placement and height needs modifying to provide optimum scenic views These analyses had to supply an ongoing picture of view possibilities in the sketching phase, and also document views of the inlet from each of the parcels in the final plan. To do this, viewsheds (grids or raster datasets generated from a viewpoint in 3-D and a surface) were created.

Basic components of the 3-D model/surface included a DEM, building themes from municipal mapping, and the parcel plan. The DEM was generated from 2.5 metre contour lines determined from photogrammetry and terrestrial height measurements. With a clear idea of the scenic possibilities of each parcel, officials were able to give better guidance to potential buyers. The city of Ithaca, NY, known for its progressive urban planning, environmental preservation and advanced e-government services, adds MapInfo technology to its online portal to provide city departments, the private sector and citizens with detailed information on the city and its services, including property photographs. (www.geoplace.com/ge/2001/)

Watershed Planning and Infrastructure Modelling

Using MapInfo for watershed planning and infrastructure modelling, Portland couples GIS tools with computer modelling systems for planning and assessment. MapInfo spatial technology helps the bureau to accurately assess existing systems, identify system deficiencies and develop a comprehensive plan for implementing necessary improvements. The New York State Division of Housing and Community Renewal's use of MapInfo technology won an Explorer Award for its e-government online application. The web-based, interactive Affordable Housing Directory enables users to easily access the inter-active directory, offering a convenient, continually updated and comprehensive directory of affordable low- and moderate-income housing throughout New York State. (www.dhcr.state.ny.us)

Location-based Web System Helps People Get and Keep Jobs

The goal of an employment agency is to help people find a job, but that help often falls short when it comes to finding the ancillary support services they may also need, such as public transportation or day care facilities. To both broaden the level of service it offers job seekers and to better monitor its own performance, Connecticut's Department of Labor (DOL) is ramping up an innovative, cost-effective, Web-based mapping application that goes well beyond simply listing and describing the jobs available in a given location. With it, users can call up easy-to-use maps via the DOL Web site to identify potential employers and to locate bus lines, hospitals, schools, community colleges, Connecticut Works SM job centers, public libraries and other available resources anywhere in the state.

Needs

- Enhance the services offered to jobseekers.
- Better monitor Labor Department performance.

Problem

- Information on resources useful to job seekers is contained in separate databases.
- Lack of ability for sophisticated cross-agency analysis.

Solution

- GIS Internet application, provide job seekers with a easy-to-use location-based system implemented cost-effectively on the Web.

Benefits

- Clients have a powerful which tool they can use by themselves.
- Labour Department can better track its performance.

The idea is to do more than just help people get a job; it is to help them keep it by giving them an easy way to locate whatever services they might need, said the geographer in DOL's Performance Measurement Unit in Wethersfield, near New London. "We use web based GIS application as a locator in our GISEPI (Geographic Information System and Employment Planning Initiative) system. Once users get on the internet, they go to our gateway page where a pushpin on a map shows their address. Right now there are seven thematic layers on the map, which encompass resources such as job center offices, day care facilities, and employment and training services providers. People have more choices today, and this is our response. While the basic need is to maintain a job bank, there's been a sea change The change is so radical that the earth has moved under our feet. We are moving away from a strictly one-to-one mode where clients have to go to our job centres and meet with a counsellor." (www.mapinfo.com)

Efficiently Delivery with Location-based Analysis on the Web

Howard county is one of Maryland's most dynamic counties in both population growth and job creation. The number of residents in its 254-square mile area has grown by 26 per cent over the last decade and stands at 239,000. Its diversified employers, primarily in technology, manufacturing, telecommunications, wholesale distribution and agriculture, have created more than 40,000 new jobs during the same period. Howard county uses location-based analysis software from GIS to manage a veritable explosion of residential and commercial development; maintain and build roads and bridges with minimum disruption to traffic flows; and, amid continuing growth, provide fire, law enforcement, public health, record keeping and other essential governmental services cost effectively.

Needs

- Provide high quality services without interruption despite tremendous economic and population growth.

Problems

- Multiple departmental databases.
- Individual workstations in remote offices require support.
- User abilities vary.

Solution

- A system uses a Web-based GIS solution to produce easy-to-use maps to speed day-to-day management; to perform location-based analysis for better planning and interdepartmental coordination; and to deliver services cost effectively.

Benefits

- Cost savings.
- Better analysis and planning.

Good decision-making is critical in such an environment. But good decisions cannot be made without relevant data in a form convenient for Howard county's employees to use. That is why personnel in seventeen county departments or major divisions use GIS. With it, they choose a physical location as a reference point and then map to it all relevant information, even if the information is contained in multiple county databases. The data are viewed in the form of maps that are intuitively understandable, unlike traditional tabular or spreadsheet formats. The maps make it easier for many of the county's 1,800 employees to perform their jobs, and governmental administration thereby has become better, faster and more cost-effective. The county has scores of applications built with GIS software, used by employees in a variety of job functions.

The wide range of applications include some which enable staff to quickly identify property owners or to generate mail lists for public notifications. Meanwhile, others facilitate coordinated infrastructure planning. For example, the Bureau of Highways handles road and road-maintenance planning, while the Bureau of Utilities handles sewer and water planning and the Planning and Zoning Department handles development planning. With GIS, the county pulls all this information together on one map and coordinates construction so that, for instance, roads are not repaved then immediately dug up to replace a water line. Still others include the

Fire Department's use of GIS to determine the best sites for new facilities and where to allocate resources for optimum response times; the Police Department's use for crime analysis and investigation, including "hot spot" trending; and the Planning and Zoning Department's use to generate base data for a major redevelopment study of major highway corridors. The county has been involved with mapping software since 1990, but in 1996 it centralized its location-based analysis capabilities in a three-person GIS Department. It operates as a central resource and treats users of the GIS software as clients. (www.mapinfo.com)

Tax Parcel Development Services

The city of Norfolk has planned to use GIS for tax parcel development. When completed, the digital tax parcel map will streamline existing map maintenance procedures that are required to support the current paper-based system. In addition, the new digital tax parcel map will enhance the value of the existing GIS basemap that includes orthophotography and planimetrics. "We are excited to be working with Baker," said Norfolk's manager of geographic information. "Baker's reputation for quality work was an important consideration for us in selecting them for the project. Our communications with Baker have been great, and everyone we work with is prompt, professional, and understanding of our needs."

"We are pleased to begin this complex effort for the City of Norfolk," said the GIS manager in Baker's Virginia Beach office. "This project will require close coordination between Baker and the City and I believe the results will reaffirm Baker's ability to be a partner through the entire data development and delivery process." (www.findarticles.com)

Municipal Mapping

Municipality of Aalborg wanted a graphical basemap in the ArcInfo environment that could be used as printed information and within the daily administration and planning. The map should be used as background for the environmental planning, AM/FM, path and road planning etc. The maps have five different scales from 1:10,000 to 1:160,000 each with their own generalization. As source material for the map the technical maps from the municipality was used.

In order to find out a solution, Geovision A/S has build a cartographic database to be used as base map to all the administrative functions within a municipality By converting map data from different sources, mainly from CAD data, generalizing and recoding information a new basic map was build in an ArcInfo environment. Five different cartographic scales 1:15,000, 1:25,000, 1:50,000 1:100,000 1:160,000 was made from the basic map, each with their own generalization. (www.esri.com/library/brochure)

Library Related Activities

Kraks Forlag A/S is a publishing company with deep historical roots dating back to the year 1770 when the first Kraks Directory appeared. From inquiries at the Royal Library of Denmark and equivalent sources internationally it seems likely that Kraks Directory is also the oldest in the world. "Although we are a company steeped in tradition, I hope that this profile will show that we are also a company which keeps up with the times." During the past ten years the company has undergone a facelift which has transformed Kraks Forlag into a modern enterprise with an advanced IT platform.

Kraks Forlag is today a highly professional provider of information to and about the Danish business sector. Realizing that new thinking was required to transform the company into a competitive, future-oriented enterprise, we have developed and maintained a strategic plan, the main elements.

One of the goals used in the evaluation was 'customer-controlled choice of media'. It was important for Krak that map customers should be able to choose for themselves their map medium: (a) printed, (b) CD-ROM, (c) Internet (www.krakskort.dk), or (d) special database abstracts (GIS and KIS). In this context it is considered that Cartographics/ArcInfo would in the long term deliver the largest number of options while at the same time system development could be standardized to a minimum of products from.

Provincial GIS

Province of Alessandria decided in 1996 to develop a GIS as a support its decision making and operational management activities. Initially informatic tools are being developed and informatic support being given only in two areas of the province's activities: finances and the protocol. In other offices the use of computers was sporadic and unstructured. Data exchange and share of resources and information were not possible. In general the informatic capabilities of the employees were quite low and it is considered inappropriate for a modern public institution.

According to the plan contemplated in the feasibility study the third phase dealing with water resources is being implemented. It will be followed by the GIS applications in civic protection. There is an interest in extending the project to the road system's departments. The province administration intends to make a part of geographical data of large professional and public interest available through an internet site. A project is under consideration within the province aiming at providing technical support to its municipalities and assisting them in training, selection, introduction and implementation of GIS technology.

In comparison to the other similar projects, the province of Alesadria opted for an initial introduction of simpler GIS technology that will gradually be replaced by more sophisticated one in the future. It is thought that the loss of possibilities of generating new information in the initial phase will be compensated by the smoother introduction of the new technology, avoiding thus the traps of initial "technological shock" that had serious negative implications, experienced in the other similar projects. The initially obtained result was the large dissemination of the GIS technology within the provincial services. It has become a real tool for daily work. At the moment of submitting the report, 7 Arcview licences have been installed and more than ten users were trained. An important achievement is the fact that the GIS concept of coping with data of spatial nature by means of GIS functionalities has affected the mentality of the employees so that the implementation of the new GIS based technology is now enthusiastically being proposed for almost all the new projects. The expected paradigm shift is gradually taking place even among the hard core of the employees who opposed or were very reluctant to GIS introduction. Another important side effect was the general improvement of informatics perception at various levels including standardizing of informatic components including, both hardware and software, as well as data handling procedures (coding, storage).

At the end importance of the implementation of proper training methodology is important. It has to be flexible, personalized and extended over an appropriate period. Best results are achieved when both trainer and trainee take part in GIS introduction by solving the ongoing operational problems. The introduction of GIS raises the interests of the future users, provides the consent for its use in future projects and makes it an integral part of the whole work process. (gis.esri.com/library/userconf/europrocgs)

Property Data and Maps on the Web

Brevard County, home of the Kennedy Space Center, is situated on Florida's east coast. The County stretches 72 miles in a north-south direction and covers 22 miles east to west. Located just 40 miles to the east of the Orlando metropolitan area, Brevard county has experienced a steady growth rate of about 3 per cent a year. The population during implementation of the project was nearly 500,000. Brevard has had an active GIS programme since 1988. Like many other GIS programmes started in the late 1980s, Brevard's programme evolved not only in terms of hardware and software but also in terms of focus. The first four years were spent developing the property parcels base layer. However, those funding the project needed something to show for the money spent, so application and map product production became the primary focus of the programme.

The site went online on August 22, 1999. The mapping web site enhancement was available just in time for the annual Truth in Millage (TRIM) notice mailing that had a section devoted to the new web site features. Response from the public has been very positive, and the site now records approximately 30,000 hits a day and over 4,000 maps are served each day. The mapping enhancement added a new dimension to the web site. Now users can access any property in Brevard county through a database query or directly on the mapping interface. Users can view property information and perform radius analysis from a chosen point. Sales analysis queries performed on the database server can be automatically displayed on the map.

Eventually, additional layers of information are to be incorporated into the web-mapping site. Work continues on addressing problems caused by map misalignment. A derivative version of the existing mapping functions will be developed for internal Intranet usage. This version will incorporate the layers available in GIS that affect property valuations such as zoning, future land use, soils, and flood zones.

The goal of the project is to allow public access to the Property Appraiser's records and maps has been realized through this technology. The web has provided the means to finally get GIS data into the hands of users in a simple and affordable interface. (www.esri.com/news/arcuser/)

GIS Needs for City

Calgary is situated on the eastern slopes of the Rocky Mountains in the southwestern portion of the province of Alberta, Canada. Calgary has a population of over 840,000 encompassing 280 square miles. It is the sixth largest city in Canada with the second highest number of head offices located within a city. The populace is well educated with over 58 per cent having a post secondary education and 61 per cent having internet in their homes (the highest total in Canada).

It is a growing, vibrant city with over 9,000 new housing starts and 15,056 building permits totalling $2 billion issued in 1999.

The concept of a unified city with a population approaching one million, the above average technical knowledge and growth averaging 25,000 per year has all contributed to the necessity for the city of Calgary to take a second look at the value of a GIS to meet its business needs.

The city started to investigate the feasibility of using GIS to meet its business needs in 1989 and began preparing an RFP in November 1990. A consultant was hired in the spring of 1990 to determine The city's spatial data requirements and they concluded that the city was spending $28 million per year processing location-related data. They sold the concept of a GIS on the basis of small gains (1 to 2%) of the $28 million being spent annually. This led to the Calgary Spatial Information Management (CalSIM) project under which the GIS project fell. The CalSIM project lasted from 1991–1996 with a $15.6 million budget. The web (intranet) application developed was based on a web delivery strategy that

- Applied the ESRI out-of-the-box functionality.
- Deployed increasing functionality in each phase.
- Added more data as it became available.
- Supported reusable functionality for other similar data.
- Had a short development time.

The *intranet web* application was delivered in five phases:

- Phase 1: Lists of maps to select and display,
- Phase 2: Search by address, parcel ID and name,
- Phase 3: List of addresses within distance and mailing lists,
- Phase 4: Parcel info based on complex queries,
- Phase 5: Thematic maps and statistical information.

The *Quick Hit* applications developed were based on a strategy that

- Built upon the GIS out-of-the-box functionality (ArcView applications).
- Used replicated city data as soon as it was available.
- That supported a wide corporate interest based on the survey.
- Had reusable functions for other similar tacks.
- Had a short development time.

The web application and quick hit applications using the above strategies demonstrated to the corporation that GIS could and should be considered as part of its business strategies. It was learnt during this process that the time estimates were suspect. As it was further learnt, the ESRI tool and gained insight into its functionality, these estimates needed to be revised. When the three-year business plan was presented to the senior management, every effort was made to ensure that they understood this and as it turned out they were very supportive.

The city was fortunate that the spatial data was stored in Oracle and that it did not have to deal with a number of legacy systems, including no data maintained from an ESRI environment.

The city's roll out strategy therefore focused on the use of ArcView and its extensions to access and query of spatial data and display the results. Replication of spatial data became an immediate strategy with the understanding that future maintenance of spatial data would take advantage of the ArcInfo 8 technology.

Maintaining spatial data using ArcInfo 7 technology was not promoted and has been limited to only a few special circumstances. The ability to replicate data has gone smoothly because of good source data and the excellent functionality of the ESRI tools. This provided the city an opportunity to leave maintenance in its current environment (MicroStation). This is important to the users as it provides them the freedom to learn the new tool and determine where their spatial data maintenance should be performed (it is believed in a number of the cases they will choose ArcInfo 8 as their maintenance environment).

An organization (GIS Centre of Excellence) to support and sustain GIS technology in the corporation is now in place. The roles for the business unit are to coordinate and facilitate GIS integration into the business units (a) to ensure GIS investments are managed, (b) to coordinate and manage data standards and data sharing, and (c) to develop strategic policies and assume a leadership role in the GIS technology within the corporation. (www.esri.com)

GIS and Local Government

Local government, both municipal and county, is often caught in the middle of divergent demands. Operating budgets can dictate limited staff sizes and increased workloads. Technologies can become outdated without continued funding or internal resources to provide the tools to do business. Conversely, people dealing with local government, taxpayers and private and public sector organizations, demand improved access to government and better increased services. Combined, local governments face a range of challenges, which in many cases grow beyond their means.

The above situation still applies to many governments; for others, partially completed or departmental solutions, have addressed a shorter term need in a band-aid fashion. Innovators, those who choose to lead rather than follow, are rarer. Sarasota, Florida and Houston, Texas, are two innovators who have meet the local government challenge head on and succeeded. Their strategy to delivery service, and the focus of this paper, was to create an Online Collaborative Community using Akanda's COLATE (Collaborative Teaming Environment)—an Internet and GIS-based framework for providing a fully integrated set of services. Here an attempt was for the justification, strategy and results experienced by Sarasota and Houston in developing and maintaining their online collaborative communities.

Sarasota is a city of 65,000, perhaps best known as a winter vacation haven. While residents and visitors to the city were enjoying its beaches and sun, the city's internal operations were not as tranquil. Operationally, the city had a series of Cobal applications for departmental functions including building permits and business licensing.

The reliance on the data in these applications, compounded by the inoperability with other systems, non-Year 2000 compliance and support problems, made for a strong technical argument for system replacement. Further, City Information Systems staff had aspirations of a much more integrated and sophisticated enterprise system that involved GIS, workflow and messaging to best manage city responsibilities such as code enforcement and development tracking review. The city envisioned all departmental applications having the same look-and-feel and access to an enterprise of data. GIS would be embedded into each application, offering a unifying and visual interface to data.

Overall justification for replacing the Cobal applications with an integrated approach were:

- Ability to make integrated decisions.
- Faster delivery and higher quality of services.
- 24 × 7 access and higher volume.
- Reduced cost and lower overhead.

With a population of nearly two million, Houston is the fourth largest city in the U.S. The growth of the Houston area is the continual and strong—it is the third fastest growing of the ten largest U.S. metropolitan areas. The population base and growth rate of Houston has placed considerable demand on its internal resources and processes, particularly in the development tracking and mapping area. The previous situation isolated development tracking and GIS:

- Development tracking for land development was a largely manual process, compounded by a nine days window for an application to be approved, denied, or sent to a planning committee for a hearing. The only automation was a dial-up modem method for private sector land developers to upload the CAD component of a land development application; after upload the file would be attached to the paper application for routing and review.
- GIS was not used as part of the land approval process, nor was a plan submitted as part of a land application used by the city's GIS.

The mixture of booming land development demands and inefficient methods to track and review applications for development was a strong qualifier for implementing a new solution for Houston.

The Sarasota and Houston cases share several common threads:

- Each had outdated technologies that were impeding day-to-day departmental operations and preventing the sharing of information with other departments.
- Existing applications served a single departmental function, with no reference towards or operability with enterprise systems.
- Proponents wanted to encourage the flow of information for applications that paralleled the actual workflow in the city.
- Proponents wanted to promote a 'One Stop – Self Help' model of customer service and leverage developments in collaborative and internet technologies.
- Proponents wanted solutions that could accommodate their future growth and demands
- Both ESRI technology users with goals of spatially enabling their databases and processes.

The change process has varied noticeably between Sarasota and Houston, a result of previous systems and technology available at the time of change. Sarasota had already installed Lotus Notes as their messaging and workflow platform when the project to replace their Cobal departmental applications commenced. This situation provided to be very timely, as Notes laid the foundation for future replacement to the Cobal applications, where GIS and departmental process functionality would be added to a Notes/Domino application.

Sarasota decided to follow a hybrid development approach, where custom developed applications would be built upon their Notes platform. MapObjects would be used to add a spatial dimension to the applications, leveraging the city's mapping investment. Over a span of two years, five applications were defined by the city and designed, implemented and supported by Akanda:

- Building permits and inspections management
- Code enforcement and complaint management
- Contractor trades licence management
- Business licence management
- Property sites address management

Internet enablement was identified in the project planning phase as a longer term goal for the application suite. Further, Houston has been a comparatively late adopter of Online Collaborative Communities, with project planning starting in 1999. However, the two years between the project starts for Sarasota and Houston have seen a significant change in the role of the Internet. Accordingly, Houston's COLATE solution—also Lotus Notes based—was to be fully internet enabled. Houston's process involved the removal of the dial-up server for posting plats and the paper development tracking process. The replacement COLATE solution involves a completely internet-enabled end-to-end system to accept online development applications, manage and track the application process, and accept input from outside commenting agencies (e.g. utility companies). Highlights of the Houston implementation include:

- Land developers being able to submit land development applications online, including forms and CAD plat drawings.
- City planning department staff has a tool to manage the entire development review process. Staff can be assigned to cases, comments and approvals assigned on a rules and/or case basis, and agendas and content for committee meetings are automatically generated.
- Markups of plats can be marked up online by commenting agencies. Utilities can assign street numbers and other commenting agencies can identify changes through their browser.
- CAD format plats, once approved, will enter the City's ArcInfo database.

Sarasota and Houston represent two cities who have harnessed the enterprise wide benefits of collaborative approaches to local government management. While the role of the internet in the COLATE deployment for each city differs, each city has replaced older, inefficient processes with internet based approaches to doing business. (http://gis.esri.com/library/userconf/procoo/professional/papers/PAP933/p933.htm)

Enterprise GIS Hub and Data Warehouse

Faced with the problem of sharing enterprise data between the many bureaus and agencies within its jurisdiction, the city of Portland in Oregon has developed an innovative internet-enabled solution that allows all parties access to city information without migrating from their current mix of data maintenance systems. The city of Portland has solved the problem of sharing data by providing an enterprise data warehouse. The Enterprise GIS Hub (EGH) that has been developed by Portland is a high availability spatial database that provides online internet access to the city's enterprise data sets. In addition to the internet browsing and querying capabilities the hub provides a web-based data extraction page that permits the user to obtain a copy of the data of their choice, for any area of the city and in the format of their own GIS system. They may then import the data directly into their own systems and use it natively. The beauty of the approach is that all bureaus can use any of the city data sets without the difficulty of having to migrate to a new system.

Bureaus within the city of Portland use a rich mix of spatial data systems including ArcInfo, Microstation, AutoCAD, MapInfo and ArcView. Tabular information is managed within a variety of databases such as Access, Oracle, SQLServer, Sybase, etc.

This mix of systems has caused the exchange of data between bureaus to be difficult. This led to a number of problems as city personnel found themselves spending increasing effort in tracking down data and verifying it for accuracy and currency as well as servicing requests for their own data from other bureaus. Inevitably bureaus found themselves maintaining their own copies of the enterprise data sets for use in their own systems. Bureau data maintenance engineers had less time to spend maintaining the bureau's core data sets and this led to maintenance backlogs and data currency issues. In the face of increasing costs and duplication of data, the city's GIS Management Committee championed the notion of a multiple-platform GIS data warehouse to serve the enterprise data-sharing needs of all bureaus.

CGIS contracted Convergent Group, a spatial systems integrator in Englewood, Colorado to architect and build a pilot system to demonstrate the strengths of the data warehouse approach. The hub stores enterprise data within a Spatial Database Engine (SDE)-enabled database hosted by Microsoft's SQLServer. SAFE Software's Feature Manipulation Engine (FME) provides spatial data translation services to translate source data into the hub and also to translate hub data into the many extract formats supported by the hub. SQLServer's native replication functionality is employed to replicate the enterprise data to all participating hub servers to provide local copies of the data to all user groups. GeoNorth's MapOptix Internet Map Server is used to serve the hub data to the internet as well as to configure user access to data and hub functionality. City personnel do not require any additional software, they can view the data online using a standard web browser. Client access to the hub data is facilitated through the application development and customization activities of Portland's CGIS. The Enterprise GIS hub allows city bureaus to share information from the comfort of their own environments.

Implementing an enterprise system based on such a rich mix of vendor technologies has been a significant integration challenge that required the cooperation and commitment of the vendor parties, which in some cases has meant the release of enhancements to vendor products. Of the technological challenges that have been overcome in the development of the hub, perhaps the greatest has been the replication of SDE data between hub servers.

Data replication provides a number of advantages including local repositories for main user groups, scalability through additional database servers and availability through the use of load balancing software to automatically distribute requests based on the available databases and their workload. Data replication is a cornerstone of the hub architecture that required significant investigation of SDE before it could be accomplished in all its flavours. The lessons learned during this part of the project have been fed back to the SDE development team at ESRI in an effort to facilitate the support of replication in later releases of SDE.

The robust architecture of the hub has been designed to scale to meet the growing demands of the city and to provide data access 24 hours a day to satisfy the city's mission-critical systems. The hub's inherent replication scheme insures that the data on all participating hub servers is automatically synchronized. (esri.com/news/arcnews/)

GIS Access Application

The city of Olathe, Kansas, has implemented an ArcIMS application for general GIS query and generation of hardcopy maps. The city's objective for this application is to provide ad hoc query of the city's GIS data, with an option to produce hard copy output. The application was designed to be easy for city staff and policymakers to use with a very low amount of training. It is assumed that these users will have no background in cartography or GIS, and their hard copy output will not require highly precise spatial displays. Because of the limitations of the ArcIMS programming toolkit, the quality of hard copy output is limited to what the user sees on the screen. For users that require presentation quality, precision, and/or sophisticated ad hoc analysis, the city's ArcView licences and GIS staff expertise are available to satisfy these needs.

The primary tool for searching and zooming to a particular location in the city is an address, and the user may select whether to use a street centerline address or a parcel address. In addition to the search by address functionality, staff can search and zoom by subdivision name, government building name, and departmental map sheet designations. Users are able to transfer their compositions into Microsoft Office documents (e.g., Microsoft Word). For generation of hard copy output, users can specify a map title, map comments, resolution, size, and format. Dynamic definition of other text elements and/or editing of annotation is also available.

Other browser-based general GIS applications include mailing list generation, pin mapping or dot mapping, tagging locations, merging map layers, and customizing maps. By placing these applications on the desks of end-users, a large proportion of general spatial queries can be successfully addressed without the specific involvement of the central GIS staff.

After a series of workshops to determine public access policy, the city of Olathe, Kansas, used ArcIMS to develop a 'first generation' map server application for disseminating information and answering frequently asked questions. Because the city cannot assume that users have any familiarity with spatial information, this application was kept simple, with the objective of targeting city residents' information requirements. More extensive functionality targeted at commercial users was to be added in the future.

Access to the mapping component is from the city's previously existing web site. After a disclaimer screen, at which the user 'Accepts' the disclaimer, a city map is displayed. The list of all available layers is displayed, and the user may select which layer will be active (for display of attribute data). Addresses and government buildings are the primary tools for searching and zooming to a particular location in the city. (gis.esri.com/uc2003)

Business Process-Specific GIS Application

The state of Kentucky Cabinet for Families and Children (CFC) has used internet technology to develop and deploy a business process-specific application for supporting the staff of the Independent Living Program (ILP). CFC operates the ILP and helps foster children (clients) between the ages of 12 and 21 develop skills that support independent living. An original ArcView application combined the ILP database with GDT's DynaMap 2000 to create a searchable geocoded database of locations for clients, care providers, and community resources. The programme staff used the application to locate services (e.g., schools, hospitals, and churches) that are in close proximity to a client. The application results can be displayed as a map of the area, a list of available services and their addresses, or imported into a document for use in form letters and mailing labels.

An updated version of the ILP application was developed using MapObjects IMS. The browser look and feel has simplified the staff's training requirements, resulting in increased benefit from the application. In addition, serving maps and data over CFC's Intranet has resulted in improved security, simplified application updates and upgrades, increased consistency, and reduced cost. (http://www.esri.com/new/arcnews/summero/articles/foster-children.hym)

Municipal Government

The Nashville-Davidson county metropolitan government needed to integrate its map maintenance and parcel update processes within its Land Information System (LIS). The objectives for this integration included eliminating redundant data entry and reducing the lag time associated with parcel updates. Operational data was not capable of achieving these objectives, for the following reasons:

- Most users relied on text-based queries, not GIS,
- The long duration required for parcel update transactions would tie up the associated data,
- Constraints in the technology of Nashville's mainframe database management system.

Implementation of the data warehouse involved transforming the LIS operational data into an architecture that is suitable for supporting complex management level queries. This facilitates acceptable performance for complex queries, and eliminates conflicts with maintenance and updates. Some of the specific steps included:

- Replicating ArcInfo coverages into SDE layers,
- Extending and restructuring layers to provide the functionality necessary for the anticipated edits and queries,

- Integrating a free-standing database into the operation for tracking maintenance and time lags,
- Extracting mainframe data and restructuring it into a relational database management system,
- Developing a bilateral synchronization between the two versions of the textual data.

The resulting data warehouse is now in production, and the Nashville-Davidson county metropolitan government is achieving efficiencies and improved effectiveness from integrating its map and LIS updates. (http://www.nashville.gov/mpc)

GIS Enterprise

Enterprise GIS has the potential to support sweeping transformations in business processes, especially when multiple departments (that may never have worked together in the past) are involved. An example of an organization that has achieved benefits from such a transformation is Prince George's County (Maryland). There, the land development process involves multiple departments, including planning, environmental regulation, inspection, health and fire. An early step in implementing an integrated permitting and GIS system was the documentation of existing business processes and the identification of opportunities for improvement. As a result of this analysis, the integrated permitting system was tailored to fit the enhanced multi-departmental workflow. In the resulting operation, each review and approval triggers movement of a permit application to the next node in this complex process. The benefits include the following:

- More accountability—for whose approval is the application waiting?
- Improved service levels—reduced lag time.
- Enhanced effectiveness—applications aren't "lost".
- Greater efficiency—faster answers to questions, less time lost tracking application status.

Business process reengineering actually has little to do with GIS *per se*. It is simply about making an organisation work more efficiently by optimizing the integration of computer technology. Because it is a challenging, detailed process, many organisations are unwilling to spend the necessary time and effort. Nonetheless, optimization of the organization's business processes is necessary to realize the full potential offered by the new technology.

Going beyond the conventional definition of Enterprise GIS will generate broad support, results (productivity, effectiveness, and service improvements), and payback. Experience with local governments has proven that a well-planned, well-managed GIS implementation can result in cumulative payback exceeding cumulative expenditures by the end of the third year of implementation. Going beyond Enterprise GIS is an effective approach to assure broad support for GIS from users as well as policymakers, to remain focused on improving the delivery of services, and to achieve the expected return on a GIS investment. (http://www.goprincegeorgecountry.com/Government/Agency Index/OJTC/GIS/map-horz.asp)

GIS and Business

Changing a business process to take advantage of GIS technology can produce improved efficiency, higher service levels, and/or increased effectiveness (e.g., quality, consistency). Some examples of changing a business process to take advantage of GIS technology are a direct result of the capabilities implemented:

- The city of Raleigh (North Carolina) changed the workflow for updating parcel maps, thus eliminating two positions (i.e., more efficiency).
- The city of Winnipeg (Manitoba) changed its process for drafting legal survey plans, reducing the time required from three days to half a day (i.e., higher service level).
- The city of Minneapolis (Minnesota) has achieved greater consistency and accuracy by changing the process of generating geographic-based mailing lists (i.e., more effectiveness).

(http://imaps.co.wake.nc.us/imaps/; and http://www.cl.minneapolis.mn.us/about/maps/gis.main.asp).

GIS and Public Works

The Yuma County Department of Public Works has embarked upon the information super highway by bringing computer-based technology into the County Asset Maintenance Management System. The former manual system was cumbersome and needed modernization in order to optimize expenditures for county asset maintenance. A study was performed and a pilot project was completed in 1998, which demonstrated that an off-the-shelf computer-based software package, utilized to establish a database of the current condition of all public works physical assets, will in fact save tax dollars over the old manual system at a cost/benefit ratio of at least 3:1.

Land Base Design: In setting up specifications for the accomplishment of the pilot study, both the city and county of Yuma Mapping/GIS units were contacted to make sure that land base integration would be accomplished as a by product when the entire Asset Software Modules (ASM) project was completed. This integration is needed to make sure that all geographic-based information generated by one agency can be utilized by others without rigorous and time-consuming transformations between unlike systems. The land base being used by both the city and county mapping units is based on the United States Public Land Survey System of Sections (PLSS), Townships, and Ranges from the initial point on the Gila Meridian and Baseline, which governs land descriptions in Yuma county. The PLSS is digitally integrated to the Arizona State Plane Coordinate System (ASPCS) (NAD, 1927). This coordinate system has been used due to the fact that most cadastral data now available, without doing new control surveys over the entire county, was identified earlier by other agencies of government for mapping purposes. it is highly recommended that this project be fully integrated with the ASPCS/NAD'83 when it is adequately densified on the ground by GPS technology.

Public Land Survey Identification: In order to utilize the PLSS, the ASPCS, and the ASM requirements, all at the same time, a digital system of numbering is being utilized to identify section and quarter section corners of the PLSS that will tie directly to the ASM requirements for line segmentation and identification. The known coordinates of the ASPCS/NAD'27 are input within the ASM line segments, with node identification input fields. This is a beneficial aid in the identification of cadastral parcels on the current assessment rolls and located along any given road or highway. Also, current GIS software can easily be implemented within the ASM for specific spatial analytical tasks related to geography.

Pavement Management Module: The Pavement Management Module (PMM) provides a systematic and consistent method for determining maintenance needs, priorities and the optimal time of repair. Early detection and repair of a roadway distress are extremely important in reducing long-term maintenance costs. Most pavements exhibit a deterioration rate that accelerates as they reach advanced age. When examining deterioration curves, it becomes clear that significant maintenance costs can be saved if remedial action is undertaken prior to a roadway reaching the steep decline in condition. PMM is used to track the lifecycle of road segments in the network and to highlight segments in need of attention. In network level, in network-level management, the entire roadway network is considered for budgeting, planning, scheduling, and selection of maintenance projects. Sufficiency inspections are periodically conducted for all segments in a network. These inspections are fast and easy observations of condition, geometry, safety, ride, etc. Ideally, inspections should occur before the segment begins the sharp decline in condition and at a point where timely maintenance can still be scheduled. The Critical Pavement Condition Rating (PCR) value can be used to evaluate the need for preventive and major maintenance. The Critical PCR value occurs at the point where the condition of a segment begins to deteriorate at an accelerating rate. Typically, the maintenance costs increase dramatically once a segment reaches this point. It is advantageous to begin maintenance efforts before the Critical PCR is reached.

Today, upper management is more demanding of management techniques that consider the needs of the entire network. With money in short supply, it is imperative that the greatest benefits are achieved from each dollar expended. Network-level analysis complements project level analysis to arrive at the required conclusions.

The deterioration matrix within the ASM allows you to define a deterioration curve that is associated with each pavement class. Once defined, these deterioration curves can be referenced elsewhere in the PMM application. One can create a deterioration curve for each pavement class. A deterioration curve is a model of how the condition of segments that belong to a pavement class will deteriorate over time (on the average). With a deterioration curve defined for a pavement class, individual road segments can be compared to the curve to determine performance or life expectancy. The deterioration curves are most accurate when they are developed from local historical data.

Pavement Analysis Module: Assists in creating multiple planning models and 'what if' type scenarios and generates multi-year budgets and maintenance profiles. Also assists in the preparation of decision trees, calculating cost *versus* benefit scenarios, predicting performance of individual assets, and aids in projecting funding to achieve overall agency goals.

Safety Signage and Markings: Is a complete inventory of signs, mountings and pavement markings. It yields a complete GPS/GIS location and identification for spatial analysis and mapping of individual assets. It supplies a complete history log of events with attached images and videos. It also supplies queries, reports, and cost of materials analyses along with access to predefined sign and marking libraries.

Bridge and Culvert Module: It provides a complete bridge and culvert inventory along with inspection records by category. It provides a complete structure history with database rollback queries and reports. Also, provides full English and metric support, while maintaining a complete structure inventory and appraisal information. It also supplies complete support for FHWA, NBI and PONTIS data.

Asset Manager Module: It is the workhorse of all of the ASM for it serves as a clearing house for all managed assets. Records labour, equipment and material costs. It relates work activities to specific work orders. It projects and tracks maintenance of schedules. It furnishes customized queries and reports while tracking inventory levels. It also tracks public complaints and generates work requests/orders. It may contain standard operating procedures (Cook Book) for most all types of maintenance work. This is truly the work horse of the system.

Following new techniques can be adopted for this purpose:

(a) *Voice Activated Data Collection:* It provides vocabulary building capabilities and GPS on the fly locations. Captured data downloads data directly into predefined data bases with no more manual data input except for QA/QC. This technology speeds up the data collection and data processing by about 3x.

(b) *Voice Recognition Technology 101:* It interprets spoken words and converts it to text while associating an asset or other item with their exact physical location via GPS/GIS. It stores data in an RDBMS using key fields to relate one record with another. One simply tells the laptop field computer what you see and it will record the field data as defined in the vocabulary dictionary. In the office one downloads the data to your office computer and performs QA/QC to check the accuracy of data being exported to the proper asset management module.

When undertaking an asset management project like this, be aware that several months' research and study must be done before buying into any particular software. Unless your agency is a very large agency, an off-the-shelf software package will probably yield the best results for the time and money spent. Also, one should bench mark test the top three packages identified to see which one will perform the best in your specific environment. By doing a pilot project one can better make a more intelligent judgment on the best package to use. Once the decision is made to buy a specific package, a decision as to how to best implement the system is a major item to consider. In Yuma county, one module at a time was done because of a lack of in-house resources to do anything else. However, if the money and the physical resources are available, and can be utilized, it is strongly suggested that the most needed modules be implemented at the same time to save duplicating trips into the field to gather data for one module without getting everything needed for several modules. Inspecting roads, bridges, culverts, and traffic signs can be done at the same time, on the same segment of highway, using two technicians. By implementing only one module at a time, both time and effort is duplicated.

Further, it has been found that a stand alone asset management server is the best solution to avoiding down time. Having data base sets out on a large network, tends to slow done production and often corrupts certain types of data. This means daily backup is mandatory and become very time consuming to reset back to a certain date and time. Lastly, make sure your vendor offers great technical support. The 1-800 phone number does not usually give you the timely response needed to identify specific remedies to problems that can arise without warning at times. Using internet (e-mail) messaging can be faster when attached digital files can be immediately examined by technical support personnel. Even a faxed document showing the problem you are having is better than trying to explain everything over the phone. Sometimes a combination of all three methods is required to solve problems as quickly as possible.

One thing is clear to all who manage public works assets-maintenance funds are becoming increasingly difficult to come by. As a result, improved methods for determining maintenance needs and priorities are required. It is no longer acceptable to focus on short-term repair techniques. Greater emphasis must be placed on selection of remedies that provide the greatest long-term benefits. Today, upper management is more demanding of management techniques that consider the needs of the entire set of assets. With money in short supply, it is imperative that the greatest benefits are achieved from each dollar expended. Network-level analysis complements project-level analysis to arrive at the required conclusions. (http://www.dot.pima.gov/gis/data/contents.metadet.cfm).

Automated Information Mapping System

In the mid-1980s the Johnson County Board of Commissioners initiated the Automated Information Mapping System (AIMS). AIMS was developed to enable the county to meet the growing mapping and documentation demands of the Kansas State Legislature. In addition to meeting the requirements of the legislature the county realized the overall benefit of having accurate and current maps available to the cities and citizens of the county and AIMS has been growing ever since. One of the latest endeavors of AIMS was to compile new planimetric data. As a part of the new planimetric data, contours were developed for the entire county. This elevation information could then be used to support storm water studies. These storm water studies are being completed on a watershed by watershed basis and submitted to the county in a form that will supplement their enterprise GIS system. Upon completion of a watershed, methodologies were analysed and passed on to the next study so as to refine the process and run each one as efficiently as possible.

Burns and McDonnell is tasked with completing the third storm water study, for the Mill Creek Watershed. The study includes a hydrologic analysis utilizing the HEC1 modelling software and a hydraulic analysis utilizing the HEC-RAS modelling software. The study was to analyse the 100-year flood plain and determine where changes need to be made to the current FEMA boundaries. In addition to re-mapping the floodplain the county is using the project as an engine to supplement their enterprise GIS system.

The storm water study can be broken down into five major phases or steps. The initial steps deal with setting up a foundation on which the rest of project will operate and the last step is the delivery of the flood plain analysis for the watershed.

- The first phase of the project is to come up with a database design that will house all the data and serve as the central file cabinet for the whole project. The design of the database is partly set by the county's data standards and the need to make sure that the information could be used efficiently for this study as well as any future studies or projects.

- The second phase of the project is the collection of all the required data for the storm water study. This phase included attaining the county's current planimetric data as well as the collection of new field data.
- The third phase deals with supporting and developing data for hydrologic stream model.
- The fourth phase deals with supporting and developing data for hydraulic stream model.
- The final phase is the culmination of all the modelling into new flood plain delineations for the watershed.

The primary goal of this watershed study is the mapping of the 100 year flood plain but the real benefits of the study are much greater than just the mapping of the flood plain. Granted, whether the information management aspect of this project is present or not the flood plain analysis would still be done but the county would be short changing themselves because of all the opportunity they left out on the table. Johnson county has chosen to grab every opportunity they can to work towards a future goal of enterprise information management for all systems in the county. They may not be where they want to be right now but are using opportunities like this watershed study to work towards their goal. A little for thought in the mid-1980s and looking at all the opportunities in each project they need done has set the county up with a information management system capable of aiding in any number of activities. (http://gisesri.com/library/userconf/proc95/to050/P014.html)

Urban GIS

Belo Horizonte, the fourth largest Brazilian city, was one the first Brazilian municipal administrations to develop an urban GIS. Through the constitution of 1988 new responsibilities and powers were transferred to the municipal governments. So there was the need to increase the level of economic and administrative efficiency through the adoption of new approaches, like the technology of geoprocessing. In 1989 Belo Horizonte started to implement a full-scale GIS, instead of beginning with a pilot project. In 1992, a very large database was assembled in vector format from stereo-photogrammetry, comprising around 95 different themes, covering all kind of geographic objects. A municipal level IT-company has been responsible for the design and development of the urban GIS and for the management of the geographic database. To make the system more user-friendly the addresses in the city has been geographically localized. This is said to be an important step towards the successful implementation of the GIS. After possession of a basic set of information one started focusing on four different applications: Education (school registration system), transportation and traffic, urban planning applications and health (surveillance of infant mortality). What is said to be the main success factors of this GIS use in the municipality of Belo Horizonte are:

- One did not rely on international technology transfer but build in-house expertise over a long period of time.
- The company has set up an organization that is dedicated completely to the development of geoprocessing expertise to support municipal administration in Belo Horizonte.
- The systematic and integrated manner in which the database has been build.

(www.gisdevelopment.net/application/urban/overview/urbano.0004.html).

GIS and Geoplan

A good example, where GIS has been implemented successfully, is the Province of Cebu. From 1991 to 1994, the German Agency for Technical Cooperation (GTZ) supported in the province of Cebu the establishment of a service institution for applications in the field of GIS. In April 1995 the GIS Cebu Province was successfully transformed into the Geoplan Cebu Foundation, which runs independently from the administration of the provincial government. During the years that GTZ has been involved in this project various reports from short-term consultancies and project documentations are available. Unfortunately, these reports are unpublished documents for internal use only. This is regrettably as a lot of, for this study, interesting information was to be available in these documents. But something is said about the first years of Geoplan. The most hindering factor of its first year was the unstable financial situation and the lack of financial buffers. One expects this not to change during the following 2-3 years. Another identified bottleneck is the lack of qualified GIS personnel and the high staff turnover.

When talking about the transformation of the GIS Cebu Province into the Geoplan Cebu Foundation it is said to be successful. Reasons given for this is the increasing flexibility to offer services and products on the GIS market and that the foundation is now able to make decisions independently without government intervention. The future trend for Geoplan will be to move away from low-end digital cartography to higher-end GIS solutions for local clientele. The only thing of concern is the relatively high staff turn-over. This is partly due to the low salaries, but also to changed work conditions in a non-governmental environment.

(http://www.gtz.de/lamin/deutsch/geoplan/e_pic.htm)

GIS and City Council

A GIS pilot project was commenced in Lilongwe City Council (LCC) in September 1995. This project was under the auspices of the World Bank and ITC was the implementing institution. The focus of the study was the Town Planning and Estates Management Services Department as its nature of work foresees a need for a GIS. This department had little experience with the use of information technology. For this reason, a training program was one of ITC's activities. Besides, the other activities were: (a) analysis of the user requirements, systems analysis and design; (b) pilot system development and modification; and (c) consolidation and enhancement of systems. Before the methodology for the design and implementation of the project was adopted the (local) situation had been analysed. There seemed to be some barriers to success:

- *Technical barriers*: no prior computer skill, irregular electricity supply and limited local support for computer technology.
- *Financial barriers*: doubt about financial support.
- *Institutional barriers*: poorly paid and not well motivated staff, general organizational problems and lack of awareness and knowledge about GIS.

One of the problems faced during the project and also known as a danger for the continuity of projects is the loss of key professional staff. Staff with computer and GIS skills are targets for other employers and difficult to replace. There are some general lessons learned which should be taken into account with the expansion of GIS to other councils:

- the availability of professional staff with GIS knowledge and experiences,
- the availability of computer staff,
- the availability of good quality base maps or a sound and affordable method for map updating,
- a more gradual introduction of (spatial) information technology into the organisation
- the need for very substantial staff development programmes aimed at raising and maintaining skill,
- the feasibility of utilizing the LCC system should be considered as a model for other councils thereby reducing development costs,
- the need to identify means of cost recovery and reserving finances for future system.

(www.gisdevelopment.net/thesis/thesis1/less4.html)

GIS and Information Gates

As local governments across the country seek to improve public access to data and services, Dames and Moore's Information Technology Group is helping show the way at the Virtual Oakland CEDA site. The site, which provides a real Internet-based geographical information systems (GIS) software solution, represents a significant milestone in the group's ongoing efforts to streamline information access in municipalities and companies around the world. Virtual Oakland is maintained on a web server operating over a Microsoft Windows NT network. Citizens can access Virtual Oakland through any internet browser by downloading the Autodesk MapGuide plug-in from the Virtual Oakland homepage.

Once inside the site, users can query a growing array of databases by clicking symbols on a variety of layered vector maps. Because whole documents are available via a few simple mouse-clicks, one can expect that civic information will be clearer and more available to citizens than ever before. In municipalities, GIS departments become a bottleneck that restricts access to data. Not only do GIS staff has to process information requests, but they have to publish the information for customers as well. When you make information available through Autodesk MapGuide, the public can instantly access distributed databases through an intuitive interface that can be understood immediately. (http://usa.uutodesk.com/adsk/servlet/item)

Conclusion

Governments at different levels are increasingly becoming dependent on web based GIS for providing services, empowerment, taking people into confidence and to become more transparent in their transactions. The case of application of GIS for rural markets is slightly different. It is important for both, the government and the industry (Sagar, K., 2003). Such initiatives have become administrative necessity and political compulsion, particularly in a country like India where the governments have to deal with masses. Nevertheless, public access to information has become global phenomenon. Chapter 17 will provide more examples in this regard.

REFERENCES

Amanna, Vijay and Suruchi Raina (n.k.), GIS for rural marketing (www.gisdevelopment.net/application/business/mi03031.htm)

Sagar, K., Anand (2003), Agmarknet—rural empowerment: Data for development, *Geospatial Today*, Vol. 2. issue 2, pp. 48-49.

6

GIS in Water and Wastewater Utilities

The daily demands on water and sewer districts are tough. Managing the flow of water and wastewater to service homes and businesses is a difficult task. In doing so, a utility needs to manage many different kinds of geographic information. They must keep track of the location and condition of water mains, valves, hydrants, meters, storage facilities, sewer mains, and manholes. It requires planning and engineering, operation and maintenance, finance and administration, and also external systems. A majority of water or wastewater utilities use GIS technology to integrate all kinds of information and applications with a geographic component into one, manageable system. GIS lets you organize, manage, and distribute geographic information culled from various databases while maintaining data integrity and focusing on project direction.

Management of a Water District

The Las Vegas Valley Water District (LVVWD), this Nevada city's major water utility, supplies approximately eighty per cent of the water used by the valley's residents and businesses. The existing population of one million translates into some 194,000 accounts, and the area's booming growth adds an additional 1,000 customers each month. As the Las Vegas Valley enjoys tremendous population growth, the LVVWD faces challenges of keeping pace with water facilities infrastructure. The LVVWD completed integration of the utility's CAD drawings and GIS. The resulting database of drawings and maps, complemented by an Oracle database, resides on a server from which construction and maintenance crews in the field can access raster images and associated data on more than 12,000 parcel maps and legal documents.

The division wanted to further streamline its CAD and GIS efforts with a closely integrated software solution for generating and improving spatial locations through scan conversion as well as maintaining GIS attribute data. Nevertheless, a typical map creation process begins prior to construction. An LVVWD technician scans a developer's drawings, then digitizes and spatially references water facilities such as water pipelines, valves, and hydrants. Next, primary GIS attribute data such as make, model, and material properties are attached to the facilities on these maps. After construction, the data is revisited, its location updated based on a post-construction inspection, and a record drawing is created. The primary GIS attribute data is also updated at this time. Finally, all of the water facilities are converted into GIS features and the primary GIS data are integrated with the other existing GIS data.

AutoCAD is applied to provide a seamless connection between the division's CAD crew and its existing facilities GIS. It allows direct GIS connectivity to AutoCAD Map. This software's strengths include creating geometry and maintaining GIS attribute data. Coupled with Softdesk S8 imaging and ArcInfo GIS analytic tools, this new software solution is suppose to help in achieving greater efficiency in map creation. AutoCAD is integrated into LVVWD's complete GIS solution. AutoCAD has proved to be reliable, stable, and feature-packed. Since Windows NT–based solution is used, LWWD now has a roomy 32-bit memory for customization within its ARX object-oriented tool set. (www.directionsmag.com/article.php)

GIS Sewer Map

The New York City Department of Environmental Protection (DEP) to provide GIS mapping of the entire city's sewer system. Baker Engineering NY, Inc. is a unit of Michael Baker Corporation. The three-year, $10.4 million project involves GIS application development, scanning approximately 110,000 plan sheets, converting the data into a seamless, city-wide geo-database, field verifying attributes for 20,000 manholes, and providing GIS applications and training for DEP personnel in each of the five boroughs of New York City. "We are very excited about this challenging project," Baker's project director said. "To ensure success, we will devote our full range of GIS and engineering expertise to the effort. By developing state-of-the-art solutions for this data collection and conversion project, we will be providing New York City with not only a highly accurate sewer map, but also a digital foundation that will serve them years into the future. This is the premier GIS conversion effort in the nation today, and I am very pleased that the DEP has placed their confidence in Baker." All work was to be coordinated out of Baker's new project office in Long Island City, N.Y. (www.esri.com/news/arcnews/winter0304.articles/nyc-dep-uses.html)

Capital Improvement Project Planning and Analysis

PlanGraphics, Inc. founded in 1979, is a full lifecycle systems integration and implementation firm that provides a broad range of services in the design and implementation of information technology in the public and commercial sectors. PlanGraphics' experience with spatial information systems and assets provides a critical discriminator among other IT consulting and integration firms. PlanGraphics' is headquartered in Frankfort, Kentucky and has regional offices in Maryland, Colorado, California, Rhode Island, New York, and New Jersey (www.plangraphics.com). PlanGraphics was awarded contracts by the Metropolitan Water District of Southern California: Frankfort, Ky. and Newport Beach. PlanGraphics teaming with ESRI and Landata Airborne Systems, has been awarded two open-ended contracts with the Metropolitan Water District (MWD) of Southern California's Engineering and Information Technologies Divisions providing engineering, data integration, and implementation services. The Metropolitan Water District is a consortium of twenty-six cities and water districts that provides drinking water to nearly 17 million people in Southern California.

PlanGraphics was to assist MWD's Engineering and Information Technologies Divisions, as needed, with mapping services and consulting services. MWD plans to utilize GIS and IT for capital improvement project planning and analysis, as well as (a) to streamline their data architectures, (b) to identify system integration and data optimization procedures, and (c) to enhance their internal systems and applications and improve their use and distribution of spatial data sets. All of these activities were suppose to aid their future project efforts, including development of their current reports on water supplies, water storage and distribution, and their seawater desalination programme. The contracts permit billing for professional time and material at negotiated rates.

The practice manager for PlanGraphics' Western Region noted: "The contracts provide an excellent entree into one of the country's largest and most technically advanced water utilities. Given PlanGraphics' track record of long-term successful relationships with clients, we believe these initial assignments will be precursors to a steady and substantive working relationship with MWD." (*Business Wire*: Aug. 6, 2002 and OTCBB:PGRA—News)

Design Analysis, Risk Assessment and Remedial Activities for Water Utility

Recently, Haestad Methods announced closing a strategic sale with Sanepar (Sanitation Company of Parana), one of the largest water supply and wastewater treatment companies in Latin America, which serves more than seven million residents and over 95 per cent of the population in Parana, Brazil. The Brazilian utility selected Haestad Methods' WaterCAD and SewerCAD water resources modelling software to conduct design analysis, risk assessment, and remedial activities. In this way, Haestad Methods' technology was to be used to maintain and preserve Sanepar's 600 water distribution and 250 sanitary sewer networks in Parana.

"WaterCAD has been an excellent tool for the analysis and design of our water distribution systems," said the Civil Engineer for Sanepar. "The software's most valuable characteristics include its speed and reliable results." Sanepar has long recognized Haestad Methods' products for their ease of use, flexibility, and powerful features. Over the course of its seven-year relationship with Haestad Methods, Sanepar has purchased more than thirty individual licences of Haestad Methods' software attributing to increased efficiency and reduce costs of operating its water and wastewater networks.

"Haestad Methods is proud to have strengthened its long-standing commitment to the Brazilian market," says the Latin America Sales Manager for Haestad Methods. "We look forward to extending our reach throughout Brazil by offering technical advancements that help improve the way our clients manage their water resources."

Sanepar provides water services to the majority of the Brazilian state's population, supplying 7.6 million inhabitants with water distribution and sanitary sewer services. Sanepar was the first water utility in Latin America to achieve ISO 14,000 certification and retains an excellent reputation within Brazil for technical excellence and public education. Since a move to private sector ownership in June 1998, Sanepar has begun to focus on efficiencies and practices that have immediate environmental impacts. (http://www.sanepar.com.br.) (Versao em Portugues http://www.haestad.com/portuguese/news/08-01-02.asp; and Waterbury, Conn., Aug. 5/PRNewswire/)

Enhance Service for Metropolitan Water District

Bexar Metropolitan Water subscribes to MapCiti™ to enhance services. It is a web-enabled GIS solution to facilitate more accurate and timely communication. Syncline, Inc. announced that Bexar Metropolitan Water District has subscribed to MapCiti (www.mapciti.com), Syncline's fully automated e-government mapping service. MapCiti enables organizations to immediately web-enable their GIS and provide essential services and information to internal departments and the general public without additional investments in technology.

Bexar Metropolitan Water provides water services to 75,000 customer accounts in the San Antonio, Texas area. The utility dispatches approximately thirty servicemen per day to perform routine maintenance on the area's water systems. Currently, dispatchers communicate the details of a particular job to servicemen in the field via telephone. With MapCiti, Bexar will be able to load all of its geo-referenced infrastructure data, including aerial photography, into MapCiti's fully hosted, web-based mapping application and create regional maps of the city's water infrastructure online. Dispatchers and servicemen will access and view these maps from any location, using only a standard web browser, to communicate about jobs. By clicking on entities on the map, they were to retrieve detailed information about those entities, including hotlinks to other information sources housed at Bexar.

"MapCiti will enable us to integrate multiple sources of data about our infrastructure into maps that are easy to access and use for performing maintenance and visual analysis," says the GIS analyst at Bexar. "The service also allows us to have an offsite backup for our valuable GIS data, and to disperse it departmentally as needed." Further, the co-founder and COO of Syncline, said, "Bexar Metro recognizes the tremendous time and cost savings associated with more accurate communication between dispatchers and servicemen." He added, "In this way, MapCiti will enable Bexar to provide faster and better service to its customers."

Founded in 1997, Syncline, Inc. is the leading provider of products and solutions that allow customers to visualise and analyse spatial data using maps they create, share and manage via the Internet. The company's customers include state and local government organizations, as well as the Fortune 1000. Syncline partners with industry and technology leaders such as ESRI, Oracle, Sun MicroSystems, TeleAtlas and GDT to deliver its comprehensive solutions. A member of the ESRI Corporate Alliance, Syncline was named ESRI "Partner of the Year" in 2001. Syncline is located in Boston, Mass. (www.syncline.com). (Boston, Mass.: July 10, 2002)

Implementation of Water and Wastewater Geospatial Infrastructure Management System

Winnipeg is the capital city for the province of Manitoba, Canada, and has a population of 620,000. Winnipeg is situated, geographically, in the centre of North America making the city Canada's largest distribution centre between Vancouver and Toronto. Winnipeg is one of the few North American cities poised to handle domestic and international distribution of manufactured goods by air, rail, and highway, with direct links to seaports.

In 2002, the Intergraph Mapping and GIS Solutions (Nasdaq:INGR—News) announced that the city of Winnipeg Water and Waste Department (WWD) has successfully implemented a water and wastewater geospatial infrastructure management system using GeoMedia® technology. The department uses GeoMedia, GeoMedia Professional, GeoMedia WebMap Professional, and GeoMedia PublicWorks Manager to improve access to the city's asset information. The WWD chose GeoMedia PublicWorks Manager and GeoMedia Professional for data maintenance on the city database that serves 178 square miles and contains over 200,000 pipe segments and more than 1,000,000 additional features. The department uses GeoMedia and GeoMedia WebMap Professional to share and integrate geo-referenced information with all the city departments and external agencies such as utility companies. Winnipeg uses Oracle® and Oracle Spatial for its enterprise wide database for use with the GeoMedia-based geospatial infrastructure management system. This system enables the department to effectively maintain and update the city's water and sewer facilities, all within a single environment.

Using GeoMedia WebMap Professional and a staff-developed intranet application, iView, city employees can access all spatial information in a format customized for each department. The WWD has made great use of this application throughout the organization to enable quick and easy access to important facility information through their internal web. Using GeoMedia technology the new system met the department's three primary mapping and critical facility management requirements: (a) to preserve the accuracy and integrity of existing and new data through the introduction of rule-based data maintenance, (b) to store data efficiently in a non-proprietary format using an industry-standard database, and (c) to provide new advanced analysis functions to all engineering and technical personnel.

"The entire project went extremely well, and the department is up to speed and in full production," says the supervisor, Drafting and Graphic Services at the city of Winnipeg. "Using GeoMedia technology's ability to integrate directly with Oracle, we have improved access to our asset information for both our engineering and technical staff as well as our field personnel. With the tools available in GeoMedia Professional and GeoMedia Public Works Manager, the data entry workflow has been made much easier, and the accessibility of the spatial and attribute information has improved as well. GeoMedia's tight integration with AutoCAD® will also allow for easy digital submission of as-built information directly into the system, further improving efficiency." (*Business Wire*: July 18, 2002)

Developing a Sewer Master Plan

The city of Stockton, with a population of 250,000, averages approximately 38 million gallons of waste water inflow per day. HDR was contracted to update the City's sewer master plan, find cost-effective ways to serve the rapidly growing northern and northeastern sectors of the city, and develop a GIS interface for the sewer model. Sewer block book information had been recently converted to GIS so it could be used in the sewer master plan model development. Sewer master planning projects identify required system improvements and develop cost-effective solutions. Typically, a computer model simulates the hydraulic conditions in the sewer system. Previously, the cost to develop a data management system to support a database or a GIS interface for the hydraulic modelling of the system would have exceeded the cost of the engineering work. In most applications such a high development cost was unacceptable to clients. Costs to develop data management systems have come down, in part, because system inventory and land use data are now available in digital format. Even with these improvements, effectively linking GIS data to the sewer master plan and developing an interface presented challenges.

The development of the GIS interface using ArcView GIS and HDRLink has provided the city of Stockton with a practical, user-friendly planning tool to conduct "what if" analyses and development impact assessments. Incorporating standard data management and output processing features in HDRLink made it possible to provide this sophisticated GIS interface at a minimum development cost.

(http://gis.esri.com/library.userconf/pro98/PROCEED/ABSTRACT/A316.HTM).

Dircctional Interface

Hamilton is located at the west end of Lake Ontario, about 80 km from Toronto and 100 km from Buffalo. This city has a population of 467,000. It covers and area of 280,000 acres. In 2001 the city of Hamilton amalgamated with other towns and cities to form the new city of Hamilton. Operational maintenance is responsible for water distribution and wastewater collection systems. It is not responsible for pumping stations or treatment plants.

GIS and Maintenance Management System or MMS as of 1999 completed databases for water and wastewater infrastructure in GIS and MMS. Both databases were maintained separately, with a common linked field for future integration. The GIS contains the schematic diagrams with little asset information. The MMS system contains all asset information including work order history, costing, labour, etc.

Easy tool for field and operations staff to identify assets in the Hansen MMS system with the asset they are working on in the field. Need reliable work history information and inspection information about assets, to drive the maintenance management programmes. It is driven totally by Water Wastewater Operations department, not Information Systems department.

Field Requirements

- Need a tool to measure distance.
- Need to print the screen.
- Most importantly, the bi-directional requirement to select assets in GIS or Maintenance Management and move seamlessly between the two systems.

Using Geomedia Public Works Solution, integration of the maintenance of the GIS and MMS databases is expected in future. It is required to integrate the Geomedia GIS database with the Water and Wastewater hydraulic modelling software; and develop a network-tracing tool for the water and wastewater distribution networks (www.directionmag.com/article.php).

Public Utility

Seattle Public Utilities (SPU) has the largest water utility in the state of Washington and one of the largest in the US. The water utility supplies high-quality water to nearly 600,000 people in the metropolitan Seattle area through direct retail service and another 625,000 people through twenty-eight wholesale customers. The area served by the regional supply system covers 320 square miles. The water pipeline system consists of 160 miles of supply mains, forty-five miles of feeder mains, and 1,620 miles of distribution mains. The pipeline system ranges in age from newly installed to over a hundred years old. Various different pipe materials were use throughout the years with varying installation quality and under many different environmental conditions.

As the water distribution infrastructure ages, it becomes increasingly more challenging to assign limited capital expenditures to prioritize the repair, replacement, or rehabilitation of pipelines. There is a growing need by municipalities to find better ways to prioritize their future projects. Much research has gone into the process of water distribution system rehabilitation planning. The American Water Works Association (AWWA), American Society of Civil Engineers (ASCE), and other professional associations have produced numerous publications on the subject. Results and recommendations from these researches have been adopted by water utilities in varying degrees. Most water utilities have adopted some form of subjective ranking system to prioritize pipeline rehabilitation. A smaller number of water utilities have completed statistical analyses to predict pipeline failure and incorporate the results within a cost-benefit analysis. SPU water engineers and managers have seized upon the advantages of accessing water main characteristic information through Seattle's GIS and have significantly reduced the time spent identifying and prioritizing water mains for preventive maintenance, rehabilitation, or replacement. SPU now utilizes GIS-based information in day-to-day decision making and capital improvement programme preparation with a focus on the water pipeline infrastructure.

The backbone of the evaluation process rests on the GIS analysis and data processing to provide all the attributes of the pipelines as well as to find spatial relationships to other layers of information. To maximize the performance of the application, the GIS analysis is done as a preprocessing step. All attributes identified as parameters to be used for statistical analysis as well as for ranking the pipelines are denormalized into a single table. The following are some of the GIS analyses and processes performed to compile the information for this application:

- Identify and compile essential attributes of the water pipeline in the GIS database and various other related tables.
- Combine segments of pipelines to produce pipe segments of similar characteristics that are more representative of city blocks. The combined pipe segments are more closely related to pipeline projects. In other words, a project to rehabilitate or replace pipelines will more likely be carried out on a whole city block than on individual pipe segments as captured in the GIS. Pipe segments captured in the GIS are split at all fittings and junctions to provide the most detailed information that would meet all usage requirements of the GIS database.
- Perform overlay analysis of the pipelines to capture layers of information that are represented by polygon features such as soil, water pressure zones, land use zoning, steep slope areas, landslide areas, liquefaction zones, flood-prone areas, neighbourhood classification, etc.
- Perform proximity analysis of the pipelines to capture layers of information that are represented by point-and-line features such as soil corrosivity points, electrical vaults, underground conductors, etc.
- Perform calculations of annual water usage by linking GIS pipeline data with an external database that provides information on customer water usage.
- Perform calculations of static water pressure by relating pipelines to the elevation contour layer and the pressure zones that the pipelines fall within.
- Perform statistical summaries on the number and types of service taps to the pipeline segments.
- Perform connectivity analyses of pipelines to critical services such as hospitals, medical facilities, dialysis centres, schools, etc. Compile and create additional data from other independent studies that would be beneficial to this pipeline evaluation process, such as information about the seismic reliability of pipeline segments.

(www.autodesk.com/mini-sites/green/pdf/SPU-case3.pdf)

GIS and Hydrographic Surveys

The New Mexico Office of the State Engineer (OSE) is charged with administering the state's water resources. To accomplish this, OSE has setup a strategic plan with four priorities: (a) eliminate the water rights application backlog, (b) protect New Mexico's water and meet interstate obligations, (c) complete water rights adjudication, and (d) automate information and processing of its databases.

The OSE is using GIS and Information Systems (IS) tools to accomplish these priorities. OSE has initiated the first steps in the adjudication process to abstract its paper or document information databases and to automate its hydrographic surveys through an innovative approach using private enterprise to create a computer-based IS/GIS for the Hydrographic Surveys of the Lower Rio Grande (LRG) and Nutt-Hockett (NH) basins. How the surveys were conducted, how the adjudication process works, the IS/GIS created, and the lessons learned are to be presented.

The advent of the computer and its wide use that occurred in the early 1980s changed the way the OSE conducts its hydrographic surveys. The first steps were to computerize the massive water rights documentation, which culminated in the mid-1990s with the development of a computer-imaging and database system called Water Administration Technical Engineering Resource System (WATERS). Once completed, WATERS will allow users to abstract the water right file, which will be related to an individual farmer's parcel and the eventual perfecting of the farmer's water right.

At about the same time, OSE decided to make the transition from traditional, non-electronic methods of data collection and data management to computer-based hydrographic surveys. The LRG and NH hydrographic surveys were started in mid-1997 and were designed to interface the written documentation or tabular data (*descriptive* information), and the digital map data (*spatial* information), and make the data and information so developed available to a user through a computer.

The *descriptive* information, mostly water rights records, was analysed and abstracted records were obtained, each one constituting the subfile, a unit describing a farmer's tract or plot of land. These subfiles were assigned a unique identifying number on the computer-based system, which now became the Information System or IS.

Further, in order to create the necessary *spatial* information, ground control was established along the irrigated portion of the LRG, and aerial photography and multi-spectral imagery were obtained. The ground control was used to control the aerial photography and produce ortho-photos of the irrigated areas, ortho-photos that were then digitized and introduced into a GIS. The multi-spectral imagery, already on a digital format, was then geo-corrected using the ortho-photos and introduced into the GIS.

Parallel to these activities extant maps and geo-referenced information—including state and county boundaries, city locations, lakes and rivers, roads, townships, ranges, sections, PLSS data, satellite imagery and field-obtained GPS data—were collected and brought into the GIS. All *spatial* information was brought to a common projection, NAD83 and State Central Zone.

The IS and GIS were interfaced using Informix and ArcInfo software respectively. The IS/GIS application was housed on a computer and peripheral setup specifically built to house it under UNIX. At the same time, a series of Graphical User Interfaces or GUIs were developed to handle the IS/GIS data on the UNIX and on a PC-environment. This PC-environment was configured so Microsoft Access can access, view and query the *descriptive* Informix database and using ArcView to view and query the *spatial* database.

An integrated IS/GIS is essential to the way hydrographic surveys are now handled by OSE. It allows us to print at will the final report and all necessary maps, as well as the approximately 10,000 offers of Judgment to potential claimants of the LRG and NH basins in a matter of weeks. The Nutt-Hockett basin hydrographic survey itself was completed in about seven months and more than 90 per cent of the offers have been accepted, making this an adjudication almost completed in less than two years.

The Rincon Section, part of the LRG Hydrographic Survey and containing about 1,200 potential claimants, was completed in about eighteen months. Most of the Offers of Judgment in this section have been sent out and it was expected to have the adjudication complete in about 2-3 years. Further, the Northern Mesilla Section, the second part of the LRG and containing about 2,400 claimants, was to be completed by June 30, 2000, i.e. in about thirty months. The Southern Mesilla and the outlying areas, the other two sections of the LRG Hydrographic Survey, were to be completed by the end of the year, about thirty-six months since starting the project. These two sections contain about 6,000 claimants, and their adjudication may take close to five years.

Strong emphasis has been placed on obtaining user friendly products for both OSE immediate users: technical and legal personnel. The several GUIs produced and the fact the IS/GIS can be used on a PC-environment are the most obvious user friendly outputs so far. During the conduct of these new generation hydrographic surveys we have collected airborne multi-spectral and satellite imagery, USGS Digital Ortho Quad and Quarter Quad (DOQ and DOQQ) data, digital elevation models (DEMs), well data, and digital line graphs (DLGs). Using these data and visualization tools from ArcView 3D-Spatial Analyst and ERDAS Imagine software, we have put together 3D views of our study areas.

Besides these visualization tools, useful when presenting results to non-technical audiences and for administration of resources, Portable Document Format (PDF) files were produced. These PDF files of the final hydrographic surveys for the NH and the Rincon Valley Section of the LRG basin have been placed on the web site, www.ose.state.nm.us. The main objective of this action by OSE is to make these surveys available to the public in general.

In late 1999 the OSE initiated another action to determine the future use of GIS in the agency. A study was contracted out to the private sector to help OSE in the development of an enterprise GIS. This Enterprise GIS study has three main components: (a) develop a detailed project plan, (b) develop a detailed GIS design, and (c) implement a GIS pilot project. This study was to be completed by the end of 2001 and was to guide the use of GIS in the OSE for the foreseeable future.

It is expected that the future computer-based hydrographic survey work in the OSE will have following features:

- The complete hydrographic surveys will be placed on the internet for public access.
- Surveys will be carried in a laptop with GUIs and visualization tools for individual legal and technical users.

- Visualization will be accomplished using *fly-by* movies and 3-Dimensional models for presentation of hydrographic survey results and as tools in water resources administration.
- The next generation of hydrographic surveys will be in an electronic, computer-based environment, using interactions between GIS, remote sensed data and GPS.
- Aerial photography will no longer be needed because digital cameras or multi-spectral scanners will obtain a digitally corrected map.
- Crop identification and delineation of tracts will be obtained from remotely sensed data.
- Field verification of these tracts will be done using a hybrid GIS/GPS/Orthomap system.

GIS and Wastewater Management

The city of San Diego Water Department began the conversion for its water and wastewater GIS programme, SPLASH, in 1990. The programme is now successful beyond the highest expectations, but the path to that success was not an easy one. Although the struggle to develop and migrate to a new GIS platform was not easy, it has provided endless benefits to the city of San Diego Water/Wastewater GIS and MIS programmes. However, the benefits were realized immediately, measuring them at six months and then again at eighteen months. To date, these are some of the measured and documented benefits that were realized:

- Completion of the 900-drawing backlog in one year (the original estimate was five years)
- An 800 per cent increase in productivity, with a decrease in training costs and training time
- 99.6 per cent system reliability, measured by tracking application, hardware, and network downtime
- 50 per cent reduction in IT development and maintenance costs
- 300 per cent decrease in hardware costs
- The annual GIS budget is no longer in the "red"
- The data are trusted and the project has full management support
- GIS-analysis functionality is the focus, instead of data entry
- Ad hoc maps and queries are done in minutes instead of days, by users instead of IT professionals
- Application flexibility
- Full integration, without data lockout, to all enterprise applications

When all is said and done, it was learned that some very valuable lessons can now be incorporated when a new project is initiated, such as:

1. Thoroughly document the project requirements. This requirements document provides a clear picture and direction to the project's final destination. It supplies critical information early on to the consultants and developers so that plans, schedules, and resources can be acquired. The document establishes the project guidelines and

deliverables. Include it in the contract documents and use it to enforce the final delivery of the application.

2. Provide a dedicated team of professionals. This team should include representatives from all aspects of the programme, from IT professionals through the end-user. A full-spectrum team ensures delivery of a product that meets the requirements, as well as, the final acceptance and usability of the application.
3. Train the team. When implementing new technology, it is important for the entire team to have some familiarity with the software. This enables the team, users included, to make more knowledgeable decisions regarding issues that arise during development and implementation.
4. Document the acceptance process. Because there are documented requirements, there is a format for establishing an acceptance matrix that will allow the development progress to be tracked and rated. This is critical to keeping the project on schedule, while providing the team with necessary information regarding the performance of both the developer and software vendors.
5. Be open-minded, but demand proof of functionality. Require that proof of functionality be done on-site with the data, and then it is a real proof. The project has requirements and deadlines. Demand proof that the technology can provide what has been requested in the timeframe required. Not to allow the project to be held up for the potential delivery of "vapour-ware." If it is critical and can not be done immediately, then search to be continued for a product or vendor that can deliver. Change is difficult, but explaining to management that the project is behind schedule and over budget is no fun either, so one has to be open-minded. Change might be what the project needs to be successful.

Ten years ago the City of San Diego Water/Wastewater GIS Programme had a strategic GIS plan, and thanks to the innovation of object technology, the programme is back on track with that plan. This involves integration with the work order and modelling applications, providing water/wastewater facility data to the City/County GIS programme, looking at GPS for future data collection, and adding additional facilities and attributes along with access to additional data, such as environmentally sensitive lands, scanned improvement and replacement drawings, geologic faults, storm drains, and easements.

The three years since the implementation have brought the SPLASH programme to a level of respect throughout the department. The implementation of a GIS programme can be expensive, but with the right technology, costs can be reduced and investments can be leveraged throughout the enterprise immediately. The city is finally reaping the benefits of the hard work and investment. The process of migrating from a legacy system to new technology might not be an easy one, but it is definitely worth the effort.

(www.directionmag/pressre;eases.php).

GIS and Water and Sewage System

The Las Virgenes Municipal Water District (LVMWD) is a small to medium sized water/wastewater utility the serving 122 square miles in the western portion of unincorporated Los

Angeles County and including the cities of Calabasas, Hidden Hills, Agoura Hills and Westlake Village for potable and recycled water services. The wastewater treatment service area also includes portions of eastern Ventura County in southern California. LVMWD provide water and sewer service to a population of approximately 65,000, it maintains its own sewer treatment plant, and it provides recycled water for irrigation within its 122 square mile service area.

By early 1990s, some senior staff at the district were convinced that LVMWD needed to implement GIS. They had learned enough about GIS technology to be convinced that it would provide a significantly better tool for performing many of the mapping and information management tasks that were central to the utility's planning, demand management, and facility operations.

Recently, LVMWD has completed Phase 1 of its GIS development programme. This has included the development of the base map, the implementation of early applications, and the performance of a water and sewer system data conversion pilot. They are in the process of preparing a bidding document with which they will hire a data conversion firm to complete our water and sewer system data conversion during January/February of the following year.

With the completion of the water and sewer system data conversion, they will move into the 3rd phase of our GIS programme that will focus on achieving several key system integration goals. A primary goal will be to achieve a data-sharing interface between our Customer Information System (currently Orcom) and the GIS. This important point of integration will allows them to collect and analyse water use data in support of our conservation programme. It will also help them to quickly respond to customer information requests by providing a link between our customer database and our GIS facility database.

Another system integration priority at the district is being achieved regarding tighter data integration between the GIS and the hydraulic model. It is currently working with Haestad Methods, which is an older version of Cybernet was used and integrated with the GIS base map when the district updated its Potable and Recycled Water Facilities Master Plans in 1999, to explore the best approach for sharing data between these two information systems. It is anticipated that a tighter integration between the GIS and the hydraulic would be feasible.

(www.gis.esri.com/uc2001/sag/display/index.cfm).

Water and Wastewater Utility Map

The city of Bloomington, Indiana implemented a citywide GIS in 1996 using aerial photography to generate base map data and manual digitizing to create data layers for the city owned water and wastewater utility. Base map layers included buildings, streets, parking lots, elevation lines and points, drainage basins and water features. Cultural features such as political boundaries, planning zones and properties (maintained and shared by the county in the same GIS software) are also a pert of the system. The GIS data has been kept up-to-date through periodic import of AutoCAD drawing files showing details of new building, road and subdivision projects. Bloomington uses Genasys GIS software and has distributed a general user interface, called GENIUS, on the desktop of most of the city's office-based employees.

In the spring of 1998, the first work began using GIS for storm water analysis. This initial work was focused on modifying the extents of a drainage basin layer in the city's GIS. Using two-feet contour intervals and spot elevations, derived from 1996-air photos, the assistant city engineer re-digitized the boundaries of these basins to more accurately reflect their extents. Smaller subsets of original basins were also digitized to generate additional analysis points and isolate areas such as sinkholes. These refined drainage basins were then used to calculate their total area so that more specific calculations for runoff could be generated. These calculations were all done by hand initially, which were not only tedious but also time consuming and inaccurate. With the city exploring using these basins for long-term storm water inventory planning and the use of runoff-based storm water user-fees, a more automated process was needed.

The city's GIS contained other relevant base map data for calculating runoff such as the permeable soil types and impermeable streets, buildings and parking lots. Since the features found within these basin areas were not all closed polygons with stored area values, they were being measured manually and placed into the runoff calculations. The first step on the way to automating this process was to take the existing soil type layer, with hundreds of soil type polygons and generalize it down to the four SCS hydrological soil types. Once these soil polygons were reclassified, the like polygons were dissolved into one another to produce a master layer of hydrologic soil types.

The impermeable streets and buildings existed as closed polygons within the city's GIS and thus could be queried for area measurements. Parking lots, however, were not closed area features and modifications were made to this layer to close off the parking lots using coincident line features shared with roads and buildings. The nature of a traditional map layer based GIS requires true polygon (closed area) features to be entirely contained within one file. Area features that might share a boundary with an area on another map layer, required the same edge to be maintained in two files. Ensuring that these edges were indeed coincident, allowed for precise measurement and accurate representation. Bloomington's parking lots were never digitized or maintained as polygons. Copying these coincident line features from building and street layers proved to be the largest single effort in the use of GIS for storm water analysis.

Generating the topology of closed, impermeable surfaces would not have been possible without the aid of aerial photography and ground truthing. Careful analysis of base map data compared with photographs and field surveys was necessary to eliminate ambiguity found in the GIS. Further refinement of these layers required that holes existing in the impermeable layers, such as courtyards in buildings and landscaping within parking lots, were considered permeable. This classification allowed for a conservative estimate of impermeable surfaces, erring on the side of the storm water customer. These holes in impermeable surfaces were tagged as islands, allowing for reclassification in the future. Sidewalks were not consistently maintained in the GIS and were, therefore, not considered in this analysis.

With all the necessary base map layers cleaned up and forming closed polygons, the first application, a series of basic overlay commands, were scripted for calculating runoff at the drainage basin level. The basin polygon was used to clip the soil type layer, and the building, parking lot and street layers. With these subsets of base map data, the building, parking lot and street layers were merged together to form one homogeneous, impermeable feature. Areas outside of these features as well as within the island holes were not part of the impermeable data. This impermeable layer was then overlaid on top of the soil types. The resulting basin polygon could be queried for total impervious surface and total open permeable soil types given in both percentage and area measurements. These values were used for general basin planning and could be regenerated as new development filled in the basin.

When the city storm water utility was formed and the decision to use a runoff-based user-fee was made, these layers were considered on the smaller scale of permeable and impermeable surface within a given property.

(www.bloomington.in.gov/hand/block-grants/con-plan-final.pdf).

Water and Sewage

The Oslo Water and Sewage Authority (OWSA) have an enterprise-wide information system known as VIS. With this system they are now able to provide accurate, detailed information about the city's water and sewage systems. The thought of having VIS started in 1991, and the first modules were up and running in 1993.

OWSA has a vision for its business. It will supply clean water for all purposes. This will be achieved through aiming at these business goals: "Customers and users should all be satisfied with OWSA's products and services, including their accessibility, quality and price." It is an agency for the city of Oslo city council and is fully self-financed. Their main income is water and wastewater fees. We supply approximately 520,000 inhabitants and all the businesses in the city with water and wastewater removal.

VIS today helps in being an efficient customer oriented service provider. The customer centre can answer most requests for data (maps and installation documents) on the fly by using property-ID or address as input to VIS. The council also extracts data for sale by using VIS. This is both pipeline information and geotechnical data. All in-house map productions, both standard and one-of-a kind thematic maps, are made by using VIS. All in-house employees can make their own maps as needed, while we still produce analog maps for use in the field. All document handling is now fully computerized. Incoming documents are scanned before being distributed. All correspondence out of the house is filed in the document system which can also handled reports, email etc. in addition to traditional correspondence.

The water and sewer module is used a lot in the sewer planning and also for daily maintenance work and rehabilitation. The water department is just starting to use the system for their planning and forecasting of flooding, snow measures etc., but do use it for their daily maintenance work. VIS also provides the city's planning authority with accurate pipeline information, and geotechnical data. This is used both for planning purposes and building permits.

Nevertheless, the council has not been able to reach the ideal world where data is stored and updated in only one database. It still has to use copies of the original databases in some of the applications. And it does not have online access to some of the external databases, like the city base map database and property database. This causes some extra work, but fortunately the data updating and loading routines work quite well, so they do not have many problems with different versions of data in different databases.

The future depends on what are we at present working with? Today we are mainly focused on efforts that can utilize the investments we have already done in the system. According to VIS authorities, the future course of action is as follows:

- *Customer services*: Improvement of the notification system for businesses is a high priority. When we know there will be construction work, or after we know of emergency shutoffs, we want to be able to notify the businesses that will be affected directly. Presently we only inform the property owners.

- *Mobile units*: We have a pilot study running for field GIS solutions. The project will hopefully have ten pilot installations by the summer 2001. We will only have data viewing in the field, no updating. We hope that the field units will save work and energy, by having all the correct and available information with them at all times.
- *TV-inspection*: We have presently bought software for a pilot study on integrating TV-inspections in VIS. This will be done through the map module, and will include an integrated map based archive and viewing system for the videos from pipe inspections. This will make the videos more accessible and more easily usable for rehabilitation planning purposes.
- *Internet customer services*: OWSA is proud of being a very customer oriented agency. We try to improve our customer services all the time. We are in the process of getting a web site for data inquiry up and running. Presently, it will be restricted to our professional users. They will be able to order their maps, drawings and other documents on the web. And they can either receive the documents electronically, or by traditional methods.
- *New versions of system for scanned documents*: We have problems both with our document handling system, and our system for scanned documents. The plan is to get new VIS modules for these applications during 2001. Hopefully we can keep the data formats as they are, without undertaking a major data conversion job.
- *Integration localization in the economy system*: The new economy system in OWSA will enable the staff to localize expenses to a certain geographic position. That is, adding GIS dimensions even to the economy system. This will help us in keeping track of which part of town has the more costly rehabilitation, and get the project cost of larger rehabilitation project divided according to localization. The economy system is ready for testing January 2001.
- *Integration with laboratory database*: We plan to make a more permanent easy use connection between the lab database and the desktop mapping system. This is to enable easier presentation of lab results from field samples.
- *Organizational issues*: The VIS system today is working fine, but it is difficult to keep the system with all its applications updated at all times. We also have a problem with users creating their own subsets of data etc. Therefore, the central VIS persons need to get focus back on the core system and data, and not necessarily participate in all the VIS subprojects. We also see that the company policy of distributing the GIS knowledge to each division only has worked to a certain extent. It is very difficult to keep developing new methods and new GIS products when everybody has to start from scratch. We are, therefore, in the process of upgrading the mapping section to Geospatial Information Section, with a separate function that will coordinate GIS use and effort throughout the whole organization.
- *General system maintenance*: The system utilizes several databases. We presently operate with copies of the public property database and the city base maps. We hope to establish proper online connections to these databases in the future. We even copy our one utility map layer to a separate database used by the desktop mapping system. In the future we hope to be able to have one geographic database. Hopefully we will not need to copy any data, but can access all both internal and external data online

from the source databases. Another challenge even further ahead is changing our data model from a graphic model to a true network model.

We also hope the web technology will help with increasing the use of the system even more.

(www.gisdevelopment.net/proceedings/gita/2001/hdti/toorunnb.adp)

Water Distribution System

The city of San Francisco's Water Distribution Department uses *Gatebooks* to maintain and operate the city's water distribution system (WDS). Consisting of hardcopy published maps in a bound book, this combination of water distribution maps and supporting textural information is the primary information resource for performing their mission. CH2M HILL began a pilot project in March of 1998 to establish the same information in digital format. Specifications for the automation of both graphic and corresponding attribute information were developed and used to successfully produce two prototype map pages in ArcView. As a result of this proof-of-concept, the City's Water Distribution Department (CWDD) has requested that the pilot project be expanded into a citywide *Gatebook* GIS according to the formats and configurations defined in the pilot project.

The various components involved to establish the desired GIS include the data conversion to establish digital database design along with a Graphical User Interface (GUI) to expedite its data maintenance and the performance of key water system analysis procedures using ArcView. The other component of this GIS is the training of CWDD staff in the use of the data model and GUI to make it possible for them to improve the operational efficiency and effectiveness of city's water distribution system.

SAAE-VR is a municipal authority responsible for the collection, treatment and distribution of potable water as well collection and treatment of wastewater. Located in Riode Janeiro's state, Volta Redonda is a city with: 300,000 inhabitants, 170,000 water connections, and 168,000 sewer connections.. The project goal is the implementation of a complete GIS system including GPS field data collection, data conversion and validation as well web map publishing. The features are as follows:

- Network Tracing.
- Water: Finding closest valves to isolate a water main leak.
- Wastewater: Tracing upstream to locate the source of contamination or trace downstream to find the potential path to the outfall.
- Validate features.
- Pop up lists enforce associations and enhance productivity.
- Interactive and batch commands to populate connectivity.

(http://gis.esri.com/library.userconf/proc00/professional/papers/PAP367/p367.htn)

7

GIS for Health and Human Services

Most problems facing the world and health today—environmental, economic, political, social—exist in a geographic context and any analysis must consider these factors. Understanding issues ranging from epidemiology to access to healthcare providers requires understanding the geographic context of these issues. Issues like finance and human resources, marketing and promotion, operations and customer service, research and planning are related to it. GISs provide powerful tools for geographic analysis for almost any health discipline. GIS technology is one of the fastest growing technologies in health today, helping professionals in the even more complex health industry manage their information to make better decisions.

The UK's Coventry Health Authority, with an ethnically diverse population of some 230,000 people, uses MapInfo to map health care providers and spatially analyse providers' locations in relation to key socio-demographic indicators. The London Ambulance Service responds to some 11,000 emergency calls a week with the help of MapInfo technology to help improve call handling and response time. With MapInfo, the ambulance service plots the location of ambulance stations, hospitals and areas of high priority calls, as well as to plan for crew deployment. Further, the US Pacific Northwest National Laboratory detects airborne contaminants and visualizes safe and unsafe areas with MapInfo technology (www.mapinfo.com). There are several examples where GIS has been used for the day-to-day activities related to health and related human services.

Epidemiology Detection and Emergency Response

The US Lightweight Epidemiology Advanced Detection and Emergency Response System is a national, real-time disease detection and coordinated response system that allows enterprise spatial analysis to be rapidly deployed. The MapInfo based solution enables disease testing information to be shared; data collection and analysis; resources management; and incident response and situation awareness (Fig. 7.1).

The New York State Office of Mental Health (OMH) develops a versatile solution with MapInfo technology to manage 25 campuses throughout the state, encompassing more than 1,000 buildings and more than 27 million square feet located on some 5,000 acres. Providing services to about 6,000 inpatients and 32,000 outpatients annually, OMH distributes dynamic data on floor and plot plans, building photography, utility costs, pending bond information and space utilization. (www.mapinfo.com)

New York State Office of Mental Health

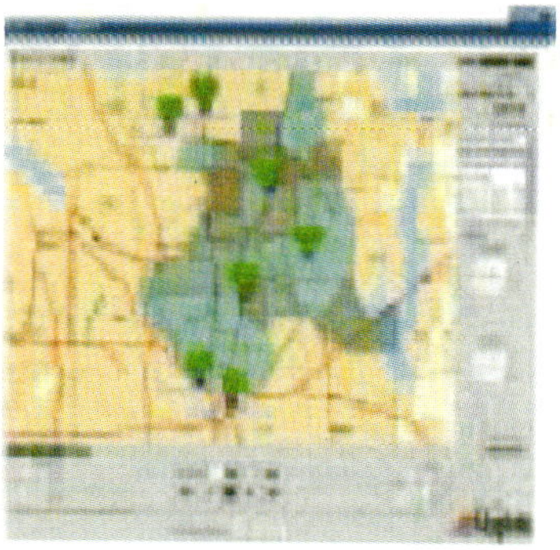

Fig. 7.1: Emergency response system

New York State's Office of Mental Health (OMH), headquartered in Albany, operates more than 1,300 psychiatric facilities located throughout the state, comprising more than 33 million square feet and located on some 5,000 acres. It is a challenge to manage such a large network of specialized facilities, both for the headquarters administrators who develop and implement long-range economic master plans, and for the field managers charged with serving a particular site's clients effectively and safely. Client, facility and infrastructure requirements change constantly, but overall OMH resources are finite: facilities budget requests total about $200 million annually, even though only about half that amount is available. The need is for a open, distributed decision-support system for better collaboration among the HQ and field personnel responsible for facilities planning. But the problem is of (a) multitude of computing platforms, and (b) diverse data types. In order to bring out a solution the SpatialWare server-based GIS technology with Oracle database, and accessed via the OMH intranet GIS Professional/custom browser desktop client was considered. The benefits were as follows:

- Allows OMH to analyse multiple "what-if" scenarios
- Easier to track Federally reimbursed costs
- Can quickly spot health and safety-related issues

The Facility Information System (FIS) has enabled OMH to improve many functions that depend on access to detailed facility information. Available on the web via the OMH intranet, the system enables managers to process and analyse multiple "what-if" scenarios at any time interval, and managers can view a final scenario of a fully implemented plan. Each configuration, viewed on a map, can help determine the best solution for short- and long-term facility operations. The system also streamlines the tracking and reporting of expenditures eligible for federal reimbursement, and has helped OMH receive about $700 million in reimbursements. Improvements in health and safety are another key benefit. Officials claim that "Health and safety issues override planning criteria for allocating budgets and scheduling work, and the FIS system helps us see on a map where immediate attention must be paid."

(http://www.omh.state.ny.us/omhweb/countyguide/)

Relief Operations in Africa

Several organisations have created web based and freestanding GIS to support humanitarian relief projects for the Horn of Africa. The United Nations, Humanitarian Assistance Information Center, and U.S. Department of State are working in concert to produce a geospatial data system that will support humanitarian relief efforts for natural disasters and political upheaval. Various data sources have been searched and pertinent information retrieved. What is lacking are land use maps and some form of physical imagery to "see" what is happening on the ground. A global mapping project sponsored by NASA and NIMA has provided LANDSAT imagery and land use maps from that imagery for the continent of Africa. LANDSAT imagery and a 13-class land use map created from that imagery are delivered through an ESRI ArcIMS site for the Horn of Africa. This presentation outlines the development and implementation of the remote sensing in support of humanitarian relief in eastern Africa.

In January 2001, the US Department of State (DOS) made a request of NIMA for aid in generating humanitarian planning maps for countries in the Horn of Africa: Djibouti, Eritrea, Ethiopia, Kenya, Somalia, Sudan and Uganda (Fig. 7.2). The maps were to assist relief workers in the planning of humanitarian efforts to relieve suffering associated with natural and man-made disasters. The data were to be accessible through an on-line mapping system utilizing ESRI's ArcIMS™ software. Earth Satellite Corporation of Rockville, Maryland was retained to create the web site and populate the mapping system with raster and vector data sets (www.esri.com).

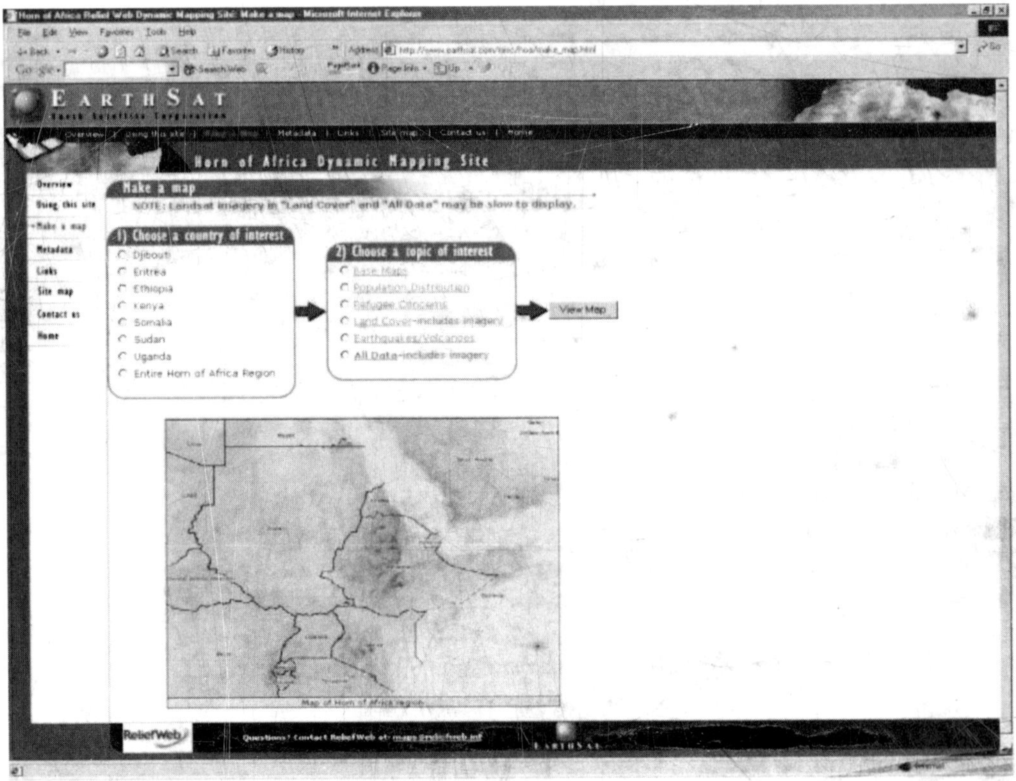

Fig. 7.2: Orientation of the Seven Countries in the Horn of Africa

Tool for Relief Planners

As indicated above, the humanitarian organisations working in the Horn of Africa use data to support aid for a wide range of needs: famine, war, refugee displacement, and natural disasters. Having data prepared in advance of a crisis allows for a more rapid response. The United

Nations Office for the Coordination of Humanitarian Affairs (OCHA) has initiated field-based information management activities through its Field Information Support Project. ReliefWeb's Map Centre provides easy access to humanitarian maps from a wide variety of sources. The Horn of Africa ArcIMS site was designed to provide a dynamic mapping capability and to support field operations through access to base geographic data.

Although all data was public, sometimes it was necessary to consider the implications of publicizing a data set, e.g. locations of refugee camps. How to display disputed boundaries was another issue. EarthSat modified a version of the political boundaries that indicated unresolved boundaries, posted on the web site with a disclaimer.

The goal of the Horn of Africa ArcIMS site was to provide a data platform, available to many, to be used as a planning tool from which basic queries could be performed and maps could be printed. ArcIMS is strong as a data display tool and a quick map maker, and offers basic GIS tools such as queries, measurement and buffering, with only internet browsing software. Since not everyone has an internet connection, and access will vary depending on the user's hardware and the internet infrastructure in a particular region, alternative data delivery formats can be developed along with an ArcIMS site, such as bundling data with ArcExplorer and distributing it on CDs (www.esri.com).

Global Food and Water Balance Process Modelling

Food and water shortages continue to threaten populations at global scales. The inadequacy of regional transportation infrastructures to match food supply with food demand punctuates the problem. Earth Satellite Corporation (EarthSat) of Rockville, Maryland worked with iSciences, Ann Arbor, Michigan to develop a GIS that assesses food and water balances at regional scales. Flow and iterative process models were applied to map food and water supply, demand, and balance to identify chronic problem areas globally. EarthSat's GIS methods for mapping was applied for food and water balance in Africa. The pilot GIS included ArcInfo scripted models, map databases, statistics, cartographic products and ArcView projects for all of Africa. GIS tools were also developed for complex interactive buffer zone analysis in ArcView.

The pilot project developed regional scale GIS models for the entire African continent at 1 kilometre resolution. The approximate scale of use is 1:60,000,000, and datasets may be used at scales up to 1:1,000,000. All models were masked through a 8901 × 8201 1 kilometre resolution grid. Earthsat developed a supply/demand and balance approach to estimating regional food balance in Africa. A demand surface is subtracted from a supply surface to produce a food balance surface showing areas of food surplus and shortage. The food balance surfaces are based on figures contained in the Food and Agriculture Organization of the United Nations' FAOSTAT database for 1998, the most current period that was then available.

The Africa Food Balance surface will benefit from additional effort including observational information for the purpose of model calibration. The inclusion of food production from urban gardens and subsistence agriculture will provide a more complete picture of African food supply. As more current data is made available, time series will be possible for longitudinal studies, and for forecasting. Work is in progress to develop the food balance models for the South Asian continental region.

EarthSat also developed a supply, demand, and balance modelling approach to mapping renewable water resources. The Africa water balance GIS model draws total average annual water demand from total average annual renewable water supply to estimate a regional scale watershed water balance. The GIS addresses critical analysis questions, identifying by watershed throughout Africa the location and extent of:

- Renewable water supply.
- Water demand from industrial, agricultural and domestic sectors water surplus and shortfall throughout the continent.
- The dependency of watersheds on inflow from exterior watersheds to meet local water demand.

The water balance surface was also benefit from additional model calibration based on observational data. While manual techniques for identifying inter watershed flow are time consuming, no reliable and accurate fully computer automated techniques for correctly producing inter watershed flow exists. It remains more time-effective to manually identify watershed connectivity, and code the relationships into a database. As with food balance, as more current data is made available, time series will be possible for longitudinal studies and forecasting. Water balance modelling is also in progress for the South Asian continental region. (www.esri.com)

Disaster Relief Operations

Disaster relief operations require the marshalling of substantial resources and their movement to the afflicted area. It is common for such operations to be undertaken in parts of the world where local infrastructures are less well developed. Frequently, the operations must be conducted in situations where the terrain and sometimes elements of the local population are extremely hostile to the aid agencies. It is not surprising, therefore, that the command and control requirements of military or civil aid agencies, are very similar to those of military forces conducting 'conventional' operations (www.esri.com). Under the auspices of the British National Space Centre (BNSC) Applications Programme, Smith System Engineering Ltd. carried out a short study to look at how GIS could be used to support aid agencies in providing assistance to disaster-struck areas. From a series of telephone interviews and more in-depth consultations with several key agencies, the study consolidated and analysed those requirements that would benefit from the use of GIS technology, particularly where this could make use of forthcoming improvements in data availability and quality.

The striking degree of similarity between different types of agency, operating under different circumstances, illustrates the enormous potential for the application of GIS to this important area. The rapid capture and import of geographic data is vital if GIS is to be used to support a crisis situation.

Social Marketing for Medicare and Medicaid Services

Balanced Budget Act mandated that Centers for Medicare and Medicaid Services (CMS) become more beneficiary friendly. CMS has refocused to become more consumer-oriented. Discover and respond to beneficiary wants and needs. Help beneficiaries understand their benefits and CMS programs. Social marketing applies commercial marketing concepts to health and human services programmes designed to influence behavior of target audiences in order to improve personal and societal welfare. CMS has adopted a social marketing strategy to better understand and communicate with their beneficiaries.

Integrating GIS and social marketing at CMS is a challenge. CMS's Center for Beneficiary Choices (CBC) is integrating GIS into an already well developed social marketing initiative to better understand beneficiaries. The project objectives are to better identify, segment and target beneficiaries so that CMS's messages will be more relevant, focused and more well received by the beneficiaries. The role involves:

- Identify relevant data sources for specific mapping requests
- Develop map templates
- Map data, by county, to segment benefit

GIS Supports a Closer Lookciary population. A major premise of social marketing is to tailor the message to different audience segments. GIS specifically supports these analyses for CMS by:

- Integrating data from disparate sources
- Analysing and displaying data at local levels of geography
- Defining and targeting segments of Medicare eligible and beneficiaries

CMS used this map to show the Advisory Panel on Medicare Education a baseline picture depicting where the medicare beneficiaries reside by county and state. Medicare covers annual screening mammograms for eligible women. Since it is less likely that African American and Hispanic women will seek screening, CMS has targeted the over 65 age group of African American and Hispanic communities for outreach and education about this benefit. The outreach focused on one or both minority groups.

CMS is working with American Indian and Alaskan Native populations to explain what medicare services are available. CMS can use maps to locate areas with high concentrations of American Indians and Alaskan Natives in order to reach beneficiaries and their providers who offer services and Medicare information to this population.

In many Asian cultures it is perceived that older individuals live with their children or other relatives when they can no longer live on their own. In addition, Asians are less likely to travel outside of their communities for health care as their communities are important and tightly knit. CMS can target elderly Asians to help them better understand the Program of All-Inclusive Care for the Elderly (PACE), an optional medicare benefit that allows elderly people to stay in their homes and communities and still receive intensive medical, social and rehabilitative services.

Regarding Hispanic population, diabetes is a serious health concern for some minority populations due to an increased prevalence of this disease in several minority populations. CMS can target specific minority communities with brochures explaining Medicare coverage for diabetes printed in not only English by in other languages such as Spanish as well.

Beneficiaries living in rural America are approximately one in four medicare beneficiaries live in rural America. These rural beneficiaries have lower in come levels and much less access to technology compared to other medicare beneficiaries. CMS can target rural communities with easy to read information specifically written to help rural beneficiaries understand their medicare benefits.

CMS can use maps to locate enrollment trends and areas where disenrollment from medicare managed care plans is most prevalent. The current uses of GIS for CMS social marketing is important. Share maps and supporting data and documentation with CMS projects and regions and internal central office staff. Use maps to prioritize and determine specific segments of CMS beneficiaries to target. Integrate data used to produce maps with other information, such as utilization of services, for analysis and decision making. (www.appgeo.com and www.esri.com)

Spatial Analysis in Tuberculosis Surveillance

A research study was made to examine the spatial distribution of tuberculosis (TB) cases by area in Harris County, Texas over a three-year period, 1995-1998, using GIS software and spatial analytical techniques. Analysis looked at the TB incidence distribution in Harris County through the use of exploratory point pattern techniques focusing on intensity measures, a spatial scan statistic and a spatial filtering method to determine whether cases fulfilled the assumption of complete spatial randomness. Results were useful for a Houston-based TB surveillance group in tracking the spatial trends in TB cases over the years.

One of the main reasons for performing this analysis was to show that there is a definite utility in the use of GIS and spatial analysis in conjunction with epidemiological analyses in public health. This project adds a benefit of performing another type of analysis that provides the researcher with a meaningful picture of the disease patterns that can be used in conjunction with output from epidemiologic studies. However, critics may be quick to point out that this benefit has a limitation as well; the fact that this current analysis is coming on the heels of prior research findings is no guarantee that these methods would have steered the research group towards their findings. This should not hinder, however, the use of spatial analytical methods in conjunction with epidemiological studies, especially as hypothesis-generating activities and exploratory exercises useful for planning future explanatory analyses. The main focus of this project was to show that the description of spatial patterns in disease events can lead to important decisions as to where interventions may need to take place or dollars spent on control efforts. In addition, some may recognize the limits of simple univariate point analysis with the methods used here, preventing one from looking for spatial relationships that may adjust for a number of covariates together as is done in traditional epidemiological studies. There are methods that will analyse multivariate point patterns, such as a bivariate $\hat{K}(h)$ function, that could be used in the future to look at comparing differences in spatial point patterns that account for locations of two or more types of events in a study region but this type of decision should be made by all interested parties involved in the research tasks, with a variety of analytical and exploratory data for background comparison. This project serves to add to that wealth of information already present in Harris County.

The importance of GIS in health research has been documented in a large number of articles during the past decade. Various peer-reviewed journals have devoted whole issues to the topic of GIS in health research (*Journal of Public Health Management*, Vol. 5 Nos. 2, 4), spatial analysis (*Statistics in Medicine*, Vol.19 Nos. 17,18), as well as, lengthy review articles on both subjects. There have even been several books written on the theme of GIS and health, as well as, exploratory analyses using spatial statistical methods and specialized software/ training modules developed to meet the needs of researchers when stand-alone GIS software is not enough for more robust statistical analysis purposes. The benefits of combining active health surveillance efforts with systematic collection and display of geographical information have also been discussed at length. GIS provides a visual component that may often be lacking in scientific studies that can provide useful information when combined with sound statistical methods. The ease of incorporating such GIS systems into already existing database structures in public health departments and surveillance systems should become the norm in an effort to promote the timely communication of disease trends to policy makers and the general public.

The main purpose of this research study was to examine the spatial distribution of tuberculosis (TB) cases by area in Harris County, Texas over a three-year period, 1995-1998, using Geographical Information Systems (GIS) software and spatial analytical techniques. In doing this, it was anticipated that this research study would demonstrate some of the valuable assets of GIS in disease mapping and surveillance. It is expected that the information gathered by this research study will assist public health workers by identifying and providing effective examples of using epidemiologic data, public use statistical software and GIS in formulating study questions, generating and testing hypotheses, and critically evaluating maps that are prepared using GIS software and spatial statistical methods. In addition, useful resources (analytical and descriptive maps) have been produced for the Houston Tuberculosis Initiative, a population-based, active surveillance and molecular epidemiology study of tuberculosis cases reported to the City of Houston Tuberculosis Control Office. It is hopeful that, in the future, public health workers and officials will see the added value that GIS can bring to an already well-established disease surveillance group. In addition, GIS can provide data that is in a form to be readily communicated to community groups and the general public if necessary.

(www.gis.esri.com/library/userconf/health01/papers/hc01-P029/hc01-P029html).

Developing GIS Enterprise in Health and Human Services

In 1999, the County of San Diego Health and Human Services Agency made a decision to establish an enterprise GIS. They conducted an enterprise needs assessment, and have been implementing recommendations that came out of the assessment. The goals and objectives of the assessment included (a) provide tools to more effectively serve the community's health and human services needs, (b) provide an atmosphere to be proactive *versus* reactive, (c) eliminate duplication of staff effort, and (d) share data across agency programmes.

What are the benefits of an enterprise GIS? To mention just a few: improved customer service, revenue enhancement, communications, cost savings and cost avoidance. In the area of customer service, GIS can play a meaningful role in targeting prevention and educational programmes. It is most effective in providing general location information. It is an excellent tool for pinpointing underserved populations and developing client support strategies.

The use of maps and trend analysis in grant applications can greatly enhance revenue generation. Additionally, GIS is most effective in demonstrating the efficacy of ongoing programs. The use of GIS for all types of communication is obvious. That's really what GIS is all about. For example, presentations to organizational management as well as the general population is much more effective when using GIS. A picture, or in this case a map, is worth a whole bundle of computer reports.

In the area of cost savings, the use of GIS makes complex data more easily and rapidly understood. It is a tool that can be used to estimate demand for services, thus highlighting areas where resources can be redeployed to improve efficiencies. GIS can be used effectively to avoid duplication of data entry and storage. Allocation of regional budgets can be based on client population and need, and it has been shown that by analysing the distribution of welfare payments relative to the poverty level households, GIS is a very effective tool in combating fraud.

Facilities planning, which results in putting resources in the right place, is one example of using GIS in cost avoidance. Using GIS to play "what if" scenarios to develop least cost plans also results in cost avoidance. Finally, identifying spatial patterns, with the use of GIS, has obvious benefit in formulating health care reform. An example of this is looking at patterns of family violence to form multi-disciplinary teams to work in areas presenting the highest incidence of violence.

(www.gis.esri.com/library/userconf/proc03/po226.pdf).

Planning Hospital Location

The Ministry of Municipal and Rural Affairs (MOMRA) of Saudi Arabia have defined a set of criteria for the planning of hospital locations. These criteria are applied on every hospital in Jeddah in order to evaluate their locations and accessibility. Public (governmental) hospitals are selected by this way as the main provider for hospital care in Jeddah. In another words, private hospitals are not included because these hospitals have different planning and management process that can be analysed by separate studies.

The MOMRA criteria for hospital planning indicate that each service should cover a 4 km catchment area with a population capacity of 60,000 or 2.5 beds per 1,000 persons (MOMRA, 1980). These criteria are applied on every hospital using the service area function of Arcview software. The output of the function is a network coverage showing the parts of the city that are within 4 km at catchment area.

It is evident from Figures (7.3 and 7.4) that the northern part of the city is located outside the hospital catchment area. This means that the existing hospitals are not serving the whole city, and there is a large part of the city, which is based on MOMRA criteria, is not within the hospital catchment areas. The next step of the analysis was to define the size of demand within each hospital catchment area. This step will identity how many people be served by the existing hospitals, and the number of beds that should be provided by each hospital. Relating hospitals demand (D) with hospital capacities is calculated using the following formula:

$$D = P \times a$$

Where P = the total number of people living inside hospital catchment, and a = 2.5/1000 person

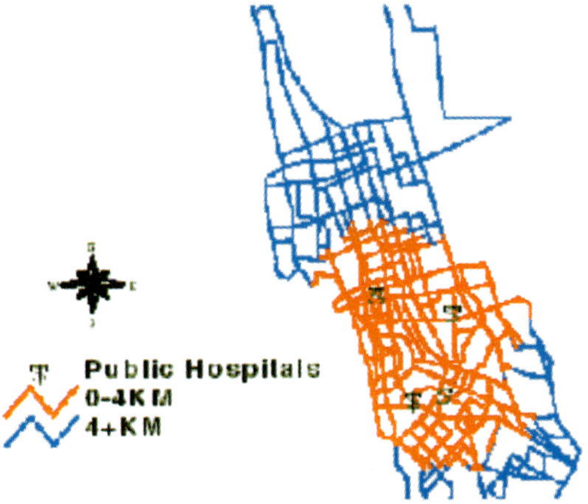

Fig. 7.3: Public hospital accessibility

Fig. 7.4: Public hospital service area

Table 7.1.: Results of 4 km hospitals catchment areas.

Hospital name	*Hospital capacity (beds)*	*Catchment area*	*Hospital demand (beds)*
King Fahad	1,001	4 km^2	974.5
King Abdulaziz	442	4 km^2	1,153.8
Al-Thugar	143	4 km^2	1,673.0
King Abdulaziz University	294	4 km^2	1,607.9
Total	1,880	16 km^2	5,410.2

Using Arc View select by them function, the present study has defined the size of people that lives inside each hospital catchment area. Theoretically, all of the people living inside these catchment areas should have free access to hospital care, but practically this is judged by the capacity of each hospital. Accordingly, if the size of demand of any hospital catchment area is more than the hospital capacity then the problem of service shortage will appear. Table 7.1 shows the results of this application, which are mainly as following:

(a) Existing hospitals capacity are greater than MOMRA criteria, e.g. King Fahad Hospital is serving 3,89,815 people where the criteria defined hospital size by 60,000 people.

(b) The size of demand within each hospital catchment area is more than hospital capacity. For example, King Abdulaziz University Hospital capacity is 294 beds while its demand is 1,608 beds.

Accordingly, it can be said that there is a great need to increase existing hospital capacities in order to overcome their existing supply shortages. The resulting shortages also clarify, why private hospitals are distributed at all parts of Jeddah City. (www.esri.com see also www.gisdevelopment.net)

Conclusion

It is pertinent to mention here that GIS has helped in human services and health care activities at all levels, starting from continent scale to a part of a city. The technology provides a wide range of flexibility however reliable data is always a problem. Remote sensing data, statistical data, vital statistics, mortality data and the like have given broad patterns which have been used for identification of priority areas and location of health care establishments.

Identifying health care needs is one of the important tasks which health authorities frequently do. Health planners keep analysing the changes that occur on health care demand. These changes usually require quick respond from health authorities, in order to keep matching health care facilities supply with the relevant demand. This is a large task that needs huge set of data including health facilities location, catchment areas, population statistics, etc. Information technology in general and GIS in particular can help the health authorities to make their required jobs in an efficient manner. For example, GIS can be used to support health planners in reaching a decision regarding increasing hospitals capacities. The present analysis has illustrated how GIS has been used for health care facilities planning. In the case of Jeddah, this application has identified the parts of the city, which require more attention regarding their health care supply. Using this application, health planners can allocate areas of the city which having poor accessibility to hospitals and according, a decision regarding improving hospital accessibility can be easily and quickly reached.

8

GIS in Retail and Commercial Business

Businesses manage a world of information about sales, customers, inventory, demographic profiles, mailing lists, and so much more. At the very core of this information is a geographic location, an address, a service boundary, a sales territory, and a delivery route that can be illustrated and interactively managed and analysed on a map. Application of GIS in retail business has increased substantially, but has a wide scope for further expansion. In other chapters, we will find several case studies where has been such applications as a part of a major business initiatives.

The impact of GIS application in retail and commercial business got a boost due to its integration with information and communication technologies. In the developed world, there is a shortage of manpower to make such applications. As a result, a lot of the projects have been out-sourced to developing countries like India. In the suburbs of the major cities we will find a new breed of people working in the Business Processing Out-sourcing (BPO) establishments. Such trend is spreading in medium size cities of developing countries as well.

Customer Relationship Management

PAC, based in Houston, Texas, offers an array of services for both offshore Gulf of Mexico and land-based hydrocarbon production operations in New Mexico, Texas and Louisiana and specializes in providing contract staff to the oil and gas industry.

The need is the ability to visualize recruits in cluster areas for productive recruiting and the filling of open positions for specific oil and gas projects. Seamless integration of technology with existing database requires minimal user training. Recruitment searches were taking days instead of minutes to complete, reducing the effectiveness of PAC to fill positions quickly and compete effectively. Further, it needed to find a way to narrow the search field. It was to base on ZIP code or county for more accurate recruitment of both urban and rural employees. With GIS PAC, recruiters were able to quickly and easily find qualified job recruits within a specified geographical area based on a ZIP code or county search. The benefits were as follows:

- Significant reduction in time spent recruiting.
- More accurate matching of job requirements with available candidates in a requested area.
- An efficient means of conducting business analysis and utilizing the full functionality of existing database.

Development of Market Strategy

Toyota Motor Corporation Australia is increasing dealer sales with the use of an online mapping system that incorporates a range of GIS business intelligence products. These solutions enable Toyota dealers to better identify and target potential new customers while also providing Toyota's national operation with a comprehensive view of the performance of state and regional offices and individual dealers.

Toyota needed a solution with ease of integration and functionality. Toyota had to implement a system via the worldwide web to disburse data to dealers across the country. Comprising a range of GIS business intelligence products, Toyota implemented an easily integrated and easy to use online mapping system. Toyota's online system maps each Toyota dealer's prime market area (PMA), the postcode-based geographical territory covered by each dealer. The system graphically depicts motor industry vehicle sales, which previously were available to dealers in monthly paper reports. Now, the vehicle sales report is mapped and delivered to each dealer online via the company's new extranet.

While each dealer will only see his individual area and the data relevant to him or her, the system provides Toyota with a national view for internal use to drill down into the various sections and subsections, including vehicle type, customer profile and sales by manufacturer. The information can be geographically based, as well as shown by sales and customer profile. The mapping system shows Toyota's national market share and how it compares to the national shares for all manufacturers. Zones, state, rural or metropolitan areas or individual dealer PMAs, may view data. It is possible to drill right down to see data on the number of doors on the car. How many three-door cars were sold nationally or in a particular suburb, it suits a very high-level view and in-depth analysis.

The benefits were as follows:

- The online mapping system enables Toyota dealers to better identify and target potential new customers.
- The system also provides Toyota national with a comprehensive view of the performance of state and regional offices and individual dealers.
- The GIS-based online mapping system decreased deployment and management costs.
- With the high level view and the in-depth analysis performed, Toyota dealers are identifying and targeting customers.
- With targeted marketing efforts, Toyota Australia is acquiring customers.

(http://www.fujitsu.com/global/casestudies/INTSTG-Toyito.htm)

Strategy for Magazine Distribution

Australian Consolidated Press (ACP) subsidiary Network Distribution Company uses a GIS mapping solution to quickly and easily visualize all the areas where ACP titles are distributed. With a map demonstrating coverage gaps, network distribution can target those areas and fill in those coverage gaps. In addition, the solution has allowed ACP to gain insight into what are needed in future targeted marketing initiatives.

The need is to quickly and easily see where Australian Consolidated Press titles are distributed and identify coverage gaps that could be filled. Magazine distributor was experiencing coverage gaps. Title distribution and agents' territories had to be easily tied in with the company's contract management network.

A GIS solution is identified to provide a clear distribution overview to ensure coverage. GIS technology integrated seamlessly into client side business applications. With the first phase implemented, ACP plans to use the solution for future marketing and subscription initiatives. Further, ACP plans to overlay geographical demographics with the magazines, which help them to use that information for marketing initiatives to increase sales. In the process it has built up a reservoir of information, and the real key now will be using that information for market analysis which will help them to get even more business benefits out of the system.

The benefits were as follows:

- Users can quickly and easily identify areas that need distribution coverage.
- By overlaying the title distribution with geo demographics, the solution provides the necessary information needed for building future marketing initiatives targeted specifically for a geographic territory, thereby increasing sales.

(http://resource.mapinfo.com/static/files/document/1074627271814/acp_casestudy.pdf)

Market Information System

When Snow Industries of America (SIA) in Reston, Virginia, decided that they needed to provide a presentation tool for their members' annual sales figures, they had only envisioned the use of their tabular data in a form-based system. SIA wanted the tool to accept a member company's annual sales information and calculate a performance and an opportunity index, enabling comparison of the firm's sales performance to that of its competitors. SIA also wanted to provide their members' sales and marketing staff a user-friendly software product that was easy to maintain while saving time and increasing overall productivity.

Snow Industries of America's has decided to use GIS technology as it would not only enhance the required presentation capabilities but also provide extensive analytical functions. Using GIS technology, SIA members could explore spatial relationships between various data sets under a variety of sales district configurations. The SnowTrak utilizes GIS technology to yield a complete map-driven marketing information system. Statistical data are easily transformed into fully interactive maps or graphs, allowing sales and marketing departments to quickly conceptualize their data on the screen. This will help in visualizing the statistical information, along with the compilation of spreadsheet data, maps, and graphs displays sales by state. In addition, comparisons of sales averages can be made on a nationwide basis that will be useful in evaluating territories yet to be established. Additional SnowTrak functions allow users to discover market potential and market share for regional or local markets and support their rapid development. SIA members are able to plot dealers and potential dealers in conjunction with a GIS database of U.S. cities containing population, age, and other demographic information. The system then allows for identification of dealers in unexplored and unexploited markets. The analysis also provides the information needed for the development of focused marketing campaigns that should ultimately increase product sales.

Electronic Bin Boards in Precision Agriculture

Vicgrain Limited specializes in bulk grain and commodity storage, distribution and loading for export. It operates storage and handling facilities at 234 locations bins throughout Victoria and southern New South Wales. Commodities are stored in silos, bunkers and sheds for customers, such as grain marketers who sell product in Australia and abroad, including millers, maltsters and stockfeeders.

Vicgrain needed to overcome its information vacuum by displaying, maintaining and updating over the web for multiple users throughout the organization what was the same manual map kept in each of the company's eight district offices. Key information about stock levels at sites were limited to personnel in individual offices, restricting Vicgrain's ability to maximize usage of its 234 bins and optimize services for customers. The solution is to have a web based GIS application that enables mapping via an intranet. The benefits are as follows:

- Enhanced decision-making and customer service.
- Increased productivity and efficiency because of faster, easier access to business-critical information and the ability to see the data geographically by site.
- Cost savings with centralized access to necessary information and reduced overhead.

The new system has achieved measurable benefits in the way information on commodity storage and site capacity can be displayed, reported and analysed by Vicgrain. This company has gained around a ten per cent increase in productivity and efficiency because of faster, easier access to business-critical information and being able to see the data geographically by site. The users can literally point and click on the map to get the information they want on any bin at any site.

(http://resource.mapinfo.com/static/files/document/1074627678 1/viegrain_casestudy.pdf)

Better Services to the Customers

Rack Room Shoes in Charlotte, NC. uses GIS to put stores in the best locations, getting message out to customers and giving the customer what they want when they finally arrive at the store. In fact, GIS application has become an integral part of store site selection, customer analysis and store performance for retailers worldwide. In successive surveys, businesses rank CRM systems as a number one priority. CRM is about identifying, analysing and understanding customers through technology.

Rack Room Shoes sees the competitive edge GIS analysis. For two years, the company has been using GIS for mapping, geographic and CRM analysis, demographic and customer analysis. By tapping into the company's customer database, Rack Room Shoes' GIS application allows executives to see what its customers look like, what they buy and where they come from. This information helps to put stores in the right locations and make sure that they are merchandised and advertised to suit the needs of the customer that will shop them.

In site selection alone, GIS applications have helped Rack Room Shoes to avoid many proposed sites which would have likely lost money had they been opened. It has also helped to identify markets and submarkets, which present the greatest opportunities. Building and operating a store is very expensive. Over the course of an average lease term, a poorly located store could cost the company hundreds of thousands of dollars. It is not unrealistic to say that by screening out poor locations the company saves millions of dollars each year with the help of GIS.

Rack Room Shoes opens about fifty new stores each year so selecting the most profitable sites is absolutely critical. To fully use location specific information, companies need to integrate spatial technology across both front office functions, such as sales, marketing, and customer support, and back-office functions, such as finance, human resources, supply chain, material management and manufacturing applications. With spatial location technology, companies can find and develop profitable customers, assess profitability of customer transactions, analyse trade areas, deliver products and services more effectively and comply more easily with local regulations and taxing.

(www.mapinfo.com/common/library/mapinfomag-summer2001.pdf)

GIS Delivery Areas for Store Locations

Domino's, one of the United States' largest pizza chains with 4,800 locations in the country, has been using GIS products for four years. Domino's based in Ann Arbor uses GIS products to map out its delivery areas and store locations. Users create multiple displays, buying patterns and demographic information and use the products for new site selection. Domino's also plans to put all its site location information on an intranet. This will gives one central location with all the standardized information. It will also give a great tool to take to the field offices.

(www.directionmag.com/pressrelease.pdf)

Commercial Real Estate

Cadillac Fairview Corporation, which is one of North America's largest investors, owners and managers of commercial real estate, uses GIS technology to combine proprietary customer data with extensive list of spatial data sets for market research. The GIS application helps Toronto-based company to gain insight into the lifestyles, demographic and purchase behaviour characteristics of any geographic area in Canada. The company also uses GIS for creating detailed maps as a powerful tool to visualize data spatially for senior management, leasing executives and tenants. GIS is used to create distinct target profiles for hundreds of products, activities and media habits. Cadillac Fairview uses the market information linked to its proprietary customer databases to be able to evaluate mall shoppers *versus* market behaviour and therefore, determine the potential for new tenants or retailers, as well as help with developing targeted programmes at best customers.

Cadillac Fairview is adept at using GIS products along with its own proprietary data to determine and monitor best customer targets for particular initiatives. In fact, in doing so, the commercial real estate company developed its first CRM programme. With it, best customer targets, identified and converted to a direct mail strategy, now can be identified and monitored for lifestyle and demographic changes.

(www.mapinfo.com/common/library/mapinfomag-summer2001.pdf)

Location of New Grocery Stores

The British Supermarket chain Tesco uses GIS applications to determine drive time polygons. The application helps the company to decide where to put new grocery stores. Tesco has been able to do a lot of different things with them especially in the area of data mining also. To look at China, people in the field can take the information from the local vendors and produce maps for the best locations. China is one of the places where Tesco is investigating at the moment. It represents an opportunity for expansion, and GIS provides Tesco with mapping systems and applications that are very useful in evaluating the marketplace. Before entering a country, Tesco uses the information to look at sales potential in various parts of the country and the country as a whole. After entering a market, Tesco uses the GIS products to aid the company in its customer analysis.

Tesco also uses GIS products to evaluate the Japanese market and map out the best sites for possible stores. Tesco puts all of its customer research into its GIS applications so it can easily evaluate markets where it is doing well and those that are under performing. Their motto is: "With a good flexible mapping system and good data, the world is yours."

Territory Identification

Anheuser-Busch in the United States uses GIS technology for dividing up its 650 territories for its wholesale beer distributors nationwide. In the past, Anheuser-Busch used paper road maps to sketch out the territories, which often caused confusion and took a lot of time. If a territory map was not clear when a new account opened in a particular area, no one knew exactly who should service that account. GIS technology allows to determine exactly who should be serving the wholesale account. It saves a lot of grief and aggravation.

Property Locator

Allied Dunbar's PropertyNow application uses GIS to allow its customers and real estate agents to search for and sell homes using straightforward and highly detailed interactive mapping tools. The Web site, www.propertynow.co.uk, reveals and analyses locations and details of thousands of homes throughout the United Kingdom. To build a sophisticated but user-friendly application, PropertyNow chose a complete spatial web based GIS solution.

Web based GIS solution delivers detailed information precisely linked to location and interactively displayed on maps at the speed and quality of service as per the expectation of the users. Without location-based intelligence, this service simply would not offer the quality of world-class service. Using online maps and data, the user may click on the area he or she wants to view. After entering housing preference and cost information, a map is created that presents an overall view of what is available that meets these preferences. Specific property details, including contact information to arrange a viewing, are available by clicking on a corresponding icon. With the zoom, detailed maps of the neighbourhood are seen.

Maps are at the heart of PropertyNow and make the service a truly useful tool for finding a dream home or marketing a property for sale by where it is located. The ability of the application to browse around and zoom into a map allows the user to see all of the options on the ground. Users may list their own property for sale on the site at no cost. The registration process will automatically locate the property on PropertyNow's mapping database, based on the postcode entered. Users can view where the system has located their property and adjust the positioning to show precisely where the property is located.

(www.mapinfo.com/common/library/mapinfomag-summer2001.pdf)

E-Business GIS Implementation for the Oil and Gas Industry

IndigoPool.com, part of the Schlumberger Information Solutions (SIS) group, brings proven M/A and D business solutions to the global oil and gas industry. IndigoPool's intelligent market place combines knowledge and experience with innovative technology, enabling buyers and sellers of oil and gas properties to reach global markets, make informed decisions, and engage in value-based negotiations with greater speed and effectiveness. Indigopool.com uses GIS application for the first fully web-compliant e-business implementation for the oil and gas industry. The implementation integrates industry-leading GIS software with IndigoPool's proven A&D technology to bring increased functionality, flexibility and speed to IndigoPool clients. This end-to-end implementation combines the advanced mapping capabilities of GIS internet map server software with the robust database benefits of Oracle technology through which enables proprietary data to be interactively incorporated into maps. The GIS implementation gives IndigoPool clients fast and easy access to a broad range of data, which includes cultural, land grid, well, production, seismic, lease and gravity or magnetic surveys. The users can intuitively move between high-quality map-based data selections, non-GIS e-business asset catalogues and E&P data rooms, enabling them to identify and investigate assets and data, and to provide visual confirmation of selections made through non-GIS means, in support of A&D decisions.

The implementation by IndigoPool provides high quality, accepted industry standard mapping data and the ability to exchange spatial information seamlessly across product suites. The GIS implementation is applicable to any SIS offering that has a web-based geographic interface to geoscience applications.

(http://www.indigopool.com)

Real Estate Investment Built on GIS

"Putting technology on the front lines for brokers" is the rallying cry of RealtiCorp Commercial Investment Fund. This unique commercial real estate firm combines the best of new and old real estate practices to provide value-added services to chain operators or multisite users (MSUs) such as the Home Depot, Target, Hampton Inn. office developers, or multifamily developers. While many real estate firms must reengineer operating procedures to take advantage of the power of GIS, RealtiCorp was designed expressly with the use of GIS in mind.

MSUs and institutional investors have reacted favorably to this mix of old and new. RealtiCorp is the only large-scale commercial real estate using a "high-tech, high-touch" GIS-based approach on a systematic basis. It has paid off handsomely. RealtiCorp, which was founded in 1995, has grown from $40 million in transactions in 1996 to over $200 million in 1998. More than 50 employees staff seven offices in South Carolina, Georgia, Florida, Virginia, Texas, and Alabama. There are plans to expand nationwide.

(www.esri.com/news/arcuser/1098/regis.html and gis.esri.com/library/userconf/proc99/proceed/abstract/9803.htm)

Affordable Space Locator Service

The South of Market Foundation (SOMF), a nonprofit economic development corporation in San Francisco, California, working with GIS planning, developed an application that assists small businesses in locating commercial space in San Francisco's South of Market area (SoMa). This application uses ArcView GIS to provide detailed spatial information for SoMa, for individual neighbourhoods within SoMa, for a quarter mile around a given space and generates a detailed space report listing. This GIS-based local economic development tool integrates information on businesses, local economics, demographics, and transportation with property information from real estate brokers to help businesses find the best possible location. This benefits the local economy as well as the business.

It provides one-stop shopping for small businesses looking for space to lease or purchase. All the information necessary to select the most suitable location for a successful small business is available through this service. As an alternative to the conventional practice of depending upon real estate brokers to select a site, it provides a much wider selection of spaces than is available from any one broker. A user-friendly interface allows users with no GIS experience to select available spaces according to desired specifications, to create a map of the area, and to produce different types of reports.

For the next step to develop this into a viable business, SOMF must increase its exposure to clients. SOMF intends to expand the service area to encompass all of San Francisco's smaller commercial districts. There are also plans to adapt the application for use on the internet.

(www.esri.com/news/arcuser/1098/soma.html)

A web-based Recycling Markets Information System

Market information is important for both buyers and sellers of recycling materials. GIS provides a way for manufacturers and market researchers to research the geographic distribution of recycling materials. Currently information on recycling markets information is not widely available to interested parties. ArcView GIS, ArcView Internet Map Server (IMS), Avenue, and HTML programming to develop the Recycling Markets Information System (RMIS) were used. This internet-based application allows interactive access to data on recycling in Texas by buyers, sellers, and recycling market researchers. It was sponsored by the Office of Pollution Prevention and Recycling at the Texas Natural Resource Conservation Commission (TNRCC).

As America is going online, a system like RMIS makes data on recycling more widely accessible anywhere anytime. It improves data sharing and is more cost-effective than other methods. RMIS provides the following four main functions:

- Assists in locating facilities that take household recyclables
- Provides locations of buyers of large quantities of recyclable materials
- Provides locations of sellers of recyclable materials
- Helps identify potential sources and markets for recyclable materials in Texas

(www.esri.com/library/fliers/pdfs/esri-microspft-texasnris.pdf)

Summary

The service sector is expanding very fast. Internet, web, telephones and other ways of communication has made this sector technology savvy. GIS has definitely a widescope of entering this sector. The links between the location of the service and its hinterland is now through the information and communication technology or ICT. Hence the distance between them is no longer an issue. As a result, India may soon become a hub for GIS related service industry.

9

GIS in Transportation

In the transportation industry geographic analysis is the key to making better decisions. Whether monitoring rail systems and road conditions, finding the best way to deliver goods and services, tracking fleet vehicles, or maintaining transportation networks, understanding these issues from a geographic perspective is crucial to deploying or spending resources wisely. It involves railroads, highways and streets, its logistics and fleet management, motor vehicle administrations, airports and aviation, seaports and waterways.

GIS technology serves three distinct transportation needs: infrastructure management, fleet and logistics management, and transit management. Transportation professionals can use GIS to integrate mapping analysis into decision support for network planning and analysis, vehicle tracking and routing, asset management, inventory tracking, route planning and analysis, and everything in between.

One of the important goals of the GIS application for transportation should be to identify accident-prone areas. Specifically this involves plotting individual accident locations, identifying accident-prone areas using GIS's spatial analysis tools, and presenting a diagnosis of accident characteristics by different types at these high accident locations. Based on the available attributes of the accident data it has been observed that the distribution of accidents varies depending on the time of a day, the type of road surface at the time of accident, the type and number of vehicle involved in the collision and so on. The following section describes how GIS software may be used to highlight specific accident types.

(a) *Accident based on type of road surface:* Pavement surface characteristic is one of the important factors, which determines the safety of a vehicle when negotiating a curve or at the time of sudden application of brakes. From the present study it has been found that major percentage of accidents occurred at both accident-prone and rest of areas when the road surface condition was dry. But comparing the percentage of accidents which occurred when the surface is wet it has been found that the accident-prone areas contribute more, leaving us the clue for viewing such accident location distributions to look for the factors and take countermeasures.

(b) *Accident based on vehicle involvement:* The accidents are also grouped based on the vehicle involvement and it has been observed that around 58 per cent of the expressway accidents were multiple vehicle accidents and the rest were single vehicle accidents. Further the collision of vehicle with stationary objects has been more in accident-prone areas. On the other hand even though most of the multiple vehicle accidents are caused as a result of head to rear type of collision, the percentage of sideswipe collision is found to be more in accident-prone areas, which needs to be focused. Hence analysis of specific types of accidents may reveal important geometric and traffic control and regulatory factors which are responsible for their causation. Further while investigating the distribution of accidents over different times of the day, it is found to be about uniform over the evening peak and night time but slightly higher during the morning peaks. Fatality percentage is clearly higher during night. When considering the speed limit there is a clear demarcation between the accident percentage contribution (around 85%) at the maximum speed limit zones (80 km/hr) and the other less speed limit zones. These types can also be visually studied for their distribution patterns.

Framing various combinations of above accident types can still improve the analysis. They help to localize accidents and thus arrive at some more useful conclusions. Further, the determination of accident-prone areas can well be approached using only accidents that result in personal injury or death, as the cause of such severe accidents cannot be reasonably attributed to randomness. There must be some deficiency with respect to the geometric features of the roadway or in traffic control and regulatory factors. Hence, by proceeding in this way and having the information about the spatial distribution of accidents it is possible to model the accidents with different geometric, traffic control/regulatory factors and environmental factors that might be responsible for a particular type of accident.

GIS is an effective tool to display different type of spatial accident distribution on digital road network. The use of GIS enables relevant accident data to be quickly processed and displayed on a map. GIS has also been used as a tool to identify hazardous locations along the expressways depending on the historical road accident data. These in turn will help to improve the safety of road by advanced planning and maintenance of the so-called accident-prone areas. For further development of the present model, the expressway database can be manipulated by adding roadway geometrics, information about traffic control and regulatory factors, flow data at different stretches of those areas. These information will then consist of the data that can be easily accessed or referred when crucial decision have to be made in relation to the type of improvement to be done at that particular area.

GIS and Railway

Belgium's TUC RAIL company was formed with one aim in mind: to design and project manage the country's ambitious and economically-important Thalys high-speed (TGV) rail link to France, Holland and Germany. The company (a venture between SNCB, the national rail carrier, the Tractbel engineering concern and others) employed software from Star Informatic from the outset for technical design and cartographic purposes. The country's dense urban environment presents a particular challenge, with licences needed at both local and regional level before construction can proceed. The ability to outline proposals in a clear and concise manner is vital in obtaining the necessary permissions.

To aid in this process, TUC Rail employed STAR software to generate DTMs for the two routes. These models were populated with buildings, trees, roads and other detail and rendered as full 3-D images, examples of which are pictured. This imagery gave politicians, officials and the public a clear idea of the impact of the lines in each locality and helped speed the process of obtaining licences. In the Republic of Ireland, public acceptance was critical to proposals that sought to upgrade the N8 (Dublin to Cork) primary road to dual carriageway standard.

A joint venture company (Parkman, Carl Bro, Punch, or PCP) was commissioned by Cork County Council to undertake the project, following which its specialist subsidiary, Parkman LandAspects, went to work preparing public consultation material. A key element in the package was a series of 3-D virtual worlds, created in Erdas Imagine Virtual GIS v8.5, that allowed 'fly throughs' of the various route options. Primarily used as a viewing and awareness tool, 3-D visualization allowed members of the public to assess the level of impact on their own localities. It also helped Parkman LandAspects respond to individual queries and concerns over the effects of the scheme. A detailed virtual world was generated using orthorectified aerial photography covering 211 sq km (commissioned from Simmons Aerofilms) and 3D DXF models of the proposed road designed using Infrasoft's Mx software and AutoCAD. A 1.5m resolution pixel-based Digital Elevation Model (DEM) was created that combined both existing and proposed elevation data. This yielded an 'as built' terrain model over which, high-resolution (1,000 dpi) orthorectified photography was draped.

This basic virtual world was enhanced by incorporating extruded vector layers (to represent buildings) and tree images (to replicate woodland), before adding 3D DXF models of the proposed road scheme. To these were added AutoCAD-generated bridge structures, street lights, traffic signs etc. Several flight paths were recorded as movies (MPEG and AVI format) and these, together with the Erdas software, proved invaluable in supporting 'live' presentations during the public consultation exercise (www.autodesk.com).

GIS and Transport Infrastructure

New Mexico State Highway and Transportation Department (NMSHTD) designs and maintains complex infrastructure projects, such as this stretch of Albuquerque interstate, using a suite of Autodesk software. In any major U.S. city today, road construction is a common scenario, with construction signs foretelling inevitable traffic slowdowns. Large and complex infrastructure projects can take state and federal agencies years to complete, and can cost billions in taxpayer dollars. The NMSHTD District 3 faces these challenges everyday. The group oversees transportation projects in the greater Albuquerque area, home to 750,000 residents, and is responsible for the state's two largest interstates—I-40 and I-25—which converge in the heart of the city. Everyday, some 200,000 people drive the "Big I," as the interstates are called. Software produced best results. Engineering productivity is vital to the efficient, cost-effective maintenance of District 3's infrastructure system.

The NMSHTD boosts productivity by using a suite of Autodesk® products, including AutoCAD® Land Development Desktop, Autodesk® CAD Overlay®, and AutoCAD LT® software, that run smoothly on its Microsoft Windows NT network. "Autodesk software is proven, cost-effective engineering technology that is far superior to other choices on the market today," declared a design specialist with the NMSHTD's Distrcit 3. "And since the software runs on our existing computers, it is the best use of taxpayers' money." "By using software early in our development process, we take the trial and error out of civil engineering," says the specialist. "For example, it is critical knowing the exact volume of earthwork when we're more thousands of cubic yards of soil. We can determine this figure with three per cent accuracy with the Civil Engineering Module in Land Development Desktop." "No matter what project comes down the road, we will execute it confidently and efficiently with Autodesk tools."

The NMSHTD completes work faster and more accurately by combining its expertise with advanced civil engineering software. For instance, prompted by a fatal head-on collision, District 3 mobilized quickly to redesign a three-mile section on nearby Tierjas Canyon. The project, which will cost $16 million, calls for widening the freeway and adding a life-saving concrete mid-section wall.

On this project, District 3 engineers and design specialists used digital survey data and AutoCAD Land Development Desktop software to design the foundation needed to support the structural improvements. They created profile, alignment, cross-section, and contour drawings of the construction site. Then using Autodesk CAD Overlay software and US Geological Survey records, they created an aerial site map. When the civil engineering design was completed, construction drawings were put out for bid. Using AutoCAD Land Development Desktop, District 3 produced these construction documents faster and better. Project completion is scheduled for fall 2000. "Using software is simply smart," adds the specialist. "We don't waste any time manually creating land development drawings that are required for civil engineering design."

Design changes are constant throughout the course of an infrastructure project. The NMSHTD completes these requests quickly by creating new drawings in the field with AutoCAD LT running on laptop computers. This action keeps projects on schedule, minimizes extra costs, and produces accurate as-built drawings. Stevens Transport Reports a 50 per cent reduction in turning accidents using Maptuit(R) FleetNav(TM) heavy truck navigation system. (www.autodesk.com)

GIS and Fleet Management

Maptuit Corporation, the only company that provides truck-to-door, dynamic navigation services, today announced that Stevens Transport has reported a significant decrease in turn-related accidents since deploying FleetNav throughout their fleet. Stevens Transport of Dallas, Texas, is one of North America's most widely respected and recognized irregular route refrigerated truckload carriers, servicing the contiguous 48 states, Mexico and Canada. In 2002, through the end of July, using the Maptuit FleetNav service, Stevens has reduced the number of turning accidents, as mentioned earlier, by over 50 per cent. "In addition to other operational savings, the Maptuit FleetNav service is providing us with important safety benefits," says the Director of Quality Assurance for Stevens Transport. "FleetNav is very valuable to drivers when navigating major metropolitan areas. Because it's on demand, drivers can utilize the service whenever necessary which helps to avoid trouble spots that may cause accidents."

Maptuit FleetNav provides automated "truck-to-door" directions from the current location of the vehicle to the consignee or shipper's location. Maptuit's advanced routing technology is designed for heavy trucks. The algorithms account for the degree of turns, vehicle turning radius and class of road, while minimizing right hand turns and decision points. Maptuit's database of the truck road network of North America is continually being augmented with a growing list of restrictions, prohibitions, low-clearance locations, and bridge characteristics to ensure that the routes generated are "truck-friendly". In addition to providing drivers with directions, FleetNav features easy-to-use reporting and management tools that enable a carrier's back-office staff to create and maintain detailed customer and delivery information, restrict roads from routing consideration, configure routing parameters, and preview routes.

Maptuit supplies transportation solutions that reduce the operating expenses of trucking companies by calculating truck-friendly routes, optimizing for miles driven, toll costs incurred and driver time. Built around the industry's most sophisticated model of the North American road network, Maptuit supplies the infrastructure needed to find addresses, calculate efficient truck-friendly routes, and determine the optimal way to assign and deploy resources. Maptuit has patents pending for its long-distance routing and alternate route generation technology. Maptuit has offices in Burlington, MA, Denver and Toronto. Maptuit is a registered trademark in the United States, and Maptuit, FleetNav and other names or logos identifying the services are trademarks or service marks of Maptuit Corporation.

(www.maptuit.com and *Internet Wire,* August 1, 2002)

GIS and Airport Management

Mineta San Jose International Airport selects GeoMedia technology to manage and support airport infrastructure and assets. Airport implements GeoMedia-based GIS to support goal of being top choice and regional gateway for air transportation services.

Intergraph Mapping and GIS Solutions announced that the Norman Y. Mineta San Jose International Airport (SJC) has chosen Intergraph's GeoMedia(R) technology to help manage and support the unique assets and operational functions of the SJC. They are transitioning from a MicroStation-based CAD environment to a fully functional GIS using GeoMedia, GeoMedia WebMap, and an Oracle Spatial database. Once all existing SJC map elements are merged into a GIS environment utilizing GeoMedia technology, airport personnel will be able to read several different GIS formats and re-project datasets on the fly. This system will provide intranet access to a computer-based system built on GeoMedia WebMap and will give greater access to all airports mapping data. The implementation of this programme will enable storage, organization, retrieval, manipulation, and graphical visualization and presentation of all airport drawings, documents, and spatially referenced asset information.

The airport's GIS programme will focus on serving three main areas, spatial data maintenance, business application development, and seamless integration of divisions using web technology. Spatial data maintenance is an ongoing challenge at SJC because of major airport expansion projects. It is imperative that GIS staff is able to accurately capture features before, during, and after construction changes occur. Business applications will be designed for the specific needs of each airport division and will include property management, security and life safety systems, utility management, ground transportation, planning or environment, and facilities management. Finally, GeoMedia WebMap will deliver maps, imagery, and data to all airport employees using the SJC intranet.

A senior geographic systems specialist for SJC comments, "Our vision is to develop easy access to all airport maps and asset information directly from any desktop. GIS is such a wonderful tool that can assist with the planning and decision-making process and can be used by any discipline, from engineering to marketing to finance. SJC and the city of San Jose are positioning to become an organization with an enterprise GIS programme and to establish SJC as a premier regional gateway for air transportation services. GeoMedia technology is providing one of the vehicles to make it happen."

Located two miles north of downtown San Jose, Mineta San Jose International Airport (SJC) is a completely self-supporting enterprise, owned and operated by the city of San Jose. SJC is the 37th busiest airport in North America in terms of total passengers. Airport passenger activity at Mineta San Jose International has grown from 6.8 million passengers in 1990-91, to over 13.9 million passengers in 2000-01. Currently, nonstop service is provided to five international cities as well as 30 U.S. destinations. Fourteen major commercial airlines operate out of the airport, which include, Alaska, American, American Eagle, America West, Continental, Delta, Frontier JetExpress, Horizon Air, Mexicana, Northwest, Southwest, United and United Express. SJC also hosts six all-cargo airlines and 180 general aviation based aircrafts. (Huntsville, Ala., May 28, 2002, www.intergraph.com)

Transportation GIS

This GIS application utilizes ArcView 3.2 Avenue programming to perform Highway Capacity Manual (HCM) capacity and levels of service analysis methodologies for freeways, multi-lane highways, and two-lane rural highways on the entire or selected segments of a regional traffic model network. Using the ArcView and TP+ Plus/Viper software interface capabilities, a traffic model is converted to a shape file and analysed in the ArcView environment. This application demonstrates a GIS/Traffic Model/HCM Interface capability that streamlines the process of impact analysis on highway projects and local development.

The California Department of Transportation (Caltrans) GIS/Level of service application was initiated as a pilot project funded by a federal grant through Caltrans, to the Merced County Association of Governments (MCAG), a San Joaquin Valley Metropolitan Planning Organization (MPO). The project focused on the development of interface capabilities between the MCAG Regional Transportation Model, and their extensive GIS programme. The immediate applications of this interface were to enhance the ability of Caltrans to use the model in the development of traffic studies, in support of State Highway System (SHS) transportation project development in Merced county.

The traditional traffic forecasting and analysis procedures performed by Caltrans staff, entailed the use of the MCAG model as the source for future Average Daily Traffic (ADT) volumes, but the more detailed portion of the traffic studies required post-processing the majority of the remaining forecasting and operational analysis work. This was necessary because the regional priorities of the MPO model, and associated performance measures, did not translate directly to project level measures of effectiveness, that included very specific hourly traffic information, and adjustment factors required for the design of SHS projects. In addition to this, the model was based on a 25 year forecast consistent with MCAG's Regional Transportation Plan (RTP), while Caltrans projects, although based on a 20 year design, had varying starting construction years, depending on transportation funding cycles.

The Caltrans GIS/Level of Service Application (CTGLA) utilizes ArcView 3.2 Avenue programming to perform capacity and levels of service (LOS) analysis methodologies for freeways, multi-lane highways, and two-lane rural highways on the entire or selected segments of a transportation model network. These methodologies are based on procedures documented in the 1997 *Highway Capacity Manual,* Special Report 209, Third Edition (HCM). This GIS/LOS application specifically accepts a shape file created in the Viper (Visual Planning

Environment) programme developed and distributed by the Urban Analysis Group/Citilabs company. The GIS Tools (Optional Add-On Module) in the Viper programme consist of a number of functions that allow the integration of the transportation model network with a GIS map. The CTGLA accepts the shape file created by Viper, utilizing all available attributes of the transportation network, as the information source to perform LOS analysis in the ArcView environment.

Capacity analysis is described in the 1997 *Highway Capacity Manual,* Special Report 209, Third Edition (HCM) as a set of procedures for estimating the traffic carrying ability of facilities over a range of defined operational conditions. The concept of level of service (LOS) is also described in the HCM, as qualitative measures that characterize operational conditions within a traffic stream and their perception by motorist and passengers. Levels of service are generally given letter designations with LOS A representing the best operating conditions, and LOS F the worst. Since this project was initiated, the HCM has been updated with minor changes to the freeway and multi-lane highway methodologies, and a completely new methodology of evaluating two-lane highways by direction of travel or by both directions combined. Although the process has started to update the Avenue programming, the CTGLA as presented at the 2002 ESRI International User Conference, has not been modified to reflect the new truck equivalency factors for freeway and multi-lane highways, or the new methodologies for two-lane highways.

For the purpose of the CTGLA, the analyses of three classes of roadways were selected. These were basic freeway segments, two-lane highways, and multi-lane highways. These roadways represented approximately 90 per cent of the SHS in Merced county. Signalized intersections were not included in this pilot project because of the complexities associated with the analysis of facilities with interrupted flow conditions, and signal timing procedures.

The following examples identifies the roadway information, including notes on items specific to the MCAG project model, used in the Avenue programming, consistent with HCM methodologies for the specific CTGLA project application.

The California Department of Transportation GIS Level of service application demonstrated the benefits of an existing and available integration of two types of systems, transportation network models and GIS. The "user friendly" environment of the ArcView programme allows the transportation analyst to import a transportation model with all associated network attributes and utilize the *Highway Capacity Manual (HCM)* methodologies to review, analyse, modify, and display calculations and results with no additional post-processing activities between the three programme platforms. Although an intermediate knowledge of transportation modelling, ArcView GIS, and HCM methodologies is helpful, ArcView users using this project application can immediately utilize transportation model data and begin performing HCM analysis, with minimal training in both programmes. The update of the LOS methodologies to HCM 2000 is in progress, with consideration of also including the analysis of interrupted flow facilities (www.esri.com).

Area Transportation Study

Chicago Area Transportation Study (CATS) is for the Metropolitan Planning Organization for northeast Illinois. As such CATS uses significant amounts of geographic and demographic data that is often hard to find within the computer system and once found, poorly documented. A study was made to document how ArcGIS and ArcIMS were used to improve data documentation and information distribution. It describes three initial projects, their design, implementation and reception. The projects include an application to distribute Census and Tiger2000 data, a Transportation Improvement Plan (TIP) application and a Regional Transportation Plan (RTP) application. (www.esri.com)

Decision Support for Navigation

Canadian Ice and Marine Services Branch (IMSB) presents a scenario of the current and future application of GIS technology. IMSB provides timely and accurate ice information for day-to-day decision support in Canadian waters. The acquisition, processing, and integration of real-time data are essential for this task. IMSB is also responsible for an extensive archive of the Canadian maritime cryosphere; specifically sea ice, lake ice, and icebergs. This information is an integral component for monitoring global climate variation since the cryosphere is especially sensitive to minute changes in environmental conditions. (www.esri.com)

Creation of an Enterprise GIS

Beginning in 2001, the NTTA undertook an initiative to manage their substantial data sets by implementing an Enterprise Geographic Information System (EGIS) in order to make better decisions and to more efficiently administer their resources. During this session the successful methodology utilized in the implementation of the NTTA's EGIS was to be presented. Topics such as essential needs assessment and implementation plan elements, project team structure, project prioritization, data quality and acquisition, key deliverables, standards, and project management were to be discussed. In addition, integration strategies with the agency's financial, toll collection, asset and facility management, and construction management software were to be addressed. Examples of milestone applications were also to be demonstrated. (www.esri.com)

GIS-Enabled Internet Site

The DTLR is the United Kingdom Central Government body with responsibility for national transport policy. Its primary objective is to maximize the contribution that trunk and local roads make to the efficient operation of the nation's transport system, promoting integration across different forms of transport and supporting wider government policies. For the first time, the DTLR has created the National Databank of Origin and Destination Surveys to provide a central repository of spatial information to help identify data for transport modelers.

Within the UK, there is an increasing interest in 'regional models' (i.e. those covering an area larger than a county) to examine the effects of long distance re-routing over a wide area. Large models require a lot of data that is expensive to collect. For this reason, whenever possible, traffic modelers were to ensure that they make the best use of existing survey data. However, it is often time-consuming identifying potential sources of movement data and can cause duplication of effort amongst interested transport modelers.

With a growing number of on-going and planned multi-modal studies, it will become increasingly important for the consultants who are developing models to assess alternative options and inform decisions on transport matters to have access to timely sources of existing movement data. As a result, in 1997 the Integrated Transport Economics and Appraisal (ITEA) Division of the Department for Transport (DfT), in joint working with the UK Highways Agency invited interested organisations to tender for the creation a database that would contain road-side interview information. Halliburton KBR was subsequently awarded this commission.

The project has resulted in the establishment of a definitive dataset that will provide valuable assistance to transport planners. The database contains the location and attributed information with respect to origin-destination transport interview surveys and currently covers the following survey types:

- Roadside interview surveys.
- Public transport surveys.
- Home interview surveys.

The purpose of the study is to:

- Explain the aims and deliverables of the National Origin Destination Survey (ODS) Databank Project.
- Discuss how the project has evolved and the stages undergone to date.
- Focus on the development on the web-enabled GIS, the issues this posed and the solutions found.

In summary, the National Origin Destination Survey Project:

- contains information about transport origin-destination surveys that have been undertaken in the UK;
- is the first time a central source of information on transport interview surveys covering England, Scotland and Wales has been successfully developed for use by transport consultants and other interested parties;
- is an initiative developed by the ITEA Division of the Department for Transport, with support from the Highways Agency;
- was developed to provide information to assist interested parties, in particular those wanting to develop transport models, to readily identify potential sources of origin-destination information;
- contains roadside, public transport and home interview transport survey origin-destination surveys;
- contains information from some 230 organisations covering approximately:
  —400 RSI Surveys;
  —7,200 RSI Sites;
  —65 PT Surveys;
  —230 PT Sites;
  —30 HI Surveys;

- consists of information about the surveys and associated locations where origin-destination data has been undertaken. It includes information about the surveys and sites, but does not contain the individual trip data. Any persons or organisations wanting trip data will need to approach the individual contacts contained in the databank.

The databank project has also demonstrated the significant benefit associated with the use of GIS technologies to spatial represent transport-related information. Without GIS technology, the ease of retrieving information and identification of the sites would be a time consuming exercise. Within minutes, consultants can now assemble the suitability of sites for use in transport models. (www.esri.com)

GIS and Residential Parking Permit Programme

An attempt was made to develop Residential Parking Permit Program Management System (RPMS) using MapObjects 2.1. This system consists of Parking Lot Design Support System and Parking Lot Management System. It is able to analyse parking lot supply and demand on the local street. And it can estimate parking demand on future. Also it can manage, update, and analyse data about parking. Officer can install parking lot on the local street using this system. Moreover this system can assign parking lots to applicants and inform the applicants about assignment result automatically. The administrator can query the information about parking lot, number, administrator, user, permit time, and so on. It is expected that this system will help in efficient management of RPMS, and it can help to solve parking problem on the local streets or residential area in the metropolitan city like Seoul in Korea. (www.esri.com)

GIS Implementation at International Airport

This case will be an overview of GIS application in an airport environment. GIS is widely used at McCarran to support the various divisions of the Department of Aviation including facilities and land use planning, business, real estate, property management, and airport operations. A study has been made to demonstrate some of the applications that were developed in-house to disseminate the available GIS data throughout the organisation. The applications include the Terminal Management System (TMS), which is a MapObjects software-based desktop application. An ArcIMS application to manage airport properties and an ArcPad application to support airport operations. (www.esri.com)

Integrating GIS into a Traffic Data Management System

This study presents the Collision, Traffic and Road Inventory System (COTRIS) that is used to track traffic collisions, traffic volumes, and road conditions in Strathcona County, Alberta, Canada. It was developed through a joint partnership between the Royal Canadian Mounted Police (RCMP) and Strathcona County. This example will show how COTRIS was designed to build "smart" data sets that are explicitly linked to a GIS street network file. The result is that every intersection and road arc segment can be queried to determine traffic collisions, volumes, or road conditions at that location. It will also present the development process from problem identification to implementation.

COTRIS is an integrated system that uses a SQLServer database, a Visual Basic interface and MapObjects development components to link GIS attributes to collision, traffic volume and road condition data. GDS and Associates Systems Ltd., a leading Canadian systems integration company, was commissioned to develop this system by Strathcona County in Edmonton, Alberta as a joint initiative between the County's Engineering and Environmental Planning Department and the RCMP (Sherwood Park Detachment).

COTRIS is designed to build "smart" data sets that are explicitly linked to the GIS. All data collected in COTRIS are linked to the GIS through a feature ID code. Users are able to query the data using a GIS tool such as ArcView or ArcInfo and instantly see the results in a map view.

One of the primary goals of the project was to improve upon the "pin map" that the RCMP created each year. This "pin map" was essentially a large map of the county hung on the wall in the RCMP Sherwood Park Detachment offices where data entry staff would place a pin for each collision that occurred during the year. At the beginning of each year, the RCMP would photograph the map on the wall, send it over to the county offices for data entry, remove all the pins and start the process again. County staff was frustrated with the lack of accuracy on the pin map. Sometimes they were not sure if the collision was supposed to have occurred in the centre of the intersection but had been moved to the side because there were too many pins in the middle of the intersection.

COTRIS is essentially a data entry system with thirteen canned reports. Because COTRIS uses the MapObjects development environment, users are able to enter both attribute and location data on the same screen. MapObjects lets users locate collisions, count stations and segments of road for storing road conditions directly on to a map of Strathcona county. However, COTRIS is not a GIS. Its main job is to help users create GIS-ready datasets.

The most important feature of COTRIS is that the SQL server database is related to the GIS database directly through the arc IDs of the streets and the node iIDs of the street intersections. This means that all records relate back to the shape files through a primary key or foreign key relationship. To get the full value of COTRIS, a GIS that can read shape files and connect to a SQL server database must be used. For example, users can load the street network file in ArcView and have available to them all the attributes stored in the SQL server database. This means that users could create maps with ArcView showing such statistics as:

- A map of all collisions where the driver was a male over 75 years old;
- A map of all collisions at an intersection where the incoming traffic volume was greater than 15,000 vehicles per hour;
- A map of all roads where the surface rating was greater than two and the traffic volumes were higher than 15,000 vehicles per hour;
- A map of all roads where there was a low surface rating and a large number of collisions (www.esri.com).

Integrating GIS Technology with Transportation Applications

Transportation studies typically include a wide variety of data. The data needs to be stored in multiple formats, assessed and edited by many team members. For a project to be successful this data must be managed and stored intelligently. This example presents the integrated use of GIS, CAD and Microsoft Access to manage, analyse and map data for a transportation access management study.

Virginia Department of Transportation (VDOT) hired the consulting firm of Vanasse Hangen Brustlin, Inc. (VHB) to develop a long-term access management plan for US Route 13 and for a portion of Route 175. The total study area was approximately 70 miles long. In addition to addressing existing safety and operational problems, the plan also took into consideration expected traffic increases through the year 2020. Due to the overwhelming amount of data that needed to be organized, a GIS needed to be established to manage the data. Microsoft Access was used to store all tabular related data. All roadway design files were prepared in CAD (MicroStation). All impact analysis, cost analysis and thematic mapping was to be performed using a suite of ESRI software.

A route-system of the study roads was created. The route-system allowed for dynamic segmentation of the events in the access databases. This allowed for conventional GIS analysis and thematic mapping of all the resources along the corridor. Resources such as median width, shoulder width, accident locations, driveway location and others were used to help determine what necessary roadway improvements were needed. Also, since the data was stored in Microsoft Access it allowed for non-GIS users to view and modify the data.

Highway engineers created all their roadway design files in MicroStation. There were five different alternatives for the corridor. Each alternative needed to be analysed and thematically mapped. Using ArcMAP all the CAD files were overlaid over the GIS base, along with all the GIS resource layers and thematically mapped. Since the highway design files were kept in there native environment (CAD) this allowed the highway engineers to modify there alternative throughout the project without having to perform any file conversions.

Another challenge was printing all five alternatives out for internal review and reports. There were a total of eighty 11×17 sheets for each alternative, with a grand total of 400 sheets. An Automated Printing Extension (APE) was written and used to produce the maps. The APE allowed the user to select a layout, scale and how many sheets to print. This extension then automatically processes each sheet and creates pdf files. Not only did APE streamline the printing process, it also eliminated the chance for user error.

The final challenge was the impact and cost analysis. Since all the highway design files were in CAD and all the resource data was in GIS, a data conversion was performed. After all the highway design files were finalized they were converted to geo-databases for analysis. First the impact analysis was performed. Water resources, wetlands, historic and archaeological areas among other resources were analysed. Second was the cost analysis. Since the geo-databases contained all the levels (layers) from CAD files we were able to associate a cost for each recommended improvement. For example, a unit cost for removal of pavement, closed medians, proposed medians, new pavement, etc. was used in the analysis. This allowed for comprehensive cost analysis for each alternative. The results where put into a Microsoft Access database for dissemination to the project team.

Using GIS it was possible to store, manage, analyse and map thousands of records of information. Since we were able to map and analyse the highway engineers CAD files duplicate efforts between disciplines were minimized and production was more efficient. This allowed for cost savings and collaboration for all team members on the project. (www.esri.com)

Highway Maintenance Planning and Network Management

Many local highway authorities in the UK are now undergoing major organizational changes and severe budget restrictions. This, when linked to increase in traffic, environmental pressures on new roads and public expectations for travel, demands a system that gives maintenance planning engineers the means to manage this complex process. There is a need for highway maintenance and the maintenance must be planned, progressed and recorded. Computer systems over the last one and half decades have tackled this problem in a textural way and many gave sterling service. To derive the greatest benefit from highway survey data an integrated GIS/RDBMS is essential.

Many local highway authorities in the UK are now undergoing major organizational changes and severe budget restrictions. This, when linked to increase in traffic, environmental pressures on new roads and public expectations for travel, demands a system that gives maintenance planning engineers the means to manage this complex process. Kent's Highway Maintenance Management System (HMMS) uses ArcInfo for this purpose. It enables the many aspects of highway network management, i.e. surveying, planning and design to be integrated into a single package. Analysis of data from various sources can be used to help balance these pressures and produce statistical information in support of budget bids. The key to all this is highly intelligent network coverage with multi access points for interrogation which is understood and used properly by staff with the appropriate skills.

The network started life as a text based node and section network design, instigated by the DoT, which used the nodes as the connectivity point. Data was captured by measuring from a node as chainage and cross-sectional position. This system worked well to create text reports and computer output of the data but not everyone, however, understood the referencing system, including the data collectors in many cases.

It became obvious that a geographical system was needed to best display the data now available. In order to give the network spatial value, the highways were digitized and the existing chart reference made a prime attribute of the new spatial arc. The network has been refined over the last eight years and now is a highly accurate representation of all adopted roads in Kent. Upgrades to the network take place using both the GIS and Chart Referencing System (CRS) modules. GIS maintains the spatial integrity and CRS ensures that node and section IDs are unique and a historical record is maintained. This record becomes part of the version control used by GIS to display historical data against a valid network.

The CRS as previously said, is not readily understood by all. With the advent of the National Street Gazetteer (NSG) it is envisaged that in time the NSG number will be the prime attribute along with relevant nodes. The connectivity between nodes would be by Elementary Street Unit (ESU) making the network even more usable and allowing the system to be operational anywhere in the UK. Use of GPS to set up a network would enable the system to be used anywhere in the world.

The data collected and processed in the past six years has indicated that there is now much to be done. Trends in personal mobility, larger and heavier lorries and the number of journeys undertaken have all shown an increase in those six years. Road usage has been a growth area and in all likelihood will remain so in future as well. The increase in traffic and corresponding decrease in funding will, inevitably, lead to a deterioration of the highway network. This in turn will lead to an increase in the demand for maintenance and consequently a better, more effective way of allocating scarce resources. Accepting the fact that a modern growing economy requires an efficient, well maintained integrated transport network, it is imperative that systems are in place so that managers have the wherewithal to plan, direct and monitor the network's well-being. Decisions need to be taken on which part of the network should receive priority. Should one maintain the nationally important strategic network to the detriment of country lanes and estate roads, or satisfy the people in their homes to the detriment of commerce ? The strategic network by its very nature of traffic size and volume will inevitably deteriorate faster than a less well used estate road but the balance of funding is the challenge the maintenance engineer must meet in the coming years.

The economics of maintenance and the whole life cost of repairs must be questioned and decisions taken that will not only protect the present but plan and strengthen the future. Kent HMMS is a system that targets maintenance schemes, gives control by auditing works and ensures delivery of management objectives. (www.esri.com)

Using Network Models to Determine Locations for Rail Terminals

In order to develop a multimodal transport system, locations must be selected where freight can be transferred between train and truck. Location-allocation models may be used to determine sites for these transshipment terminals. Such network models, derived from GIS ArcInfo, are applied to find the optimal point between minimizing the number of terminals and minimizing before-and-after transport by road. The result of these calculations is a pattern of terminals covering the whole or a part of the catchment area. The location-allocation model does not generate chains; these are calculated afterwards with the aid of a travelling salesman algorithm.

At present, shops are mainly stocked by truck deliveries. However, the growth in automobile traffic and trucking has led to increasing congestion. This puts timely delivery of goods in jeopardy. Moreover, the growing environmental pressure is unacceptable. An alternative to current practice may be found in combining several modes of transport for a single shipment. We call this multimodal transport. The project Rail Distribution Netherlands (RDN) envisions the development of a multimodal distribution system for goods destined for the retail sector. Trains provide long and medium distance transport, while trucks take care of the local distribution to the supermarkets. The goods are transferred at rail terminals. Transhipment takes place as close to the final destination as possible. In this way, the share of freight being moved by road can be minimized. The combination of rail and road transport takes advantage of the punctuality of the train and the added value of the truck's flexibility and door-to-door accessibility.

The task of determining the number of terminals needed and selecting the right locations is crucial to an efficient operation. There are two opposing driving forces that exert an influence on the optimal number of terminals. On the one hand, cost control calls for minimizing the number of terminals. The investment outlays and operating costs are inversely proportionate to the number of terminals. On the other hand, the optimal service provision and the reduction of before-and-after transport by road call for maximizing the number of terminals. The task is thus to determine the optimal number of terminals and their optimal locations. At the same time, certain principles have to be taken into account:

An optimal balance should be found between minimizing the number of terminals, minimizing the amount of kilometres driven in road transport, and maximizing the number of users to be served. The catchment areas served by the terminals should cover the entire country. The prospective sites for the terminals are tied to the locations of existing terminals and stations. The service radius of a terminal is limited to 50 km.

In order to develop a multimodal transport system using train and truck, locations for transhipment terminals have to be selected. With the aid of location-allocation models, a method has been worked out to find the optimum between minimizing the number of terminals and minimizing the before-and-after transport by road. Location-allocation models have been developed to determine the optimal location for services. The optimum is the point at which a service is made available to the users in the most efficient manner possible. The models optimize the relation between the potential services (location) and the identification of the users (allocation). ArcInfo has three types, each of which has been designed to optimize the location of facilities on the grounds of a given criterion. The *Mindistance* model minimizes the number of transport kilometres. The *Maxattend* model maximizes the number of users. Further, the *Maxcover* model tries to achieve an optimal coverage of the catchment area.

In the framework of this application, the main goal is to minimize transport costs. This model performs best in reducing the number of kilometres between the catchment areas and the terminals. On the other hand, the model has difficulty in covering the entire catchment area within a range of 50 km with a limited number of terminals.

The location-allocation models are valuable additions to the field of GIS. The main drawback that has emerged in this attempt is that such models make use of an origin-destination matrix, whereas freight transport takes place in chains. The transformation into chains in retrospect is a somewhat disappointing solution. However, the integration of chains in the location-allocation algorithm requires very time-consuming calculations. Furthermore, in a GIS, there is always a chance that errors may be replicated during transformations. In that light, the model could be greatly improved if the units of study were not spatial concentrations of population but actual locations of retail establishments.

A drawback in this application concerns the capacity to compare costs. The number of terminals and number of kilometres are dissimilar entities. How many kilometres of distance reduction would justify the construction of an extra terminal? In order to answer this question, the number of kilometres covered by truck and train would have to be given a monetary value and the extra investment and operating costs would have to be determined for each terminal. These calculations call for accurate figures on costs, and such information was not available in the framework of this study.

(www.gis.esri.com/library/userconf/europroc97/2transport/TL2/tl2.htm)

Using GIS for Transport Management

Copenhagen Transport (better known by the Danish abbreviation HT) is by far the largest bus transport authority in Denmark serving one-third or 1.7 million of the country's inhabitants living in and around the Copenhagen Metropolitan area. This area is in rapid transformation with workplaces relocating from the centre of Copenhagen to the surrounding areas and with the construction of the bridge between Denmark and Sweden. Further, HT is constantly competing against people's choice of using cars instead of public transportation. To meet these challenges, HT has set out a plan known as *Vision 2005* to strengthen public transportation within the region. *Vision 2005* incorporates application of modern information technology including GIS which is seen as an important technology for planning, operating as well as monitoring the bus transportation. A study was performed during 1996 identifying needs and application areas. Today work is in progress on compiling data and developing applications, e.g. for line planning, passenger information services and journey planning, environmental impact assessment, GPS-location of buses, and route optimization for tele-buses.

(www.gis.esri.com/library/userconf/europroc97/2transport/TL4/tl4.htm)

Spatial Analysis of Waterways

In the late 1995 shipping and waterway data was made assessable with GIS tools. As a result, it was possible to conduct specific analyses using GIS. The analysis of traffic (cargo)-flows and the consequences of a treaty that deals with cargo residues is one of the component projects of the Spatial Analysis of Waterways project.

This project explores the possibilities of integrating and analysing the numerous types of data relating to shipping. In short, the aim of the project is to conduct spatial analyses of shipping and the waterway network, investigate their applicability, and improve our understanding. The analysis of traffic flows and the consequences of unloading of inland vessels in the Netherlands make use of information from the IVS90 system.

IVS90 is the information and tracking system for shipping. Its aims include making the circulation of shipping safe and smooth, dealing with shipping accidents effectively, and compiling statistics. When vessels enter the IVS90 area, they report their presence and are tracked as they sail through it. The IVS90 area consists of 90 per cent of the main waterways in the Netherlands. This IVS90-network is made up of a collection of IVS90 "sections" which are linked to the Transport Research Centre's digital waterway network.

Some of these IVS90 sections are locks, where data is recorded on every commercial vessel that passes through. This data relates to the voyage, the cargo, and the vessel itself. Locks in the IVS90 network provide comprehensive information, and have served for many years as an important source of shipping statistics.

As is stated above, a great deal of data is available on transport by water. However, it always relates to one point (a lock) in the network. Collating data from all the IVS90 sections generates information on every voyage's chosen route. This information can be used to obtain shipping statistics on all IVS90 recorded voyages undertaken during a particular period. It is known which IVS90 sections have been passed through on every voyage. Since the IVS90 sections have been placed in a GIS environment, it is possible to represent the voyages on a map. By making selections from the total collection of voyages, specific flows of traffic on the IVS90 network can be represented in a visual form.

(www.gis.esri.com/library/userconf/europroc98/proc/idp22.html)

Track Events for Railways

Railway planning and operation relies on 150 years of experience in the use of thematic maps and spatial referencing. Today visualization and analysis is aided by a digital route network in combination with data about the track, landscape, weather, vegetation, human population, animal collisions and other events along the track.

Connecting geographical locations is the basic motivation for railways. Planning, operating and maintaining railways require extensive use of geographical and environmental information. Ever since the Irish National Railroad in 1838 produced a series of geological and property related thematic maps that may be considered a first manual GIS; the railways around the world have used maps as tools in their progress. Geology and property information is no less important in the railroad projects today, but added requirements for speed, capacity, interoperability and environmental factors brings forward new tools and methods. Computers and GIS are increasingly important.

The new high-speed railway between the center of Oslo and the international airport at Gardermoen is the largest land based project in Norway in recent decades. An early application of GIS here supported the property transactions between landowners and the state regarding areas required temporarily or permanently for the project. Digital maps were produced and connected with administrative and economical data relating the properties. Different types of land were evaluated and alternative locations of comparable quality could be offered to some of those who had to move. The geological and environmental challenges of the project include unstable sediments in the ground, water leakage in tunnels, and noise reduction efforts in residential areas. The Gardermoen trains will start operation simultaneously with the opening of the new airport soon, but use of the longest tunnel is postponed due to the water leakage problems. The announced 19 minute-ride to the airport will thus get a delayed start.

(www.gis.esri.com/library/userconf/europroc98/proc/idp101.html).

Time Travel Information Using GIS

The volume of traffic in European cities continues to increase. In the United Kingdom, traffic is expected to double by the year 2020. Government planners have realized that simply building more roads, apart from any environmental considerations, will not solve this problem. To minimize congestion and maximize use of the transport system, the existing infrastructure needs to be managed more effectively and alternative means of transport explored.

In 1992 the historic port city of Southampton, England, teamed with two other port cities, Cologne, Germany, and Piraeus, Greece, to bid for European Commission (EU) funding for a project that would integrate traffic and travel information. The resulting project created a GIS-based traffic monitoring and analysis system called the Road Management System for Europe (ROMANSE). This pioneering use of intelligent transport systems is part of a larger project known as EUROSCOPE (Efficient Urban Transport Operations Services Co-operation of Port cities in Europe) and will be used as a model for transport management across Europe.

ROMANSE is run by a consortium of partners from public and private sectors led by the Hampshire County Council and the Southampton City Council. The system, designed around the use of advanced transport telematics, addresses immediate problems and long-term solutions. It provides real-time information on traffic volumes, incidents, and parking availability to travellers before and during trips so they can make informed choices not only as to route but as to method of transport.

The information used to advise travellers can be accumulated and studied to make better policy decisions about transportation. This system also promotes the use of alternate methods of transportation such as buses and bicycles. Though the initial implementation was based on ArcInfo, ROMANSE now uses ArcView GIS and Siemens' Urban Traffic Control (UTC) system. ESRI (UK) developed an ArcView GIS extension, Strategic Information System Client that displays an overview of transport environment using information from the UTC system. Various types of transport data are displayed as themes in ArcView GIS. The basemap used for the system originated from the United Kingdom's Ordnance Survey. Road network condition data is displayed against this base. Data is collected from a variety of sources. Closed-Circuit Television (CCTV) is installed in thirty key locations in and around Southampton to monitor traffic.

Operators can pan, tile, and zoom cameras. Automated Road Traffic Event Monitoring Information System (ARTEMIS) software constantly scans the CCTV pictures to gather and process information on traffic flows, average speeds, and incidents. Additional data from roadside detectors and GPS units on buses along with the CCTV-derived information is fed into the UTC system. It is then processed and integrated by ArcView GIS and displayed at the Traffic and Travel Information Centre (TTIC) in Quay, Southampton.

Instead of trying to interpret text-based data, TTIC staff can quickly assess the current state of the road network by looking at maps created by ArcView GIS. They use the information to analyse existing conditions and spot developing problems as well as to dispatch emergency services. Information on traffic flow per minute, traffic speeds, congestion, and car park occupancy can be immediately transmitted to travellers via a system of variable message roadside signs and radio broadcasts. Free-standing computer terminals, located at transportation, shopping, and tourist information centres, supply information on public transport routes and times, help route car trips, and provide information for tourists on hotel accommodations and points of interest. Bus stops feature screen displays that show the arrival times of buses calculated from updated information on bus location obtained from GPS units on the buses. An interactive web site that provides real-time information on road conditions also allows users to check the current and predicted occupancy in car parks before travelling includes an application that helps users plan trips using the public transit system, and contains information about cycling. MapObjects Internet Map Server (IMS) software provides current traffic information through maps created on demand by visitors to the ROMANSE web site. Users can customize how the various modules are displayed when they visit the site. "ROMANSE is one of the most advanced traffic control systems in the UK. By putting this detailed integrated information on the internet, it is now at the leading edge of traffic and public transport information systems throughout Europe," says a GIS project manager for the Hampshire County Council.

The high level of system integration made possible through GIS is one of the unique features of ROMANSE. The various aspects of the project data collection from many sources; analysis using the Strategic Information System Client; dissemination of transport network information to the public via signage, information screens, radio, and the internet; and the use of that information in decision making and planning work together to minimize congestion and make the best use of resources. ROMANSE was recognized in 1995 with an award from the British Computer Society (BCS) for innovative IT projects. BCS IT awards are given not only for excellence in computing, but on the basis of the social benefit and business value to society created by the project.

The system has moved rapidly from a concept in 1992 to a fully operational system that has met with positive response from the public. Current plans call for delivery of working systems that can be marketed commercially to other cities throughout the world. This comprehensive GIS-based traffic control and travel information system can be used to alleviate traffic problems by helping the public more efficiently use the existing transport infrastructure and improve policy decisions on transportation issues in the future.

(www.gis.esri.com/news/arcuser/0799/romans.html).

GIS-based Warning System for Business Promotion

All modes of transportation—plane, train, automobile, or ship—can be adversely affected by weather conditions. For railroads, weather-related derailments can result in injury or death to workers, destroy tracks, and cause losses that range from tens of thousands to millions of dollars. The ability to briefly delay trains threatened by weather events can minimize hazard and loss. A weather-related accident involving a moving train can be ten times more costly than one involving a stationary train.

The Union Pacific Railroad (UPR), the largest railroad in North America, operates 6,847 locomotives and manages 38,654 miles of track in 23 states. UPR links every major West Coast and Gulf Coast port, serves four major gateways to the east, and is the primary rail connection between the United States and Mexico and interchanges traffic with the Canadian rail system.

In order to keep down costs and protect people and property from weather-related train accidents, UPR needed reliable, reasonably priced weather data that covered its vast area of operation. UPR also needed to be able to focus on specific locations and weather events. Decisions to stop or delay trains require real-time weather information and the ability to determine the proximity of weather events to track segments and rail cars. A GIS-based system that processed this type of weather information could fulfil many of UPR's needs.

DTN Weather Services, a division of Data Transmission Network (DTN), is the world's leading supplier of weather solutions, providing GIS-ready data, forecast services, accurate real-time weather, and display systems to more than 20,000 customers. In 2000, DTN Weather Services, an ESRI business partner, offered customers the ability to convert up-to-the-minute weather information directly into ESRI GIS formats on-the-fly using MetWork-GIS FileServer. This file server is a weather data ingest and file management system that runs on Windows NT. It supports standard internet communications protocol and various network configurations. Using MetWork-GIS and ArcView GIS 3.2, DTN Weather Services created a custom GIS-based track-warning network, the Union Pacific Railroad Weather Alert Notification (UPWAN) system that helps UPR protect its employees from injury and its equipment from damage while ensuring timely service to its customers.

"I worked very closely with the DTN development team during every stage of the process," said the manager of Systems Development and Support at UPR's Harriman Dispatch Center. "They listened to what we had to say and incorporated it into the design. The end result was a product that not only matched our current needs, but also will grow and expand with Union Pacific. We recently had a situation in Oklahoma where DTN alerted us to approaching high winds. The dispatcher stopped the train, and a short time later we received notice that 21 of the cars had blown over and were lying on their sides. If the train had been moving, the loss would have approached $1,000,000 and the track itself would have sustained severe damage. Because the dispatcher was alerted in time, the track was still usable and the overall damage was minimal."

With DTN GIS-based weather data, UPR can monitor specific geographic locations for important weather conditions. When very exact, predetermined weather conditions take place on a particular section of track, the dispatcher responsible receives an immediate, fully automated alert. The dispatcher must then acknowledge that alert within a short time or the alert moves to the next stage. At that stage, the supervisor on duty is notified. Ultimately the system can be shut down. Such steps may sound drastic, but winds, torrential rains, flash floods, and tornadoes can cause significant damage, delays, injuries, and even deaths. These hazards can all be minimized or even eliminated with advanced warning. The networked PC-based system can accommodate more than hundred separate users with text and imagery to monitor the entire UPR system. The data can also be archived for a permanent record for future review or reports.

Stopping trains before a weather-induced accident occurs is the primary goal of the UPWAN system. The specific weather events that threaten trains fall into three main categories: (a) conditions that produce high winds, (b) inundation of a track segment with water, and (c) temperature extremes that could expand or contract rails causing them to warp. After identifying the types of events that create these threatening conditions, UPR outlined the following specific criteria for generating an alert message:

- A tornado warning, flash flood warning, or severe thunderstorm warning issued by the National Weather Service.
- Temperatures lower than 0° F or higher than 105° F.
- Wind speeds in excess of 50 miles per hour.
- The presence of a tornadic vortex signature identified by National Weather Service Doppler Weather Radar (NEXRAD) that is likely to impact a Union Pacific track within seven minutes.

The ability to merge GIS-ready weather information using MetWork-GIS FileServer with ArcView GIS provides the framework for a powerful decision support system. The automated weather alert notification system developed for UPR, while a specific implementation, contains system components that can easily be reused or extended to provide solutions for other transportation clients or other industries.

(www.gis.esri.com/news/arcuser/0101/dtnweather.html)

Methodology for Spatially Based Urban Trip Distribution

The basic activities of the Ministry of Public Works and Land Planning of the Catalonia Government (PTOP) include planning and managing of geography related components: mobility, transportation, urban, land, regional, territorial. Different systems managed by different services with heterogeneous development levels and data models. Spontaneous co-ordination or gradual reforms are unfeasible in practice because of the rigidities and inertias created in all the different systems:

1. Collect and classify static geographical digital data arriving PTOP.
2. Re-structure and tune the information (tables, fields, relationships, indexes).
3. Translate to the improved Geomedia format (Smartstore, GeoMedia Access, MrSID).
4. Store and reply to static data servers.
5. Publish to the GIS Systems via metadata (SAM).

To assess spatial trip distribution, by using GIS, remote sensing and analytic hierarchy process, in order to define a displacement origin or destination matrix, without the realization of an extensive door-to-door research, following steps were carried out:

A. Integration: GIS, RS and AHP (all processes spatially analysed)

- systemic variable analysis
- complimentary GIS + AHP
- raster spatial modelling
- completes vector modelling
- pixel-level data disaggregation

B. Contributions to transport planning

- avoids door-to-door household research
- allows scenario studies (variations)
- better impedance assessment
- integrates personal technical experience (AHP)

Web applications with the most common GIS need for all vertical projects. Open for all users in PTOP, offers GIS culture to everyone connected. It shows the metadata contents related to every user profile: datasets, symbologies, pages:

- Usefulness
- Everyday work
- Decision support
- Publish broadcasting

Three layer structured:

- SAM database (metadata).
- SAM librarian (technology).
- SAM managing applications (interface).

Two object families:

- Abstract object definitions: projects, data roles, functional roles, user profiles.
- Real object references: datasets, workspaces, legends, functionalities, users.

(www.geospatialworld.com/gsw2004/proceedings/result.asp)

Road Management Information System

A system for collecting and storing road-related data requires auditing quality and currency of information. It should converts data into information to assist decision-making in main road. Australian Road Management Information System (ARMIS) operates across the state out of fourteen regional offices and 550 accredited users. It has following infrastructure:

- Operational systems to support district management
- External data capture
- Database audit that applies business rules
- Roads information data centre (with linkages to other corporate systems). Provided the first capability of displaying ARMIS data over geographical layers that were available from various sources.
- Not a true GIS representation of ARMIS data (as it did not attach spatial reference datum to each piece of ARMIS data).
- Used the proven concept of dynamic segmentation.

ARMIS GIS

- Provides a true GIS representation
- Provides a strong data management discipline
- Open architecture
- Oracle spatial cartridge
- Utilises Intergraph Geomedia

Features of the 1999 design included—

- Open architecture to provide seamless data integration with all major GIS vendor formats.
- Oracle Spatial Cartridge allowed management of spatial data by RDBMS: single unified Oracle database using standard tables; utilised existing access and security protocols and had the effect of making GIS data access transparent.

Being used in ARMIS GIS to inform users as to what data they have access to. ARMIS GIS allows users to register their own datasets: (a) user level, (b) district level, and (c) state level. Further, ARMIS GIS manages the storage of all its datasets so registered users at any location can retrieve them. The new ARMIS Online, web-based, uses following four layer model:

- Database Oracle running Unix
- Map engine in VB calling Geomedia web map and web enterprise functions
- Interface layer between user and map engine using PHP on Unix
- Browser layer (Active CGM)

The journey of ARMIS and GIS has been long and challenging and an important milestone in our ARMIS GIS development has been achieved. It now provides Main Roads Queensland with:

(a) a cost effective method of exposing many users to GIS displayed data, and
(b) it fulfils the promise of achieving full spatial integration of road-related data.

(www.geospatialworld.com/gsw2004/proceedings/result.asp)

10

GIS for Telecommunication

The telecommunications industry is changing rapidly, resulting in tough competition and an ever-increasing scope of services offered to customers. Solving the many business problems of a telecommunications company requires a good understanding of where customers and facilities exist and good information about those locations. In telecommunication industry GIS technology can be used in any of the following areas:

- Outside plant engineering
- Wireless engineering
- Operations support systems
- Marketing and sales
- Location services
- Customer relationship management
- Network planning

GIS technology enables telecommunication professionals to integrate location-based data into analysis and management processes in network planning and operations, marketing and sales, customer care, data management, and many other planning and problem-solving tasks. A GIS can integrate location-based data from databases all over the world to help in resolving and streamlining everyday business issues.

Web-enabled Mapping to Boost Customer Service

Arch Communications Group, Inc., the second-largest provider of paging services in US. The combined company was to have more than seven million subscribers and revenue of more than $800 million. With just a few clicks from within a web browser, anyone was to be able to pinpoint whether paging services from Arch Communications are available in a given geographic area. Arch Communications requires an easy solution for pager customers to determine service areas, which will also support for web-based e-commerce strategy. The problems were as follows:

- Generic maps
- Unfamiliar user interfaces
- Proprietary operating systems

Web enabled customer service system based on distributed spatial (geographic) information. Arch's customer service representatives use the system to establish areas of coverage for existing customers and sales prospects. By building its customer service infrastructure, Arch's call centres and web site have the accurate location information needed to answer customer questions immediately. In addition, Arch sales force uses GIS's customer relationship management products so customer care can match offers to customers in real-time and sales may target individual customers. The benefits were:

- People of all skill levels and all types of computers can easily access information.
- Supports buying products and service over the internet, enhancing sales by allowing Arch to target individual customers and enabling customer care to match offers to customers in real-time.

(http://internetweek.cmp.com/news1098/news102298-7.htm)

Boosting Customer Service and Identify New Customer

AT&T Corporation is among the world's premier voice and data communications companies, serving more than 80 million customers, including consumers, businesses and government. With annual revenues of more than $53 billion and 1,26,000 employees, AT&T provides local, long distance, data, and internet-access and wireless services to customers worldwide. This needs:

- Increase the capture of new revenues rather than rely on recurring revenues as its source of growth.
- Advance each sale more quickly by understanding each customer better.
- Increase the speed with its service get to market.

With business customers receiving on average three calls a day from communications service providers, AT&T recognizes that each customer interaction must be focused and meaningful. To do this, the company needed to present a vast amount of internal and external information to field sales personnel in a way that illuminated individual customer solutions. The system needed to identify and prioritize solutions based on revenue potential. AT&T also wanted to be able to advance its sales more quickly and increase the speed with which it gets its services to market. The issues were:

- No centralized source of customer data and AT&T network data for sales people to reference.
- Existing spreadsheet applications not as intuitive to the sales force.

AT&T has developed a data base called Attack using GIS technology. This helps the growth markets sales force to create maps from customer and lead source data so that revenue opportunities can be quickly identified and targeted. By colour coding the opportunities by revenue thresholds account executives can see which ones to work first. The GIS technology gave the sales representatives the tools necessary to quickly identify and target new opportunities across a multi-billion dollar market. This solution also makes it a lot easier for sales team to identify prospects and ultimately close business quickly by providing every possible cross-reference. The sales people can focus not only on their initial sales target but can also uncover other opportunities as well. The great thing about it is that it pulls all of these disparate resources together in one location. By putting customers on the map, Attack allows growth sales to locate opportunities in assigned territories and allows global sales to locate all its customers' sites nationwide. The customer location relative to the network facilities is a critical part of the sales picture in AT&T. By this process AT&T customers do not come to them but AT&T is connected to the customer order to provide services.

(http://resource.mapinfo.com/static/files/document)

Identify Additional Customers for a New Service

British Telecommunications (BT), owner and operator of the United Kingdom's dominant telephone network, sells local, long-distance, international, wireless and internet services as well as equipment. Competitive pressures in its markets continue to push BT to develop new products and services that satisfy customer's communications needs and to compete effectively with available alternatives. BT recently launched in the United Kingdom a new asymmetric digital subscriber line (ADSL) service that allows ordinary home and business phone lines to be converted into high bandwidth digital links. ADSL represents an important opportunity for BT to compete head-on with the cable companies and other high-speed data service providers for a share of the growing market for broadband services. The problems were:

- BT wanted to find a way to profile and pinpoint the "ideal" high-bandwidth customer.
- BT needed to prioritize ADSL deployment in locations with greatest potential demand.
- BT also needed a tool to perform cost/benefit analysis essential for a cost-effective nationwide roll out of the service.

GIS technology of mapping, data visualization and analysis software combined with data of postal, demographic and road has produce a powerful ADSL market analysis and network planning tool for BT (Fig. 10.1) The benefits were:

- BT has been able to pinpoint more than 400 high potential exchanges covering six million households and businesses for its initial ADSL roll out.
- The system has enabled a ranking of all exchanges offices for ADSL upgrade.
- System is helping BT plan the most cost-effective nationwide ADSL roll out.

GIS technology enabled BT to plan an extremely focused and highly targeted ADSL launch for London and surrounding areas like Cambridge, Coventry and Manchester, where demographic factors most closely correspond with the profile of ADSL subscribers and significant potential for high bandwidth services has been identified. Deployment of capability into the appropriate areas to maximize the early adoption of new high bandwidth services into the UK is a critical factor in the business planning for implementing ADSL on the national network. With such an enormous programme to be undertaken over a period of time, profiling and pinpointing ideal customers with GIS is helping BT achieve the most effective roll out of ADSL. This ADSL market analysis and network planning solution is the latest addition to BT's portfolio of GIS-based applications. The company is one of the largest users of GIS technology in the world with more than 1,000 software licences for a wide variety of key applications in network planning, operations and sales and marketing.

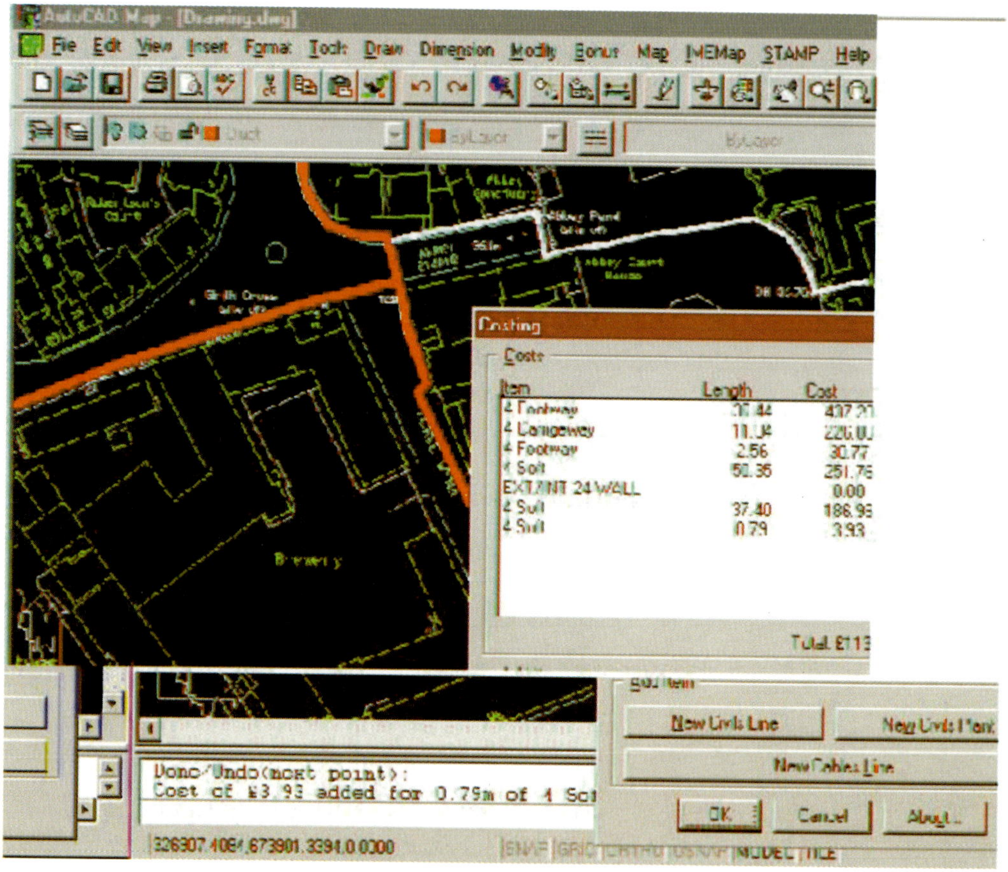

Fig. 10.1: Print Screen of BT's Web Site

Further, in order to improve customer service, BT retail equipped 17,000 field service engineers with a comprehensive MapInfo location-based solution. Aimed at reducing time spent locating customer premises and pinpointing faults, the solution will result in a predicted savings of $33.4 million over the next five years. When BT's field engineers relied on paper-based maps and diagrams, they lost valuable time searching for information needed to respond to customers' needs. Then, BT replaced field terminals with Panasonic laptops for the engineering workforce, and built a MapInfo solution enabling sharing of maps and diagrams.

The solution, built using MapInfo MapX and incorporating MapInfo, MapMarker and MapInfo. DriveTime TM provides simple yet tremendously beneficial access to fundamental information. MapInfo MapX, a mapping engine, allows for highly visual display of location-based data. MapMarker enables engineers to locate the customer on a detailed map. DriveTime delivers the quickest route to the customer or fault.

With the solution, engineers first locate a property and then progressively zoom into view road networks and local plant and network details; the location of underground ducts; and, finally, the individual cables within them. Engineers can see the proximity of the network to the customer, search by customer's address and postal code and plot the shortest route to get there, all saving further time and effort. "With this new technology, we knew that we could provide engineers with accurate information on where the customer was, as well as providing detailed maps of the ducts and cables supplying our business and residential customers," said BT's programme manager for customer service. The solution provides engineers with access to extensive plant diagram electronic records—held on 32 databases throughout the United Kingdom—along with local and national road maps for each engineer's territory. (www.gis.esri.com/library/userconf/telecom05/papers/enterprise-gis.pdf)

Assesses Customer Needs Quickly and Accurately

Nortel Networks specialist Network Systems and Solutions Group (NSSG) is not only helping the company win business throughout Asia Pacific, it is delivering real benefits to customers using GIS. NSSG's small 12-member team based in Australia, Singapore and China, uses GIS tools to support Nortel's new business bid responses throughout Asia and assist telecommunications carriers, alternate operators and service providers with strategic network planning, detailed network planning and implementation. GIS technology help Nortel's Asia Pacific operation understand market segmentation, assess demand for communication services and identify the most suitable end-to-end solution for any given customer distribution be it in a central business district, suburban or rural area. A powerful analysis tool that quickly and accurately supports businesses throughout Asia was needed which could take care of:

- Accuracy lacking with examinations of numbers just on a spreadsheet or static information.
- Turnaround speed needed to be upgraded to Internet speed.

GIS products delivered via GIS are used in demographic analysis, market segmentation, business planning, target marketing, competitive analysis, traffic engineering, network optimization and network roll out which lead to:

- Having mapping dimension for powerful and advanced analyses
- Enhanced customer service with thorough market analysis.

The NSSG selected GIS for its flexibility, portability and automation features. The planners in the group do consulting work throughout the Asian region, and they need the flexibility to use it on their desktop wherever they are on-site with the customer or overseas. With the team's input required for up to 100 bid responses and consultations a year, speed of turnaround is also critical. Before, each project might have lasted one to two months each, but now they have to be responded within weeks or even days. The clock is always ticking. Now everything that planners do is that being physically be there, they can remotely use GIS tools and data. Before these GIS tools were also available, but people were using things like overhead transparencies on their monitors to trace maps. NSSG eventually aims to make the information it has collected available throughout the company using GIS. The GIS application built specifically for mapping on the Web.

(http://resource.mapinfo.com/static/files/document/1074617546735/nortel_casestudy.pdf)

Delivery of Exceptional Network Performance and Customer Service

Orange is the largest and fastest growing digital mobile phone network in the United Kingdom with more than 2.5 million subscribers and more than 5,300 base stations worldwide. The company's strategy is to provide superior network quality and coverage thereby attracting and retaining high value customers. To this end, Orange has made significant investments in the United Kingdom and abroad in its state-of-the-art customer service centres and in the expansion and enhancement of its network. GIS technology is playing a key role in helping the company delivers on its commitment to delivering exceptional network performance and customer service. This needs

- Superior network quality and coverage, thereby attracting and retaining high value customers.
- A system that is business process driven, automates routine tasks and integrates seamlessly with other systems.

The company developed a coverage management system (CoMS), built on GIS technology, to provide accurate and consistent coverage information and coverage maps. It became the most widely used system within Orange's more than 1,200 users across customer service, network management, engineering, sales and marketing and at Orange stores and dealers used CoMS's but there were several problems. There was limited ability to customize data. The system did not integrate well with other systems. Maps were not entirely accurate and CoMS only worked in United Kingdom coordinates. Orange needed to replace the original CoMS with a system of task-driven applications modelled on its business processes that would integrate seamlessly with other systems:

- Former coverage management system offered limited ability to customize data.
- The system did not integrate well with other systems.
- Maps were not entirely accurate and worked only in United Kingdom coordinates.

GIS mapping software enables Orange's coverage management system to provide accurate displays of network coverage on detailed, interactive maps that also show Orange base transmitter sites, road networks and other key features. The benefits include most up-to-date coverage and network performance data can be delivered to customers immediately. Two primary business processes were identified as key to this system:

- Passing coverage queries from the call centre to engineering so that the company could plan for and provide better network coverage and quality; and,
- Passing network quality and operational data from engineering to the call centre so that customers could be provided with the most up-to-date network performance and coverage data.

The integration of GIS technology in CoMS and the integration of this system with the company's engineering systems enable Customer Service Representatives (CSRs) to answer queries immediately and consistently, making each customer contact productive and meaningful. For instance, when a customer calls to report a signal problem, the CSR can determine whether the caller has entered an area of variable coverage or whether the local transmitter is under repair and provide this information to the customer. Feedback of subscriber hits on coverage problems can then be sent to engineering for analysis and input to network planning. The system's accurate and current maps provide information about planned new sites; date of their activation and details of how these sites will improve and expand coverage. All of this information is regularly updated, enabling CSRs to satisfy the customer's requirements for the latest information, and available on a 24-hour basis. GIS is helping Orange to effectively manage its relationship with each existing and prospective customer. The Orange GIS team is planning to deliver CoMS functionality to subscribers via the internet using internet mapping GIS software. Intranet rollout, which will allow roaming information to be delivered to the Orange call centre via the company's intranet, will also be made possible. The company plans to build the CoMS ActiveX interface into customer information points Orange stores and dealers. (http://www.collageconsulting.co.uk/AI-GIS.htm).

Planning, Building and Maintenance of an Integrated Network Inventory System

Telkom SA Ltd. (Telkom), based in Pretoria, South Africa, has selected its Smallworld Network Inventory TM software. The Smallworld Network Inventory™ software will help Telkom with the planning, building and subsequent maintenance of a fully integrated network inventory (information) system. The new system was to replace numerous existing, disparate applications currently being used for network planning within Telkom's seven regional offices. The creation of a single, fully integrated network planning system will enable Telkom to streamline its business processes and provide immediate, nationwide access to up-to-date network information. As a result, planners will be able to more quickly and accurately assess future demand for new network and optimize upgrade plans, thus reducing the cost of new build and the maintenance of existing network. Around 1,200 users were to have access to Physical Network Inventory™ and Logical Network Inventory™ for the planning and design of both the physical network and the logical, service layer. Program Manager for Telkom commented, "We were looking for a solution that would best fit our business requirements. GE Network Solutions software provided the perfect combination of optimum functionality and value for money. The Smallworld

Network Inventory™ suite of software will enable us to integrate all our existing technologies into one, seamless system which will result in significant operational improvements and cost savings."

Director of The IST Group, GE Network Solutions' official distributor in South Africa remarked, "Basing its network planning and design on Smallworld Network Inventory™ software has distinct commercial benefits for Telkom SA. By improving the efficiency and cost effectiveness of its network planning, it is inevitably reducing the cost of network ownership—which is critical in today's competitive environment." When fully operational Telkom SA plans to integrate the Smallworld Network Inventory software system with other operations support systems such as service activation, service assurance, and fault management. President and CEO of GE Network Solutions concludes, "This is a significant order for GE Network Solutions and we look forward to working with Telkom SA and IST over the next few years to bring about a successful project."

(http://www.telkom.co.za/minisites/utilities/index.html)

Managing the Physical Infrastructure of its Fiber-optic Network

Atlanta (March 2002), GE Network Solutions, a leading supplier of strategic network solutions to the communications and utility industries, announced that Hughes Tele.com has chosen its Network Inventory software, working with GE Network Solutions world-class partner InfoTech Enterprises Ltd., to manage the physical infrastructure of its fiber-optic network. Based in Mumbai, India, Hughes Tele.com is one of India's major competitive voice, data and internet telecommunications providers, as well as the first provider of fiber-based broadband services. The company is planning to offer advanced, intelligent network services such as calling card, premium rate and freephone services. It currently offers Indian businesses a one-stop shop for local and long distance voice, data, Internet and other value-added service solutions.

Hughes Tele.com was looking for a solution that could capture and maintain all of its inside and outside plant equipment in one seamless system. Currently, that data is managed in a number of disparate systems around the company, making it difficult to keep accurate, up-to-date and manageable records. Working with technology partner and project contractor Lucent Technologies Hindustan, Hughes Tele.com selected Network Inventory software and has now begun to implement the system, which will facilitate the planning and design of its new network facilities and track the as-built information, enabling the company to more efficiently manage its fiber-optic network. "This system has been one of the key enablers in our business planning process and will continue to remain one of our focus areas in the future. Working with GE Network Solutions and its world-class partner, InfoTech, has ensured that we received the functionality we needed."

InfoTech Enterprises Ltd., as implementation partner of GE Network Solutions, primed the project from data creation, application development and interface, to other critical operational support systems such as CADTel Fault Management system, AllTel Customer Care billing system and SAP R/3. Chairman and Managing Director of InfoTech Enterprises, remarked, "InfoTech's ability to win several contracts within the Telecom industry, especially from prestigious companies such as Hughes, proves that GE Network Solutions provides an ideal product for this industry, and InfoTech is an ideal partner for the customization and implementation of services in India. We look forward to further implementation projects in the next six months."

"GE Network Solutions was the solution of choice for Hughes Tele.com because we were able to deliver a fully integrated, open system that meets the needs of the organization," said Vice President, GE Network Solutions Communications division. "We are delighted that Hughes Tele.com now considers the Network Inventory system a core decision-making module for network-related functions."

(http://www.directionsmag.com/press.releases/index.php)

Telecommunication Services

"Pronto" or "ready," the Italian salutation when answering the telephone, has particular significance in the battles currently being waged across Europe for mobile telephone market share. Telecom Italia Mobile SpA. (TIM) is the largest mobile telephone operator in Europe, serving more than 10 million Italian customers with its Total Access and Communication System (TACS) and Global System for Mobile Communications (GSM) network. Headquartered in Rome, TIM operates 20 regional offices that are responsible for such strategic tasks as deployment, operations, and maintenance including seven service centres devoted to customer relations. Its aggressive technological development and marketing services have allowed the company to add more than three million new users during the last 18 months alone.

One of the key operational systems devoted to customer care that serves this intricate wireless transmission web is TIM's Sistema Informativo Territoriale (SIT), a powerful GIS built on an integrated set of ESRI software products including Spatial Database Engine (SDE), ArcInfo, and MapObjects Professional. In addition, TIM has deployed ANGI, an SDE preprocessor developed by ESRI Italia. With ANGI, multiple transactions from different clients can be simultaneously served for access to the same data. SIT supports about 3,000 operator call stations throughout its entire territory, providing real-time answers to customer questions related to GSM and TACS network coverage. Service is provided 24 hours a day, seven days a week. Approximately 100,000 calls are answered each day, 10 per cent of which are coverage concerns with questions regarding specific areas of coverage, loss of signal, or interference. In addition, SIT provides the location of public authorities, such as police stations and the location of dealers selling TIM's range of equipment and services. A variety of tabular information is also available to the customer service operators including dealers' names, phone/fax numbers, credit card information, business hours, and so forth. Those staffing the call centres need instant answers from SIT because customer response time is measured in seconds.

TIM's Network Engineering Department has developed algorithms for its GIS to incorporate both natural and man-made topographical features that could adversely affect mobile telephone transmission. This allows the optimization of equipment installation so that coverage areas are neither over-nor under-serviced, which provides substantial cost savings to the company in a very competitive marketplace. To stimulate the growth of mobile telephone use, TIM has introduced a prepaid card that allows pay-as-you-go mobile phone service. With an in-credit card, a mobile telephone user may use his phone until his credit is expended. Then he can extend the use of the card by transferring funds at an Automated Teller Machine (ATM) to his prepaid card. The machines are located throughout the country, and with the help of its GIS,

TIM is providing ATM location maps to its customer base to facilitate their use. TIM is also developing a system that will use GIS to analyse call types based on customer habits such as location, time, and type of service: mobile to mobile, mobile to fixed, etc. This will help them provide better, more cost-effective service to their clients.

Looking to the future, research analysts predict rapid, sustained growth in Italy's mobile telecommunications market. Forecasts indicate that in a few years there will be nearly the same number of mobile and residential telephone customers in Italy. Presently, there are approximately 22 million residential customers, which constitute about 40 per cent of the Italian population. Analysis indicates that there will be about 20 million mobile telephone users two to three years from now. That figure could rise to about 30 million by the year 2005. Italians often have vacation homes that have fixed telephones, and it is thought that many of those phones will be replaced with mobile ones, which will provide the customer with greater flexibility. Equipment prices are coming down, and presently the cost of using a mobile telephone can be the equivalent of less than 10 cents per minute during off-peak hours and on the weekend. All of these factors will the cost of mobile telephone use and in turn stimulate growth in the industry.

TIM's network engineering department has developed algorithms for its GIS to incorporate both natural and man-made topographical features that could adversely affect mobile telephone transmission. This allows the optimization of equipment installation so that coverage areas are neither over nor under serviced, which provides substantial cost savings to the company in a very competitive marketplace. To stimulate the growth of mobile telephone use, TIM has introduced a prepaid card that allows pay-as-you-go mobile phone service. With the card, a mobile telephone user may use his or her phone until the credit is expended. Then the customer can extend the use of the card by transferring funds at an ATM to the card. The machines are located throughout the country and with the help of GIS, TIM is providing ATM location maps to its customer base to facilitate their use. TIM is also developing a system that will use GIS to analyse call types based on customer habits such as location, time, and type of service (mobile to mobile, mobile to fixed, etc.). This will help them provide better, more cost-effective service to their clients.

(http://www.telecomitalia.it)

AutoCAD Map–Based System Tracks Underground Utilities

Woolpert LLP ranks among the nation's top design and consulting firms. This multidisciplinary professional services firm provides a wide range of services, including architecture, engineering, and geographic information systems (GIS). AutoCAD Map® and AutoCAD® software are used extensively for engineering and GIS projects. Woolpert is committed to excellence through creative, diverse, and flexible solutions developed with the support of advanced technologies. One client, Little Rock Air Force Base, is installing new underground fiber-optic cables and hired Woolpert to develop a system for locating, mapping, and maintaining records on underground facilities. The tools were perfect for creating AutoCAD Map-based utility maps that gave Little Rock a clear picture of the underground infrastructure prior to digging. Such information helps avoid costly, and possibly dangerous, gas, utility, or water line breaks during excavation projects.

First, Woolpert used aerial photography and Global Positioning System (GPS) surveying methods to create a base map. Next, GPS surveyors collected underground line locations and, while in the field, mapped them directly into an AutoCAD Map-based GIS running on a pen-based PC. A custom AutoCAD Map-based application that links utility maps to a Microsoft FoxPro database was created next. "The AutoCAD Map-based system offers an easy user interface," says Woolpert CAD systems administrator. "New staff rotate into the AFB every few years, and training them on this system will be easy."

Ninety to 100 per cent of the visible or documented underground utility lines at the Air Force base are now identified, making it safe to dig. Not only that, the AFB has an efficient GIS and facilities management system for maintaining all above-ground and underground utilities and facilities. "We invested our time and resources into an Autodesk software solution because it enables us to provide our clients with quality solutions at a competitive price," concludes the administrator. "I am also confident that Autodesk will be there with the tools to meet our future needs."

(http://www.directionsmag.com/press.releases/index.php)

Ho Chi Minh City Post and Telecommunications

In May 2002, Intergraph Systems South East Asia Pte Ltd., an Intergraph customer since 1998, has signed an expansion contract for the supply, implementation, and deployment of a complete Telephone Facility Management System (TFMS). The Business Cooperation Contract (BCC) Board, a partnership between a subsidiary of France Telecom and Vietnam Posts and Telecommunications Corporation, officially endorsed the contract. Working in conjunction with its exclusive distributors in Vietnam, Intergraph South East Asia was to manage the project and be responsible for supplying and commissioning hardware and software, data conversion, training, technology transfer, and system maintenance.

Based on Intergraph's flagship FRAMME geofacilities management solution, the TFMS project will help France Telecom and HCMC P&T achieve a high return on their infrastructure investment. As part of the BCC agreement, France Telecom was to construct 540,000 new telephone access lines in Ho Chi Minh City. The project was to address key initiatives, such as demand forecasting, network design, investment document preparation, and construction. Further, TFMS was to have the capability to forecast demand by computing the requisite number of telephone lines for an area based on property information and penetration factor tables. It was suppose to provide a complete set of tools to assist in planning, designing, and managing the civil (underground ducts and above-ground structures), copper, and fiber networks as well. Additionally, TFMS was to speed up tender and construction activities through a module that facilitates the preparation of investment documents.

Managing director of Intergraph South East Asia, said, "This award represents an expansion of Intergraph's geofacilities management solution commissioned in 1998 and demonstrates HCMC P&T's confidence in our solutions and local expertise. We are delighted that both Hanoi and Ho Chi Minh City, the capital cities of Vietnam, use our solutions. We hope to extend this solution to all provincial cities in the country."

Ho Chi Minh City Post and Telecommunications is a subsidiary of Vietnam Post and Telecommunications (VNPT) Corporation. VNPT is the sole provider of postal and telecommunications services in Vietnam. VNPT divisions provide all local and international telephone services to sixty-one cities and provinces (web: www.vnpt.com.vn). Further, Intergraph Systems South East Asia is the distributor of Intergraph Corporation in the ASEAN region. Based in Singapore and with representative offices in Singapore, Malaysia, Indonesia, Vietnam, Myanmar and the Philippines, Intergraph SEA develops, markets and supports specialized applications based on Intergraph core products for GIS/Mapping, Utilities, Telecommunications, and Process industries, catering to the specific requirements of the South East Asian market.

(http://www.highbeam.com/doc/1G1:90682460/Intergraph+Corp).

Streamlining Work Processes

Again in May 2002, Intergraph Hong Kong Limited, a subsidiary of US-based Intergraph Corporation, announced today that Companhia de Telecomunicacoes de Macau S.A.R.L. (CTM) has purchased G/Comms, Intergraph's next-generation geospatial information management industryware for telecommunications companies. CTM maintains and operates one of the most modern and efficient telecommunications systems in the world with approximately 375,000 mobile and fixed-line customers. Additionally, CTM has become the biggest internet services provider in Macau and has provided broadband service since 1999. By upgrading its GIS, CTM strengthens its ongoing commitment to efficiently and economically provide customers with high quality, readily available local and international telecommunications services.

The company's goal is to achieve an integrated Geospatial Resource Management (GRM) environment, in which pre-integrated solutions allow a seamless workflow among applications that provision and maintain its service-delivery network. G/Comms and its family of products are based on standard, non-proprietary programming tools and represent commercial off-the-shelf (COTS) software. These characteristics set the stage for ongoing application extensions and easy maintenance and upgrades, thereby supporting CTM in maintaining the system over the long term. G/Comms will also easily interface to their existing Access Network Management System (ANIS) and Network Information Management System (NIMS). According to CTM's Head of Network Management, "In reviewing alternate solutions, we realised that Intergraph Hong Kong presented an interesting cost-benefit proposal, supported by a dedicated professional team. This solution will contribute to our goal of being one of the best telecom service providers in Asia."

"Following closely on our contract with CEM, the sole electricity provider on Macau, CTM is our second major win in this progressive and growing Asia Pacific region," said Intergraph's Regional Director, Greater China. "The CTM solution builds upon our ongoing commitment to provide the advantages of open and non-proprietary COTS software, so that our customers can respond to evolving market demands." Intergraph Hong Kong Limited will provide software technology, consulting and implementation services, and data migration for the project. The project is on a fast-track schedule, with completion an anticipated in approximately twelve months.

(www.gismonitor.com/news/newsletter/archive)

Broadband Internet Mapping of Downtown

Downtown San Diego is one of the most "wired" cities in the United States, with more than 70,000 miles of broadband backbone and end user fiber optic and copper cable. The ability to connect to such a comprehensive broadband network can be an important factor considered by businesses or potential new residents when relocating. Charged with providing information and outreach on the virtues of locating to downtown, the Bandwidth Bay project is providing information about the area's broadband network. A project partner, the City of San Diego, has developed an interactive, ArcIMS internet mapping application for viewing, querying, and printing key information about downtown and the broadband network. The result is a more complete picture of the broadband infrastructure and opportunities in downtown San Diego. (www.gis.esri.com/library/userconf/proco1/professional/papers/pap963/p963.htm).

Centralizing Corporate Assets

Southern California Edison's (SCE) telecommunications division Edison Carrier Solutions (ECS) is developing a pilot GIS/GPS asset management system. The implementation of this proposed system will aid the design, construction, and maintenance of ECS's fiber-optic network. Data collection will take place with GPS units, and data storage will be on a central server running Oracle8i. GIS users can access data through ArcView and CAD users access through AutoCAD. This project is suppose to cover GPS-based geographic data and storing the data in a centralized server which will benefit ECS's analysis and design of its network. (www.gis.esri.com/library/userconf/proco2/abstract/a0947.htm)

Cell Phone Network

T-Mobil Online with T-Map T-Mobil, the mobile telephone subsidiary of Deutsche Telekom AG and operator of one of the largest cell phone networks in Germany, has implemented T-Map, a call centere solution utilizing a mapping component based on ESRI's GIS technology. Using ArcIMS™ and ArcSDE, the web-based solution visualizes spatial information such as network coverage and service locations at a very quick response time, which significantly cuts operating costs. The GIS is charged with geodata by ArcInfo preprocessing about every two months in order to provide the most recent basemaps to the browsers.

The T-Mobil database is rich with six zoom levels available for German data, 88 foreign countries with three zoom levels, and attribute data for 350,000 cities and points of interest throughout the world. International information is available to locate foreign roaming partners and their tariffs. Response time in Germany is on average within two seconds to provide location and quality information on service dropouts at a spatial resolution of approximately twice the size of a football field for the entire country. T-Map includes all spatially related information content that is important to its customer care agents. In addition, it is embedded in a web-based information system that provides technical and product information on mobile phone devices. This includes photographs and specifications so the agent can easily explain basic maintenance, such as battery replacement and special functions, to the customer. The long-term goal of T-Mobil is to introduce spatial technology as a key service in the new generation of mobile handheld devices that offer multimedia-based information services in addition to telephony. Potential applications include e-service, e-commerce, and e-management.

(www.gis.esri.com/news/arcnews/summer03articles/t-mobile.html)

Sales Support

In response to customer requests to purchase equipment directly from the operator, Bouygues Telecom in France has committed to its own brand distribution channel. It recently opened 33 boutiques that feature GeoKiosk™, a geographic data display station that allows current and prospective subscribers to discover the Bouygues Telecom product range. GeoKiosk was developed by ESRI France for applications of public access to geographic information. Its simplicity and ease of use attracted Bouygues Telecom, which is using 100 GeoKiosk solution licences in a business application in 33 boutique stores in France. Bouygues Telecom uses GeoKiosk to enable the public to select location information from a database of 36,000 towns and cities and display available network coverage for the selected area. It is also possible to browse all of the boutique locations and their business hours, as well as obtain maps on how to get to them.

Bouygues Telecom's objective is to provide its customers with friendly surroundings where they can discover personal telephone surfaces and test the digital high resolution sound. At the heart of each boutique is a multimedia workshop console that combines pictures, sound and computer systems so that anyone can interactively discover the components of the Bouygues Telecom product offering. A recent study shows that 54 per cent of the time is spent with the GeoKiosk application. The Bouygues Telecom sales staff use the GeoKiosk 70 per cent of the time to show the coverage data to prospective customers.

The application uses ArcInfo and ArcView GIS software from ESRI to create the coverage data that forms the basis for the surface. The vector data that is stored on a protected server includes more than 160,000 polygons. The shapefiles are encrypted for security during distribution. Also used for display is raster image data from the well known Michelin maps. These images are scanned in at scales of 1:100,000 and 1:200,000 and are stored on the hard drive of each workstation. Each boutique offers three kiosks connected on a local area network running under Windows NT.

The distribution of geographical data is currently a central issue for government, business, and organisations. While the web makes it possible to reach a worldwide audience, there is much value in the GeoKiosk concept, which allows a local audience to gain access to data in a structured way. This is especially helpful for computer novices who do not know what a keyboard or a mouse is. GeoKiosk lets people browse geographic areas in a most intuitive way and learn to answer simple questions such as "Where is this located?" or "What is this map feature?"

(www.gis.esri.com/library/brochures/pdfs/telecomeuro.pdf)

Conclusion

We have noticed that GIS has become an essential component of any major initiative in telecommunication, government or industry. GIS technologies enable the internet service providers (ISP) to analyse the spatial relationship between a customer and the electric wave range from an antenna, in order to evaluate whether communication services can be provided to a customer. The client-server system is built under the combination of any industry standard software like ArcView 3.2, ArcSDE, and Visual Basic. ArcSDE stores the basemap and location data for customers and the antennas.

Internet provider business for personal clients has grown dramatically all over. Until a few years ago, the internet connection environment was largely based on dial-up connection via telephone lines. However, dial-up connection had two major problems, one being the high cost in that it required communications charges in addition to the provider fee, and the other the extremely slow speed of information communications. These problems were solved by internet constant connection environments such as CATV, FTTH, wireless LAN and DSL, leading to the dramatic spread of the internet to private homes. Under these circumstances, internet providers are exposed to severe competition with each other to acquire new clients. The key to acquiring new contracts lies in how swiftly and accurately information about the services of a provider can be transmitted to clients. Due to the competition, K-Opticom Corporation whose internet provider business is based on wireless LAN and FTTH, decided that the use of GIS was the best solution to such problems and in 2001 introduced the Area Checking System for checking the Internet service area on receipt of a new contract application.

Before accepting an application for subscription to the internet constant connection service, it is necessary to check whether the applicant is located within a serviceable area. This process is called the "area check". After the area check has been conducted, data on applicants has to be managed as the client information of each base station. If GIS were not used, the area check made on receipt of an application would be a long drawn process. If the above work were carried out for each applicant, a great deal of working time would be required to deal with many applicants and it would be impossible to respond to the dramatic increase in the number of clients demanding internet connection services. The importance of the new GIS-based "Area Checking System (ACS)" lies in enabling applications to be handled more efficiently. The requirements for this system are as follows:

(a) The base station serves as a map database and manages information on the clients allocated to it.
(b) The applicant's address is entered from the application form and is matched with the address point data containing the coordinates of each house to create point data on the map.
(c) When the applicant's location is established, the neighbouring base stations are automatically retrieved on the GIS and a list of candidate base stations is displayed. At this time, the distance from the applicant's location to each candidate base station is indicated. It is also possible to refer to the information on existing clients.
(d) Approximately 1000 applications per day can be processed.

Organizations get a return on investment by integrating GIS into their information systems. By combining up-to-date coverage data with demographics, trouble tickets, call history, and revenues, they are creating and maintaining successful marketing and customer care programmes. Successful marketing efforts require that one knows where the customers are, know what the revenues are by market segment, and can determine where the opportunities for growth are. GIS enables one to see this kind of information. This in turn allows to equip the sales force with the network data they need to sell additional services or analyse their prospects' installations. Use GIS to see how the resellers are doing, visualize market share, view market penetration, select priority areas, and more. Use of GIS can dramatically enhance the quality of your customer care by allowing call centre operators to dynamically view and query the operational status of the network. Customers with questions and comments will have an answer in seconds. The service representatives can also enter the customer information into trouble ticketing and job management applications, giving other departments access to important data.

11

GIS for Mobile Location Services

Location services combine GIS applications with easy-to-use mobile devices to provide information wherever and whenever it is needed. The GIS technology behind these services will empower an increasingly diverse range of applications, putting even more valuable information in the hands of mobile users. MapInfo announced a project agreement with the Siemens Information and Communication Mobile Group (IC Mobile) in which MapInfo miAwareTM GeoToolBox is being integrated by Siemens IC Mobile in its Location Enabling Server (LES). Siemens' LES is being implemented into Vodafone's mobile networks globally and will enable Vodafone to offer customers value-added mobile location-dependent services and applications. MapInfo miAware TM GeoToolBox provides mapping, routing and geocoding functionality. Further, Vodafone D2 selected MapInfo to provide the underlying technology platform to its location-based D2-NightGuide Premium WAP service. MapInfo's location-based services (LBS) technology platform, MapInfo miAware TM will provide the D2-NightGuide navigation and routing functions.

Continuous Navigable Routing

Recently, Navigation Technologies, a leading provider of digital maps for vehicle navigation, internet/wireless applications and business solutions announced availability of navigable digital map coverage of the Kingdom of Saudi Arabia. The completion of Saudi Arabian project represents a critical development in Navigation Technologies' digital mapping of the Middle East. The availability of navigable digital maps of Saudi Arabia enables continuous routing from Kuwait to Bahrain to the United Arab Emirate (UAE), all of which were released as NAVTECH maps in 2001. Oman will follow soon. (www.aldaleelexplorer.com, www.navteg.com and www.itsa.org)

The NAVTECH map of Saudi Arabia provides detailed coverage in Greater Riyadh and Jeddah, enabling in-vehicle navigation systems and internet/wireless applications to provide turn-by-turn directions in these areas. The launch includes more than 15,000 miles of digitally mapped roads and detailed physical attributes such as speed limits and road sign text. In addition to the standard points of interest included in NAVTECH maps (e.g. restaurants, hotels, schools, gas stations), the Saudi Arabia map was customized to include local areas of interest such as mosques. Al-Daleel Information Systems, a leading Jeddah based digital mapping and solution provider, was integral to Navigation Technologies' rapid entry into the market given their comprehensive understanding of the cities and terrain in Saudi Arabia (Fig. 11.1).

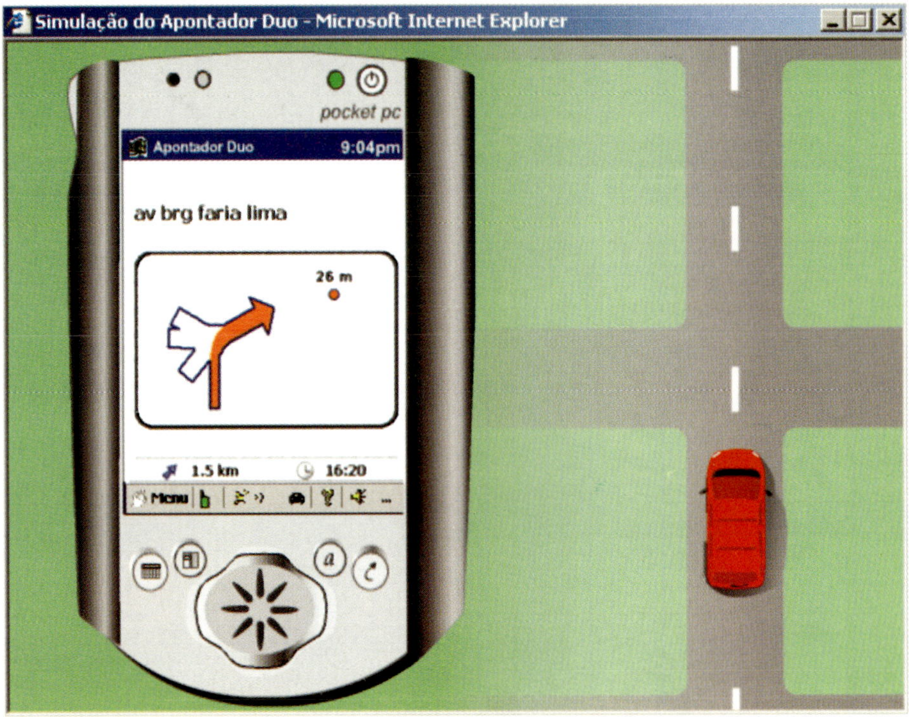

Fig. 11.3: Help in navigation

Revolutionary Safari Network

In July 2002 NavCom Technology, Inc. announced the introduction of its Safari Network SR-7100 System. The Safari Network product line utilizes NavCom's wireless communications expertise to provide mobile users in harsh environments with local wireless networks. The SR-7100 System offers a revolutionary data communications solution for systems integrators looking to expand their customers' network communications systems to the remote mobile environment. "The SR-7100 is a private, dedicated system that seamlessly integrates fixed and mobile assets into a single flexible network," said NavCom Product Manager. "The rugged design of the SR-7100 components creates an always on communications system for vehicles in harsh environments such as farm fields, construction sites, and mines."

The Safari Network SR-7100 System handles multiple sessions of varied data types simultaneously and has a 20 mile range that can be extended through its repeater capability. It builds on NavCom's communications systems expertise to provide the necessary features and capabilities to wirelessly meet most fleet management, data acquisition, machine control and local area networking application requirements. Systems integrators who are developing these applications benefit by providing greater savings and efficiencies to their customers. The innovative SR-7100 System, with its ability to simultaneously receive and transmit data in real-time, brings the sophistication of multi-service, multi-session communications networks to a new level by expanding it to the mobile wireless environment. (www.direction.mag.com)

11

GIS for Mobile Location Services

Location services combine GIS applications with easy-to-use mobile devices to provide information wherever and whenever it is needed. The GIS technology behind these services will empower an increasingly diverse range of applications, putting even more valuable information in the hands of mobile users. MapInfo announced a project agreement with the Siemens Information and Communication Mobile Group (IC Mobile) in which MapInfo miAwareTM GeoToolBox is being integrated by Siemens IC Mobile in its Location Enabling Server (LES). Siemens' LES is being implemented into Vodafone's mobile networks globally and will enable Vodafone to offer customers value-added mobile location-dependent services and applications. MapInfo miAware TM GeoToolBox provides mapping, routing and geocoding functionality. Further, Vodafone D2 selected MapInfo to provide the underlying technology platform to its location-based D2-NightGuide Premium WAP service. MapInfo's location-based services (LBS) technology platform, MapInfo miAware TM will provide the D2-NightGuide navigation and routing functions.

Continuous Navigable Routing

Recently, Navigation Technologies, a leading provider of digital maps for vehicle navigation, internet/wireless applications and business solutions announced availability of navigable digital map coverage of the Kingdom of Saudi Arabia. The completion of Saudi Arabian project represents a critical development in Navigation Technologies' digital mapping of the Middle East. The availability of navigable digital maps of Saudi Arabia enables continuous routing from Kuwait to Bahrain to the United Arab Emirate (UAE), all of which were released as NAVTECH maps in 2001. Oman will follow soon. (www.aldaleelexplorer.com, www.navteg.com and www.itsa.org)

The NAVTECH map of Saudi Arabia provides detailed coverage in Greater Riyadh and Jeddah, enabling in-vehicle navigation systems and internet/wireless applications to provide turn-by-turn directions in these areas. The launch includes more than 15,000 miles of digitally mapped roads and detailed physical attributes such as speed limits and road sign text. In addition to the standard points of interest included in NAVTECH maps (e.g. restaurants, hotels, schools, gas stations), the Saudi Arabia map was customized to include local areas of interest such as mosques. Al-Daleel Information Systems, a leading Jeddah based digital mapping and solution provider, was integral to Navigation Technologies' rapid entry into the market given their comprehensive understanding of the cities and terrain in Saudi Arabia (Fig. 11.1).

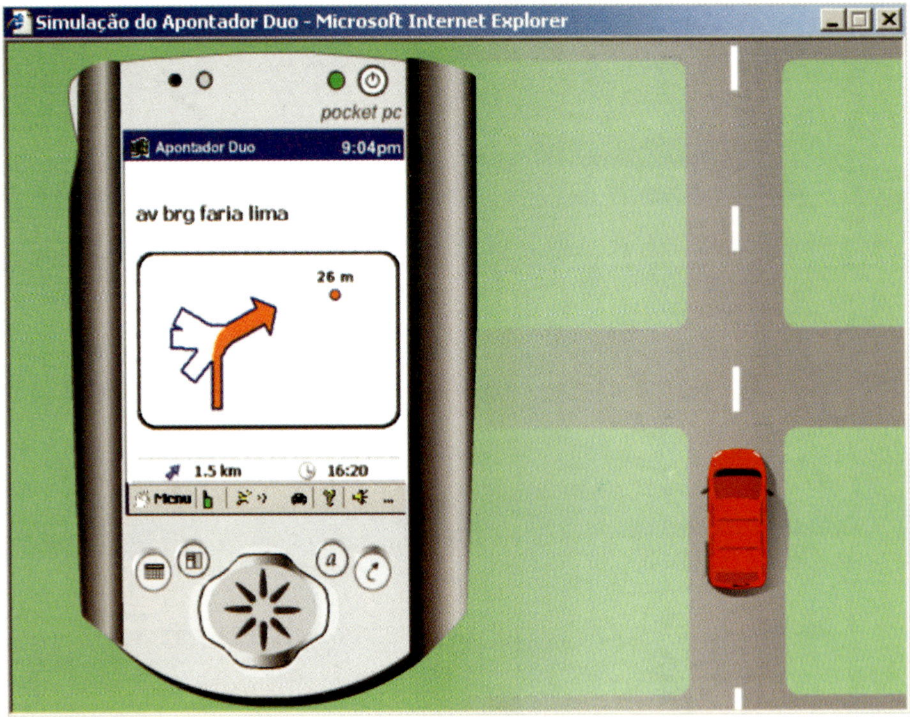

Fig. 11.3: Help in navigation

Revolutionary Safari Network

In July 2002 NavCom Technology, Inc. announced the introduction of its Safari Network SR-7100 System. The Safari Network product line utilizes NavCom's wireless communications expertise to provide mobile users in harsh environments with local wireless networks. The SR-7100 System offers a revolutionary data communications solution for systems integrators looking to expand their customers' network communications systems to the remote mobile environment. "The SR-7100 is a private, dedicated system that seamlessly integrates fixed and mobile assets into a single flexible network," said NavCom Product Manager. "The rugged design of the SR-7100 components creates an always on communications system for vehicles in harsh environments such as farm fields, construction sites, and mines."

The Safari Network SR-7100 System handles multiple sessions of varied data types simultaneously and has a 20 mile range that can be extended through its repeater capability. It builds on NavCom's communications systems expertise to provide the necessary features and capabilities to wirelessly meet most fleet management, data acquisition, machine control and local area networking application requirements. Systems integrators who are developing these applications benefit by providing greater savings and efficiencies to their customers. The innovative SR-7100 System, with its ability to simultaneously receive and transmit data in real-time, brings the sophistication of multi-service, multi-session communications networks to a new level by expanding it to the mobile wireless environment. (www.direction.mag.com)

Findings the Best Places to Eat, Shop and Play

Vindigo is the leading service provider for delivering location-based content and services to wireless and mobile devices. Vindigo's proprietary technology and network of content partners enables it to offer compelling and comprehensive location-based applications to carriers and consumers, via Palm Powered(TM) handheld computers, Pocket PC computers, and Internet-enabled mobile phones including BREW handsets. Vindigo allows people to explore, enjoy, and connect to the world around them.

Verizon Wireless, the nations largest wireless service provider, recently announced the availability of Vindigo, a personal navigation application for wireless phones that can be downloaded over the air onto select phones. Verizon Wireless, the first carrier in North America to offer downloadable applications to consumers nationwide, chose Vindigo because it offers its customers the ease of one-stop shopping. In more than 25 major metropolitan areas, Vindigo has comprehensive local information on where to eat, shop, and play. (Logos: http://www.newscom.com/cgi-bin/prnh/20000822/VERILOGO-a http://www.newscom.com/cgi-bin/prnh/20000822/VERILOGO-b)

Fig. 11.4: Navtime application

Fig. 11.5: MapQuest application

The service is complete with full-colour maps, ratings and reviews from brand-name content providers, and includes driving directions to any Vindigo listing (Fig. 11.4). The service is available to Verizon Wireless customers for $3.75 per month and will appear on their monthly phone bill. "Applications consumers can download over the air on their phone are available nationally and we believe these services will help us reach our goal to morph the wireless phone into a valuable resource for consumers who want up-to-the minute information to help them manage their life," said the vice president of wireless data and Internet services for Verizon Wireless. "It is important that we join with content providers and developers that can give our customers the same superior service we provide and that are experts in their field (Fig. 11.5). Vindigo provides a compelling consumer application with brand-name content and integrated mapping and directions, complete with hosting and maintenance."

Each productivity application published by Vindigo for Verizon Wireless runs on QUALCOMM's Binary Runtime Environment for Wireless(TM) (BREW(TM)), an advanced platform that allows users to download the applications over the air. Verizon Wireless subscribers using a BREW-enabled wireless phone that is compatible with the application will be able to select, purchase and download productivity applications published by Vindigo. "We founded Vindigo in anticipation of wireless phones as powerful and compelling as the new handsets from Verizon Wireless," said CEO of Vindigo. "Today, hundreds of thousands of consumers rely on Vindigo to find what they need, wherever they are, and our relationship with Verizon Wireless allows us to deliver a similar user experience to the largest audience of wireless phone users in the country."

Vindigo is initially available on the Verizon Wireless Z-800 with full colour display ($399.99 with a two-year service agreement). Availability on other handsets was to follow shortly. Customers can purchase the BREW-capable Z-800 at Verizon Wireless Communications Stores throughout the country and online at http://www.VerizonWireless.com. Additional BREW-enabled phones was to be available soon. Vindigo is currently available in the following U.S. cities: Atlanta, Houston, Phoenix, Austin, Las Vegas, Portland, OR, Baltimore, Los Angeles, San Diego, Boston, Greater LA (Burbank, San Francisco, Brooklyn, Glendale and Pasadena), San Jose, Chicago, Miami, Seattle, Dallas, New York, Twin Cities, Denver, New Orleans, Orlando, Detroit, Philadelphia, Washington, DC, and Fort Worth.

Verizon Wireless is the nation"s leading provider of wireless communications. The company has the largest nationwide wireless voice and data network and 30 million customers. Headquartered in Bedminster, NJ, Verizon Wireless is a joint venture of Verizon Communications (NYSE: VZ—News) and Vodafone (NYSE: VOD; LSE). Find more information on the web at http://www.verizonwireless.com.

SporTrak Map Handheld GPS

Thales Navigation is one of the world's leading developers and manufacturers of positioning, navigation, and guidance equipment with operations throughout the U.S. and Europe. Thales Navigation markets its Magellan brand GPS products in the consumer electronics, recreation, and automotive markets, and its Ashtech GPS and GPS+GLONASS professional solutions in the survey, GIS/Mapping, and OEM markets. Through its joint venture with Hertz, Thales Navigation has developed the Hertz NeverLost vehicle navigation system. Thales Navigation's key innovations include the first U.S. commercial hand-held GPS receiver for positioning and navigation, and the first handheld GPS with industry standard Secure Digital Memory Card capabilities.

Thales Navigation, a leading global provider of Magellan consumer and Ashtech professional GPS solutions, has announced the availability of the new Magellan SporTrak(TM) Map, a compact, lightweight GPS handheld with WAAS-enabled three-meter accuracy. The Magellan SporTrak Map delivers a unique set of advanced features at a new level of affordability, including a 2 MB map database, plus five times more memory for downloading additional navigation data than any comparable GPS receiver. Currently available with North American map data, SporTrak Map will also be available in Europe with a European base map. Whatever your adventure—wherever it leads you—SporTrak Map was to guide from doorway to the great outdoors and back again.

The rugged, SporTrak Map is a versatile navigation tool for activities and travel on wheels, on water and on foot. The North American database contains highways, major roads, parks, waterways and airports, as well as 6 MB of available memory for adding street detail, points of interest and 3D topographic maps from Magellan MapSend(TM) software. A new nine-key keypad provides fast, direct access to nine navigation screens, which are easily visible on the large, backlit display. Weighing just 6.1 ounces, the go-anywhere SporTrak Map is also impact-resistant, waterproof and even floats.

Fig. 11.6: Magellan Data Sets

SporTrak Map is one of three handhelds in the Magellan SporTrak series, which includes the recently introduced SporTrak Pro and the affordable SporTrak. All SporTrak products ship with a PC cable, user manual and quick start guide. Optional accessories include vehicle, bike and swivel mounting brackets, a canvas carrying case and a variety of power and PC data cables, including a cigarette lighter adapter. Details about the SporTrak Series and Magellan MapSend software products are available on their web site. (www.magellangps.com)

Thales Navigation, headquartered in Santa Clara, Calif., with European headquarters in Carquefou, France, is a subsidiary of Thales, a leading professional electronics company headquartered in Paris, France, with activities in aerospace, defense and information technology and services. Magellan and Ashtech are registered trademarks, and SporTrak, MapSend and DataSend are trademarks of Thales Navigation. (www.thalesnavigation.com)

LBS Middleware and Applications

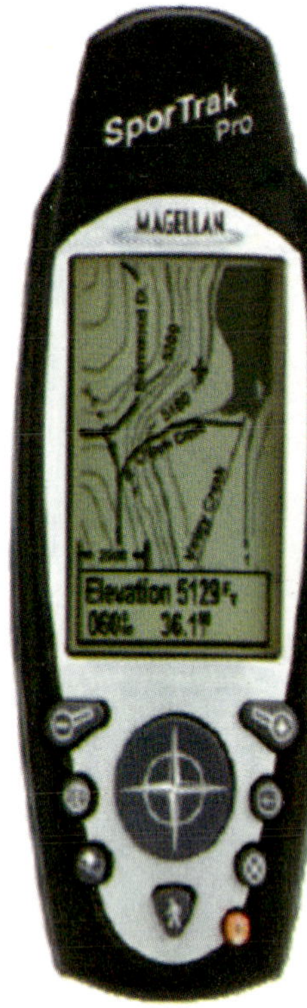

Fig. 11.7: Magellan SporTrak Pro Handheld GPS

In May 2002, LocatioNet Systems Ltd. announced that it has received two contracts from Austrian mobile operator ONE (Connect Austria Gesellschaft fuer Telekommunikation GmbH). This operator has contracted LocatioNet to supply it with its location-based services (LBS) middleware platform, along with its best-of-breed GIS toolbox and open set of APIs. The second contract is for the development and supply of two LBS applications designed to run on top of LocatioNet's middleware platform.

Under the terms of these contracts, LocatioNet has already supplied and integrated its LocatioNet Platform with ONE's cellular network, as well as with Nokia's mPosition GMLC for position determination. The LocatioNet Platform allows ONE to offer its subscribers an ever-growing array of LBS applications, starting with current 2G and 2.5G networks and handsets, and seamlessly migrating to 3G networks. "This is another step made by ONE to extend our portfolio of value added services by offering our subscribers new and advanced location-based services. LocatioNet has offered us the best technological solution for the middleware platform and applications and enables us to offer the richest services in this new area", says the CEO of ONE. Further, "We are extremely pleased that LocatioNet LBS solutions were chosen by ONE", says President and CEO of LocatioNet. "ONE is among the first mobile operators to realize that the future of LBS is not in providing a few vertical, tailor-made applications. LBS success is only possible through a one-time implementation of a comprehensive middleware platform that will manage all the LBS aspects required to provide rich service suites. We are confident that this approach will result in successful LBS-oriented business for ONE."

"LBS is one of the most exciting service areas in the mobile services industry", says segment manager for LBS at Sun Microsystems. "Everybody is looking for services that will attract end-users and generate revenue for the operators and other service providers. LBS make the difference and provide value to users who need information and access to services independently on where they are, and whatever device they have in hand at that moment. This is an essential starting point and we will certainly see an increase in user friendly services based on Java."

The LBS market is extremely complex. An effective LBS solution requires all service elements (end-user devices, content providers, GIS engines, position determination technologies, operator infrastructure, etc.) to be properly integrated. The LocatioNet solution takes the pains out of service implementation, by providing a comprehensive middleware solution, which integrates content, applications, positioning and any GIS engine to quickly fulfill user requests over any type of mobile handset. An open API and a user-friendly software development kit (SDK) make LocatioNet's middleware platform a 'must have' product not

MapObjects for cartographic data management, MICROSOFT Visual Basic and C++, Sun Java for user interaction and e-mail handling and Microsoft Access and SQL Server RDBMS for data base management. (for details contact: Sandro_Severoni@telespazio.it and: Fabio_D'Amico@telespazio.it)

Providing Instant Response

Telecom Italia Mobile S.pA. (TIM) is the largest mobile telephone operator in Europe, serving over 12 million customers with its Total Access and Communication System (TACS) and Global System for Mobile Communications (GSM) network. The service is provided 24 hours a day, seven days a week. Approximately 120,000 (140,000 peak) customer calls are handled each day, 10 per cent of which are coverage concerns, such as questions about specific areas covered by the radio signal and complaints of loss of signal and interferences. TIM is headquartered in Rome and presently operates seven regional offices that are responsible for such key tasks as deployment, maintenance, and operation of the cellular network, as well as relationships with their existing and potential customers.

The TIM's SIT project (Sistema Informativo Territoriale, or Geographic Information System) is aimed at providing better service to existing and prospective customers. SIT or GIS is serving about 2,000 operator stations. SIT provides real-time on-line answers to customer questions related to coverage of the GSM and TACS network for the whole Italian territory. The customer care operators in the call centres need instant response from the SIT because the customer is at the other end of the phone line and must get the correct answer in seconds. Quick response times mean lower costs to TIM, a higher number of customers served per unit of time, and higher customer satisfaction. SIT accepts several kinds of operators' input to select the area of interest including the name of the city, street name and number for the urban areas, and name of the nearest town/village for the rural areas.

SIT presents the output information in graphical form. The system extracts the map of the specified area from the database where the major Italian cities and the continuous coverage of the Italian territory is stored. The maps of the urban areas are 1×1 km in size, while the other maps are 30×30 km. The system has zoom in/zoom out and panning functions. All information generated by the operators' interaction with SIT is stored in the SIT's database and is earmarked with the operator's ID code. In addition, SIT provides

- Geographic location of covered areas derived from the theoretical computational models used by the Network Engineering Department and the real coverage determined by field measurements.
- The geographic polygonal coverage area of each cell including the cell ID and the associated information.
- The location of the dealers selling TIM's range of equipment and services including a variety of tabular information such as name of the dealer, phone/fax numbers, Internet address, credit card information, business days/hours, and so forth.
- Location of public authorities such as police stations.
- Location of ATM machines and post offices for purchase of TIM-Card credits.

(For details: p.paganucci@tim.it)

Innovative Services to its Customers

Bitë GSM Ltd. in Lithuania has used GIS since it began operations in 1995 and has been steadily increasing its use of the technology in recent years as the company continues to expand its wireless network. "GIS technology is critical to the success of the planning process in a cellular network operation," comments network-planning manager at Bitë GSM. "One of our primary uses of the technology is for GSM radio coverage propagation predictions. We also do demographic and customer problem analysis on top of our predictions."

Bitë GSM has implemented CellularExpert, a radio network planning application developed by HNIT–Baltic GeoInfoServisas, ESRI's GIS software distributor in Lithuania. It is built on ESRI's ArcView GIS, ArcView Spatial Analyst, and ArcView 3D Analyst software. This specialized application provides a number of critical capabilities including the management of information about network components, free-space propagation modeling, three-dimensional visualization of radio links, and advanced statistics and map query functions using coverage and interference calculation results.

Bitë GSM's technical department is using GIS for site selection and evaluation, microwave line planning, signal strength measurement mapping, network coverage prediction and analysis. Mapping signal strength helps maintain high standards in quality control by improving the predictive aspects of the implementation process.

Customer care locates coverage or quality-specific customer claims by placing reported signal problems on an intranet web page. Transmission engineers analyse the incident reports and produce terrain profiles for transmission planning. The Bitë GSM marketing department is responsible for maintaining up-to-date coverage maps of the network on an intranet page. The maps are prepared from materials originating in the technical department. "The open nature of ESRI's GIS software together with its powerful analytical and modeling abilities allows an easy integration with related applications," concludes the company representative. "This dynamic combination promotes the development of new applications that are not commercially available but provide needed capabilities to the telecommunications industry."

There is another case of the application of ESRI softwares in providing services by integrating different technologies. Waar is the name of the new mobile telephone service developed by ESRI Nederland B.V. for *m.info* subscribers of Dutch mobile telephone company KPN Mobile. By using a WAP telephone, users can view maps on their phone display of every given address in the Netherlands, with Europe soon to follow. The maps can be zoomed in or out and panned. The next step, which will soon be implemented, is a route map with accompanying text based on travel origin and destination. WAP is the worldwide standard for accessing the internet by mobile telephone. Based on ESRI's ArcSDE and ArcIMS, Waar provides quick reference map availability in a generic format. It is expected that with the continued technological developments and the integration of global positioning system, real-time navigation capabilities based on actual traffic information will be available by mobile telephone in the near future.

In a typical scenario, an operator at Neydarlinan receives an incoming call. A telephone switch electronically identifies the caller's automatic identification number and that number is matched in the Oracle database. The database provides the operator with the caller's address and postal code information, and this information is then georeferenced to locate the incident and to determine the appropriate agency to respond. In moments, the operator has all the necessary information to dispatch a police, fire, medical, or ambulance unit. "Our GIS has proven to be an extremely powerful solution for a mission-critical information system like the one at Neydarlinan," comments the project manager. "It enables the implementation of a fault-tolerant system with standard computing tools and opens up a new dimension in GIS."

(www.gis.esri.com/library/userconf/europroc97/10amfm/AM2/am2.htm)

Electricity Utility Service

FPC is the second largest investor-owned electric utility in Florida. Founded in 1899, the company serves approximately 1.4 million customers in central and northern Florida. Its service area encompasses approximately 20,000 square miles, including the cities of St. Petersburg and Clearwater, as well as the rapidly growing area of central Florida surrounding Orlando. FPC's effort to expand customer relationships is to provide competitively priced energy, excellent reliability, and outstanding customer service. The Delivery 2000 (D2K) project is a package of new technology and delivery systems that is streamlining operations and improving customer service:

- A mobile data plan was part of FPC's initial business strategy in planning the D2K project.
- An analysis was performed on who would be using the mobile application to determine the number of units and training requirements. A review summarized where the users would be located, how many would use the system, how often they would be in the field and for how long.
- All potential sources of data used and applications that would share the same data with the mobile system were identified—along with where the data would be stored, how frequently the various sources would be accessed, if remote access to the data would be required from the field, and whether the data would be required in real-time or some other form.
- As part of the overall D2K evaluation of technology, FPC identified software and hardware providers and the support resources. FPC selected off-the-shelf solutions, which were less expensive and easier to implement than custom solutions.

Mobile solutions have automated the distribution of crew workload at FPC, enabling dispatchers to focus on emergency situations and serve larger workforces and territories. By transmitting job status information to the office, dispatchers and supervisors can quickly monitor progress and reprioritize orders. Field crews stay informed through interactive messaging with dispatchers and access to critical corporate information systems. Mobile deployment at FPC has optimized crew efficiency Œ mobile units provide field crews with mobile links to the dispatchers and electrical facilities data using electronic map.

Rural Fire Service

On December 25, 2001, bush fires burned out of control and looked like they would engulf many densely populated areas and national parks on the outskirts of Sydney, Australia. Using MapInfo spatial technology, the New South Wales Rural Fire Service (NSW RFS) produced a sophisticated and extensive data analysis of the affected area, gaining a tactical and strategic view of the situation. MapInfo spatial technology includes MapInfo Professional and MapInfo. (www.mapinfo.com)

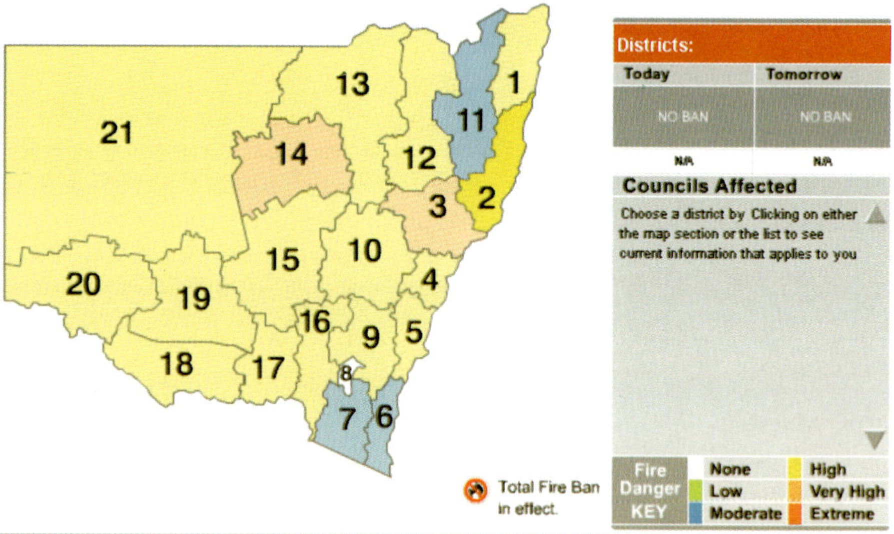

NSW Fire Areas							
1	Far North Coast	7	Monaro Alpine	12	Northern Slopes	17	Eastern Riverina
2	North Coast	8	ACT	13	North Western	18	Southern Riverina
3	Greater Hunter	9	Southern Ranges	14	Upper Ctl West Plains	19	Northern Riverina
4	Greater Sydney Region	10	Central Ranges	15	Lower Ctl West Plains	20	South Western
5	Illawarra/Shoalhaven	11	New England	16	Southern Slopes	21	Far Western
6	Far South Coast						

Fig. 11.8: A Total Fire Ban Symbol in the NSW Fire Areas

MapXtreme, MapInfo Easimaps, MapInfo Mapshop, MapInfo Proviewer and MapInfo Encounter TM —powers a data portal accessed by NSW RFS. This spatial data portal has allowed the NSW RFS to make better decisions regarding risks to people and the environment. In the case of the bush fires, spatial analysis was used to determine the types of houses to be built and the materials to be used in bushfire prone areas. Through 3D and hazard and risk analysis, data was transformed into pictures providing a better understanding and a powerful perspective of the situation. Via the portal, NSW RFS studies highlighted certain patterns and trends that otherwise had been impossible to see. NSW RFS is even able to investigate the likelihood of arson. Based on area demographics, NSW RFS maps historical bush fire areas with current data to determine whether the fire was possibly the result of arson or natural causes. The MapInfo-powered portal provides NSW RFS with a clearer view of information held on its corporate database, resulting in improved business decisions for future planning and management. And with GIS technology, NSW RFS better understands the state's environment, enabling it to apply for extra federal funding.

Conclusion

From the above case study it is apparent that the GIS business model is not restricted to software only. The integration of hardware and data with software is also an essential component for bringing out solutions for location based GIS applications. Recently we have seen considerable innovation in the use of geographic information coming from outside the established GIS industry as Google, Microsoft and Yahoo (GMY) and continue to develop their map-enabled search tool. The combination of these major IT players is bringing a whole different perspective to the industry as GMY embed geospatial tools into their applications (Parson, 2006). Embedded GIS facilitates the integration of geographic information with IT infrastructure, makes intelligence about location accessible to more business applications, and better information, and agile decision making (Wu, 2006). This trend is more apparent in providing location based services. Mobile mapping systems (MMS) have opened new avenues in data acquisition and updating and substituting conventional methods by providing fast, efficient, cost-effective and complete data (Anderson *et al.,* 2005).The technologies are already becoming digital and further integration will lead to better services. The limits and options are endless. Nevertheless, the world is becoming more and more organized, predictable and functional with the help of spatial technologies.

REFERENCES

Anderson, Teri, Ali Roshannejad and Naser El-Sheimy (2005), MMS on a roll, *Geospatial Today,* Vol. 4, issue 4, pp. 26-32.

Parson, E.D. (2006), Embedding geospatial technology into main stream, *GIS Development,* Vol. 10, issue 2, pp. 18-20.

Wu, Jian Kang (2006), Embedded GIS – An overview, *GIS Development,* Vol. 10, issue 2, pp. 28-30.

12

GIS for Electricity and Gas Utilities

In today's competitive world, a successful utility must take maximum advantage of each of its considerable resources, from people to equipment to information. Integrating geographic and other corporate data to make maximum use of this valuable asset has become absolutely vital. Utilities require a spatially enabled technology infrastructure on which their operational, customer, and decision support systems can leverage the information they have about their assets, employees, customers, and their relationships. In addition to solutions that are scalable, supporting work for individuals, work groups, and the entire enterprise, utilities need access to their information from the office, the field, or anywhere the user may be.

Energy has become pivotal for the modern society. In developed countries the per capita consumption of energy in the form of electricity and gas is highest. On the other hand, the developing countries are trying to conserve and find out new sources of energy, including the non-conventional ones. Nevertheless, the management and distribution of energy has important to both, the developed and developing countries. Here the role of GIS has been considered for better efficiency in the distribution, design, networking and safety.

Design, Construct, Operate and Maintain Distribution Network

North Carolina Natural Gas (NCNG), a subsidiary of Progress Energy (NYSE: PGN), provides natural gas and related services to more than 176,000 residential, commercial and industrial customers in south-central and eastern North Carolina. The company is headquartered in Raleigh, N.C. NCNG was to replace its paper mapping system with GIS Technology in order to design, construct, operate and maintain its gas distribution network. It was also to use the web based GIS application to distribute spatially enabled applications across the enterprise. NCNG chose this solution to empower a tool to make smart business decisions while exceeding customer expectations. GIS applications offer flexibility and scalability to grow with its business needs. Ultimately, NCNG will be more efficient and effective not only in engineering and operations, but also in other functional areas such as marketing and accounting. (GE network solutions; www.ncng.com)

Generating Portfolios, the Location of Transmission Bottlenecks, Congestion Costs

Midwest Independent Transmission System Operator, Inc. (MISO) has decided to use GIS for analysing critical power market issues such as the value of generating portfolios, the location of transmission bottlenecks, congestion costs, and the attributes of different market structures.

The GIS application incorporates the entire fundamental requirements for a computer simulation tool to accurately forecast prices, congestion, and system utilization under the Locational Marginal Pricing (LMP) market system that is contemplated in the Federal Energy Regulatory Commission's Standard Market Design Giga-NOPR. The software integrates detailed representations of a power system's load, generation, and transmission into a single simulation. This enables the user to calculate hourly production costs, while accounting for the impact of transmission constraints on the economic dispatch of generation. The software provides strategic data for making informed market decisions:

- Hour-by-hour nodal or bus spot prices of energy
- Hourly line flows and congestion costs
- Physical and financial transmission rights modelling
- Unit revenues based on megawatt output and bus spot prices
- Hourly emission quantities as well as removal and trading costs
- Identification of companies and generators responsible for power flows on lines.

(GE network solutions; www.utilitiesproject.com)

Standardize the Management of Electric Distribution Asset Lifecycle

Hydro-Québec was created in 1944 by the Québec government, its only shareholder, supplies electricity to over 3.5 million customers in Québec. With assets of nearly $60 billion, annual sales of $11.4 billion, and a work force of over 20,000, Hydro-Québec ranks among the leaders of the North American energy industry. Hydro-Québec's Distribution division is responsible for an electrical distribution system with 106,448 km of lines. Hydro-Québec has decided to use GIS technology for the standardization of the management of their entire electric distribution asset lifecycle. Hydro-Québec is undertaking this major GIS integration project, in conjunction with an on-going complete business process reengineering initiative in the distribution area. It was also to use GIS to automate planning, design, construction, maintenance, project management, engineering, and operations.

The proposed GIS solutions was to integrate with SAP's Plant Maintenance and Project Systems (PM/PS) application modules. The solution will be deployed to 3,500 end-users, including many web and mobile computing users, within a period of thirty months. The GIS application will be integrated with its own outage management system (CED) and Cyme's distribution engineering analysis package. Hydro-Québec relies on SAP PM/PS to perform maintenance management and distribution work management. This integration will allow Hydro-Québec to fully leverage its SAP investment. The system was to roll out in its entirety by 2004. Further, it was to use a phased approach for rolling out the solution, beginning with a pilot implementation in the summer of 2003. (GE network solutions; www.gesmallworld.com)

Standardization of Management of its Distribution Networks

Xcel Energy was formed in August 2000 from the merger of Denver-based New Century Energies and Minneapolis-based Northern States Power Co. Xcel Energy has selected the GIS technology to standardize the management of its distribution networks throughout the company.

Xcel Energy, the nation's fourth largest natural gas and electric utility company, was to implement the GIS technology across its entire merged service territory.

Minneapolis-based Xcel Energy Inc. (www.xcelenergy.com) operates in twelve western and midwestern states. The utility provides a comprehensive portfolio of energy-related products and services to 3.2 million electricity customers and 1.7 million natural gas customers through its regulated operating companies. Xcel Energy owns nearly 90,000 miles of electricity transmission and distribution lines, and more than 31,000 miles of natural gas pipelines (www.gesmallworld.com).

Outage and Service Order Management System

TXU (NYSE: TXU) is one of the largest investor-owned holding companies in the world, with assets of US$40 billion. TXU serves more than 11 million customers in the US, the UK, Continental Europe and Australia with electricity and natural gas services, energy marketing and trading, telecommunications and other energy-related services. TXU Networks, Australia has implemented a major new outage and service order management system using GIS technology. To comply with the newly deregulated electricity market requirements in the state of Victoria, TXU Networks replaced its existing outage management and trouble call system with a GIS technology complied outage and service order management system.

The GIS solutions has been integrated with the new TXU deregulated billing system and meter reading applications. Initial steps included migration of all legacy customer information into Oracle database. This is now the primary customer information system that retailers can interface to for all customer connection point, premise, and meter information. A web based GIS application has also been developed to manage service orders for new connection installations and for alterations to existing connection points, premises and meters. Outsourced service providers can now view their task workload for dispatching work to the field crews via the internet. The web based solution can also display all the trouble calls, planned and unplanned, that exist within the GIS application, and enables TXU Networks to distribute the GIS data and functionality across the entire TXU Networks business (www.gesmallworld.com).

Managing Electrical Networks

Light Serviços de Eletricidade is an electrical distribution company that provides power to 3.4 million residential, commercial and industrial customers in 30 cities in the Brazilian state of Rio de Janeiro. Light is owned by Electricite de France (EDF). Light Serviços de Eletricidade (Light), an electric utility in Brazil, has implemented GIS Technology to manage its electrical distribution networks. Light, which distributes power to the city and parts of the state of Rio de Janeiro and one of the largest privately owned electric utilities in Brazil, is putting together an integrated solution that includes SCADA, commercial, operations, and analysis systems. GIS technology is at the heart of this solution, providing a common geographic data repository for Light's distribution area urban mapping, electrical network and dispatch control center data.

Light has chosen GIS Solutions to achieve the highest level of integration and to provide the customers with the benefits of state-of-the-art distribution network management. Light is a progressive company and is using GIS solutions technology to become more competitive and to respond to the Brazilian regulatory agency requirements. GIS technology will feed Light's distribution management system, which will integrate with a variety of corporate systems including: SAP/R3, PRAO (Planification des Reseaux Assistée par Ordinateur), UE (Utilities Expert) and Pegasus. Light was to use GIS application for network planning and design as well as for distributing spatial and non-spatial information via the internet (GE network solutions; www.sainco.com; http://www.lightrio.com.br).

Outage Management System

American Electric Power Company is a multinational energy company based in Columbus, Ohio. In the US, it provides electricity to 4.9 million customers in eleven states, with more than 38,000 megawatts of generating capacity. American Electric Power Company (AEP) has successfully gone into production with GIS technology to manage their network operations and quickly restore electric service during outages, reducing their internal development costs. Further, this will also expect to improve reliability and customer satisfaction. A web enabled dispatch tool has also been implemented, providing AEP with the ability to assign and dispatch work crews from any dispatch centre throughout the organization (GE network solutions; www.gesmallworld.com).

Enhancing Business Practices

GIS technology has been a strong management tool for large utility companies. Until recently, many small utilities have stayed away from incorporating a GIS because they believed that at the small business level GIS is not cost effective. However, GIS coordinator for Truckee Donner Public Utility District says that cost is not really a factor in deciding whether or not to employ a GIS. With the declining cost of GIS software, off-the-shelf application and data model implementations, and application sharing, a utility of small or medium stature can get as much or more out of a GIS as a large utility without bearing high cost burdens. Truckee Donner Public Utility District (PUD) is a small utility in Lake Tahoe that uses GIS in a big way. With 13,500 electric customers, 15,000 water customers, and a new developing business in broadband telecommunications, the company has been able to implement many off-the-shelf GIS applications to enhance business practices that were often thought of as unattainable by small utilities.

In order to keep costs down, Truckee Donner performs its own data conversion, software configuration and data modelling. The PUD has developed custom applications for small fees and has used these applications to trade with other utilities, again enhancing what it already has for no additional cost. GPS is used as a data entry foundation upon which the land base was developed. The PUD, sanitary district, town, county, state, department of transportation and regional engineering firms now use the created base maps, allowing for easy integration of shared data among the different agencies. Also, by sharing *as-built* information, the PUD no longer has to maintain information not pertinent to the district's utilities; instead the data is maintained through the proper agency and shared through a data exchange agreement, cutting down on a lot of time and material that is often directed at land base maintenance currently seen at many larger utilities.

Truckee Donner is located in California's Sierras, not far from Lake Tahoe where only two seasons impact the utility company: the winter season and the construction season. These two seasons last approximately six months each. In the winter season, the average snowfall is 400 inches, but the community can receive as much as 900 inches annually, burying much of the district's underground facilities. During this time, the GIS coupled with a GPS is needed to locate the facilities and provide navigation to them. Most facilities management in the winter season is restricted to maintenance activity. Depending on weather patterns, the construction season usually starts about mid-May. During the winter, the ground is permafrost, so as soon as spring arrives, everyone begins digging at the same time. Business starts happening fast, and the utility company's GIS is actively coordinating efforts. It locates joint projects and trenching projects. The company matches its maintenance and facilities development schedule with other construction agencies such as CalTrans, California's road maintenance agency. If CalTrans is widening a highway, the utility company uses its GPS/GIS to maintain its facilities or plan new projects at the same time. By knowing where and when construction will occur, the district can build out its network with reduced cost and labour. As a seasonal town that must build out networks to serve, both seasonal residents and vacant lots, this reduction of cost is critical. ArcFM also has been effective in supporting the company's joint construction projects with tits conduit system management application. Recently, the district, in a joint project, constructed a large conduit system through the town. The Miner and Miner Conduit System Manager has been helpful in describing what conduits exist and where they exist and tracing out what facilities are available for lease or sale. It also became useful for relocation and maintenance of buried conduit and pipe.

Prior to implementing the company's GIS, the majority of the knowledge about facility location existed in the field crews' *walking maps*. These *walking maps* were prepared by the planners and foremen who had constructed a majority of the system. Without traditional maps or designs either in paper or electronic format, newer field crews had insufficient reference tools. The need for GIS was obvious. Another testament was during an outage in the dead of winter in which the electric superintendent, with only four hours of training, was able to come in at 2 a.m., turn on the GIS, locate the address of a phone call, and trace the network to locate the outage. It must have saved at least a half hour of driving the circuit. With GIS, it was possible to go right to the problem. This shows that GIS not only the benefits a small utility but also the software makes it even easier for training and application. A small utility that has a GIS can have a great impact on its own community. Starting the GIS for this sized venture takes just one or two quality people to get it going. (ArcGIS and ArcFM; www.esri.com)

Geofacilities Management System

AnShun Electric, located about 970 miles southwest of Beijing, serves 50,000 customers in the city of AnShun in Guizhou province of mainland China. This initial implementation by AnShun is the pilot system for an integrated Geospatial Resource Management system that will eventually be deployed by AnShun City Electric Company of Guizhou province. AnShun Electric Bureau has purchased GIS software for the geospatial information management system of its electric utilities. The utility's investment in advanced technology was to position it to provide efficient, economic, and high quality service in an increasingly competitive environment.

The pre-integrated GIS solutions help to streamline day-to-day activities that provision and maintain a service-delivery network. GIS software was seamlessly integrated with AnShun's new work management system, helping to optimize facility management, maintenance and inspection activities, and power service analysis and management.

The GIS application was considered to help to establish an enterprise facility database capable of using real-time data to monitor the electrical network. Additionally, it was also help to streamline workflows, strengthen network security, provide high quality electric supply and service, and fully leverage this advanced technology by sharing information across the organisation.

(www.gismonitor.com/news/newsletter/archive/091902.php)

Providing Integrated Asset Management System

PowerNet serves more than 63,000 customers with a distribution network covering more than 12,000 km. As a key component of PowerNet's corporate information infrastructure, GIS was to extend the company's integration capabilities and maximize data usage, improving the company's overall service standard and competitive market presence. The integration of open architecture facilitates of GIS technology with PowerNet's Asset Management System, provides a seamless workflow among applications that plan, construct, and maintain the company's supply network.

(http://www.intergraph.com.au/sgi/customers/g-electric-suite-products.pdf)

Geofacilities Management System

Niagara Falls Hydro, located in Ontario, Canada, distributes electricity to approximately 33,000 customers over an 80 square mile area. The GIS system was to integrate Niagara Falls Hydro's existing computer-based and manual systems that manage and maintain the company's electric distribution network. The initial implementation is expected to show noticeable benefits and time savings by reducing redundant processes and enabling more efficient capture, design, and analysis of network asset information. The GIS technology has been selected to streamline operational workflow and satisfy future information demands. The application will provide a data warehouse and common repository for all asset information required to support planning, estimating, work management and engineering network analysis. The project includes interfacing to Niagara Falls Hydro's Customer Information System (CIS) and work-estimating systems and was to allow office personnel, and eventually field personnel, to access and view current asset-related data.

(http://www.intergraph.com./customerstories/3324.asp)

Network Analysis

Nashville Electric Service (NESA) is one of the ten largest public electric utilities in the United States. Nashville Electric serves more than 300,000 customers in Middle Tennessee. The NESA service area encompasses 700 square miles, which includes all of Davidson County and portions of six surrounding counties. The functionality of the application includes tools that change automatically to streamline the process of adding and moving features. Over 200 commands and tools were created yet the end-users enjoy a simple and easy to use interface with which to perform the database maintenance. Extensive use of the new feature-linked annotation was also used while maintaining the existing information.

In 1990, Nashville Electric Service (NESA) decided to implement a GIS-system to secure integration between utility and customer data, which is called Giselle system. The Giselle system at present is basically a system for data conversion. But it is clear for everybody that great competitive benefits can be obtained if the converted data are integrated fast in other operational procedures. The GIS technology facilitates a major redesign of the business processes, based on integration between administrative, technical and geographical data. With Giselle, IT-department of NESA has got a strong basis for internal marketing of additional GIS systems, and NESA has already approved two new projects (www.gis.com/speciality/utilities/electric.hmtl).

The data model contains a geometric network, feature class subtypes, attribute domains, validation rules, connectivity rules, and feature linked annotation. The GIS application follows NESA's real world electric features including network connectivity with connectivity rules through the geometric network functionality (www.esri.com).

Geospatial Network Management

Alagasco, a subsidiary of Birmingham-based Energen Corporation, is responsible for distributing natural gas from the pipeline directly to their 475,000 customers. Alabama Gas Corporation (Alagasco) use GIS technology for a complete geospatial network management solution. The GIS application selected by Alagasco to manage and maintain their gas distribution network is designed to meet the requirements of the gas distribution industry.

The implementation of GIS facilities in the Alagasco system will take place in two phases. Phase one includes the deployment of GIS software, the design of the gas model, the importing of the Alagasco network data, and subsequent development and management services to enhance the implementation and provide for unique user requirements. System deployment throughout Alagasco's seven division offices was to take place during the second phase of the project. Additional phase two requirements include the integration of GIS application with Alagasco's gas analysis application. GIS application was to provide the powerful facility management capabilities. This will help to manage the inquiries and permits related to gas distribution networks, and support superior service delivery and customer satisfaction. As the largest distributor of natural gas in Alabama, Alagasco needed a system that could keep pace with the company's continued growth and functional requirements.

(www.directionsmag.com/pressreleases.php)

Geospatial Network Management Solution

Southside Electric Cooperative (SEC) currently provides electric services to some 47,000 consumers in south central Virginia. Incorporated in January 1937 and headquartered in Crewe, SEC is the largest geographic rural electric cooperative in Virginia, with more than 7,000 miles of distribution line. This phase includes deploying the supporting GIS application and a facilities management application. The data model was to be developed and the legacy mapping system of SEC was to be migrated to the GIS environment. The resulting solution includes integrating the GIS with an Outage Management System (OMS), Automated Meter Reading (AMR), Customer Information System (CIS), and an engineering analysis application.

For the better preparation for the planning, design, and implementation of energy choice requirements, SEC is aggressively working to improve cost efficiencies while continuing excellent record of service reliability. Realizing current technology can produce the fastest and greatest return, applying GIS solutions is identified as a key factor by each division of our operations.

(www.directionsmag.com/pressreleases.php)

Electric Circuit Mapping with Inventory Data

Faced with the need to use large scale electric facility data for smaller scale schematic circuit mapping, the City of Riverside Public Utilities Department employed many unique techniques to create these critical output products. Elegant approaches to scaling, placement and page management usher several Arcplot commands including Overpost to the "next level".

The initial objective of the City of Riverside GIS implementation was to replace all manually maintained maps with a common base and department specific data layers to support the historical as well as enhanced map products. One of these map products is the electric circuit map. Autocad and manual methods maintained these maps. They were at various page sizes, orientations, and map scales. The map scales ranged from 1" = 200' to 1" = 800' and the page sizes ranged from A to D. The primary reason for the different scales, page sizes, and orientation was to minimize total page counts while maintaining clarity. Several inset maps clarified congested areas. The electric department automated their facilities with a sophisticated database design using ArcInfo coverages and Oracle to support 1" = 100' inventory mapping. This scale, while very clear and easy to read, did not lend itself well to circuit mapping due to the number of pages required to cover each circuit. The engineering staff attempted various scales and page sizes before standardizing on one page size (D) and map scale (1" = 200'). Even at the standard scale, inset maps are required to detail congested areas.

By using a creative approach to processing the data in a specific way, by understanding the limitations of Overpost, and by tenaciously pursuing the course, the public utilities department was able to reuse GIS data developed by them for inventory mapping at one scale and use it at another scale for circuit mapping. Many people at the city who saw the early attempts doubted the efforts. It was nice to prove it could be done. (www.esri.com/news/arcuser/798/electrical.html)

Tools for a Changing Environment

The Convergent Group of Engelwood, Colorado, has developed a decision support tool for a gas utility marketing department based on ArcView GIS. Five customized ArcView GIS applications were developed using Avenue, Dialog Designer, and PresenTable from Applied GIS, Inc. to allow the client to market more effectively in both regulated and deregulated markets. The Client's Goals Convergent Group's client must market products and services for a gas utility that works in both regulated and unregulated market areas. This client wanted tools to analyse existing residential, commercial and industrial customers and to determine the products and services that were sold to those customers.

In order to better analyse existing and potential customers, the client also wanted to create segments based on common characteristics to aid in targeting specific products and services to customers. The client also wanted data and tools to look into other geographic areas outside regulated service areas in search of potential customers as well as to protect existing contracts from competitors. In particular, the client wanted to prevent customers from bypassing their company and tapping into interstate pipelines. Finally, the client wanted a system for tracking the success of its sales force, displaying results and reporting in tabular, chart, and map formats for corporate presentations, mailing lists for marketing pieces and leads lists for direct sales activities.

The five applications that are used for market planning integrate the spatial components of external data sources with the client's internal data. Using these applications, the client can learn more about existing and potential customers. Customer and potential customer base data are segmented so specific products and services can be targeted to customer groups based on spatial and aspatial criteria. Actual sales are tracked and compared with marketing goals. Also, customers whose location could allow them to connect directly to interstate pipelines are identified. The applications have customized query, analysis, thematic mapping and reporting functions. Customized maintenance utilities allow for additions to the data model. (www.esri.com)

Management of State Energy

In 1992, the state government in Western Australia commissioned the Energy Board of Review to investigate ways to improve the efficiency of the energy business in the state. The Board came back with a recommendation the State Energy Commission of Western Australia, which supplied gas to the area, be split into a number of competing businesses. Alinta Gas of Perth, Australia, was created as a result of state government moves to deregulate the market. Alinta, which manages one of the world's longest natural gas distribution pipelines, is using ArcFM to develop an integrated GIS solution to remain competitive while effectively serving its 350,000 residential, business and industrial customers. (www.esri.com/news/arcuser/798/utilities.html)

Integrated OMS, CAD and GIS Solution

The workflow to restore facilities crosses many system boundaries, thus requiring tight integration to reduce restoration time by critical minutes. The integrated OMS, CAD, and GIS solution provided by Intergraph meets the Florida Power Corporation or FPC corporate goal of reducing power interruption time through tightly integrated optimization of each step in the restoration workflow. A fast, smooth flow through and between each step is paramount in significantly reducing outage time. Functionality from each component is used as the work progresses through the restoration cycle. Although the attempt here is primarily on OMS, certain design and work management integration issues are also critical to keeping the facilities model current and accurate. Therefore, the procedure incorporates the design or work management workflow and identifies component engagements for reference.

The FPC project team viewed GIS as the foundation for the D2K project. The FPC asset database requires a central repository that is spatially oriented and capable of integrating with WMS and OMS as well as other third-party applications. For FPC, GIS is the keeper of the assets and is responsible for providing asset information to the user community. Further, FPC and Intergraph are confident that integrating GIS, WMS, and OMS applications in Delivery 2000 will provide an extremely effective solution for meeting FPC's future challenges. This confidence is based is based on the following factors:

Committed Team: The "Z" team approach enhances the probability of corporate-wide approval, acceptance, and ultimate project success.

EVA Financial Analysis: This application requires a project plan so that the timing of expense and capital costs and benefits can be modelled precisely. This financial statement based methodology ensures that investments are actually benefiting the project including close project monitoring and evaluation. The analysis methodology used will continue to be extremely detailed in developing all avenues of costs and benefits. In addition, it will provide a complete quarterly plan for deployment and maintenance over the full lifecycle of the solution.

Off-the-Shelf Solutions: Using off-the-shelf capabilities will reduce customization and affect the total solution cost favourably. In addition, properly developed pre-integration can greatly reduce future costs for maintaining interfaces. Geospatial technology plays a critical role in reducing restoration time.

Restoration Time: The integrated GIS/OMS/CAD solution will provide restoration benefits that will affect the entire corporation, including the SAIDI results. (www.intergraph.com)

Operations Management via the Web

PacifiCorp is one of the West's largest utilities, with a six state electricity service area consisting of both urban and rural regions. In the fall of 1997, PacifiCorp was in the midst of a major reorganization of distribution management personnel, consolidating the trouble-call input and management function from nearly hundred district offices into two call centres. As the call centres and dispatch managers became more removed from the local dispatch operations and field personnel, a system for distributing trouble call and related information to local workers became a critical missing component. The company had recently started using a new automated Distribution Management System, but that system was not designed to be installed on hundreds of desktops over low speed communications lines. Even so, the re-deployment of dispatch centres was underway and could not be delayed while technical support managers searched for a method to distribute trouble-call data to remote field locations.

PacifiCorp's Operations Support Management responded to the urgent requirement with an extremely aggressive project to deploy operations, customer and outage data to every desktop in the company. The chosen strategy was to use very lightweight and easy-to-deploy web based tools, delivering active maps and dynamic access to time-critical operations data to any company desktop that supported an internet browser and minimal connectivity to the corporate intranet. The team also decided that they wanted to make the map a key element of the interface to the data, allowing users of the system to navigate through the map, thematically display trouble call information on the map, and tightly couple the map to the tabular data. In the application it was realized that the new system could not follow the approach of many traditional

GIS systems. Although GIS data and thematic maps would be a key element of the new system, the system could not become a "GIS System" with the typical licensing, training, complex GIS functions, and development time of such systems. Instead, the system focus would remain on deploying tabular data and spatial data with a map based interface that would be as easy-to-use an interface as could be devised. Also, it was critical that the system deliver absolutely superior performance; slow response times typical of so many GIS systems could not be tolerated in this application.

To display maps in the web browser, the team chose the MapGuide product from Autodesk. MapGuide was selected because of the ease of integrating data from multiple sources, the good performance of the product, and the advantages of a vector based display. In Map Guide, maps are not just pictures or images, but consist of live objects that the user can select and can be acted on under the control of a Java programme running in the browser.

The Operations Visualization System is not a traditional GIS or AM/FM system, but instead uses an easy-to-use map interface to allow users to view and query operations and facilities data. Using a rapid development methodology and internet tools, a powerful spatially enabled system for viewing electric utility operational data and facilities was deployed to hundreds of users in just four months. The system has minimal requirements at the desktop, needing a web browser and modest network bandwidth, and has thus proven quite easy to deploy. Training and support requirements have also been modest. The map viewing technology used, Autodesk MapGuide, has proven powerful and robust, and yields good performance and flexibility.

The system has enabled PacifiCorp to address customer service needs faster and continue a high quality of service with fewer staff. OVS is well situated to become a key platform to deploy operations and facilities data to every employee in the company. (www.autodesk.com)

Reengineering the Field Technician

A few years back a well known cable company analysed how their field technicians were spending their time. The analysis revealed the following utilization statistics:

—57'XOn-job time—time spent installing and servicing customer accounts.
—Thirty-one per cent job-related overhead—time spent driving to customer sites, communicating with dispatch, and waiting for assignments.
—Seven per cent management-related overhead—time spent in team meetings, training, and reporting.
—Five per cent vehicle-related overhead—vehicle refuelling and maintenance time.

Further investigation revealed field staff had difficulty in getting through on the radio, and were sitting in their vehicles waiting for instructions to the next service call. With over 40 per cent of a technician's average day spent on activities other than installing and servicing customer accounts (i.e. overhead), any technology that can assist in reducing field technician overhead time would have a positive effect on operational productivity. The inability to contact a dispatcher because of a high volume of traffic is time wasters. For the mobile worker, increased productivity is all about saving time. Automated Vehicle Tracking (AVL) is one technology that can save time. For most organisations, an increase of one customer service call per day by each field technician would significantly improve the bottom line.

Integrating GPS positioning, AM/FM databases and wireless data communications, AVL offers opportunities for better management and utilization of a utility company's mobile workers. If the benefits are there, why has the industry been slow in adopting this technology? Perhaps, the main reason that AVL has been slow to take hold is the fact that vehicle tracking alone does not yield the business improvements required for a company to invest in the technology. AVL is a piece of the technology that needs to be integrated into work management processes and software applications. As the utility industry heads toward deregulation, many companies are investing in business process reengineering. Products that incorporate AVL into their solutions, addressing not only newer technologies such as AVL, but go hand-in-hand with recommended improvements from reengineered work flows, will be in great demand as competition forces utility companies to reduce their costs and improve service.

Today, utility companies are increasingly evaluating AVL technology in the form of pilot projects. As the industry leaders roll out and successfully integrate these AVL products into company operations, other utility companies will quickly follow suit. The day is fast approaching with AVL imbedded solutions projected to grow exponentially over the next few years.

The capabilities of AM/FM and GPS based vehicle tracking are rapidly being combined to offer new and exciting applications for improving the efficiency of the mobile workforce. New GPS/GIS systems, such as GeoLink@ PowerTrakTM from Baker GeoResearch, provide sophisticated AVL, fleet management and GIS field mapping capabilities in a single open systems environment. These combined systems make it easier to build and maintain AM/FM databases that can support robust vehicle tracking and map based computer aided dispatch applications. The creation of AM/FM databases that meet the needs of a wide range of emerging and increasingly popular GPS based vehicle tracking applications will be the real challenge facing our industry as we move into the 21st century. (www.directionsmag.com/pressreleases/index.php)

Rural Utility Service, United States Department of Agriculture

The Rural Utility Service (RUS) is a United States Department of Agriculture programme to assist co-operatively owned electric, telephone and water utilities with loans, design standards and quality assurance. Cooperatives are businesses owned by the members or users of the services, the business provides. Electric utilities of this type range in size from two hundred to over one hundred thousand members. Aiken Electric Cooperative (AEC) has 35,000 active meters, making it a mid-sized RUS utility.

The following advertisement may fit the search for the perfect AM/FM/GIS coordinator at a cooperative. Wanted: *One GIS coordinator skilled in logical and physical database design, computer aided design, GPS fundamentals and equipment, GIS projection and mapping standards, budget creation and management, project coordination, workstation and server hardware and software, and local and wide area networking. Four-year engineering graduate preferred for additional duties to be assigned.* There is very seldom one expert available in all the areas listed; therefore, one has to learn to adapt to the changing demands and make sound decisions using available information. The internet becomes an invaluable tool in collecting the required background material.

AEC is migrating from a paper system developed in 1959. This is a map or grid-based system that is employed at many cooperatives. The current map numbers are the basis for customer accounts and pole locations. The migration from a 40-year old unchanged system brings with it both cultural and political challenges. Management at AEC began researching the feasibility of replacing the existing system in 1996. The initial goals and budgets were derived with assistance from South Carolina's state owned generation, transmission, and distribution utility, Santee Cooper. It has more than 15 years of electronic mapping experience, and along with the cooperatively owned G&T, Central Electric, provides generation and transmission services to 15 distribution cooperatives in South Carolina. The resulting project plan and budget was submitted to AEC's board and the project received approval in 1997.

AEC has made a commitment to purchase corporate systems that have existing interfaces between them whenever possible. Systems that it plans on incorporating include Interactive Voice Response (IVR) system, Trouble Call Analysis (TCA) system, Automated Meter Reading (AMR), and Voice Over 1P(VOI) integration. Each of these systems may have integration benefits and must be reviewed closely for inclusion into the existing corporate projects. The cooperative felt that they must be aggressive in their view of the future, but not bind or blind ourselves with complexity.

With focus and active interest, AEC has come a long way in a short time towards incorporating a new mapping system. However, time frames are tricky entities, and no one at a cooperative is allowed to wear just one hat. If the coordinator has the background to manage the issues identified, it is highly likely that they will be drawn into additional technology based tasks. It is prudent to anticipate that some or all project phases may require "elasticity" at best, or "take more time" at worst, than initially forecast. This is not necessarily a negative finding; on the contrary the positive side effects of permitting more time and interaction for input from staff and facilitating understanding of the cooperative way of business is a true bonus perk. (www.directionsmag.com/pressreleases/index.php)

AM/FM/GIS Implementation

As part of their AM/FM/GIS implementation project, Allegheny Power directed the work of a GeoResearch Inc. to perform electric distribution system field inventory and mapping services along 24,500 circuit miles in a 12,000 square mile service area. Over 500,000 distribution utility poles needed to be mapped in a 13-month period, representing one of the largest and fastest GPS/GIS field inventory and data collection efforts ever undertaken by an electric utility. The project required the rapid staffing and effective management of over 40 field data collectors—a challenging task.

This project was to involve the management and planning issues involved in completing such a large-scale, fast track field data collection project on schedule. Project planning considerations, recruiting and training, management of a large field staff, field data post processing issues, and QA procedures was to be addressed. The Electric Distribution mapping was part of an initiative at Allegheny Power to build a geospatial inventory model of their electrical facilities for their AM/FM system. This initiative also includes paper-to-vector conversion and attribution of other division's data. Once the database is complete, Allegheny Power will have a connectivity model of their network for load analysis, design, and outage management processes. The following statistics of this project are provided. Allegheny Power's System includes:

—23,000 distribution circuit miles
—520,000 distribution poles
—13 month deadline
—95 per cent - 100'/40attribute accuracy
—3-5 meter positional accuracy

GeoResearch

—5 field offices
—35 data collectors at one time (58 total collected poles)
—5 supervisors
—8 GIS/data processors

The field methodology was simple enough, they would walk to each pole once, starting from the substations and follow each circuit outward until it ended its electrical flow. They would carry a GPS receiver integrated with a personal computer running a DOS software which allowed the collection of the required attributes, and tagged the attributes to the latitude and longitude of each pole.

This project was completed on-time, though not without considerable effort, overtime and costs. As with most large scale projects, many mistakes occurred, particularly in the first couple months, most of those mistakes resulted in lost productivity (poles per day), and we were constantly scrambling to adjust our staff numbers to keep pace with delivery schedules, and to replace staff lost to turnover. One of the largest mistakes from the beginning was overestimating the number of poles the collectors could map each day. The rugged terrain in per cent of the collection area prevented reaching the estimated productivity goals on most days. However, in the relatively flat, rural area of Mercersburg, PA, the collectors averaged well over the target on most days. In addition to the overestimating of daily productivity, it was difficult to overcome the productivity losses created from numerous equipment failures throughout the length of the project (especially at the beginning). Additionally, since the company was constantly adding new staff, they almost always had some number of new collectors "in training" which required personnel resources that would otherwise have been utilized as supervisors, auditors, GIS processors, and collectors.

The company feels that this job could have been performed much more efficiently and profitably in two years rather than in one year. It takes a few months after training to really develop a smooth and efficient working environment, and it takes time to absorb the significant start-up costs. The technology for this type of project is here, and it works very well. The key to success is good management starting from the estimating and proposal stage through recruiting, training, planning and production. (www.gisdevelopment.net/proceedings/gita/1993/mobile/mobile001pf.htm)

Area Resource Management (ARM) System

Illinois Power Company (IPC) uses an enterprise wide Area Resource Management (ARM) system for managing essential infrastructure data that helps to reduce operating costs and improve customer service. As part of this solution, widespread deployment of the AM/FM/GIS data asset and technology to the field provides a strategic business advantage in an increasingly competitive industry. IPC currently supports the following three mission-critical field applications using a single digital map publishing service product delivered on CD-ROM.

—Distribution/Records Maintenance
—Engineering Design
—Storm Damage Assessment and Restoration

Capable of running on virtually any Microsoft Windows based PC, the ids CD-ROM product provides a simple-to-use, fast, powerful and low cost tool that can be readily integrated with the technology and tools already used every day in Illinois. IPC is successfully exploiting the use of mobile technology to increase the quality as well as quantity of digital information available to field staff. In addition, corporate and field office staffs are finding new uses for the AM/FM information, increasing the value of this corporate data asset. Widespread distribution of the Illinois Power AM/FM data asset coupled with automated work management and the other key corporate database systems through ARM truly empower field personnel to leverage this valuable AM/FM data asset.

One of the lesson learnt is "simplicity is key to success". When implementing a solution to a workforce that is not computer literate even though it is technician based, keep it simple. Technology solutions must be powerful, easily supported and scalable within the organisation. The solution must take advantage of "best-of-breed" technology using out-of-the-box and of the shelf-technology. Technology enables and supports the process and workflow. IPC is devoted to using technology as a work flow enabler. The original core capabilities are still fundamental to the remote mapping solution and subsequent versions. As importantly, the remote mapping solution continues to be functional and relevant because the solution is user driven. By developing internal owners, user input can be leveraged to improve systems and applications. For example, every six months, a new WIMS release is developed in house and map viewing software updates can be provided with each data update on CD-ROM. (www.esri.com/library/brochures/pdf/gis-sois-for-power-generation.pdf)

Gas Distribution Maps and Facilities Records

New York Energy East Corporation (NYSEG) is a wholly owned subsidiary of Energy East Corporation and provides energy delivery and related services to 811,000 electric and 250,000 gas customers to more than one-third of New York State. It operates a gas storage facility and 7,000 miles of gas pipeline to serve its gas customers in 230 cities, towns and villages. Gas distribution maps and facilities records are the basis of the GBU's asset documentation and engineering, construction, operations and maintenance activities. The automation of this information and making it available throughout the organisation is seen as key to effective communication and to automating and improving critical engineering and operations workflows.

Gas franchise expansion projects beginning in 1987 provided positive change in gas mapping and records methodology but glaringly identified inefficiencies and out of date existing maps and records. The positive changes were the impetus needed to establish the need for change based on proven productivity increases based on advanced technologies. A major requirement was an efficient, expedient methodology for base map preparation supporting fast paced construction and documentation for final as built maps and records. After several successful expansion projects, a plan based on solid cost and schedule projections was prepared, presented and approved in late 1992 with more than 1,650 activity items identified in the project. The AM/FM/GIS project consisted of several primary goals:

—The development of mapping standards.
—The development of a detailed land base for geographic referencing and control of gas facilities.
—The automation of mains related information from as: builts and work orders.
—The automation of service related information from existing manually maintained service cards.
—The linkage of mains and services network with associated physical and operational information (attribution).
—The reconciliation of customer, street addressing and premise location.
—The production and electronic distribution of map books to the field.
—Continual incremental improvement of the facility data and access.
—Use of off the shelf software.

Upper management support throughout the project has been excellent due to solid communications, timely progress reports, acceptable incremental product deliverables and an under-budget project. Management of a large multi-year project is critical to success and can be very demanding but extremely rewarding if managed correctly. In-house conversion of a time consuming all encompassing project is not for everyone, but in our case we were willing to be responsible for the entire conversion process. Continuing their spirit of community involvement with many joint partnership projects, the company has associated with which became an added benefit and helped with critical aspects of the funding process of a very long and complicated project. The New York State Digital Orthophoto Quarter Quadrangle (DOQQ) programme was another joint project completed as a partner. By embracing change through technology the company was able to provide better customer service, respond quicker while making our work easier. Leveraging the investment is on-going and continues to pay dividends. (www.advantica.biz/PDF/electri-NYSEG-Solution-Brief.pdf)

Pipeline Management

El Paso Natural Gas Company (EPNG—a subsidiary of the El Paso Energy Corporation), together with its sister company, El Paso Field Services, owns and operates over seventeen thousand miles of pipeline throughout Oklahoma, Texas, Arizona, Louisiana and New Mexico.

Information about these pipelines is maintained by the drafting group of Rights, Records and Land Services (RRLS) in a series of paper alignment sheets that are copied and distributed on an as-needed basis. With increasing market sector competition and the need for immediate data access, EPNG soon realized that a significant change in their graphical data management techniques was required. To this end, EPNG solicited software vendors for turnkey solutions. After evaluating several turnkey proposals, and recognizing that the company had just invested significant capital in both, hardware and software solutions, the project team decided to investigate the possibility of taking advantage of these investments and still address the need for improved data management.

With this in mind, EPNG sought out to evaluate the existing software and hardware, the services of MOORE Resource Systems (MRS) to base at EPNG and determine if these solutions could be used to improve the overall management of the pipeline data set. EPNG, together with MRS, embarked on a pilot project. This project set out to demonstrate the viability of utilizing the existing client-server and CAD (computer-aided drafting) infrastructure to maintain and distribute the vast data stores for the pipeline and ensure its long term availability by moving it from a paper environment to a digital format.

Niagara Mohawk Power Corporation

Niagara Mohawk Power Corporation (NMPC) is a public owned electric and gas utility with a service territory of 21,000 square miles reaching throughout upstate New York. It serves 1.5 million electric and 500,000 gas customers. Its main office and computer centre is located in Syracuse, New York. The geographic area spans 37 counties including 670 cities, towns and villages. A sampling of the electric facilities profile shows: 131,000 circuit miles, 2,200 feeders, 1.2 million poles and 345,000 transformers.

NMPC in 1999 was to complete a three-year effort to convert data from various disjointed systems into one seamless GIS database. As data for a region is converted it is made available in production where GIS users have access to a suite of applications such as work order design, estimating, records management, etc. An extensive QAQC plan was developed to ensure network connectivity throughout the electric model down to individual customer locations. This intensive effort continues realizing this connectivity is the foundation for the Outage Management System (OMS).

NMPC, like most utilities, is under transformation with re-engineering efforts addressing the way the business can respond to the competitive market place. Several of the outcomes from these efforts can have a significant impact on the architecture decisions to support the OMS. For instance, when performance feasibility studies were done at NMPC in early 1997, there were four Regional Control Centers (RCC) that managed outages.

—Central area RCC located in Syracuse.
—Western area RCC located in Buffalo.
—Eastern area RCC located in Albany.
—Northern area RCC located in Watertown.

The plan then was to have a distributed environment with each centre being capable of stand-alone operation. Since that time the Northern RCC has been closed and the management of its geographic area consolidated with the Central RCC. Now there is a strong indication that over the next two years all centres will be consolidated into the centralized location. However, a requirement surfaced that suggests ten to twelve district offices may require access to the system during certain storms. Additionally, NMPC is planning on implementing a new Customer Information System (CIS) in 1999 which is the entry point for customer trouble calls, so the company needs to plan the handling of calls from the old and new CIS systems. Hence, the company noticed the outage management business processes demanding the need for an adaptive outage management technical architecture that can support the changing requirements of the business, yet integrate with the enterprise wide technical architecture. (www.esri.com/news/arcnews/fall02articles/elpaso-corp.html)

GPU Energy

In 1997, GPU Energy, Reading, PA 19640 USA made a decision to change the way the company operates. The proposed changes would affect virtually everyone in the organisation. This task seemed quite daunting in most respects, and GPU needed everyone's help to make it happen. It was determined early that the technology could provide the necessary tools to make this transition successful. While evaluating systems during the initial stage, GPU determined that SAP@ R/3TM could provide most of the necessary components for the new organisation. However, this type of system was missing a key component that could help GPU manage assets effectively while streamlining the record keeping process. A GIS could provide this missing component while giving users the ability to link work activities, estimates, and design data. Many of the new work processes being developed required this type of information, so it became very apparent that integrating GIS and SAP R/3 was one of the keys to making this a successful transformation.

GPU Energy looked to technology to help accomplish the goal of its enterprise wide change programme. The key ingredient for this change was the people of GPU Energy, equipped with the proper tools to get the job done. The integration of SAP R/3 with Smallworld was proven early on to deliver bottom line benefits while enabling the technology to have a combined capability not offered by each system individually. It was imperative that business workflows be provided with adequate and timely data to complete work orders. For this to be successful, two main integration pieces had to come together. First, virtual database integration allows the GIS and SAP R/3 data models to be treated logically as a single model, enabling sophisticated planning features for SAP R/3 data. The second piece, business process integration, offers complex business processes freely spread over GIS and SAP R/3. In many situations it makes sense to extract data in GIS and pass them as input data to SAP R/3 (and *vice versa*). GIS features such as network tracing can be utilized as special pre-processing tools for SAP R/3, allowing both systems to concentrate on their strengths, and eliminating the system's other weaknesses by cooperation at the business process level.

(www.gisdevelopment.net/proceedings/gita/1999/enterplan/cnterp001pf.htm)

West Kootenay Power

West Kootenay Power is a small vertically integrated electrical utility servicing company having about 120,000 customers in the southern interior of British Columbia. It has four small

(50 MW) hydroelectric generating plants, 950 miles of transmission lines and 5,700 miles of distribution lines (both overhead and underground). Electric utility plant is geographic by nature. Unlike other industries that have all the facilities in a small area, utilities have plant that is spread out over hundreds or thousands of square miles. Thus, the logical means of tracking facilities is through maps. A database linked to a mapping system allows information about the assets to be represented graphically. Inserting information through the mapping system allows for improved data management.

West Kootenay Power must control the vegetation under and near its distribution and transmission facilities to ensure that electricity is safely distributed to its customers without interruption. Tall growing vegetation near power lines is a major cause of electrical outages. Tree line contact can also create a public safety hazard. The company required a means of recording and tracking tree trimming and removal activities so that the encroaching trees can be plotted on maps for brushing crews. Also, by utilizing information on vegetation growth rates and vegetation management costs directly from the database, West Kootenay Power staff was able to create annual vegetation management plans for various budgetary scenarios. The business model presented here provides an overview of the business processes that will support the following:

—provide historical information for vegetation planning purposes
—generate annual vegetation management plans
—execute tree trimming and removal activities
—monitor progress and generate progress payment supporting documentation
—evaluate effectiveness of work and contractors

Although, the processes can be grouped into planning, execution and close phases. In this case, much of the data retrieval is in the form of database reports rather than maps, especially in the planning stage. Maps are only required in the execution stage to provide a visual display of problem trees so that contractors can find the sites. The functional requirement is indicated below.

Source Data: Provide a database to capture and store information related to hazardous trees, slashing and pruning. The system will support extraction of sub-databases, by feeder, which will be provided to contractors for entering vegetation data.

Specific Attributes: Hazard tree rating information will be stored in the database to permit West Kootenay Power staff to perform vegetation management scenario analysis by hazard rating. Work conducted was to be indexed by year to permit historical analysis. The primary key in the database will be a span *id* which will remain consistent and will provide the master link to the AM/FM mapping data. Autocad Map based functions will be utilized to update feeder and management area fields in the database. The existing AM/FM plotting functions will be utilized, to the extent possible, to provide the graphical interface to the vegetation management database.

Displays/Reports: The system will generate maps identifying the span *id* and the work specified to be conducted using symbology approved in advance by West Kootenay Power. The existing West Kootenay Power vegetation management sheet surround, modified with final symbology will be employed. The system was to generate the reports required by the contract documents to be provided to the contractors. The format of this report was to be defined in advance as part of the detailed specifications.

Computer Hardware and Software: The system was to be included an Autocad Map utility which will permit West Kootenay Power staff to add new poles and spans and automatically update and maintain a master span *id* table. The system will utilize a Microsoft Access interface, forms and wizards which will navigate non-technical users through the process of entering and maintaining database information. The system will automatically extract a replicated database which will be supplied to the contractors for updating as part of their work programme.

Maintenance and Support: General maintenance of the computer hardware was to be provided by the IS department. Further, support for the AM/FM system was to be provided by the Technical Services Department. A more complete and extensive list of specific features was used by Kanotech to develop the vegetation management application.

A well developed mapping system is an invaluable tool for supporting operations in an electric utility. Given the geographic nature of utility plant, specific work to be performed on the electrical system can be planned, monitored and reviewed using map and database information provided by the mapping system. The above are only two applications where the mapping system is utilized to support operations. They were presented to give an understanding of how an application can be developed, starting with the business process and working through to the final system features. In addition to the above, other applications are either developed or under development at West Kootenay Power, including: streetlight inventory, shared pole (with telephone and cable) monitoring and control, property tax analysis and pole test management.

(www.nwcouncil.org/fw/subbasinplanning/kootenai/plan/Assessment-07/Reference.pdf)

Cemig, Minas Gerais

Cemig is the main power company in the state of Minas Gerais in Brazil and its purpose is the generation, transmission and distribution of power within its concession area. Its main focus is electric power with a basically hydraulic generation matrix. The company has about 4.9 million customers, distributed in an area of 586,000 square km—about the same size as France—286,000 km of distribution network, and 365,000 transformers installed. About 10,000 km of distribution networks are built every year to supply new customers, growing at an average annual rate of 7 per cent. Their customers are spread in agglomerations ranging from a big one, with 1.6 million customers—the metropolitan region of Belo Horizonte, to medium one—seven cities over 150,000 customers, and many small centres about and over 3,000 customers, performing a total of 793 cities.

To manage this asset, the company has been investing since 1970s in computer systems. The first successful one was a mainframe based system to control the low tension secondary network, from the transformer to the customer. This system has been very necessary because of the secondary network topology used, which sometimes encloses many street blocks and hundreds of customers. The system was dubbed CONSEC, which stands for Secondary Control, and was put into operation in 1982, using since the beginning of the UTM geographic coordinate system.

Further, Gemini is an integrated project for mapping and management of the electricity distribution system. The main goals are the production and update of mappings (Cartor Module) and the storage, update and use of the electrical network data to support the planning, design, maintenance and operation activities (Prolux Module). The main emphasis is on the operation of the distribution system. The system emphasizes geographical functions and modeling of the distribution networks, in the form of the AM/FM, running under Windows NT, on a client-server architecture. It is a low cost platform that allows a perfect matching between resources and the size of the database, being easily scalable: the same software is installed at small computers serving small cities and NT clusters for reliable operation at larger centres. The system has been completely developed by using object oriented methodologies, and supports the main databases in the market, making it easily adaptable to cities of several magnitudes. They are now using Oracle database for the major cities and Interbase for all others, due to price and support issues.

The Gemini Project has two main modules named Cator and Prolux. The Cartor Module corresponds to obtaining and updating the cartographic bases in digital media, and is a customization of the existing CAD software. The Prolux Module, which corresponds to the management of the distribution network, has been entirely developed by the company's team to better meet the electrical distribution needs.

Since its first release, the Gemini Project has achieved great reliability and trust, eliminating all conventional paper and wall maps used to support the operation centres' main tasks. The quality of the operation centres' work was greatly improved, reducing the average outage service time, because of the easy analysis of outage problems, and better specification of long manoeuvers involving many switches and equipment. Users can get all needed information at the computer. One important fact is the improved environment at such centres, which are now quieter and less stressing, leading to better analysed solutions during emergency outages.

This project release has introduced great changes to the way operations are processed are conducted at the company, making the whole scenario more reliable and clearer. It has definitely changed the profile of Cemig Operation Centers and their associated activities (mvaa/99).

(www.prudente.unesp.br/rbc/-pdf-54-2002/54-06.pdf)

Savannah Electric

Savannah Electric (SE) had a need and a desire to implement a new outage management system in a short period of time and to minimize the costs. Their previous outage management system was tabular and only analysed meters, transformers and the immediate upstream device of the transformer. They wanted the new system to be graphical and capable of understanding and analysing the entire distribution electrical model. Throughout the project, SE made decisions to ensure success. These decisions enabled SE to complete the project from start to finish in less than twelve months as opposed to many other OMS implementations that have taken three to five years. The company's previous outage management system, written in 1987 and modified in 1993, was not millennium compliant. In addition, it was implemented on a mainframe that was also not millennium compliant. SE wanted to retire this mainframe in 1999, rather than make it millennium compliant. Thus, they had significant motivation to achieve the goal.

SE's first GIS system was implemented in 1984. In 1991, Southern Company started developing the Southern Electric Geographic Information System (SEGIS). SE became part of Southern Company shortly afterwards. They implemented the new system in 1995. The goal of the SEGIS project was to procure and implement a full-featured GIS system supporting the business and engineering needs of all five of the wholly owned electric utilities that are subsidiaries of Southern Company. Specifically, the design of SEGIS included these goals:

- It must contain sufficient data and be organized to support application development.
- It must meet reasonable performance expectations related to specific application requirements.
- It must be easy to use and the design should minimize complexity of operation.
- Its data redundancy should be minimized.
- Its design must be flexible to allow currently designed applications to change and it must be extended to facilitate new applications.
- It must support critical business requirements unique to the nature of electrical utilities business functions.

Although SEGIS has grown from the initial procurement, its roots are firmly planted in solid ground. The SEGIS design evolved from a pilot project led by two groups. Environmental Systems Research Institute (ESRI) was primarily responsible for the land base design. Miner and Miner Consulting Engineers (Miner and Miner) was primarily responsible for the electrical model. Southern Company Services enhanced the pilot applications significantly before they were implemented into production. The SEGIS implementation is based on ESRI ArcInfo, Oracle RDBMS, and is installed on a Sun server using Hummingbird to emulate a Sun workstation on standard PCs (with large 21" monitors).

Nearly a decade later, much of the initial SEGIS design and implementation is alive and well. Two of the five Southern Company's utilities (Savannah Electric and Mississippi Power Company) have SEGIS fully implemented and in production; another, Alabama Power, is partially in production; and a fourth, Gulf Power, is currently evaluating and contemplating a SEGIS implementation. The SEGIS design and implementation is ideal for supporting projects like the SE's OMS.

Minor changes have been implemented since SE has gained on-the-job experience with the application. The company met their goal of implementing the system in time for the millennium, with enough time to spare to work on other auxiliary applications that also had to be removed from the mainframe. Furthermore, they had everything in place just in time for both their second storm of the century within five years that occurred in June, and for their brush with Hurricane Floyd, that caused much of Savannah to evacuate in September. Their previous OMS would have been flooded by either of these storms.

(www.esri.com/news/arcnews/winter0506articles/southern-company.html)

The Salt River Project

The Salt River Project (SRP) is one of Arizona's largest electric utilities and water providers. SRP serves more than 700,000 residential, business and industrial customers with electric power throughout a 2,900 square mile area in central Arizona. The electric side of SRP is a political subdivision of the state of Arizona. In response to an increasingly competitive business environment, SRP is moving towards integration of data and management functions through the use of GIS. SRP, as are many other electric utilities, is placing increased emphasis on controlling operating costs. The functions typically supported by GIS systems, work management, outage management, inventory and asset control reduce operating costs by providing personnel with timely information that results in increased accuracy, efficiency, productivity and communication.

Taking this philosophy one step further, SRP has implemented an integrated solution of distribution analysis that addresses the most common concerns of distribution operations and planning namely voltage, loss and overload. Future extensions were to address concerns in the areas of quality, protection and reliability.

The transition to GIS has focused primarily on the elimination of tedious paper databases and multiple analysis programmes as well as centralization of management functions. The extension of these systems to support engineering analysis can provide additional benefits. At present, most implementations of distribution analysis tools have focused on the use of the tool's analysis capabilities as simply one step in the solution of recurring problems like voltage, loss, quality, protection and reliability. This approach can be tedious and time consuming. This approach also requires a high level of expertise in the use of the tool.

The approach of integrating analysis functions, as has been done at SRP, goes one step beyond the simple embedding of analysis tools. The suggested approach de-emphasizes the use of analysis tools in favour of a focus on problem solving. Detailed knowledge of the use of the analysis tool is not required. Rather staff can be directed to a problem area and offered a choice of actions (solutions) that will reduce or eliminate the problem. When integrated like this, analysis functions become available to a broader range of staff, increasing efficiency and productivity, but more importantly, permitting better decisions to be made by more staff. Problems are identified and solutions quickly offered. Implementation of a solution is more rapid and the operation efficiency of the utility is enhanced. This case study has presented a set of function libraries that encapsulate the algorithms needed for the analysis and optimization of distribution systems. The availability of these algorithms with a well defined application programme interface enables system integrators and other solution providers to truly integrate distribution analysis into the GIS environment.

(www.gis.esri.com/events/index.cfm)

Baltimore Gas and Electric Utility

Baltimore Gas and Electric (BGE) Utility's Substation and System Protection Department consists of 293 people responsible for the design and building of BGE's three to five new substations each year, and also handle the on-going updates and modifications to existing substations. One of the department's tasks is to produce the construction drawings for these substation projects. Their deliverables include the wiring drawings necessary to document the electrical system as well as the structural drawings for the physical plant. (www.autodesk.com)

Nevada Power

Technological innovation and peak customer service are longstanding traditions at Nevada Power. A time tested combination, they have helped deliver power to the people for nearly a century. Nearly 600,000 customers in Las Vegas, North Las Vegas, Henderson, Laughlin, unincorporated Clark County and small parts of Nye and Lincoln counties depend on Nevada Power for electricity. In 1997 the company's customer growth rate jumped to 6.4 per cent, making it the fastest growing electric utility in the United States. Official estimates place Las Vegas Valley's population at 2 million by 2005.

Before Nevada Power moved to AUD, designs were produced manually or with a less powerful CAD system. Designers were often forced to estimate what material to order. Many times orders were short and material was missing. Even if the designer ordered the correct material, errors were still made when a ruler was used to take measurements off the drawing and devices and boxes were counted manually. Today, after designers and coordinators for the utility's Las Vegas North and Las Vegas South design departments identify the power source for a new project, they use Autodesk Utility Design to lay out the design and create cost estimates, a bill of materials, and the final construction documents.

Again, Nevada Power will be breaking new ground as the first United States utility company to integrate an AutoCAD Map based Utility Design solution with Autodesk GIS Design Server. The unique configuration was to provide the desktop, web based, and mobile tools needed to manage the utility's entire workflow including design and planning, data storage and management, reporting and analysis, and web and mobile communication. Data stored in a standard Oracle database will flow from the Oracle data warehouse to the GIS application server, which will serve it up through the company intranet or the internet to browsers on mapping and engineering desktops and to mobile field locations. (www.autodesk.com)

Conclusion

The modern society is heavily dependent on electric power hence any problem in the supply of this energy source will directly effect the population, industry and the government. From the above examples it is evident that GIS has been used for the management of power, its distribution, designing, planning and also to reduce paper work. Different GIS softwares have been used in combination with the data base management software and hardware. AutoCAD, ArcInfo, Oracle and other softwares are very popular for such AM/FM applications. In fact, one of the widest users of GIS is in the electrical utilities.

REFERENCES

CSDMS (u.d.), *Power on Demand by 2012: The Role of GIS, GPS and Remote Sensing in Power Sector,* Centre for Spatial Data Management and Solutions, Noida, India.

Marutish, V. (2004), An enterprise GIS system for electric utilities: New approach to rofitability, *Geospatial Today,* Vol. 3, Issue 3, pp. 30-33.

13

GIS in Public Safety

GIG plays an important role in public safety. Response capabilities often rely on a variety of data from multiple agencies. The capability to access and process information quickly and deploy resources where needed can be mission critical. Information about the location of an incident or disaster is often crucial in knowing how to respond. Fire, disaster planning, emergency medical services are a few examples where GIS technology can play a vital role. GIS software uses geography and computer-generated maps as an interface for integrating and accessing massive amounts of location-based information. GIS allows public safety personnel to effectively plan for emergency response, determine mitigation priorities, analyse historical events, and predict future events. GIS can also be used to get critical information to emergency responders upon dispatch or while *en route* to an incident to assist in tactical planning.

Dynamic Flood Warning System

Floods are the major disaster affecting many countries in the world year after year. It is an inevitable natural phenomenon occurring from time to time in all rivers and natural drainage systems. It causes damage to lives, natural resources and environment as well as the loss of economy and health. The impact of floods has increased due to a number of factors, with rising sea levels and increased development on flood plain. However, information is required for the development of warning system.

With the increase of population and industrial growth, more and more settlements and development are encroaching in flood plains. Hence floods in Bangladesh have been causing the most severe economic damage among all kinds of natural disasters around the world. In order to formulate flood management and control policies, a thorough understanding of flood problems and their solution is required. This can be done through data collection and analysis along with flood impact assessment. A study was attempted to develop an integrated approach of a new model for warning of floods rather than flood mapping for flood risk assessment in flood prone countries for regular monitoring of damages.

Hydrodynamic model MIKE 11-FF (NAM, HD and FF) was successfully integrated with GIS in the Arc view GIS environment. MIKE 11 FF has been designed to predict the variation in discharges and water levels in a river system as a result of catchment rainfall and inflow and outflow through boundaries in the river system. GIS has been used intensively to create watershed models from digital elevation (DEM) data to trace flow paths to get a complete surface for identifying the actual flood flow, flood affected properties, agricultural areas, population and infrastructure.

Flood inundation maps were created by overlaying land use features and infrastructures to delineate flooded areas using dynamic spatial modelling. GIS was used for the analysis to determine hazard zones in the maps, which serve as risk zone identifiers for the general population. This information leads to locate safer location to evacuate the affected population utilizing the optimum route. Complete statistics will also be released for the predicted time stamps in advance by 72 hours, 48 hours and 24 hours. Finally, the information of population, land use and infrastructure are likely to be affected was to be made available in a user-friendly interface for timely decision for flood preparedness. Mostly, the northern part of the study area is always in the flood affected region. Flood inundation maps were prepared using simulation results. As the simulation results at every hour for the next 72 hours are available, so flood inundation maps can be prepared at any of those time using the forecasting water levels. Water levels in the flood phase maps (F0-F4) were classified into five classes—dark blue showing higher water depth.

There is considerable crop damage almost every year of moderate flood. Overlaying land use map over the flood map, the area of affected crops was found out. The area was then reclassified according to the types of crops. Possible percentage of affected crops in the next 72 hours was worked out.

Information of predicted populations and infrastructures affected were also established using overlay operation. Those maps indicated the accessible route to evacuate people during emergency. It is important for the warning system, to develop and find out the efficient route to provide direction to the vulnerable people to move out during emergency. Before doing this, the need was to come up with a plan that will enable to choose the best possible way to minimize travel time to reach to the destination in an effective manner. Arc view Network Analyst helped in using networks more efficiently. It assists in finding out the quickest way to get the safest location and to determine the best route to reach. Overlaying this information with population data, it is easy to understand the situation, from where most of the people getting service and where to be provided for additional facilities (Fig. 13.1).

Warning is crucial in emergency management. Adequate flood warning is a crucial element in emergency management, and operations of all of the systems around flood forecasting that actually deals with. Being able to give information on what is going on, and provide an adequate warning time to all of the public and emergency responders is necessary. The aim was to develop a user friendly system to support flood warning in advance at 72 hrs, 48hrs and 24hrs. Maps of the flood affected land use features, such as roads; agricultural lands, population and households can be released for above warning time stamps. In addition, complete statistics can also be available about the affected region. The information of population, land use, infrastructure etc. to be affected will be available in term of user-friendly queries for timely decision for flood preparedness.

Perhaps floods cannot be avoided, but by implementing effective flood prevention schemes, one can reduce damages from severity, if sufficient information for flood forecasting is given timely. Dynamic use of GIS integrated with hydrodynamic model provides useful measures towards disaster preparedness and planning. This dynamic model can be automated fully to be accomplished with the existing flood warning system to provide warning for the people without the knowledge of GIS to identify the possible inundated areas to take initiative to evacuate people to the safe places in time. It can help to promote public awareness in disaster management activities as a part of focusing the dissemination of forecast at the grassroots level. Therefore, new flood control and management schemes based on dynamic model can be implemented for important disaster management aspects like prevention and mitigation, preparedness, response and recovery and also planning for operational activities, immediately before, during and after the floods.

(http://www.gisdevelopment.net/application/natural_hazards/floods and http://www.dhigulf.com/upload/developmentofafloodforecastingsysteminbangladesh.pdf)

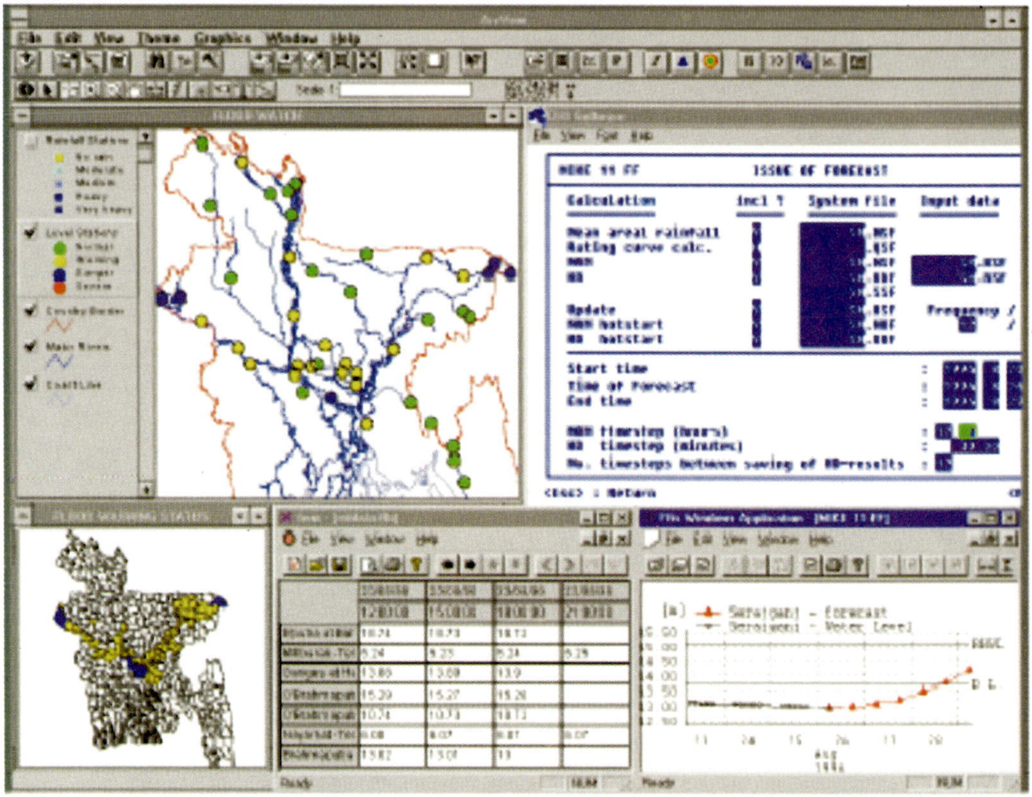

Fig. 13.1: MIKE 11 Flood Watch System

In developed countries the problems are also similar, but with the help of technology, management of flood hazards is better. In UK, one of the wettest years on record, Simmons Aerofilms acquired vertical aerial photography of some of the worst hit areas, including the River Medway in Kent, which burst its banks at Yalding. On its own, such photography is a valuable resource for many applications, especially those concerned with hydrological monitoring. Its value is further enhanced when orthorectified and combined with related geographic data sets in a GIS. (http://www.simmonsaerofilms.com/links/index and http://www.simmonsaerofilms.com/links/index)

City Police Department

In the US, orthorectified photographs have been used to circumvent crime. With accuracy an important consideration, ER Mapping employed the MAPS DTM to orthorectified the Simmonds photography and provide a basis for 3D visualisation. The use of DTMs for risk assessment is illustrated here in a sequence of images. The New York City Police Department uses MapInfo technology in its next-generation agency wide crime analysis system, enabling all city precincts, transit districts and housing services to analyse crime patterns by location and frequency—yielding better crime prevention. This system enables police to view and analyse crime in surrounding areas and command centres. (www.cwhonors.org/case-studies/NYPD.pdf)

Web-based Crime Mapping and Analysis

For tackling crime, MapInfo software has been used to develop GIS for mapping and analysis. Suffolk County, NY, has been implementing MapInfo technology with the aid of MapInfo partner TSI for more than 10 years. Relying on MapInfo technology for public safety, the Suffolk Police Department uses location-based intelligence to fight crime with web based crime mapping and analysis, web-based automatic vehicle locator of its patrol cars, and tracking of officers with wireless devices. Attracting international recognition, Suffolk's innovative use of MapInfo technology ensures the public's safety, as well as its police officers' safety.

(www.tsimapping.com/aboutus.html)

Emergency Information System (EIS)

It appears that MapInfo GIS software has been found to be more popular while dealing with different public safety measures. The case of Turkey may be mentioned here. Its governorship of Sakarya employs MapInfo technology and 3D mapping capability in its Emergency Information System (EIS) to help it analyse, plan for and forecast situations—particularly important for the earthquake-prone nation. EIS immediately provides maps and data of an affected area in times of crisis. EIS also assists rescue teams, ambulances and fire and police departments. In addition, EIS helps responders and government officials establish temporary shelters and health units, plan evacuation systems and collection donation points for clothing and equipment. Sakarya also provides information to the public and other government organizations via the world wide web with vector and raster based maps and interactive GIS maps.

(http://www.isprs.org/istanbul2004/comm2/papers/139.pdf)

Making Better Decisions Regarding Risks to People and the Environment

On December 2001, bush fires burned out of control and looked like they would engulf many densely populated areas and national parks on the outskirts of Sydney, Australia. Again using MapInfo spatial technology, the New South Wales Rural Fire Service (NSW RFS) produced a sophisticated and extensive data analysis of the affected area, gaining a tactical and strategic view of the situation. MapInfo spatial technology—including MapInfo Professional, MapInfo MapXtreme, MapInfo Easimaps, MapInfo Mapshop, MapInfo Proviewer and MapInfo Encounter TM—powers a data portal accessed by NSW RFS. This spatial data portal has allowed the NSW RFS to make better decisions regarding risks to people and the environment. In the case of the bush fires, spatial analysis was used to determine the types of houses to be built and the materials to be used in bushfire prone areas. Through 3D and hazard and risk analysis, data was transformed into pictures providing a better understanding and a powerful perspective of the situation. Via the portal, NSW RFS studies highlighted certain patterns and trends that otherwise had been impossible to see. NSW RFS is even able to investigate the likelihood of arson. Based on area demographics, NSW RFS maps historical bush fire areas with current data to determine whether the fire was possibly the result of arson or natural causes. The MapInfo powered portal provides NSW RFS with a clearer view of information held on its corporate database, resulting in improved business decisions for future planning and management. Further with GIS technology, NSW RFS better understands the state's environment, enabling it to apply for extra federal funding. (www.regional.org.au/au/gia/12/382bunker.htm)

Emergency Response Faster and More Accurate

Cumberland County, N.C. knows a fast and accurate response to an emergency team's question "How do we get there?" can mean the difference between life and death. That's why E911 planners for this North Carolina County use GIS technology to swiftly pinpoint the location of every emergency and guide response teams to its scene. Cumberland County is home to more than 300,000 people living in areas as varied as the city of Fayetteville, Fort Bragg army base and scores of rural outskirt communities. The county's two public safety answering points (PSAPs) respond to about 10,000 calls each month dispatching police, ambulance, fire or sheriff to a third of those. Several years ago, planners realized that increasing the accuracy of location data was essential to improving response time to countywide emergencies.

In order to improve response time to emergencies countywide, the county needed a way to quickly, easily and accurately calls up maps identifying the locations of emergencies. Addresses were collected from city records and government data, which sorely lacked precision and detail, forcing dispatchers to assume and approximate where emergencies were and how to get the response team to them

The county exhaustively implemented a door-to-door geocoding and mapping operation that yielded a complete and accurate picture of Cumberland County. The county shares maps and analyses with anyone seeking current county location data and maps. Further, increased accuracy in identifying addresses and particularities of the location, such as long driveways or those located indirectly off a street, enables 911 dispatchers to: immediately locate exactly where a 911 call originates, see the movement of the emergency team in real-time and, specifically direct emergency technicians—all yielding increased response speed. With GIS technology, Cumberland significantly reduced cost and time required to programme and integrate its data.

Cumberland County is looking to the future with an eye on: impending E911 legislation; making location data universally accessible for editing and adding functionality like hot linking data and images to maps. This county is preparing for FCC-mandated regulation to take effect that requires wireless carriers to provide location data for mobile 911 callers. It expects a seamless integration of this latitude and longitude data—delivered to 911 networks from wireless carriers—with the address and mapping solution, thanks to GIS's proven data integration superiority. County planners are also evaluating GIS Spatial Ware,data management software that enables businesses to store, manage and manipulate location-based data within a relational database management system. This will allow multiple users to edit county location data files and maps. It also plans to exploit GIS professional's capability of hot linking a variety of data to images on maps. This way, additional key location information such as subdivision drawings can be scanned, hot linked to maps and accessed with one click on a map. Users of county's location data and maps will get the critical back-up and added detail needed to make important decisions, especially the life and death ones. "GIS pulls it all together for us," said one of the official. "It is head and shoulders above other systems." (www.mapinfo.com/static/files/document/cumberland-casestudy-optimized.pdf)

Fire Fighters Monitoring Raging Fires

In August 2002, the City of Prescott's Emergency Operations Center in Arizona found Earth Resource (ER) Mapping's image processing software, ER Mapper, instrumental in managing and monitoring raging fires that destroyed the Prescott National Forest and surrounding areas at rapid speeds. Due to rough terrain and steep timbered ridgelines fighting the aggressive wildfire became extremely difficult. The City of Prescott used ER Mapper to firstly seamlessly mosaic twenty 1-metre IKONOS satellite images into a single 3.2GB file. This image was then compressed to a manageable size of just 12.8MB by using ER Mapping's ECW (Enhanced Compressed Wavelet) technology.

The City of Prescott used the compressed image to create a set of image based maps. These image-maps then acted as backdrops to multiple vector coverages including street centre lines, road outlines, property boundaries, water resources, electrical, gas lines and building footprints. These detailed maps could then be distributed to all staff at the emergency operations centre, police and fire station to assist as a visual perspective tool to help manage and monitor the fire. GIS Coordinator for the City of Prescott said, "It is evident that we were extremely lucky. The use of fire maps and imagery had provided the response personnel a means of attacking the fire throughout the night. They were able to visualize the terrain and vegetation, the access to the area on roads and trails, the location of resources, and the location of residences that were in danger. Being able to see this throughout the night gave them the advantage of operations during many hours of cooler temperatures and no-wind conditions."

(www.gismonitor.com/news/newsletter/archive/080102.php)

Ground Zero Exhibit

Denver event provides a unique look at the site of the 9/11 World Trade Center attacks through technology used in the rescue and recovery effort. Event is open to the public from May 23-29, 2002. On Wednesday, May 22 at 6:00 p.m., some of Colorado's most respected and influential public safety officials were gathered to view and discuss an educational exhibit that shares an important perspective on recovering from the events of September 11. "Charting Ground Zero—Before and After" features maps and geospatial imagery that were created following the World Trade Center attacks as part of the rescue and recovery effort. The technology used to create these images continues to be used today by local, state and federal agencies to assess damage, monitor the progress of recovery, and safely deploy personnel and equipment in the disaster zone.

The imagery represents a convergence of technologies that will have far reaching effects on how the country can respond to attacks, natural disasters, and other emergencies in the future. "These images are the result of new and existing technology and have provided tremendous assistance to the recovery efforts at Ground Zero," said Dr. Sean Ahearn, Associate Professor at Hunter College and Director of the Center for Analysis and Research of Spatial Information Lab. "This is the same technology that our country will depend on, in many ways, to respond to any future attacks. The exhibition is meant to familiarize people with the technological advancements and share with them how these advancements can benefit us all." An original exhibit in partnership with the City of New York and the Woodward Gallery was also created. The exhibition includes operational maps created for New York City's Office of Emergency Management, as well as aerial imagery, laser ranging images and other images from the site. Two companies with a presence in the Denver area, LinksPoint and GE Network Solutions, whose technologies were involved in the recovery efforts were instrumental in bringing this exhibition. San Jose Fire Department selected STM or Solid Terrain Model for training, pre-fire planning, and fighting wildfires state-of-the-art, full-colour, and 3-D scale model. It includes San Jose's east foothills and the Alum Rock area.

(www.directionsmag.com/pressreleases.php)

Solid Terrain Modelling (STM)

Solid Terrain Modelling (STM) is again used for different types of applications. STM, the maker of the world's most accurate solid terrain models, announced that, as indicated above, the San Jose Fire Department has purchased an STM model for use in training, fire pre-planning, and fighting wildfires. The state-of-the-art, full-colour, 3-D solid scale model of San Jose's east foothills and Alum Rock area includes the area where a dangerous and costly wildfire occurred in October 2000. The four-foot by four-foot model is an exact replica of the entire area, and includes access roads, hazards and jurisdiction boundaries. "This model will help us provide firefighters with continual strategic and tactical training for fighting fires in the geographic area where wildfires are most likely to occur," said the battalion chief, San Jose Fire Department. "The model represents this area far more accurately than any maps or photos we have. During an actual wildfire, we can transport the model to the incident area command post and use it to plot the fire as it burns. The model will give the incident commander a real-life picture of the area, where the fire is and where it is going. In addition to access roads, jurisdiction lines and possible hazards, fire officers will be able to see the geography, scale, distances, terrain, points of view, and sight lines. "We're receiving a lot of interest about our 3-D models from a number of fire and police departments," stated the president of STM. "We've already sold a number of STM models to the National Forest Service for training and pre-fire planning, and for field use."

STM models provide a level of comprehension not found in any other mapping technology. People can gather around an STM model and view it from any angle. They can discuss it, touch it, point to specific features on it. As they do, everyone in the group "gets" it—the scale, distances, terrain, colouration, points of view, sight lines and more. To produce a model, STM uses satellite digital elevation and photometric data for the selected geographic area. The digital elevation data is used to control STM's highly calibrated cutting machine as it carves the model into high density plastic foam. Then, the photometric data is used to guide STM's printing machine as it applies a satellite photograph directly onto the model's surface. The result is an exact replica of the geographic area—with breathtaking detail and in brilliant colour. About 17 STM models are now on display at the National Geographic Society's Explorers Hall museum in Washington, D.C. Currently, STM models are used by law firms; emergency response groups (fire, police); land management offices (construction, civil engineering, real estate, recreation); developers; transportation agencies; utilities (oil, gas, hydro); architects; museums; military, federal, state and municipal agencies; the U.S. Department of Defense and others (*Business Wire,* May 21, 2002).

Risk Assessment Tool

Maps do empower owners of property near flood zones. For the first time, real estate owners will have access to information that helps them more accurately assess the likelihood of their properties being flooded, thanks to Nation Wide Flood Research, Inc. The company today announced the launch of TruRisk (SM), a proprietary mapping and risk assessment application that shows where properties are located in relation to special flood hazard areas. Beginning June 2002, Nation Wide Flood Research's clients will have the option to receive TruRisk maps in either English or Spanish, initially at no cost, with their flood zone determinations. Owners and potential buyers of property in special flood hazard areas must often purchase flood insurance to obtain a real estate loan. Traditionally, the only information provided to property owners has been a Federal Emergency Management Agency (FEMA) Standard Flood Hazard Determination and a notice to the borrower of property stating whether a building is in a special flood hazard area and whether flood insurance is required. Property owners wanting more specific information had to obtain maps from the FEMA Map Service Center, but FEMA maps do not provide distances to the nearest flood zone, making it difficult for property owners to assess their risk.

"TruRisk allows property owners to more accurately evaluate their need for flood insurance," President and CEO of Nation Wide Flood Research, Inc. (NFRI) said. "Many residential and commercial property owners are unaware that flood insurance can cost as little as $9 per month or that the likelihood of a property being flooded is equal to or greater than their car being stolen or property burglarized." According to the FEMA, about 25 per cent of all flood insurance claims are for properties located outside of identified high risk areas. TruRisk helps property owners whose real estate is in close proximity to a special flood hazard area to self-assess their risk and evaluate their need for flood insurance.

NFRI, a premier provider of flood zone determinations, was founded in Dallas, Texas in 1995. It provides standard and life-of-loan flood zone determinations, census tract data, and risk assessment information for all fifty states and US territories. Lenders, insurance companies and real estate owners across the country rely on NFRI for flood zone risk assessment (*Business Wire,* May 20, 2002).

Emergency Management of Seismic Event

"Sige on line", representing the evolution of the current info-system, is dedicated to the management of an emergency in case of a seismic event (SIGE), currently on service at the National Seismic Survey. "Sige on Line" goes over the local operating condition, task previously done by SIGE, becoming a new concept of distributed activity and different field of operativity, through the integration of different technologies: GIS, wEB, line-links etc. After a simplified description of the SIGE is illustrated, the new project, Sige on line (SOL) and an experimental application with the contribution and participation of the National Seismic Survey (NSS), the Department of "Protezione Civile" and the province of Modena, Italy has been mooted.

The NSS, as member of the Civil Defence body (law 225/92), has to provide the specific activities for the aftermath of the seismic event and its following post-event management. SIGE is an integrated tool used for supporting civil defence operations in case of seismic events in Italy with the aim to do prevention, minimise impacts and facilitate emergency intervention. This task is fulfilled through the production of real time tabular reports, technical maps and relevant scenarios.

The system uses the data (location, magnitude, time) of ING or Istituto Nazionale di Geofisica for monitoring seismic network in connection with the integrated Seismic Risk Information System. Just after the event the NSS gives an informative report called SIGE. Such report gives an homogeneous information content, allowing to value the global square area of the seismic phenomenon regarding its entity, land extension and social and urban involvement.

After 45 minutes from the signalling of the event, the paper report is transmitted from NSS employees to Civil Defence Department. In this way it is possible to overcome knowledge blackout occurring immediately after "post-event". Report contents cover the following areas: demography, services, networks, and historical seismology and damage scenarios (property and people).

Risk analysis and response GIS and public safety is linked with such a project. Concern for public safety has grown in the wake of numerous and costly emergencies in recent years. Events such as Hurricane Mitch have been costly not only in dollars but in human suffering. Deemed one of the worst natural disasters in the western hemisphere in this century, Hurricane Mitch roared into Central America on October 26, 1999, causing six billion dollars in losses to the region, killing more than 10,000 people, and leaving another 1.5 million people homeless.

GIS was instrumental in recovery efforts for this area. ESRI joined with the United States Geological Survey (USGS), Silicon Graphics, Hewlett-Packard, Motorola and Microsoft in sending computer systems equipped with GIS software and databases to the region. These systems were used to coordinate and prioritize rescue and relief efforts. The Central America Disaster Atlas, created by a team of volunteers from the USGS and ESRI, was distributed to Honduran and international agencies so they could create accurate maps to guide rescue, relief, and reconstruction operations.

(www.gis.esri.com/library/userconf/europroc98/proc/idp47.html)

GIS-Based Tools

There are other initiatives of the application of GIS in disaster management in the US. GIS has proved a powerful tool for assessing risk and prioritizing needs. An earthquake loss estimation software application called HAZUS has been developed as part of Project Impact and is available at no charge to state and local governments. HAZUS works with ArcView GIS to produce detailed maps and analytical reports that describe a community's potential losses.

As part of Project Impact's programme to educate the public about natural disasters, FEMA added online mapping of tornadoes, hurricanes, hailstorms, earthquakes, windstorms and flood information. ESRI developed and hosts the hazard maps site where visitors can build maps that show the history of natural disasters for ZIP Code areas, towns, or congressional districts.

The Project Impact Community Challenge Grant Programme, sponsored by ESRI, supports the use of GIS as a tool for effectively creating disaster-resistant communities. Grants were to be awarded to communities with innovative proposals for using GIS to develop or implement damage prevention strategies. ESRI was to provide software and training grants totalling approximately \$100,000 and FEMA will provide training and technical assistance. Selected communities were to be featured in the public safety sessions at the 2000 ESRI International User Conference and the 2,000 Project Impact Summit. (www.esri.com/news/arcnews/winter0203articles/south-caroline-deviser.html)

Fire Agencies Improve Response with GIS

Fire agencies on all levels strive to preserve lives, property, and natural resources. Many agencies have embraced GIS as a tool that helps them balance needs, uses, and hazards to promote sustainability of the environment while identifying and limiting vulnerability. The Cedar City Interagency Fire Center (CCIFC) in Utah, dispatches fire personnel for a number of agencies including the Bureau of Land Management, the National Park Service, the Bureau of Indian Affairs, and the State of Utah. When responding to a fire, CCIFC dispatchers plot real-time GPS positions and reference them against the GIS, which contains data on ownership, roads, wildlife habitats, slope classification, fire fuels, zones of moderated suppression, and other related themes that consolidate text, tabular, and other data previously available only from wall maps, manuals, and separate databases. Now dispatchers need only query an area to obtain concise instructions on the constraints and objectives for the area in question. (www.gis.esri.com/library/userconf/proc98/PROCEED/T0700/PAP653/p653.htm).

The Winston–Salem Fire Department in North Carolina implemented a system that uses ruggedized PCs mounted on emergency vehicles that run a custom GIS application built using MapObjects and NetEngine. The application displays a map showing the optimal route to the emergency location. This route can be instantly modified to take into account streets that are blocked. The application also provides information on all available fire hydrants, any known hazardous materials in the area, the locations of special needs populations, and other pertinent information. A touch screen lets firefighters pan and zoom on the map and use icons that link the GIS application to additional information such as floor plans, diagrams, or other pre-fire survey information.

(www.gis.esri.com/library/userconf/proc01/profesional/papers/pap512/p512.html)

Similarly, the City of Laguna Beach, California, has used GIS to pursue both policies. In 1993, a fire driven by high wind conditions caused extensive damage to the city and resulted in the loss of 286 homes. The city has used GIS to help rebuild the portions of the community that were destroyed in the fire. GIS proved so useful that it has been adopted as a tool to improve city operations. As a result of the 1993 fire, Laguna Beach extensively updated the safety element in the general plan. The availability of mapping, aerial photographs, and analysis tools in GIS helped speed and improve the revision process. The new safety element includes an emergency routing programme and fuel modification specifications and requires an emergency preparedness plan that is updated on an annual basis. Zoning Analyst, a custom ArcView GIS extension developed by ESRI business partner Geographic Information Services, Inc., assists planners in providing public information, processing cases, generating public notification letters, and producing maps.

(www.gis.esri.com/library/userconf/proc00/professional/papers/PAP875/p875.html)

Fire Hydrant Maintenance Using GPS and GIS

A joint project developed by the Fire Department and GIS staff of the City of Euless, Texas, leverages existing GIS data and GPS technology to benefit both groups as well as other city departments. Euless, a midsize city with a population of 44,000, is located midway between Dallas and Fort Worth.

Fire hydrant maintenance is a chore performed by the Fire Department. Fire fighters visit each hydrant in the city to check its general condition. Sometimes they trim the weeds, grease the caps, or exercise the valve. However, because not all of the hydrants appear on the map, and not every hydrant is correctly located, the real purpose of this exercise is to familiarize fire fighters with hydrant locations. To improve incident reporting and computerize response preplanning, the Fire Department acquired a new management software called Firehouse. This software allows external data to be imported. Address data that had been tested and used by the city's GIS Department for the last ten years was imported into Firehouse. Addresses for all of the city's homes and businesses were now available from a pick list on the incident report screen.

Not only was useful data added to the Fire Department's software but the GPS location of the hydrant helped increase the accuracy of the water map. New hydrants were added, bogus hydrants were removed, and most of the other hydrants were shifted slightly in order to correctly represent their true location. The same procedure was to be performed on the valves to increase the accuracy of the city's water maps. In addition to these benefits, the project has the potential to help in other ways:

- The Public Works Department describes the location of hydrants that are being taken out of service for water line repairs to police and fire dispatchers. Prior to the hydrant survey these descriptions were often limited to "the one in front of the third house on the left north of the intersection." Dispatch was often confused about exactly which hydrant was out of service. Now hydrants are referenced by the unique identifier. It is clear which hydrant is down.

- The Public Works Department is also using ArcExplorer to look up hydrant ID numbers by street address or graphic selection. This test project lets staff identify hydrants that require maintenance that the Fire Department would not normally perform.
- The latitude/longitude coordinates for locations in the city can quickly be approximated based on the nearest hydrant. The Fire Department may provide hydrant coordinates to the Care Flight helicopter ambulance so that it can find locations more quickly.

(www.gis.esri.com/news/arcusers/1000/hydrant.html and www.gis.esri.com/news/arcusers/1000/umbrellag.html)

Critical Support for Wildfire Response

More than 5.9 million acres in the western United States were burned by wildfires as of September 2000. According to the National Interagency Fire Center (NIFC) in Boise, Idaho, at one point during the summer of the same year, 84 fires, involving more than one million acres, were active in Arizona, California, Idaho, Montana, Nevada, Oklahoma, Texas, Utah and Wyoming. In response to these devastating wildfires, federal firefighting agencies and private industry created the Geospatial Multi-Agency Coordination (GeoMAC) group.

This project developed an internet application based on ArcIMS and ArcSDE that provides decision support for prioritizing the use of wildfire suppression resources and critical information for public and firefighter safety. The GeoMAC team was assembled quickly. Business and functional requirements were developed on the fly concurrently with the actual application development efforts. In a period of less than two weeks, the GeoMAC project progressed from conception to full implementation.

The initial goal of the GeoMAC project was to implement a strategic level internet-based mapping application to aid in the allocation of firefighting resources to the most critical fire incidents in the Great Basin region of western Montana, Idaho, Oregon and Utah, said the incident commander for the GeoMAC incident team. "These maps will enable fire personnel to obtain a quick snapshot of near real-time fire hazards."

The GeoMAC project rapidly evolved and a variety of different data sources were identified, prioritized and processed into suitable formats and made available over the internet. The data available over the web included course resolution fire location data from several sources including NIFC's Incident Management Situation Report (Sit Rep), thermal patterns representing potential fire hot spots from processed Advanced Very High Resolution Radiometer (AVHRR) data and additional potential fire locations identified from visible light data extracted from Defense Meteorological Satellite Program (DMSP) sources.

As the value of the GeoMAC Web application was realized, the project rapidly expanded from the initial strategic level incident mapping application to include other levels of functionality. Almost immediately, the value of disseminating the information to the general public became apparent. The geographic area covered by GeoMAC was expanded as fire communities in other western states recognized the value of providing data via the web.

Other web based mapping applications were subsequently developed. One of these applications gives fire managers more highly detailed data at the incident level. Links to fire weather stations, 1:100,000 scale digital line graph (DLG) roads and hydrographic data, hydrologic units, aviation hazards, and digital raster graphics (DRG) 7.5-minute USGS quad data are included in this application.

Another application supports tactical aircraft use. The Temporary Flight Restrictions (TFRs) application uses the Internet to disseminate and update TFRs from the fire incident command center. It is a significant improvement over current operating procedures. The TFR application uses NOAA flight navigation sectional maps to provide background images.

As the initial wild land fire situation eased, the GeoMAC team shifted the focus to supporting Burned Area Emergency Rehabilitation (BAER) teams by providing spatial data online. With this data BAER team members could see old fire perimeter data and the 1:24,000, 1:100,000, and 1:250,000-scale quad indexes at the same time and evaluate which quads were needed to perform standard BAER analysis. BAER team members could use the web application to directly order the appropriate data via the internet. (www.gis.esri.com/news/arcusers/0101/geomac.html)

Fire Hydrant Inventory Made Simple

The Elsinore Valley Municipal Water District (EVMWD) in Southern California was to update data on approximately 4,000 fire hydrants using ArcPad, a handheld computer and a GPS unit. By using custom forms created for ArcPad, data collectors will not only verify the location of the district's hydrants but will also gather information on the make, condition, and capacity of each hydrant. In addition to the need to simply locate all the hydrants to link the test results, the local fire authorities wanted detailed information about fire hydrants such as the type of hydrant, number of outlets, size of each outlet, and other information. Requests for hydrant information originating from two sources—within EVMWD and from local fire agencies—made the GIS staff decide that it was time for a fire hydrant inventory.

The solution proposed by the staff used a Sokkia GIR 1000 GPS unit, a Compaq IPAQ PocketPC, and ArcPad from ESRI. The existing fire hydrant database, street rights-of-way, street centerlines, and miscellaneous layers are be loaded on the IPAQ and set up in ArcPad. A custom form was created for the hydrants for data collection and update. For each hydrant, the data collector refers to the hydrant database, verifies the hydrant in the field, and inputs data into the form. A hydrant that does not show up in the database is located using GPS. At the end of the day, data is downloaded from the IPAQ and the GPS receiver. Points taken with the GPS are loaded in the ArcInfo database, and attribute information is loaded into a Microsoft SQL Server database.

(www.gis.esri.com/news/arcnews/winter0102articles/eisinore-valley-mwd.html)

Virus Surveillance

For the first time avain virus with mosquito vector was seen in US in 1999 in New York City. The crows are especially susceptible. This disease is usually mild in humans, but can be deadly in young and old. But found in birds or mosquitoes in thirteen states; and 18 hospitalized human cases and one death were reported. This activity relates to inform the activities of the state and local health departments in the Washington metropolitan area regarding larviciding and adult mosquito spraying: (a) risk and possible cases in human, and (b) informing the public. The other dimensions of the project involve:

- Two clusters of WNV+ birds in 2001
  Major cluster at DC/Maryland border
  Secondary cluster in Virginia
- Evolution over time
  First bird in early July at centre of major cluster
  Cluster spread during the season but stayed centered at this spot
  Empty centre at end of September

The above inference may be due to testing as the clusters are not due to reporting patterns. Mapping allowed to see patterns not otherwise apparent. Geographical, especially at jurisdictional boundaries, and temporal evolution of clusters evaluate risk to humans based on density. Obviously, an integrated and automated GIS could have made this information timely.

(www.gis.esri.com/news/fall03articles/new-york-city.html.)

Icelandic GIS Used for Emergency Services

The European Union (EU) is implementing a process to standardize the varying infrastructures among the countries within its organisation. This includes the standardization of all emergency response systems. Members of the EU voted to make all emergency response numbers 112. For Iceland this meant abandoning its disjointed network of independent agencies and reorganizing them so that one organizational body would handle all forms of emergency response. Iceland's entire emergency response system is now operated by Neydarlinan HF, a privately held organisation that has been contracted for service by the Iceland government. Shortly after its foundation, Neydarlinan approached HNIT Limited, a leading consulting firm in the field of GIS in Iceland, for the development of an emergency response computer system that would include computer telephony, relational database, GIS, and protocols for short messaging and pagers.

When calling 112 in Iceland, responding telephone operators can now dispatch assistance from more than 200 different response agencies across the country. They include fire, police, and ambulance units, and each has its own service area. Prior to the new system, each agency had a separate number to call if emergency aid was required.

(www.gis.esri.com/whitepapers/pdfs/gis-and-emergency-mgmt.pdf.)

Search and Rescue (SAR) Mission

Even the most experienced hikers and climbers may find themselves in need of rescue and, if so, can breathe a sigh of relief knowing that GAF of Munich, Germany, has developed a mission support tool to locate them and co-ordinate the Search and Rescue (SAR) Mission. Known as SAR Land, it is specifically designed for mountain rescue, although application in humanitarian aid missions and for emergency service support during big events is also envisaged.

14

GIS in Law Enforcement and Criminal Justice

Geography plays an important role in law enforcement and criminal justice. Response capabilities often rely on a variety of data from multiple agencies and multiple sources. The ability to access and process information quickly while displaying it in a spatial and visual medium allows agencies to allocate resources quickly and more effectively. In the *mission-critical* nature of law enforcement, information about the location of a crime, incident, suspect, or victim is often crucial to determine the manner and size of the response.

GIS software uses geography and computer-generated maps as an interface for integrating and accessing massive amounts of location-based information. GIS allows law enforcement and criminal justice personnel to plan effectively for emergency response, determine mitigation priorities, analyse historical events, and predict future events. GIS can also be used to get critical information to emergency responders upon dispatch or while en route to an incident to assist in tactical planning and response.

GIS for Crime Information and Analysis

The UK's Leicestershire Constabulary, with more than 2,000 police officers and 900 support staff working in 26 police stations throughout the county, uses MapInfo in its Crime Information System, enabling all police officers, from bobbies on the beat to the chief constable, to access crime data analyses and quickly identify crime patterns. It appears that not a very high end GIS is required for such applications. In fact, easy to use software which can be customised for organising information with quick response time is better suited. Further, the London Metropolitan Police, based in New Scotland Yard, also uses MapInfo technology to map street crime throughout the greater London area, achieving within the first few years of usage thirteen per cent increase in crime solving. They continue to use MapInfo in its crime fighting today, working to ensure the safety of 10 million people within the London area. (www.macrimeanalysts.com/articles/historyofcrimeanalysis.pdf)

Similar to above, the New York City Police Department (NYPD) uses MapInfo technology in its next-generation agency wide crime analysis system, enabling all city precincts, transit districts and housing services to analyse crime patterns by location and frequency — yielding better crime prevention. This system enables police to view and analyse crime in surrounding areas and command centres. On the other hand, the NYPD's agency wide intranet-based crime analysis system provides crime analysis capabilities to every precinct, transit district and housing service area in the city. The cornerstone of this crime solution is MapXtreme, MapInfo's powerful Java-based internet mapping server, allowing the nation's largest police department—with 76 patrol precincts and 40,000 officers—to access and share crime data. In the same country, the Yale University Police Department, with a served population of 20,000, uses MapInfo-based CrimeInfo, developed by MapInfo partner Charles River Technologies, to analyse incident patterns for improved campus security. Since the university spans several city neighbourhoods, knowledge of incident patterns within and outside the boundaries of the university is essential to meeting the department's mission of reducing and preventing crime, in partnership with the city of New Haven. MapInfo technology is implemented in the Minneapolis Police Department's crime fighting and deployment application, known as CODEFOR. This year, in CODEFOR's first year of operation, serious crime dropped 16 per cent. Portland, OR, also uses MapInfo for day-to-day law enforcement in crisis analysis and management, resource deployment and threat assessment. Portland also employs MapInfo analysis capabilities for its homeland security efforts. With the help of MapInfo partner, John Lowe and Associates, Portland uses mapping and thematic functions to visualize potential and actual threats to its public safety, particularly performing on-the-fly spatial analysis for evacuation, emergency notifications and exposure to hazardous materials. California Department of Justice's MapInfo-based crime fighting and analysis strategies include forecasting, statistics, area profiles, trend identification, pattern analysis and tactical planning.

(www.solimserver.geography.wisc.edu/axing/teaching/geog578/references/nytimes-01-20-2000.html and www.resource.mapinfo.com/static/files/document/1104948703881/atlanta-pd-geoworld.pdf)

Australia's New South Wales Police Service's Mapping Unit, recognized as a model of best practice, and provides a specialist mapping service in support of police investigations, operations, intelligence and administration throughout New South Wales. Using MapInfo technology, the unit also supports the state emergency operations centre during emergency incidents and disasters and the police operations centre during major policing operations.

(www.mapinfo.com/common/library/mapinfomag.spring2002.pdf)

Track Criminal Activity in Mushrooming Atlanta Community

The customer for tracking criminal activity in Atlanta is the Bureau of Police Services in Henry County, GA, crime analyst uses GIS technology to keep pace with changes in one of the fastest growing areas in the nation. Fifteen miles from downtown Atlanta, Henry County's population jumped from 58,741 in 1990 to more than 119,000 today. According to crime analyst, tracking criminal activity in the midst of mushrooming communities and commercial developments requires keeping pace with the proliferation of new roads and streets so crime incidents can be tied to specific addresses and locations. There is a tremendous growth problem. The county is gaining up to thirty-five new streets a month. This requires intelligence mapping technology to automatically geocode addresses and periodically updates the county roads, streets, developments and continually expanding city boundaries. The problem is that the Bureau's paper-based crime mapping procedures unable to track events in rapidly expanding county. The solution considered appropriate was to have crime mapping unit using GIS Professional for mapping and analysis and GIS. This helped in keeping street and address database updated. Combined technologies enable crime-mapping unit to stay current with county growth and provide precincts with crime incident data in near real-time. The ability of the unit to identify crime patterns as they are developing allows precincts to respond more quickly than was possible with the previous pin-mapping system. Crime rates in the county are decreasing significantly while the population is increasing dramatically.

GIS Professional was used to build a multi-layered county map of police precincts, districts, categories of criminal activity, dates, times, etc. MapMarker PLUS automatically geocodes the addresses, allowing to track the geographic locations of crime events and connect them to specific addresses. Incident reports filed by officers on patrol go into a Police Track Database. The analyst pulls the reports from the database, loads them into GIS Professional, and analyses them for patterns of specific crime types. With the resulting information, he develops thematic maps and related data for distribution to precincts via the county intranet. This is near real-time crime mapping. Criminal activity can be plotted as it is occurring and show how it is developing, how it is moving across the county boundaries. For special department studies or briefings, PowerPoint presentations were used. Crime mapping is carried out on a stand-alone PC tied into a network environment. The analyst runs GIS Professional and MapMarker PLUS on Windows NT 2000. The speed and facility of the system enables the department to be much more proactive in tracking crime events. If an individual is breaking into vehicles, or if there is a series of burglaries, we can see the cluster start, and actually watch it migrate into other areas. From each event, we can collect pieces of information, including date, time, and location and add these to be used. Collectively that information can tell us whether we are dealing with one or more individuals and where to more accurately place patrols to catch or at least deter offenders working in a particular area. The facility and speed with which GIS Professional and MapMarker allow us to stay current with the county's rapid growth have enabled us to provide precincts with accurate information in near real-time. According to the crime analyst, "since we can see crime patterns developing, our response times are now much faster than they were before we had this capability. We are already seeing the benefits of this system; crime rates in the county have been coming down significantly at a time when our population is increasing dramatically." (www.co.henry.ga.us/HomePageMisc/NewcitizenGuide.htm)

New York State Division of Criminal Justice

New York State Division of Criminal Justice Services (DCJS) relies on MapInfo for state wide crime analysis system. MapInfo MapXtreme powers first state wide, cross-jurisdictional, web-browser based crime mapping system in Troy, New York. For the first time, all New York State law enforcement agencies will be able to analyse crimes across city and county borders on a map, enabling them to better isolate patterns and trends. MapInfo worked with the Vera Institute of Justice in developing this web-browser application.

With MapXtreme, law enforcement officers in the office, including patrol officers and commanders, will be able to map crimes by type, date, time and other parameters. They are able to produce maps, view the information as bars on line charts and identify crime density through hot spot analysis. Law enforcement agencies can also produce maps of crime occurrences and arrests within their jurisdiction and across the region. Computerized crime mapping has transformed law enforcement, enabling agencies to track crime rates, develop solutions and reduce crime more efficiently than ever before. Vice President and Adviser, homeland security said "We are thrilled that the DCJS, one of the nation's leading developers of criminal justice technologies and information systems, selected MapInfo to develop this state-of-the-art application. The system, which is the first of its kind in the nation, will be instrumental in bringing city and county law enforcement agencies together to reduce crime and protect the residents of New York State."

(www.mapinfo.com/common/library/mapinfomag-spring2002.pdf)

Border Patrolling

Border Patrol is the mobile uniformed enforcement arm of the Immigration and Naturalization Service. Their mission is the detection and prevention of the illegal entry and smuggling of aliens into the United States. Border Patrol agents perform their duties along and in the vicinity of 8,000 miles of international boundaries by automobile, boat, aircraft, horseback, snowmobile, jet ski and afoot. Most of the areas patrolled by the Border Patrol are remote with varying types of terrain along with sub-terrain such as storm drains and tunnels. These areas are normally accessed using unnamed and unmaintained roads or trails. The Border Patrol does not have the luxury of tracking events using the conventional geocode methods with street maps, as do other law enforcement agencies.

With the revolution in GIS and desktop computing, this scenario has changed. The Border Patrol now utilizes ArcView GIS and the global positioning system to map out operational areas, track events, and analyse smuggling patterns. They have combined GIS with other technological advancements, such as biometric identification and remote sensing to conduct unprecedented analysis on smuggling activities. These advances have also opened the door for multi-agency cooperation between the Border Patrol and other federal, state, and local law enforcement agencies. (Ref: Smuggling Operations Revealed Using GIS Track: Law Enforcement and Criminal Justice; by Daniel T. Isenberg; also http://gis.esri/library/userconf)

Crime Analysis and Mapping

A Crime Mapping and Analysis Programme was presented for Tactical Crime Analysis Using ArcView. The presentation was to cover methods for performing tactical crime analysis using ArcView GIS 3.2. The participants were to be given a conceptual framework for tactical crime analysis, discuss alternative map layers and data to incorporate into a tactical analysis, and be shown several GIS methods and practices to analyse the sample series. Point symbol or "pin mapping", proportional symbol mapping, as well as polyline to point file techniques were to be shown and explained in an attempt to determine the hunting ground, develop possible suspects or investigative leads, and predict future targets spatially and temporally. (Ref: Tactical Crime Analysis Using ArcViewTrack: Law Enforcement and Criminal Justice Author(s): Sean M. Bair).

Further, in October 1999, the Hudson-Webber Foundation awarded a grant to the Detroit Police Department and Wayne State University (WSU) to train 2,000 Detroit police officers in a new computerized crime-mapping system. The system uses ArcExplorer software (in combination with a web browser) to provide daily information, in map form, to police officers across the city over the department's wide area computer network. Using the mapping system, officers can view crime patterns and access crime data, and supervisors can maximize resource allocation. In other cities, computerized crime mapping is credited with helping to dramatically reduce crime by supporting better problem solving. In Detroit the technology is being deployed so that it is accessible to all police officers—from top executives to patrol officers. As such, the goal of this project is to effect systemic change that will result in improved policing and safety in the city of Detroit. This project features a rigorous evaluation. Key research questions focus on the use of the system by police officers at all levels (executive, command, investigative, patrol, special operations), officer data needs, officer opinions on the utility of mapping, and the actual impact of crime mapping on resource allocation and crime levels. (Ref: The Impact of Crime Mapping in Detroit, Michigan Track: Law Enforcement and Criminal Justice by David Martin; also http://gis.esri/library/userconf).

Youth Violence

An attempt was made to describe a police-medical collaboration to better understand youth violence in Springfield, Massachusetts. Police data on violent incidents involving youths with spatial data and socio-demographic data from the American Community Survey in a GIS was combined. The geographic distribution of incidents and identified police sectors at increased risk for youth violence was examined. Seventy per cent of all incidents occurred in the offender's home sector. Regression analysis identified race and several economic indicators of the home sector as significantly related to volume of incidents. This will help in demonstrating how this approach can be applied to improve the design of intervention programmes and to plan resource utilization in preventing youth violence. (Ref: Using GIS and the American Community Survey in Understanding Youth Violence Track: Law Enforcement and Criminal Justice by Jessica Renae Schueler, Jane Leslie McCall Garb, Colleen Diane Flannery, Amy K. Pasini, Richard B. Doneit, Richard B. Wait; also http://gis.esri/library/userconf).

False Alarms

Linking GIS capabilities with an automated software system such as CryWolf for the tracking and processing of false alarm calls has resulted in a more efficient use of time and resources and a reduction in false alarms. The Charles county, Maryland (111,050), False Alarm Reduction Unit (FARU), in collaboration with Public Safety Corporation (PSC), has served as the initial site for the operation and evaluation of the CryWolf software. Alarm data is imported from computer-aided dispatch and used to send letters and prepare reports. Data and statistics can also be graphically presented on local county maps using MapObjects programming. Additionally, graphs and charts can be prepared and exported to other programs for use in reports to management.

The number of municipalities adopting alarm legislation is rapidly increasing due to the growing concern over police officer safety and the costs related to the reduction in manpower and resources due to false alarms. Unfounded alarm calls reduce law enforcement effectiveness and undermine their ability to respond to more critical calls. For residential and business property owners who are concerned about crime, alarm systems offer an added sense of safety and security. They can be an effective crime deterrent, but if they are unreliable and result in false alarms, they pose a major problem for everyone.

Due to the requirements for new legislation and the resulting delays in trying to create a quick fix in-house, there is a growing need for a sophisticated software system to assist in the tracking and processing of alarms. PSC has created a GIS-based application that brings the power of geo-spatial analysis to the false alarm arena. Utilizing ESRI's MapObjects v2.0 allows CryWolf to be translated into a comprehensive programme planned, designed, and created specifically to track false alarms. Unlike current software solutions, CryWolf is a very user-friendly system that maps alarm locations, allowing for the visual interpretation of alarm patterns. Agencies can customize multiple programme features, such as clearance codes, letter formatting, and custom letterhead design and utilize a variety of data collection modules for alarm companies, alarm registrations and alarm call entry. CryWolf is designed to use enhanced TIGER data 'shape' files for the current jurisdiction, and offers the mapping capabilities that geo-spatial users expect. Truly customized software for tracking false alarms can cost up to several hundred thousands dollars. The more elaborate the requirements are and the more customization requested by the user, the more costly the software will become. CryWolf is designed as a generic false alarm application with a variety of customization features. This allows the software to accommodate the needs of many jurisdictions without having to resort to custom code.

Linking GIS capabilities with an automated software system such as CryWolf for the tracking and processing of false alarm calls has resulted in a more efficient use of time and resources and a reduction in false alarms. During the initial testing and verification phase, almost 7,000 false alarm calls have been processed through the programme. Test notifications, and invoices have been mailed to hundreds of county residents to date. The receipt of these letters has already impacted the county by accelerating the number of registrations being received each day. Director of FARU remarked that with the CryWolf application we have been able to accomplish more in the past four months than what we were able to do in the preceding sixteen months.

The reduction in false alarms is essential to address the rising costs to tax payers, raise additional funds to be allocated towards community education and crime prevention, increase response time and presence in the community and free up personnel from manual tasks by providing the billing mechanism to generate fees, which will more than pay for the initial investment in the CryWolf application. Furthermore, alarm users will gain a better understanding of the impact of repeated false alarms, with more alarm businesses and owners inspecting and updating their systems to prevent further false alarms. (Ref: Using Maps to Track the Incidence of False Alarms Track: Law Enforcement and Criminal Justice by Edward Chicca, Denise Shaw; also http://gis.esri/library/userconf).

Community and Public Safety

Winston-Salem, North Carolina, was the first community to develop a community safety information system (CSIS) as part of the Department of Justice (DOJ)-sponsored Strategic Approaches to Community Safety Initiative (SACSI). The successful focus of the Winston-Salem initiative has been on the implementation of research-based strategies to reduce and prevent juvenile violence. CSIS, an internet map server application that integrates diverse community data from multiple agencies, was developed as an analytical tool to assist the local SACSI efforts. CSIS has been enhanced to expand its functionality. The local SACSI has been awarded a $1.8 million grant from a local charitable trust to establish a Center for Community Safety at the Winston-Salem State University campus. The work to reduce juvenile violence and improve community safety however continued. (Ref: Community Safety Information System in Winston-Salem, North Carolina Track: Law Enforcement and Criminal Justice; by Julia B Conley)

In public safety, geography is playing an important role more so than ever. Most 9/11 incidents often generate a variety of data from multiple agencies. The capability to access and process locational information quickly and deploy resources where needed can be extremely critical. Decisions based on this flood of information are about to wash over the continent's public safety agencies. More and more Public Safety Answering Points (PSAP) are receiving E911 (Enhanced 911) calls from cellular devices. The FCC 99-245 mandates for wireless is slated for October 2001 holds PSAPs accountable for locating cellular E911 calls geographically. Plant equipment's ORION solutions based on ESRI core technology are on the job. (E911-GIS Applications in Public Safety Plant Equipment Inc. Track: Law Enforcement and Criminal Justice by Jeff Freeland; also http://gis.esri/library/userconf)

Law Enforcement

A presentation showcasing GIS applications being used by the Greenville, South Carolina, police department that impact daily operations was made. Topics include crime analysis, traffic collision analysis, tactical/SWAT applications, court prosecution presentations, critical event planning, and how GIS supports community-oriented policing, and accreditation. Actual GIS projects are used, for example, to show how GIS was employed to determine three different patterns of auto breakings involving professional auto breakers, vagrants, and juveniles. Traffic collision analysis and coordinate mapping from patrol vehicles were demonstrated. This presentation is intended to highlight real-world GIS uses in law enforcement that may be easily adopted in other agencies. (Ref: GIS and Law Enforcement: Real-World Applications Track: Law Enforcement and Criminal Justice by Robert Gamble)

In another case, the Spokane County Sheriff's Office and the Spokane police department used GIS in their two-year investigation of the Spokane Serial Killer. Victims were geographically classified using ArcView GIS. Suspect tips were geocoded for spatially relevant events that coincided with known activities of the serial killer. The USGS Animal Movements extension was used to generate the Jenrich-Turner home range of victims and known key *modus operandi* evidence. The centres of the calculated ranges were within twelve blocks of the suspect's residence. Using ArcInfo, information from the arrested suspect's GPS transmit/receiver was calculated into a common projection and datum and combined with statewide data downloaded from federal web sites. The resulting map has been distributed to law enforcement agencies in Washington state and several unsolved cases have reopened. (Ref: GIS in the Spokane Serial Killer Task Force Investigation Track: Law Enforcement and Criminal Justice by John Bottelli, Ariane Schmidt, Laurie Anderson)

Further, methamphetamine continues to be a major threat to public safety in Illinois. The reported number of labs seized rose 925 per cent from 1997 to 1999. To stay ahead of the epidemic, a methamphetamine risk model was created to identify potential hotspots in the state at risk to the methamphetamine epidemic. By integrating known characteristics of methamphetamine users and Illinois county demographics into a geographical model, high-risk areas having characteristics conducive to methamphetamine use and manufacture could be identified. The risk model serves as a strategic tool that assists law enforcement and communities in forming proactive strategies to combat the insurgence of methamphetamine. (Ref: Illinois Methamphetamine Risk Model: Predicting an Epidemic Track: Law Enforcement and Criminal Justice Author(s): Boyd W Butler, Lex Bitner, Jennifer Gardner).

In a different application, the National Center for Missing and Exploited Children (NCMEC) assists law enforcement agencies in locating missing and abducted children by distributing the child's photograph. Former and/or current addresses of the abductor are gathered from searching the public record databases. Using MapObjects, an interactive mapping programme has been created that allows NCMEC staff the added capability to assess leads based on chronological and geographical proximity to reported addresses of a suspect and/or a missing child. This ability to view leads, both by location and by time, will add a new dimension to how analysts determine the value of information reported to the centre.

NCMEC, located in Alexandria, Virginia, is a non-profit organization congressionally mandated under the Missing Children's Assistance Act of 1984. NCMEC's mission is to assist law enforcement, families, and social-service professionals in finding missing children and help prevent child sexual exploitation. In our goal to recover children, NCMEC has recently developed an interactive mapping application that connects NCMEC's data to geographical or spatial data and visually represents NCMEC's attributes on a map. The application assists case managers and case assistants by allowing them to visualize all the location information related to each case and then focus poster distribution to specific areas for valid leads. The mapping application will assist staff analysts by spatially scoring leads based on the sighting location and its relationship to address information that the analyst has recently incorporated into NCMEC's database.

By being able to visualise the locations of addresses and high-priority leads, it is hoped to recover more children in less time. A visual representation of a case will enable to create or modify strategies for additional poster distribution to generate more valid sightings. It is also expected to be able to predict routes of the abductor or child based on earlier sightings, and poster distribution can be focused on a localized area where the child or abductor may be headed. Finally, otherwise undetected patterns may be seen when the case is viewed geographically. (Ref: Law Enforcement, MapObjects, and the Recovery of Missing Children Track: Law Enforcement and Criminal Justice by Geraldine Kochan; also http://gis.esri/library/userconf)

Emergency Services

Here is a case on the central role of GIS in the creation and integration of an E911 database by the St. Louis county police department. The St. Louis county police have successfully implemented GIS practices that allow integration of the E911 database with a new Computer-Aided Dispatch (CAD) to provide emergency services to the more than one million citizens of the county. Additionally, the E911 database has been integrated into an in-house GPS application and has become the focus of a multi-departmental data migration to an ArcSDE environment. (Ref: Integrating GIS into Emergency Services: A Case Study Track: Law Enforcement and Criminal Justice by Paul Trudt, Mark Dougherty)

Summary

The above examples do indicate the potentialities to apply GIS in different spheres of law enforcement and criminal justice. But these case studies are from western world where the government and justice departments have sufficient funds to use GIS related technologies to circumvent crime. The conditions in the developing countries are entirely different. The day-to-day problems have over taken the long term preparations such as development of GIS data base, training of personnel, and creation of GIS infrastructure. Nevertheless, 9/11 types of emergency, terrorism and high-tech crimes have compelled the governments all over to give due attention to GIS. Even the judgments are now based on technical or scientific data generated from spatial technologies such as air photos, GIS, remote sensing and GPS.

15

GIS in Defence Applications

Geography, or terrain, matters to the military commander who must understand terrain to make informed decisions in all matters, whether administrative, operational, tactical, or logistical. GIS can prove to be useful in this regard. Though at present, most armed forces and defence organizations continue to use paper maps as the principal tools for understanding terrain. As computer systems and digital geographic information (DGI) becomes a more important decision support tool, however, they will need GIS to exploit that DGI to make decisions regarding the base or facility management, command and control, hydrographic, production and logistics. The choice for the selection of GIS software has to be limited from the off the shelf standard packages. However, it has been observed that the choice varies depending on various applications. Softwares have been made customized perhaps for which the industries have never thought. The potentialities are enormous. Some examples have been mentioned here.

Command and Control

In Sweden, GIS is a fundamental technology in the military's Command Control and Communication Information (C3I) system, which captures, stores, standardizes, analyses, and georeferences the many disparate types of data bombarding their military commanders on a daily basis. As the use of GIS grew at Swedish Air Force (SAF), it became apparent that a comprehensive battlcfield GIS was needed. Essential functions included the production of tactical situation maps, convoy planning, target analysis, war game monitoring, and a variety of related applications.

In 1996, in a determined effort to reduce costs and standardize all GIS development for its C3I system, the SAF decided to adopt commercial off-the-shelf (COTS) products for an enterprise GIS. A GIS embedded in superior C3I systems provides options to overcome some of the inherent problems in using the technology on the battlefield. It provides a lower learning curve to the end user and allows quicker response to rapidly changing battle conditions. SAF developed GeoPres API, particularly for symbols management and geographical data grouping. GeoPres is the world's most advanced deployable command GIS. It allows the distribution of information from the company command level to armed forces headquarters. It also functions as an integral part of a decision support system. The system enables the building of powerful symbology for military units including size, type, transportation component, and name of unit. It is also possible to create new symbology. Since component symbols can be built, it has applicability outside the military. GeoPres was developed for the military, but is generic enough for general use.

(www.esricanada.com/english/events/defence/files/2004/John-Day-keynote.pdf)

Geospatial News on Demand Environment

As CNN and other news sources broadcast news around the clock, MITRE Corporation in Bedford, Massachusetts, captures the footage and maps it. MITRE is an independent not-for-profit company working in the public interest. It operates federal funded research and development centres for the Department of Defence, the Federal Aviation Administration, and the Internal Revenue Service. In May 1998, the Defence Advanced Research Projects Agency (DARPA) with a project to develop a Geospatial News On Demand Environment, called GeoNODE, tasked MITRE. GeoNODE helps people navigate and perform analysis on news coverage emanating from CNN broadcast news, print media, archives, and other sources.

GeoNODE allows analysts to navigate the news both temporally and spatially so that they can see the geographic distribution of news at the time it was reported. DARPA is investing in GeoNODE to help military personnel visualize information. Personnel can look at patterns and add value based on their knowledge of common sense issues and the kinds of risks involved. The news-processing component of GeoNODE is provided by a MITRE-sponsored research project called Broadcast News Navigator (BNN). Part of the news processing involves a natural language processing system called Alembic that extracts from each news report the named entities (people, locations, organizations, dates, and currencies), which are then plotted to a map using GIS. These component technologies have in part benefited from DARPA's prior investments in natural language processing and broadcast news understanding.

GeoNODE has the potential to greatly reduce the amount of time intelligence analysts have to spend watching broadcasts and searching archives. This process operates in near real time. It takes several hours to process the CNN video, so the information is typically available the next day. A valuable source of information about weapons of mass destruction is found in the archives of the Center for Non-proliferation Studies of the Monterey Institute of International Studies, called the Monterey Abstracts. This data is collected off HTML pages, stored in Sybase, and plotted to the map just like the CNN information. (www.es.purdue.edu/homes/diffton/dowmwnt/CDE01.pdf)

Spatial Data Standards for Facilities, Infrastructure and Environment

The Computer-Aided Design and Drafting (CADD)/GIS Technology Center for Facilities, Infrastructure, and Environment was established at the Information Technology Laboratory (ITL), US Army Engineer Research and Development Center (Waterways Experiment Station [WES]), Vicksburg, Mississippi, in October 1992 to support the Army, Navy, Air Force, and Corps of Engineers. One of the major initiatives assigned to the CADD/GIS Technology Center is the development of the Spatial Data Standard for Facilities, Infrastructure, and Environment (SDSFIE). The SDSFIE is the only non-proprietary GIS data content standard designed for use with predominant commercial GIS and relational database software. The SDSFIE focuses on development of data content standards to facilitate share ability and interoperability with the enterprise GIS at Air Force, Army, Navy, and Marine Corps installations; US Army Corps of Engineers Civil Works activities; and other federal government organizations.

This non-proprietary design, in conjunction with its universal coverage, has propelled the SDSFIE into a *de facto* standard throughout the Department of Defence; other federal, state, and local government organizations; public utilities; and private industries throughout the United States and the world. The CADD/GIS Technology Center annually updates and expands the SDSFIE. (www.gis.esri.com/library/userconf/proc03/po519.pdf)

4. Providing Environmental Information and GIS Analysis Tools to Support Decision-Makers

Naval Warfare Publication 4-11 (NWP 4-11) was promulgated in October 1998, and it requires that environmental considerations be included in operational planning. It further provides Navy and Marine Corps commanders and planners with doctrine to accomplish their assigned mission while achieving the highest possible degree of environmental protection and compliance. In addition, NWP 4-11 includes guidance on the preparation of an environmental annex to an operations order or operations plan. In order to support fleet commanders and planners, a GIS application called the Navy Environmental Planner (NEP) has been created for the Navy. The NEP allows the fleet planner to combine and view many types of data including environmental data, infrastructure data, baseline cartographic data, and oceanographic data. Furthermore, the NEP provides the ability to produce an in-depth analysis of the environment of a user-specified area of interest in an easy-to-use graphic interface.

(http://isc.wes.army.mil/products/tssds-tsfms/tssds/articles/papers/EsriMiltaryBaseMgmtBrochure.pdf)

Installation Management for Regional Planning, Facilities Management, and Public Works Operations

The Commander Navy Region Hawaii (CNRH) use GIS to keep track of its shore installation facilities and infrastructure and desired the capability for information retrieval and analysis to support its facilities planning, management, and public works efforts. A component-based GIS application called the Regional Shore Installation Management System (RSIMS) has been created. It combines spatial data with tabular information to provide its users with a visual tool that allows them to intuitively perform reporting and analysis functions for regional planning and facilities management. RSIMS links the Region's spatial data to databases such as Naval Facilities Assets (NFA), which contains real property inventory for land, buildings, structures, and utilities owned or leased by the Department of the Navy. RSIMS is also linked to maintenance management system software that tracks and manages repair, maintenance, and service work orders for many military activities in Hawaii.

The integration of GIS technology with operations databases results in an easy-to-use, interactive graphical interface allowing users to navigate and retrieve information via a map. It also provides reporting and analysis functionality for work orders, equipment inventory, preventive maintenance inspections, and job plans for recurring work. Through RSIMS, multiple users can instantly access spatial data from their desks in real time and are able to overlay various data types onto each other. RSIMS has increased worker efficiency and improved response time. Users are now able to do data analyses and see relationships between different data sets—tasks that were difficult or impossible to do when the data existed as paper maps and stand-alone databases.

Another benefit RSIMS has provided is the ability to share data among various departments, thus eliminating data redundancy and fragmentation as well as streamlining data maintenance by making it easier, faster, and more cost-effective to update the maps. RSIMS utilizes a spectrum of spatial data required for facilities and installation management.

(www.esri.com/industries/defense/files/defense-esri-infrastructure.pdf)

Integrated Corporate Solutions to Installation Management

The US Army Training and Doctrine Command (TRADOC) trains the Army, develop the war-fighting doctrine, and commands assigned activities and installations. The Deputy Chief of Staff for Base Operations Support (DCSBOS), Headquarters (HQ), TRADOC, Fort Monroe, Virginia, provides mission support and facilities installation management and accesses, trains, sustains, and deploys the forces. DCSBOS supports sixteen installations, or cities, where more than 150,000 people work, live, and train on two million acres of land and in almost 170 million square feet of facilities. Further, DCSBOS is organized into functional directorates that include military and civilian personnel management, resources management, logistics, acquisition, engineering, community activities, and other processes to manage the infrastructure, services, and environmental issues for the TRADOC installations.

Substantial information resources used in decision-making are spatial data or directly related attribute data. Unfortunately, DCSBOS had no structured information library for integrating, viewing, and analysing current or historical information for various sources. All of the various DCSBOS functional IT systems manage, store, and summarize data in stovepipe hierarchies. Also, there was no repository of time-based, summary information to help make decisions. As a result of an intensive, interactive planning process involving all the eleven directorates, DCSBOS created an implementation plan to guide the development of a corporate database (CorpDb). With its purpose to "Achieve Information Dominance —The Right Information at the Right Time," the CorpDb system is a framework to plan, design, create, and dispense geo-based information technology to support decision making throughout the DCSBOS base operations organization.

DCSBOS, developed an enterprise-level, GIS-based decision support system (DSS) known as the BASOPS Corporate Database. Using the CorpDb approach effectively breaks down the stovepipe decision-making of the past. The system is a relational database model integrating spatial data (data that references a location on a map), attribute data (descriptive information about the spatial entities), and regular tabular data. The GIS technology allows massive amounts of information to be stored, accessed, managed, and analysed on the computer, then viewed geographically on the screen in map form. Data is shared across these functional domains. Both spatial and non-spatial data can be directly exchanged with the Army's other major commands, Department of the Army, and other military services.

Facing a wide range of potential users, the CorpDb's designers created a system for the entire user community. System developers access the CorpDb's capabilities and certain designated power users in each directorate through desktop GIS loaded on individual PCs. By having direct access to GIS application, users can create and add new data layers and perform advanced analyses. The remaining DCSBOS users, whether on site or in the field, reach the system through CorpDb web application (CorpDb Web), now under development. The Garrison Commanders and Directors of Public Works (DPWs) in the sixteen installations have been provided copies of their installations' version of the CorpDb. The next phase of installation support will allow installation users to reach the CorpDb web site. Also, an installation version of the CorpDb LAN system, with direct links to the DCSBOS CorpDb, is in the planning and design stage.

(http://isc.wes.army.mil/products/tssds-tsfms/tssds/articles/papers/EsriMiltaryBaseMgmt Brochure.pdf)

Bird Avoidance Model

A problem often faced by both military and commercial pilots is loss or damage to equipment when an aircraft collides with birds in the air. A bird strike at any point during a flight can cause serious equipment problems and even result in the loss of human life. Since 1985, there have been more than 30,000 birds-aircraft strikes recorded by the US Air Force alone. These strikes have caused thirty-three fatalities, thirty lost aircraft, and more than $450 million in equipment damage. Nearly one quarter of all bird strikes occurred on low-level training routes or over military operation areas, and nearly two-thirds of total equipment damage occurred on those same training routes and operation areas. Thus, knowing bird strike probability on any particular mission can help prepare pilots, planners, and schedulers in the selection of flight paths and schedules. US Air Force uses the Bird Avoidance Model (BAM), which has been created using, GIS technology to analyse and correlate bird habitat, migration, and breeding characteristics with key environmental and human-made geospatial data. The original BAM is an intuitively designed desktop application for the Bird Air Strike Hazard (BASH) team of the US Air Force. The BAM application includes interfaces for specific users such as aircrews, planners, and schedulers. The original BAM application has been transferred to the internet. The BAM internet application allows the user to identify, zoom, pan, and view on the web various models of bird strike information within specific time spans.

The BAM web site not only provides users with information, but it also provides them with access to frequently asked questions and instructions for using the models. The user can create and interact with maps that display the models and other geospatial themes.

(www.esri.com/news/arcnews/sumer04articles/usaf-uses-gis.html)

Management of Natural Lands/Assets and Human-made Infrastructure

Defence organizations are frequently being asked to make their operations more efficient using new technologies. To increase efficiency, data management needs have become a priority in defense organizations around the world. The tenth Civil Engineering Group (CEG) located at the US Air Force Academy (USAFA) started using GIS technology to create an intranet based application for their existing planimetric and orthophotographic data. The resultant application gives users the ability to query, identify, zoom, pan, and view various layers of mapping information via the web. It is also important that any intranet or internet GIS solution be easy to use and have a high level of functionality. Therefore, this intranet GIS solution for the tenth CEG features a cursor display with live geographic coordinates and a zoom to building or road function. The application also allows the user to query and identify functions. With this intranet GIS application, the tenth CEG implemented a comprehensive, integrated, base wide GIS that aids in the management of their natural lands or assets and human-made infrastructure.

(www.geoinsight.com/projects/USAF/Academy)

Mission-Critical Decision Support

Geospatial information plays a key role in our national defence. Knowing what is where is the key to mastering the battlefield, installations, and deployed areas that support the war fighter. The United States Air Force Vision 2020—"Global Vigilance, Reach and Power" recognizes information superiority as one of six Air Force core competencies required to defend the United States and protect its interests through aerospace power.

The Air Force is implementing geospatial information technologies to support mission processes at each installation in accordance with Air Force GeoBase foundations and principles. GeoBase is an Air Force initiative to develop enterprise geospatial information cooperatives to support mission objectives at Air Force installations. The GeoBase initiative prescribes a set of principles, policy, and guidance to govern the process of geospatial information management at USAF installations. GeoBase Simulator demonstrates how network-based geospatial information technologies could be used to support real-world crisis and daily mission decision making at Air Force installations, ranges, and forward mission locations.

The Air Force exercised two crisis scenarios—a bomb threat and an aircraft accident. In the first scenario the Wing Operations Center (WOC) is notified by the security forces that there is a bomb threat to a key building on base. To address the requirements of the first scenario, a blockade and building status tool has been developed that can be used by security personnel in situations where a controlled area must be cordoned and traffic control is necessary. The tool automatically centres the view to a selected structure, adds a user-defined cordon area, generates traffic control points (TCPs) for road intersections immediately outside the cordon, and creates a formatted report of the intersections available to the WOC as well as the on-scene commander via a wireless network and mobile computing.

In support of evacuation of the cordoned area, the tool enables the user to modify TCP status and location, display building evacuation status, view structure floor plans, and report which buildings fall inside the specified cordon. The formatted building list is also available to the on-scene commander. In the event of an airborne toxic material release, the tool is integrated with the ALOHA aerial plume generation application co developed by NOAA and the EPA. The specified cordon area, traffic control points, and selected buildings and roads can be modified to account for both the plume footprint and 95 per cent confidence area (established to account for sudden weather changes). The blockade and building status tool is designed to support a primary seat (at the WOC) and multiple secondary seats via ArcView RPC Client/ Server requests. This approach allows the primary seat to control the display of building and TCP status, intersection, and building status reports and the display of potential plume footprints for the entire support team. Using RPC, the views on secondary seats at remote locations can be refreshed with real-time updates.

As part of the bomb threat scenario, a call is received at the WOC indicating that a munitions convoy is *en route* from a munitions igloo to the airfield. Established munitions convoy routes intersect the cordoned bomb threat hazard area. In the second scenario, an aircraft accident unfolds as the Marine Corps contact the WOC to report an Air Force F-16 has crash-landed on their air station. The key crisis response objectives are to mitigate any hazards associated with the site and route emergency responders to the scene of the accident. The application determines drive time to route first responders and HazMat teams to the accident site.

(www.gis.esri.com/library/userconf/proc04/docs/pap1839.pdf)

Facility Management

United States Navy in Yokosuka, Japan, Pearl Harbor, Hawaii, and San Diego, California use GIS technology to improve the overall work flow processes of the Public Works Center (PWC) staff. Applications include for casual users and custom tools for comprehensive document integration of floor plans, photos, and cultural and natural resource surveys from multiple departments.

Custom tools developed for long-range capital improvement planning and three-dimensional visualization of project plans allow personnel the ability to see how their base will look years down the road. The integration of these applications creates more efficient workflow processes. Consequently, this allows the staff to make better-informed decisions.

In Pearl Harbor, GIS is used to support regionalization efforts including footprint reduction: the demolition of inefficient structures to reduce operating costs. At the request of the Commander of Navy Forces, Japan, tools were developed to analyse and correct chronic parking problems when the fleet is in port. A tool has also been developed for the Environmental Department of Commander Fleet Activities, Yokosuka, which links existing water sample databases to the GIS base map. In October 1998, CADD/GIS manager, PWC Yokosuka, embarked on an ambitious project of the creation of accurate base mapping of their facilities. As the project developed, the US Air Force saw the financial benefits of the efforts. Under the direction of PWC Yokosuka, many DOD installations in Japan are developing planimetric maps, digital orthophotos, and three-dimensional models.

(www.esricanada.com/files/software.GISDIC-v1.pdf)

Cultural Resource Management

The cultural resource management effort at Edwards Air Force Base (AFB), California, makes extensive use of a Geospatial Information System (GIS) for a wide variety of tasks. GIS is used to document the cultural resources at Edwards AFB. Cultural resources sites and isolated finds are recorded and are linked to a detailed database of information about each resource. In addition to this, information has been created for data such as cultural resources survey areas, early homestead patents, potential historic districts, and cultural resources management planning regions. Because several different contractors are involved in collecting cultural resources data, ensuring that the contractors all collect the same data and record it in the same way is a very real concern. To ensure that contractors collect artifact data in the proper data format, and to ensure that the artifact data required by the base is collected for each artifact, a custom database was created. All of this descriptive data are available in pull-down pick-list menus, helping to eliminate typos and misspellings from the database. A 'remarks' field is provided for contractors to include any additional artifact information not available in the lists.

GIS allows the use of remote sensing data to assist in the recording of cultural resources. Some of the historic period cultural resources at Edwards AFB were identified from old USGS maps and land records, and have not been formally recorded. Recording these sites is an ongoing project at Edwards. To get a better idea of the size and features present on these sites, the existing location information is overlaid on aerial photographs of the base using GIS technology. In addition to the many benefits to managing cultural resources and project planning, GIS also supports the study of the many prehistoric sites on base. GIS provides the ability to overlay the locations of recorded prehistoric sites with other data on the physical environment. Site or artifact types can be quickly correlated with prominent vegetation or soil types, land slope, or proximity to water sources. As more of Edwards AFB is surveyed and more prehistoric sites receive detailed examination, these types of GIS analysis may provide new insight into the lives of the people of the Mojave of long ago.

(www.mojavedata.gov/docdata/docs/EAFBINRMP2002.pdf)

GIS Software Options

The Australian Army has selected ESRI software for the GIS component of its Battlefield Command Support System (BCSS). Both the Australian Navy and Air Force have implemented similar systems, and BCSS was to provide the Army with a long overdue command support system. ESRI's ArcView GIS software provides the basis of the GIS module for BCSS, together with its ArcView 3D Analyst and ArcView Spatial Analyst software extensions. The software was to eventually provide users with tools to perform sophisticated terrain and mobility analysis as part of their command support role. The initial rollout of BCSS will consist of 140 ArcView GIS licences as well as ArcView 3D Analyst and ArcView Spatial Analyst extensions.

BCSS is a modular system based on commercial off-the-shelf (COTS) technology. The BCSS project team has been watching industry trends closely and only tested those COTS products that can realize at least eighty per cent of their solution and have the potential for further customization. Initial development of ArcView GIS has already been carried out in the engineering and intelligence departments within the Army, and this work will be incorporated into the GIS module for BCSS. The development process has involved close consultation with current and potential users. The Army's First Brigade was to provide testing services as the system is being rolled out. The plan is to gradually expand the use of the BCSS in a structured manner. According to a project team member, "We are planting the (GIS) seed by providing access to the technology." He hopes the first stage of BCSS will whet the appetite of Army users and provide a taste of what the technology can offer.

The project team is well aware of the rapid nature of change in the information technology industry and is consequently operating on a timescale of six monthly software reviews with hardware reviews every two years. It is anticipated that this approach will minimize the problems encountered with lengthy development periods, which could conceivably produce outdated applications. CelsiusTech Australia, prime systems integrator for the BCSS project, is incorporating other software into the system including the Command Data Network System-Army (CDNS-A), Lotus Notes, and the Microsoft Office suite. ADI has been contracted as the GIS developer for the project. The system will operate under a Windows NT environment on commercial and industrial workstations.

(www.esri.com/industries/defenses/files/defense-esri-army.pdf)

Australia's Defence Topographic Agency Deploys Nationwide Production System

In April 2000, in a move towards total automation, Australia's Defence Topographic Agency (DTA), formerly the Army Topographic Survey Establishment begins full-scale production of digital military geospatial data and hard copy maps using their new PARARE system. PARARE is the term coined for a larger programme within Australia that, in addition to data conversion outsourcing contracts, includes the implementation of a nationwide, enterprise topographic production system within DTA, the Australian equivalent of the National Imagery and Mapping Agency (NIMA) in the United States. DTA supplies maps and digital data to the Australian Defence Force and also unclassified products to the wider public through AUSLIG. ESRI–Australia's Defence office began work on the GIS component of the $18.4 million (Australian) PARARE project in November 1998 as a subcontractor to RLM Systems Pvt. Ltd. "The new system will provide the ability to produce military geospatial products required by the defense services of today and tomorrow," says DTA General Manager. Further, he said, "The new system provides greater flexibility to produce defence-specified products, and as other products come online in the future, we will have the capability to produce those as well." Increasingly, the DTA has been tasked to produce digital geographic products. Despite some forecasts of the digital age completely replacing the need for paper products, the General Manager says he has seen no evidence of this. He says close to 1,000,000 maps have been provided by DTA for the peacekeeping operation in East Timor recently.

(www.esri.com/news/arcnews/spring00articles/astralias.html)

Digital Data and Map Production Facility

The PARARE system provides a complete digital data and map production facility and will provide the largest map production capability in Australia with 113 cartographers, GIS specialists, image analysts, and technicians involved in production. It is believed that one of the key advantages of the system is the ability to produce digital products in ESRI format, which is what many defence users require. The PARARE installation includes several licences of ESRI's Production Line Tool Set software, forty licences of ESRI's flagship GIS software ArcInfo, forty-eight licenses of ArcView GIS, and three licences of ERDAS IMAGINE.

"The most important aspect of the new system is a GIS database as the backbone for all digital and hard-copy map production," says the DTA strategic development manager responsible for the evaluation of new technology and systems. In the past digital data was collected purely as a means to create the end product, a map. "Over time, CAD data became useful as digital data, and the demand has continued to increase," says the manager. As the use of digital data increased it was found that the CAD data was not of sufficient quality for further manipulation. The new GIS database will be the source of data for all products from quality hard-copy maps to digital products such as Digital Terrain Elevation Data and Vector Map (VMap). Data will pass through strict quality controls before it is loaded into the GIS database, and the structure of the database will ensure consistency and quality.

DTA will make extensive use of PARARE's map production capabilities to produce a number of hard-copy maps such as Topographic Line Maps (TLM) at four scales (1:10,000, 1:25,000, 1:50,000, and 1:100,000), Joint Operations Graphics (JOG), Operational Navigation Charts (ONC), Tactical Pilotage Charts (TPC), Jet Navigation Charts (JNC), Navigation Plotting Charts (NP), Air Staff Planning Charts, image maps, orthophoto maps, and special purpose maps.

The transition has involved enormous change for DTA staff. The staff was involved in production using the old system and has undergone intensive training in the use of the new software and hardware. The manager says that the implementation of PARARE has been a "big cultural change for people from a CAD system to the GIS; (however) almost everyone has adapted to the change and moved forward." The production process is broken down into components beginning with feature extraction from digital aerial photographs. DTA task contractors fly for aerial photographs that are loaded directly into the Helava SOCET SET system for feature extraction. The vector and elevation data extracted from the photography is then loaded into ESRI's Production Line Tool Set applications to create the digital and hard copy products.

Each component of the production process has a team of people assigned to it. For example, the Feature Extraction Cell staff will take the aerial photography and produce vector data, the Hard-copy Conversion Cell staff will load the vector data into the GIS database, and the VMap Production Cell staff will use this data to produce VMap Levels 1 and 2 maps. Hard copy map production also makes use of the data in the GIS database with staff in the Finishing Cell using the Production Line Tool Set to manufacture a digital map product. The process is completed with the digital version of the map being sent to the BARCO film writing system and then printed on the Roland offset printing press, which can produce up to 10,000 maps per hour.

(www.esri.com/news/arcnews/spring00articles/astralias.html)

Visualization and Intelligence

Applied Analysis Inc., one of the leading remote sensing firms, is implementing a web site for one of its military customers that will be accessed through SIPRNET—a classified version of the internet. The unique capability of the site includes the ability to interact with very large image files, covering an extensive geographic region of the world. The insights and lessons learned in adapting ArcIMS to this environment and mission will be presented.

This case study describes a geodata visualization system based on ESRI's web mapping solution. Geodata visualization refers to the means of visually correlating georeferenced imagery and vector data in a map context. Satellite-image exploitation software is used to derive ground cover classification products from imagery and the results are served up in map form where they can be visualized along with the base imagery and geographic features stored in a GIS. ArcIMS software enables this visualization using standard web browser technology, thereby imposing minimal hardware requirements and minimal learning curve on the user's part.

Applied Analysis Inc. has a long history of developing software tools to exploit remotely sensed geographic imagery. The current version of the subpixel classifier spectral image processing tool allows users to develop spectral signatures from multispectral and hyper spectral imagery and apply them to locate materials of interest within a scene, as well as scene-to-scene. The unique subpixel detection capability of this software allows the user to detect materials that occupy as little as 20 per cent of a pixel and correctly distinguish spectrally similar targets, such as a particular vegetative species, interspersed with other vegetation. The software has been used in a wide number of applications, including urban land cover studies, wetland tree species identification, crop identification, environmental monitoring, wetlands identification, and pest infestation location.

A US Government customer contracted with Applied Analysis Inc. to use our subpixel classification technology to create land use analysis products from commercial satellite imagery. The customer's area of interest covers millions of square miles, and dozens of Landsat Thematic Mapper image scenes were required to be processed. Additionally, the customer was interested in how land usage changed over time which dictated that archival satellite imagery be processed to develop a historical record.

From the user's standpoint, GVS is a set of dynamic web pages that allows them to focus in on an area of interest, select the products (overlays) they are interested in, and create a printed output. They typically begin with the GVS Products web page. This page lists available products by country. Hyperlinks within this page take you to the GVS web page containing products for that area. They can also jump to the GVS Global Entry page and browse for product areas graphically. Links to pages containing more information about GVS and the sponsoring project are also provided on the product's web page (*Source*: Gene V. Roe).

Analyzing Sonic Boom Footprints of Military Jets with GIS

Testing and training exercises involving supersonic US military aircraft can produce sonic booms. By using ArcView and PCBoom3, the U.S. Air Force's sonic boom modelling tool, the effects of sonic booms in a particular environment can be studied. Currently partially integrated PCBoom into ArcView, with a view to full integration of sonic boom modelling into a GIS is to follow. Its implementation allows users to either ingest actual flight data or design flights. Using real flight data helps review past supersonic impacts. Creating model flights allows the user to predict future impacts and determine how they may be reduced. Using this modelling capability, coupled with ground reference data, we are performing a parametric study that will be the basis for an interactive flight planning tool (*Source*: Andrew S. Rogers).

Creating a GIS Community: DoN GIS Communications Network

The Department of Navy GIS community is creating a communications network to centralize their GIS information and knowledge. The goal of this effort is to set up a seamless, interactive process for sharing GIS information, preserving institutional knowledge, and bringing together members that are isolated by geography or organizational divisions. At present the community's online tool, DoN GIS Communications Network (CommNet) allows members to share GIS data, project information, contract language, lessons learned, and contact information. With this approach, community members define content, duplication of effort is minimized, and investments are maximized (*Source*: Linna S. Manomaitis).

GIS in Armed Forces

GIS in an Armed Forces organisation is not a matter of choice of a certain software product or a matter or a choice of a specific platform, but is a systematic way of working with a number of new tools to handle the geographic information as one of the basic factors in the assessment and decision process in a headquarters (HQ), and it does not matter if one talks about a battalion HQ or an armed forces HQ. This implies that the use of GIS, necessarily, will affect both HQ organisations and their way of working. This awareness has grown in the Swedish Armed Forces (SAF) since the end of the 80s. This has also led in many way of trying to consider all aspects on the development and implementation of GIS which SAF wants to try out.

In 1989, there were more brigades in the Army, more combat vessels in the Navy, more fighter squadrons in the Air Force than today. Sweden had a rather static planning resulting in training of their units again and again in the same geographic area (the area in case of a war) and through this training the security forces got an extensive geographic knowledge—about the area where they were supposed to execute combat, but nowhere else. At the same time digital geographic data began to be available and the software packages as well to reasonable prices. A number of computer enthusiasts started to develop simple GIS applications.

In 1990 the Armed Forces and the National Land Survey signed an agreement concerning a shared GIS-centre, with people from the Land Survey organisation as well as with, initially, one officer from the Armed Forces. At the same time they tried to get an idea about what was going on in several places and they tried to get all these development activities to face a common objective. Further, in 1991 they had our first tests on GIS-applications during a divisional exercise in the northern part of Sweden. They also had the first agreements with the civilian producers of digital geographic data about availability for the Armed Forces. In order to establish some sort of umbrella organisations for the GIS efforts, the Armed Forces GIS Steering Committee was established with following mandate:

- to give entitlement to start new development activities.
- to co-ordinate GIS development efforts.
- to allocate the resources at the GIS-centre.
- to allocate other resources for GIS development as SW, data sets, test facilities etc.
- to encourage the initiation of new GIS efforts.

In 1993 they started developing the present GIS strategy which was signed by our Supreme Commander in 1995. Thus it has continued, more and more rapidly.

One of the most important tools for establishing a consistent GIS development and implementation in a large organisation, independent of which type of organisation it is, is the presence of a strong GIS strategy. It is not enough to have the strategy developed, signed, distributed and read. A strategy always consists of a number of outer frames for what is accepted and what is not accepted. This is often considered limitations for the developer but is absolutely necessary in order to achieve the desired co-ordination with the overall context for a C3I-system. If that strategy, with its limitations, is to be complied with, it has to offer something very beneficial for the individual who has been given the task to develop a GIS application. The GIS-strategy offers:

- GIS consulting resources
- Geographic data
- Software recommendations
- Hardware recommendations
- Development equipment
- Application test facilities
- Project manager training

Further, in addition to above, it offers a couple of other facilities without costs for the responsible project manager or his commanding officer. These assets are funded centrally in the Armed Forces. In this way the strategy will become close to loved.

One core issue in the strategy is the mandatory development process, with a limited local prototype phase initially, where the main responsibility is with the project manager, who always has to be an end user of the application under development. And this emphasis on the end user will be a central part of the entire development process until the implementation of the application. After the local prototype phase, where real trial and error is allowed we have a coordinated project phase aiming to the implementation. During this phase attempts can be made to widen the project organisation, involve reference groups and external test groups and start preparing the training of future users. Another core issue is the careful separation of common GIS tools as—(a) projection handling, (b) positioning, (c) best map functions etc. which are required by every user, and also for activity related applications which are requested by representatives for a specific branch, service or function. Such activity related applications can be—(a) artillery sites, (b) radio link communication establishment, (c) bridging, (d) low-flying paths for attack helicopters and so on.

The responsibility for development of those common functions is a matter for a central project organisation. On the other hand is the responsibility for the development of activity-related application for a particular issue for those who are in the middle of a specific activity, with both feet in the mud. Those persons are the only individuals able to describe the specific activity careful and systematic enough to become subject to an activity-modelling process subsequently followed by a data modeling process. And here it is vital to be extremely careful. Otherwise one can not get the desired results.

(www.gis.esri.com/library/userconf/europroc99/proceed/papers/paper904/P904.html)

GIS-Support for Peace Keeping Operations

Again Swedish Armed Forces (SAF) together with GIS-centrum takes an active part in the development of Military Geographic Information Systems (MGIS). A GIS can in many respects provide support for activities at a peacekeeping unit. The project, GIS Support for International Missions (GIS IM), was initiated by the commanding officer of one of the Nordic UN battalions in Bosnia. The GIS IM concept includes training of operators, application development, field verification, hardware and software. GIS IM is developed within the SAF's GeoPres concept.

The application development is based on prototypes at three sites, one for the Swedish battalion in Bosnia, one for training at SWEDINT and one for further development at GIS-centrum in Stockholm. The objectives are as follows:

- To develop an operational GIS support for different branches involved in international missions.
- To establish routines for the supply of geographic data.
- To plan and carry out training of GIS IM operators.
- To integrate GIS IM in the training and simulator environment at SWEDINT.

The extensions for ArcView is for (a) Military geographic information—support for documentation of military and civilian objects of interest for the mission, (b) visibility analysis—support for reconnaissance in order to establish observation posts, and (c) military symbols (situation maps)—handling of the most important NATO symbols and different types of unit boundaries.

A GIS-based UK Army Command and Control System have slightly different application. The Battlefield Management system (BATMAN) was in the final stages of development and was to be deployed in November 1996. This system was supposed to demonstrate the integration of commercial GIS and Oracle RDBMS with a modern development environment. BATMAN is a client/server application. It comprises ArcView and a bespoke Delphi client accessing the active data resident in an Oracle database running on an NT server (patrik.bremborg@lm.se).

GIS at Defence Research Establishment

The Danish Defence Research Establishment (DDRE) is both doing research and handling project for the defence. One of the interest areas for the Human Computer Interaction Section (HCI-section) is GIS.

Military operations have always required knowledge about the geography. They take place in time and space and require knowledge about the terrain. Therefore, the military has always been large-scale consumer of maps. When GIS made it possible to handle maps more effectively, and update and store digital maps more easily, GIS became an effective tool for the military. Today when GIS is moving beyond two dimensions, it has become even more essential to the military. Today most GIS are still just toolboxes that make it possible to store, display, and analyse geographical information. They often require extensive educational exercise to run the programme. Further, the analysis requires effective digital representation of the terrain, and one should be able to work together with other systems including other GIS. This gives the following interesting topics or reasons for promoting GIS-related activities:

- Generation of geographical information
- Visualization of geographical information

- GIS Interoperability
- Analysis in 3 and 4 dimension

The goals are to (a) require and pass on information on GIS, (b) improve the usability of future GIS-products for the defence, and (c) to improve the visualization of the large amount of geographical information required in military operations.

(www.gis.esri.com/library/userconf/europroc97/miltary/M2/m2.html)

Conclusion

The defence GIS market offers some of the greatest potentials for new applications and integrated GIS technology and systems. This market has traditionally been one of the larger consumers of geospatial technology, information and related services, although most production activity so far has been conducted in-house. As we move ahead, the defence and security market remains large and strong with an emphasis on maintaining global capabilities that enable the mobilization of operations over land, sea and the air. Overall, the defence GIS market spans a range of different requirements which are diffused, vary greatly and tend to be user driven. The nature of the market has moved from customized large-scale developments to commercial-off-the shelf (COTS) products. Similarly, the pace of technological development and changing operational environments have resulted in a change to meet the requirements toward flexible components, interoperability, standard tool kits and data and integrated systems development (Baijal, R., n.k.).

Future market opportunities in this sector are of course numerous. Key areas include advanced communications and secure networking to meet the demands for on-line, just-in-time delivery of information and decision support capabilities. Development of experimental, soon-to-be operational mobile warrior and integrated systems for in-field application continues to dominate research and early pilot phases. These and associated technologies such as advanced guidance, high resolution imagery, operational picture and precise location determination remain critical areas for GIS technologies, often within larger operational command and control systems. Along with the technological components getting integrated with the defence needs of the country, digital mapping, remote sensing, soft photography, GPS and GIS have made deep in roads. The field operations, un-manned aircrafts (UAVs), fly-through the difficult terrains and missile trajectories require spatial data and GIS has a role to play.

Nevertheless, defence systems require well supported access to geographic information. These requirements fuel an ever-expanding demand for innovative geospatial information discovery, access, integration and management tools to support the provision of responsive, accurate and timely global GIS. Delivery of the information from internally produced, discrete hard copy products towards interoperable, seamless, on-time delivery of services via national and global infrastructures, in essence, mobilizing technology support the delivery, storage and accessibility of information from within and outside the defence sources.

Defence forces all over the world have diversified their activities. It is not only the combat operations which they have to be involved, but terrorism and management of disasters have been added to their responsibilities though occasionally. Nevertheless, they have to be prepared for any eventualities: natural, man-made or geopolitical. As a result, maintenance, facility management, visualization, peace-keeping operations, research, 3 and 4 D analysis, command and control, decision support system and the like have made in roads in the defence operation. Such initiatives are not only limited to headquarters operations, but the field and lower formations do require such preparedness which they have do depend on GIS related activities.

REFERENCE

Baijal, R., M.K. Arora and S.K. Ghosh (n.k.), A GIS assisted knowledge-based approach for miltary operations (www. Gisdevelopment.net/application/military/overview/miltaryo0001.htm).

16

GIS for Decision Support System

The concept of Decision Support System (DSS) is based on the seminal work by Simon and associates in 1950s and 1960s (Simon, 1960). In the 1970s the concept of decision support systems (DSS) began to develop in the Information Systems (IS) community, notably with the work undertaken at the Massachusetts Institute of Technology (Gorry and Scott-Morton, 1971, Little, 1971). By the early 1980s there were many books and papers published in the DSS field and DSS had become a recognized part of IS (Sprague, 1980; Bonczek *et al.,* 1981). DSS had evolved out of the business data processing tradition and usually dealt with the financial and operating data associated with business use. The volumes of data involved with such systems were relatively small compared with those in the geographic domain. As computer systems became more powerful, some DSS type applications evolved that used map display or employed spatial information. A good example is the Geodata Analysis and Display System (GADS) which was used for routing applications (Grace, 1977). Nevertheless, the technology it used had limited graphics and inadequate processing power to exploit the full potential of spatial applications.

While these developments in DSS were taking place in the IS community in the 1970s, a largely separate trend of development took place in GIS, with developments largely concentrated on geographic data processing applications (Nagy and Wagle, 1979). Spatial applications had placed heavy demands on the technology, and this slowed the progression from data processing to decision support applications. However, over time improving computer performance led to increasing interest in spatial *what-if* analysis and modelling applications. The idea of a Spatial Decision Support System (SDSS) evolved in the mid-1980s (Armstrong *et al.,* 1986), and by the end of the decade SDSS was included in an authoritative review of the GIS field (Densham, 1991). This trend was evident in the launch of research initiative on SDSS in 1990 by the US National Center for Geographic Information and Analysis (Goodchild and Densham, 1993).

Consequently, by the early 1990s, SDSS had achieved a recognized place in the GIS community and was identified by Muller (1993) as a growth area in the application of GIS technology. The delay in the recognition of SDSS, compared to other DSS in other domains reflects the greater demands of spatial processing on IT. Nevertheless, despite these developments SDSS does not occupy a central place in the GIS field; and many introductory GIS textbooks do not mention SDSS at all (Clarke, 1997; Bernhardsen, 1999). This may reflect a feeling among many in the geographic disciplines that SDSS applications involve a diversity of techniques from different fields largely outside the geography domain. SDSS is an interactive, computer-based system designed to *support* a user or group of users in achieving a higher *effectiveness* of decision-making while solving a *semi-structured spatial decision problem.* Three terms, i.e. semi-structure spatial problems, effectiveness, and decision support capture the essence of the SDSS concept.

Less attention was paid to SDSS within the DSS research community, until the mid-1990s when some work in this area began to appear (Wilson, 1994). One of the first papers in an IS related publication illustrated the effectiveness of SDSS technology (Crossland *et al.*, 1995). Recently the benefits of SDSS for both inexperienced and experienced decision-makers (Mennecke *et al.*, 2000) were discussed in *MIS Quarterly*. The field has now grown to the point that it is made up of many threads with different, but related names, such as collaborative SDSS, group SDSS, environmental DSS, spatial knowledge based, expert systems and the like. The main characteristics of spatial decision problems include:

- a large number of decision alternatives,
- the outcomes or consequences of the decision alternatives are spatially variable,
- each alternative is evaluated on the basis of multiple criteria,
- some of the criteria may be qualitative while others may be quantitative,
- there are typically more then one decision-maker (or interest group) involved in the decision-making process,
- the decision-makers have different preferences with respect to the relative importance of evaluation criteria and decision consequences,
- the decisions are often surrounded by uncertainty.

Basic Decision-Making Process and GIS

Simon (1960) suggests that any decision-making process can be structured into three major phases (Fig. 16.1):

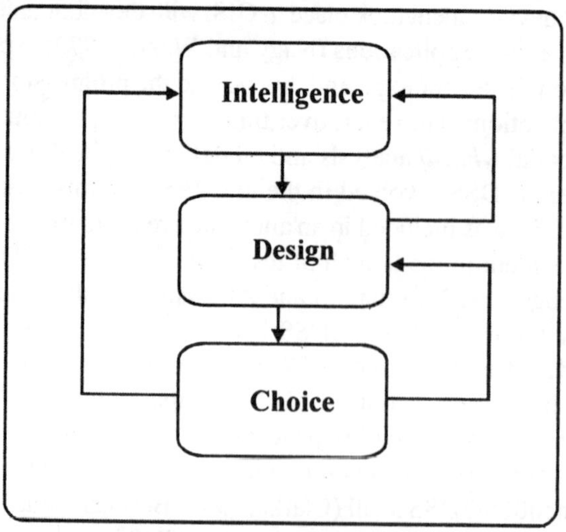

Fig. 16.1: Phases of decision-making

- *Intelligence*—is there a problem or an opportunity for change?
- *Design*—what are the decision alternatives?
- *Choice*—which alternative is best?

It is important to find out how and to what extent GIS technology provides support required in each of the three phases of decision-making:

In the *intelligence phase,* which involves searching or scanning the environment for conditions calling for decisions, requires an exploratory analysis of the decision situation. GIS can play a vital role at the initial stage of spatial decision-making by helping in coordinating decision situation analysis through its ability to integrate and explore data and information from a wide range of sources. GIS can also effectively present information in a comprehensive form to the decision-makers.

The *design phase* involves inventing, developing, and analysing a set of possible decision alternatives for the problem identified in the intelligence phase. In this stage a formal model is typically used to support a decision-maker in generating the set of alternatives. While an increasing number of GIS systems are described as systems for supporting the process of designing and evaluating spatial decision alternatives, but most commercially available GIS lack the kinds of spatial analysis and modelling required by decision-makers. The capabilities of GIS for generating a set of alternative decisions are mainly based on the spatial relationship principles of connectivity, contiguity, proximity and the overlay methods. In current GIS environments, models for generating decision alternatives operate in the background, detached from users' insights and qualifications.

The *choice phase* involves selecting a particular decision alternative from those available. In this stage each alternative is evaluated and analysed in relation to others in terms of a pre-specified decision rule. The decision rules are used to rank the alternatives under consideration and the ranking depends upon the decision-maker's preferences with respect to the importance of the evaluation criteria. The critical for use of GIS in the choice phase is the capability of incorporating the decision-maker's preferences into the decision-making process and in general, GIS systems do not provide a mechanism for flexible incorporation of the decision-maker's preferences into the decision-making process.

It can be observed that GIS systems have limited capabilities of supporting the design and choice phases of the decision-making process. The systems provide a very static modelling environment and thus reduce their scope as decision support tools—especially in the context of problems involving collaborative decision-making.

Problems Semi-structured Decision

Any decision problem falls on a continuum that ranges from completely *structured* to *unstructured* decisions (Simon, 1960). The structured decisions occur when the decision-making problem can be structured either by the decision-maker or on the basis of relevant theory. The structured decisions are programmable and can be solved by computers. On the other hand, the unstructured decisions occur when the decision-maker is unable to structure the problem and the problem cannot be structured on the basis of a relevant theory. These decisions are non-programmable and must be solved by the decision-makers without any assistance from a computer. Most real-world spatial decision problems, if not all, can be found somewhere between these two extreme cases of completely structured and unstructured decisions; these decisions are called *semi-structured* ones. The SDSS concept has major application this area. The structured (programmed) part of the problem may be amenable to automated solution by the use of a computer, while the unstructured (non-programmed) aspects are tackled by decision-makers (Fig. 16.2).

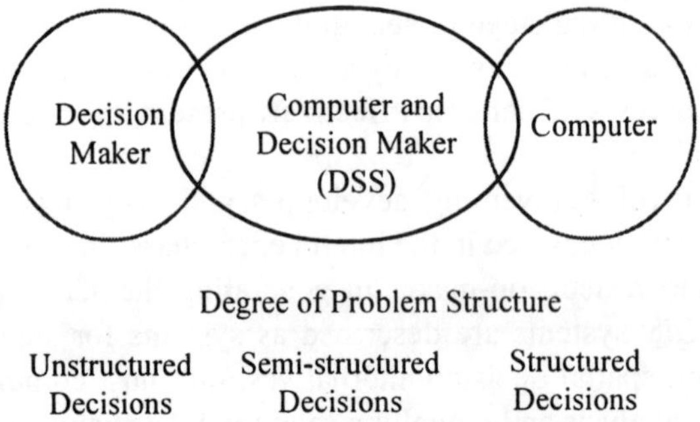

Fig. 16.2: Degree of decision problem structure.

The main aim of the system is to improve the effectiveness, rather than efficiency of the decision-making process. The higher effectiveness is achieved by incorporating the decision-maker judgements and computer-based programmes into the decision-making process. To be effective the system must be easy to use. The system also helps the users to explore the decision problem in an interactive and recursive fashion in all phases of the decision-making process.

Principles of Spatial Decision Support System (SDSS)

The technology for a DSS must consist of three sets of capabilities in the areas of *dialog, data,* and *modelling* (the DDM paradigm) (Sprague and Watson, 1996). A well design SDSS should have balance among the three capabilities (Fig. 16.3).

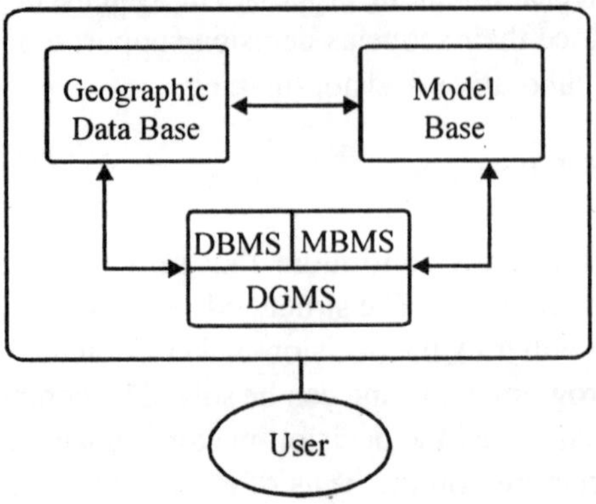

Fig. 16.3: Components of SDSS

- The *Data Base Management System* (DBMS) contains the functions to manage the *geographic data base;*
- The *Model Base Management System* (MBMS) contains the functions to manage the *model base;*
- The *Dialog Generation and Management System* (DGMS) manages the interface between the user and the rest of the system.

Technologies for Developing SDSS

DSS tools facilitate the development of either a DSS generator or a specific DSS (Fig. 16.4).

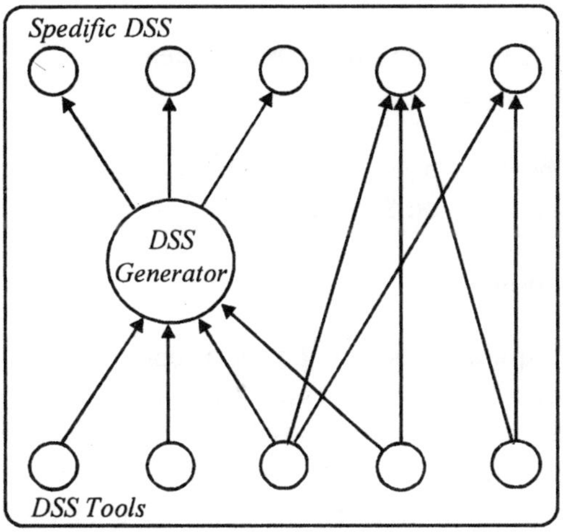

Fig. 16.4: Three levels of DSS technology (*Source*: Sprague and Watson, 1996)

Following are examples of some programming languages:

- The procedural programming languages and code libraries (e.g., Arc Macro Language (AML) scripting tool of ARC/INFO, Avenue—ArcView GIS software's built-in object-oriented scripting language, TransCAD—Caliper Script macro language, MapInfo—MapBasic);
- Visual programming language (e.g. STELLA II, Cantata and Khoros);
- Inter-application communication software (e.g. dynamic data exchange (DDE), object linking (OLE), open database connectivity (ODBC));
- Simulation languages and software (e.g. SIMULINK, SIMULA);
- Application programming interfaces (API) (e.g. the IBM's geoManager API, Java Advanced Imaging API, TransCAD's API);
- Applets (e.g. GISApplet, Microsoft Visual J++),
- Visual interfaces, graphics and colour subroutines (e.g. graphical user interfaces—GUI).

DSS generator is a package of related hardware and software which provides a set of capabilities to quickly and easily build a specific SDSS; examples include:

- GISystems (e.g. ARC/INFO, ArcView, ARCNetwork, Spatial Analyst, MapObjects LT, GRASS, IDRISI, MapInfo, TransCAD);
- Database packages (e.g. dBase, Access, Paradox);
- Decision analysis and optimization software (e.g. LINDO, EXPERT CHOICE, LOGICAL DECISION);
- Statistical and geostatitical software (e.g. S-PLUS, SPSS, SAS);
- Simulation (e.g. Spatial Modelling Environment).

Specific DSS are systems devoted to the analysis of a particular set of decision problems; the systems which actually support the decision-makers in tackling semi-structured problems; examples include:

- Active Response Geographic Information System;
- IDRISI Decision Support;
- GeoMed;
- Spatial Group Choice; and
- winR+GIS Spatial Decision Support.

Decision Support in Planning

Any planning initiative requires considering a lot of options before taking a final decision. Whether the pros and cons have been assessed ? In what way the proposed action is most viable ? What are the priorities ? Is the proposal financially sound in immediate and in long terms ? What are the alternatives ? Finally, whether such action is politically acceptable ? Issues similar to above are likely to crop up whenever a planning exercise is taken up. Answers to such questions are essential. A flow diagram (Fig. 16.5) is included here to show the role of GIS in such complicated decision-making process. Further, decision-making process in heterogeneous multi-user environment is shown in Fig. 16.6.

DSS *versus* SDSS

A SDSS is different from a DSS in the fact that it is used to support decision processes where the spatial aspect of a problem plays a decisive role. Despite ongoing discussion, there is no agreement on a single definition. Some compare this to describing a car where there seems to be general agreement on what it is, but not on what are the necessary attributes for giving it such a label. This explains why most authors simply end up with a list of characteristics of what a SDSS might comprise. For this study we prefer the statement by Densham (1991) that SDSS are ''explicitly designed to provide the user with a decision-making environment that enables the analysis of geographical information to be carried out in a flexible manner''. The lack of widely supported and unambiguous definitions of DSS and SDSS do not seem to hamper the development of such systems. On the contrary: there are innumerable systems under development for supporting a variety of decision questions. Most popular application areas are, at least in The Netherlands, coastal zone management, river management, water resource management and land-use planning.

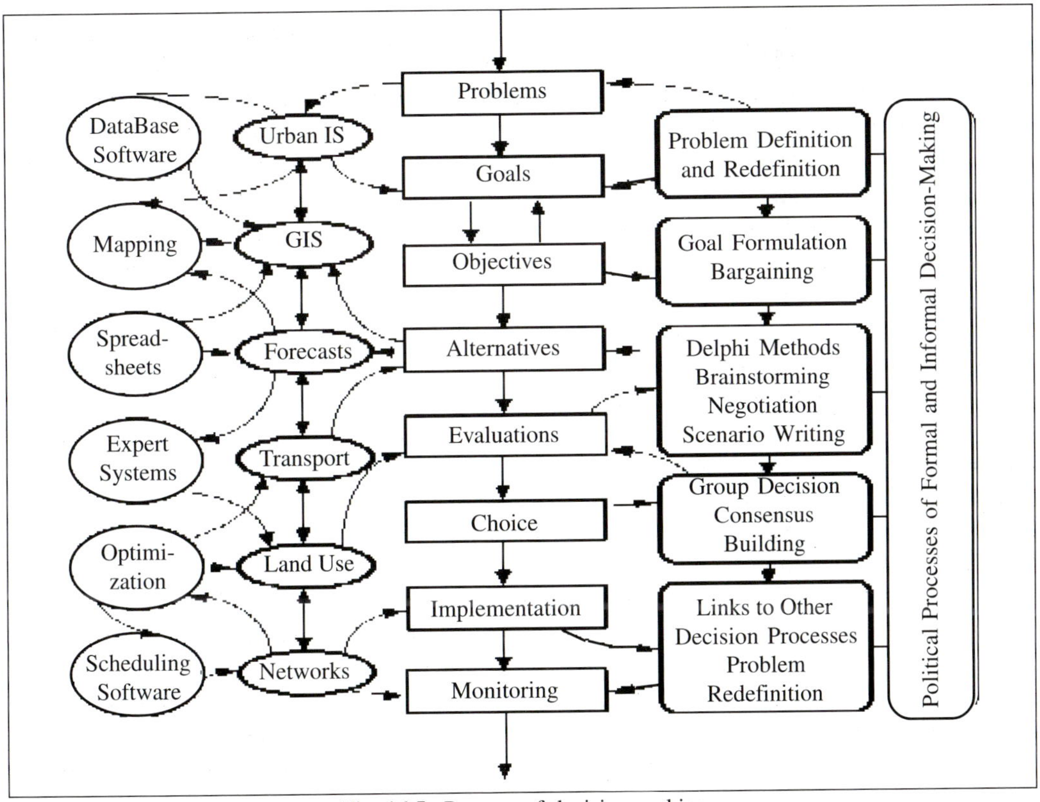

Fig. 16.5: Process of decision-making

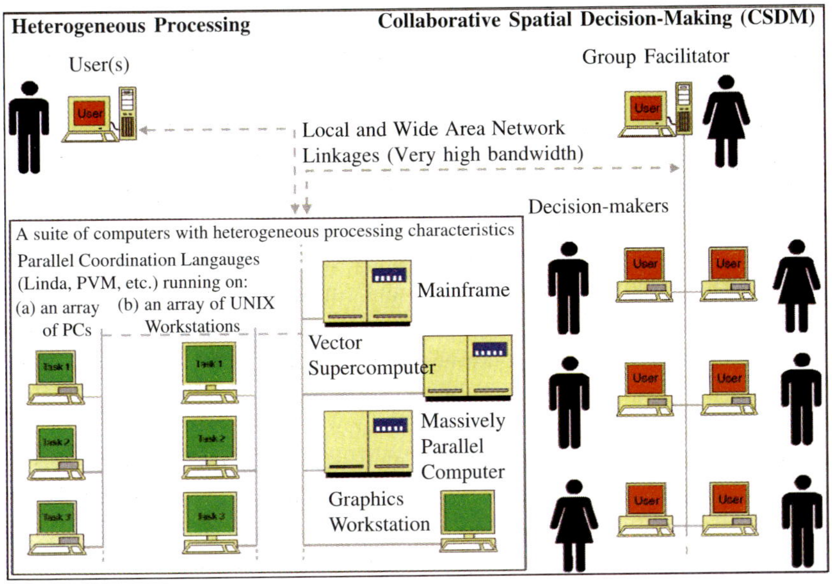

Fig. 16.6: Heterogeneous processing environments of SDSS

Despite the high effort and large amounts of money spent on developing DSS, there are many that are never or hardly used (Adelman, 1992). There are many reasons for this lack of success of decision support systems. One reason can be that the users find the system too detailed, time consuming and costly to use (Ubbels and Verhallen, 2000). Other reasons are related the general complexity of the systems, while still other reasons are related to the uncertainty of the model output and on the appropriateness for solving the decision question. It is stated that the limited involvement of users in the development phase can lead to unsuccessful DSS. In addition, the need for training in the use of each particular DSS has been emphasized. Only then can a DSS be used effectively, the results free of errors and useful for decision-making. Four categories of factors which constrain the use of analytical tools in decision-making has been described.

These are too little, too much and wrong information, time and resource limitations, lack of access, and the hurdles of communication and trust. Some of the reasons for unsuccessful DSS mentioned earlier are strong indications that users are not always able to take systems into use as intended or expected by developers. In other words, the functionality of the systems is not in all cases optimal. Functionality is a very wide term and can include almost any aspect of a Spatial Decision Support System (SDSS).

There are probably just as many reasons for poorly functional SDSS and, therefore, unsuccessful SDSS as there are systems built. There are several suggestions to be found in the literature describing the characteristics of a DSS on which to base a comparison or an evaluation of the systems. Ubbels and Verhallen (2000) evaluate a set of DSS according to their suitability for collaborative planning processes using the characteristics user-friendliness, prerequisite for collaborative processes, transparency, flexibility, and the way the effects of possible actions are estimated. Using these characteristics the authors determine the suitability of tools for specific user groups and decision-making phases. Crossland, Wynne and Perkins (1995) investigated the impact on the decision-makers performance when using SDSS in a decision-making process. The performance was measured by the accuracy of the solutions and the time it took to come to the solutions.

It was concluded that SDSS users have a shorter solution time and fewer errors compared to those not using SDSS. Adelman (1992) presents three methods for evaluating a DSS in technical, empirical, and subjective evaluation terms. These methods of evaluating a DSS are intended for implementation during the development phase of the DSS, to keep the development on track. However, the methods can also be used when evaluating a system after completion. The technical evaluation method focuses on system characteristics and includes, among other criteria, the adequacy of the selected analytical methods, software development costs, software tests and verification, and adequacy of the knowledge base. The empirical evaluation method on the other hand focuses on obtaining objective measures of the system performance. An important question in this type of evaluation is whether decision-makers perform better with a system than without. The subjective evaluation method focuses on assessing the system from the perspective of the potential user in order to identify whether the end-users find the system useful or not.

This study focuses on the functionality of SDSS. Because the emphasis is on characteristics of the system this approach can be labelled as a technical evaluation. It is very difficult to perform an empirical or subjective evaluation because of the inevitable dependency on end-users. As stated earlier many SDSS are not taken into use and if they are, documentation of end-users experiences is rare. Five examples of Spatial Decision Support Systems (SDSS) for coastal zone management are used to test the research questions. These examples can be seen as representative for the state-of-the-art in The Netherlands in this field.

1. *Nature Development and Valuation (NDV Module)*—which supports the design, development and valuation of nature on the Dutch coast. The potential for nature development is thereby quantified, analyzed and presented. The application is the land reclamation project Maasvlakte 2, located South of Rotterdam (Ruijgrok, de Vries, Baptist, Meulen, and Zaadnoordijk, 1999).
2. *Ecotope evaluation DSS (ECOPEIL)*—which supports questions concerning the expected ecotopes and the consequences of water level changes on the naturalness and biodiversity in the Jsselmeer area (Jans, Platteeuw, Tosserams, and Schiereck, 2000).
3. *Shoreline DSS (Shoman)*—which supports suitability assessment of locations for large-scale nature development projects on the Dutch coast. The application is the development of a wet dune valley in the Province of North—Holland (Heuvel, Katwijk, Triest, and Goedhart, 1993).
4. *Wadden Sea DSS (WadBOS)*—which supports the analysis of current policy and management issues in the Wadden Sea. It permits assessment of the e.ects of policy decisions and the feasibility of management measures (Uljee and Englen, 2000).
5. *Eco-morphological DSS (EMM)*—which supports the analysis and evaluation of morphological and ecological effects of different dredging and deposition interventions. The application area is the Western Scheldt estuary (Baptist, de Vries, and Wang, 1998).

One of the main objectives for taking a SDSS into use is that it must be possible to test alternative solutions. In creating alternatives, users can incorporate ideas and knowledge of the decision problem and possible solutions can be tested in terms of their feasibility. It is important that de.nition of alternatives can be managed by users. When making an alternative gets too complex for a user, the user can easily feel overwhelmed by the task and it is less likely that the SDSS will be a success. The complexity of making an alternative depends, among others, on how many parameters need to be changed and how easy it is to make changes to the parameters.

An example of a situation in which some users can feel overwhelmed by the task of making alternatives can be found in the eco-morphological DSS (Baptist *et al*., 1998). Here only two decision variables play a role, which implies that it should be an easy task to make an alternative. However, this is not the case. The decision variables are the amount of dredged material and the amount of deposited material, both in m^3. These are entered using two separate maps, one for deposition and one for dredging.

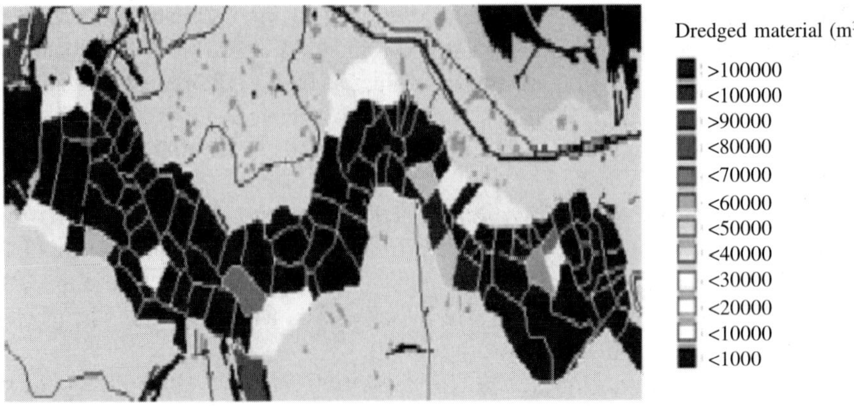

Fig. 16.7: Standard input in the eco-morphological DSS for decision variable dredging for one time step (*Source*: Baptists *et al.*, 1998)

Fig. 16.7 shows an example of a dredging input map. Input data must be provided for 112 possible locations for each time step. Not only does the user, therefore, have to think of the amount of material either to dredge or deposit when making an alternative, but also the spatial pattern of the input data has to be taken into consideration. The time step of the eco-morphological DSS is thirty days and because the DSS calculates long-term morphological and ecological effects on the estuary, putting in data for a time period of twenty years is not inconceivable. In addition to the challenging task of putting in data for one time step, the user has to repeat the input task a large number of times, both for the dredging and deposition decision variable. Not only is then the task very time consuming but also users easily lose the overview of the spatial patterns of the input data that very likely change in time. Generating alternatives, therefore, becomes a very complicated task.

In the four other SDSS, generating an alternative is different from the eco-morphological DSS and is more directed towards entering and changing parameters. In the Nature Development and Valuation DSS users need to design a land reclamation area and add management practices and functions of the designed area. A similar option is also available in the Ecotope Evaluation DSS where users can change the elevation of an area and add the management practice of this area. In the Shoreline DSS users must delineate an area which represents the form of a wet dune slack. The elevation of this new area must then also be filled in. In the Wadden Sea DSS, contrary to the Shoreline DSS, users are expected to filled in many, highly diverse, parameters. The emphasis on putting in data using maps is much less apparent, with the only input of data using maps being when users wish to open or close areas for functions like fishery or recreation.

Acceptance of a DSS to a large extent dependent of its ease of use. An important element in this respect is how well a user is guided though the system, for example whether the next step is self-explanatory or not. The organisation and number of windows at any one time is also important as is navigation through the system and its facilities. An example of a screen that is self-explanatory is shown in Fig. 16.8 and is taken from the Nature Development DSS (Ruijgrok *et al.,* 1999).

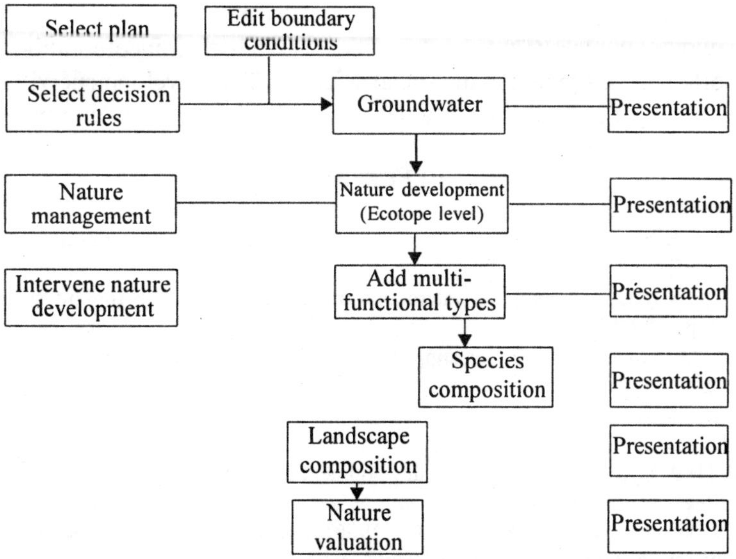

Fig. 16.8: Nature Development DSS (*Source*: Ruijgrok *et al.*, 1999)

The same type of guidance is found in the eco-morphological DSS. However, the developers of the Wadden Sea DSS have chosen a very different approach, where the path through the system consists of many levels. Because of this it is easy for users to lose overview of which level they are on and what task to perform next. Another disadvantage is that the number of windows open at any given moment can be very large. Systems like the Shoreline DSS and the Ecotope Evaluation DSS also possess.

Conclusion

Currently, most SDSS are tailored for specific problem domains. Users often will not recognize that SDSS in quite different application domains have similar architectures and combine similar elements. The applications use the attribute language of the domain, even though the spatial data inputs, manipulations, and outcomes may be identical. SDSS represent the integration of ideas, methods, and techniques from many of the social, behavioural, and management sciences. Their usefulness in the policy and planning sciences is based on the degree to which they successfully incorporate the findings and approaches of these sciences. Their focus on the spatial and the geographic data lead to a very important link with the evolving field of geographic information science. The rapid development of national spatial data infrastructures provides a rich and expanding database to which attribute information can be attached.

Spatial Decision Support System (SDSS), at their inception, were a derivative science that built upon the successful development of models of behavioural, social, and political processes. However, with the expanding use of SDSS in a wide variety of decision situations, theories and models of these processes will have to account for the use of SDSS. When normal decision-making patterns involve SDSS use, they will become an important element in social science theories and models to explain decisions in these areas. This evolution from the use of social and behavioural models as exogenous units in SDSS to their incorporation as endogenous units is an important development in the social sciences. This development will mark the more explicit incorporation of geographic variables in social science theory and method.

REFERENCES

Adelman, L. (1992), *Evaluating Decision Support and Expert Systems*. Wiley Series in System Engineering. New York: Wiley.

Alter, S.L., (1977), Why is man-computer interaction important for decision support systems? *Interfaces*, Vol.7, No.2, pp.109-15.

Alter, S.L. (1977), "A taxonomy of decision support systems", *Sloan Management Review*, Vol.19, No.1, pp. 39-56.

Alter, S. (1980), *Decision Support Systems: Current Practice and Continuing Challenges*, Addison-Wesley, Reading, USA.

Armstrong, M.P. (1994), Requirements for the development of GIS-based group decision support systems, *Journal of the American Society for Information Science,* Vol. 45(9); pp. 669-77.

Armstrong, M.P., Densham, P.J. and Rushton, G. (1986), Architecture for a microcomputer based spatial decision support system, *Second International Symposium on Spatial Data Handling*, Vol. 120-31, International Geographical Union.

Armstrong, M.P., De, S., Densham, P.J., Lolonis, P., Rushton, G., Tewari, V.K. (1990), A knowledge-based approach for supporting locational decision-making. *Environment and Planning B: Planning and Design,* Vol. 17, No. 341-64.

Baptist, M.J., de Vries, M.B., Wang, Z.B. (1998), *Ecomorphological Decision Support System for the Western Scheldt (software and manual)*. Delft, The Netherlands: WL/Delft Hydraulics.

Bernhardsen, T. (1999), *Geographic Information Systems: An Introduction*, Wiley.

Bhargava, H., and D.J. Power (2001), *Decision Support Systems and Web Technologies: A Status Report,* Proceedings of the 2001 Americas Conference on Information Systems, Boston, MA, August 3-5.

Bonczek, R.H., Holsapple, C.W. and Whinston, A.B. (1981), *Foundations of Decision Support Systems*, Academic Press, Orlando.

Carver, S.J. (1991), Integrating multi-criteria evaluation with geographical information systems, *International Journal of Geographical Information Systems,* Vol. 5(3); pp. 321-39.

Church, R.L., S.R. Loban and K. Lombard (1992), An interface for exploring spatial alternatives for a corridor location problem, *Computers and Geosciences,* Vol. 8(9); pp. 1095-1105.

Crossland, M.D., Wynne, B.E. and Perkins, W.C. (1995), Spatial Decision Support Systems: An overview of technology and a test of efficacy, *Decision Support Systems*, Vol. 14, No.3, 219-35.

Clarke, C., Clarke, M., (1995), The development and benefits of customized spatial decision support systems. in P. Longley and Clarke, G. (eds.), *GIS for Business and Service Planning*. Geoinformation International, Cambridge, pp. 227-54.

Clarke, K.C. (1997), *Getting Started with Geographic Information Systems*, Prentice Hall, Upper Saddle River, NJ.

Davis, G. (1974), *Management Information Systems: Conceptual Foundations, Structure, and Development*, McGraw-Hill, New York.

Dhar, V. and Stein, R. (1997), *Intelligent Decision Support Methods: The Science of Knowledge.* Prentice-Hall, Upper Saddle River, NJ.

Dickson, G.W., M.S. Poole and G. DeSanctis (1992), "An overview of the GDSS research project and the SAMM system", in R.P. Bosttrom, R.T. Watson and S.T. Kinney, *Computer Augmented Teamwork: A Guided Tour*, New York: Van Nostrand Reinhold, pp. 63-79.

Densham, P.J. (1991), Spatial decision support systems, in D.J. Maguire, M.S. Goodchild and D.W. Rhind (eds), *Geographical Information Systems: Principles and Applications*, London: Longman, pp. 403-12.

Densham, P.J. and M.F. Goodchild (1989), Spatial decision support systems: A research agenda, in *Proceedings GIS/LIS'89*, Orlando, FL., pp. 707-76.

Eastman, J.R., P.A.K. Kyem, J. Toledano and W. Jin (1993), *GIS and Decision Making,* UNITAR, Geneva.

Ferguson, R.L. and C.H. Jones (1969), "A Computer Aided Decsion System", *Management Science*, pp. B550-B561.

Gerrity, T.P., Jr. (1971), The design of man-machine decision systems, *Sloan Management Review,* Vol. 12, No. 2, pp. 59-75.

Grace, B.F. (1977), Training Users of a Decision Support System, *IBM Research Report RJ1790*, IBM Thomas J. Watson Research Laboratory, May 31, 1976.

Gray, P. (1981), The SMU decision room project, *Transactions of the Ist International Conference on Decision Support Systems*, Atlanta, pp. 122-9.

Gray, P. (1987), *Guide to IFPS (Interactive Financial Planning System),* McGraw-Hill Book Company, New York, 1987.

Houdeshel, G. and H. Watson (1987), "The management information and decision support (MIDS) system at Lockheed-Georgia", *MIS Quarterly*, Vol. 11, No. 1.

Huber, G. P. (1982), Group decision support systems as aids in the use of structured group management techniques, *Transactions of the Second International Conference on Decision Support Systems*, pp. 96-103.

Goodchild, M. and Densham, P. (1993), *Initiative 6: Spatial Decision Support Systems (1990-1992),* National Center for Geographic Information and Analysis, Santa Barbara.

Gorry, A. and Scott-Morton, M. (1971), A Framework for Information Systems, *Sloan Management Review,* Vol. 13, Fall 1971, 56-79.

Grace, B.F. (1977), Training Users of a prototype DSS, *Data Base,* 8, 3, 30-36.

Little, J.D.C. (1971) Models and managers: the concept of a decision calculus, *Management Science*, Vol. 16, No. 8, pp. 466-85.

Keen, P.G.W. and M.S. Scott-Morton (1978), *Decision Support Systems: An Organizational Perspective*, Addison-Wesley, Reading, MA.

Heuvel, T. van, Katwijk, V.F. van, Triest, H. van, and Goedhart, K. (1993), *Shoreline DSS (Shoman) (software and manual),* The Netherlands: National Institute for Coastal and Marine Management.

Inmon, W.H. (1990), *Using Oracle to Build Decision Support Systems*, QED Press.

Jans, L., Platteeuw, M. Tosserams, M., and Schiereck, M. (2000), *ECOPEIL decision support system (software and manual)*, The Netherlands: Institute for Inland Water Management and Wastewater Treatment.

Jankowski, P., T.L. Nyerges, A. Smith, T.J. Moore and E. Horvath (1997), Spatial group choice: a SDSS tool for collaborative spatial decision-making, *International Journal of Geographical Information Systems*, Vol. 11(6), pp. 566-602.

Keen, P.G.W. and M.S. Scott Morton (1978), *Decision **Support** Systems: An Organizational Perspective,* Addison-Wesley, Inc., Reading, MA.

Kimball, R., W. Thornthwaite, L. Reeves and M. Ross (1998), *The Data Warehouse Lifecycle Toolkit*, John Wiley and Sons, New York.

Klein, M. and L.B. Methlie (1995), *Knowledge-based Decision Support Systems with Applications in Business*, John Wiley and Sons, Chichester.

Little, J.D.C. (1971), "Models and managers: The concept of a decision calculus". *Management Science,* Vol. 16, No. 8, pp. B466-85.

Malczewski, J. and W. Ogryczak (1996), The multiple criteria location problem: 2, Preference-based techniques and interactive decision support, *Environment and Planning A,* Vol. 28(1); pp. 69-98.

Mennecke, B.E., Crossland, M.D. and Killingsworth, B.L. (2000), Is a map more than a picture? The role of SDSS technology, subject characteristics, and problem complexity on map reading and problem solving, *MIS Quarterly*, 24, 4, 601-4, 625-9.

Muller, J.C. (1993), Latest developments in GIS/LIS, *International Journal of Geographical Information Systems*, Vol. 7, No. 4, pp. 293-303.

McCosh, A.M and M.S. Scott Morton (1978), *Management Decision Support Systems,* Macmillan, London.

Nagy, G. and Wagle, S. (1979), Geographic data processing, *Computing Surveys*, Vol. 11, No. 2, pp. 139-81.

NCGIA (1990), Research initiative 6: Spatial decision support systems, *NCGIA Technical Paper 90-5*. UC, Santa Barbara.

NCGIA (1995), Report from the specialist meeting on collaborative spatial decision-making, *Initiative* 17, National Center for Geographic Information Analysis, UC Santa Barbara.

Powell, R. (2001), DM review: A 10-year journey, *DM Review*, URL www.dmreview.com.

Power, D.J. (2000), *A Brief History of Spreadsheets*, DSSResources.COM, World Wide Web, http://dssresources.com/history/sshistory.html, version 3.3, 03/11/2000.

Power, D.J. (2000), Web-based and model-driven decision support systems: Concepts and issues, *Proceedings of the 2000 Americas Conference on Information Systems*, Long Beach, California, 10-13 August.

Power, D.J. and G.L. Rose (1976), Improving decision-making behaviour using the Hewlett Packard 2000/Access System. *Proceedings of the American Institute for Decision Sciences*, pp. 47-49.

Rockart, J.F. (1979), Chief executives define their own data needs, *Harvard Business Review*, Vol. 67, No. 2 (March-April), pp 81-93.

Ruijgrok, E.C.M., M.B. de Vries, M.J. Baptist, F.V.D. Meulen, and W.J. Zaadnoordijk, (1999), *Nature Development and Valuation Decision Support System (software and manual)*, The Netherlands: WL/Delft Hydraulics.

Simon, H.A. (1960), *The New Science of Management Decision*, New York: Harper and Row.

Sprague, R. (1980) A Framework for the development of Decision Support Systems, *MIS Quarterly*, Vol. 4, No. 1.

Sprague, R.H. (1989), A framework for the development of decision support systems. In: Sprague, R.H., Watson, H.J. (eds.), *Decision Support Systems: Putting Theory Into Practice*, Prentice-Hall, London, pp. 9-35.

Sprague Jr., R.H. and J.E.D. Carlson (1982), *Building Effective Decision Support Systems*. Prentice-Hall, New Jersey.

Scott Morton, M.S. (1967), Computer-Driven Visual Display Devices—Their Impact on the Management Decision-Making Process, Doctoral Dissertation, Harvard Business School, 1967.

Scott Morton, M. S. (1971), *Management Decision Systems; Computer-based support for Decision Making*. Boston, Division of Research, Graduate School of Business Administration, Harvard University.

Scott Morton, M.S. and A.M. McCosh (1968), Terminal costing for better decisions, *Harvard Business Review*, Vol. 46, No. 3.

Scott Morton, M.S. and J.A. Stephens (1968), *The impact of interactive visual display systems on the management planning process*, IFIP Congress, pp. 1178-84.

Sharda, R., S. Barr and J. McDonnell (1988), Decision support systems effectiveness: A review and an empirical test, *Management Science*, Vol. 34, No. 2, pp. 139-59.

Sprague, R.H., Jr. (1980), A framework for the development of decision support systems, *Management Information Systems Quarterly*, Vol. 4, No. 4, pp. 1-26.

Sprague, R.H., Jr. and E.D. Carlson (1982), *Building Effective Decision Support Systems*. Prentice-Hall, Inc., Englewood Cliffs, N.J.

Sprague, R. H., Jr. and H. J. Watson (1979), Bit by bit: Toward decision support systems, *California Management Review*, Vol. XXII, No. 1, pp. 60-68.

Sprague Jr. R.H. and H.J. Watson (1996), *Decision Support for Management*, Upper Saddle River, N.J.: Prentice Hall.

Turoff, M. and S.R. Hiltz (1982), Computer support for group versus individual decisions, *IEEE Trans. Communications*, COM-30, Vol. 1, pp. 82-90.

Ubbels, A.A., Verhallen, A.J.M. (2000), *Suitability of decision support tools for collaborative planning processes in water resource management*, Report No. 9.067, The Netherlands: Institute for Water Management and Wastewater Treatment.

Uljee, I., and Engelen, G. (2000), *Wadden Sea Decision Support System (software and manual)*, Research Institute for Knowledge Sytstems, Maastricht.

Wilson, R.D. (1994), GIS and decision support systems, *Journal of Systems Management*, Vol. 45, No. 11, pp. 36-40.

Web References

- Batty M. and P. J. Densham (1996), Decision support, GIS, and urban planning, http://www.geog.ucl.ac.uk/~pdensham/s_t_paper.html
- *Journal of Geographic Information and Decision Analysis* (JGIDA), http://www.geog.uwo.ca/gimda/journal/journal.htm
- Malczewsk, J. (1997) The internet resources for geoinformation-based decision analysis, http://www.geog.uwo.ca/gimda/intres.htm
- Larry, D. (1992), SDSS for location planning, or the seat of the pants is out, The Geographer's Craft Project, Department of Geography, University of Texas at Austin. http://www.utexas.edu/depts/grg/gcraft/notes/gisapps/sdss.html
- NCGIA Initiative 6: Spatial decision support systems (SDSS), http://www.ncgia.ucsb.edu/research/initiatives.html#i6
- NCGIA Initiative 17: Collaborative spatial decision-Making, http://www.ncgia.ucsb.edu/research/i17/I-17_home.html
- Power, D.J. (1997) DSS research resources, http://dss.cba.uni.edu/DSShome.html
- Watkins, Jr. D.W. and D.C. McKinney (1996) Recent developments associated with decision support systems in water resources, http://earth.agu.org/revgeophys/watkin00/watkin00.html

17

GIS for Internet and Other Applications

Geography plays an important role in sustainable development and humanitarian affairs. Planning and implementation of complex development goals and humanitarian response efforts often require spatial analysis and map creation capabilities. Data for these efforts is often in a variety of formats (maps, databases, spreadsheets, images, and so forth) from multiple agencies and multiple sources. The ability to access and process information quickly while displaying it in a spatial and visual medium allows organizations to improve decision-making and promote better organizational integration and knowledge management. Geographic information system (GIS) software uses geography and computer-generated maps as an interface for integrating and accessing massive amounts of location-based information. GIS allows international agency staff to integrate data for planning and implementation purposes. Analysis of geographic features with a GIS allows the analyst to view new patterns, trends, and relationships that were not clearly evident without visualization of the data. GIS can also be used to get critical information about a humanitarian crisis to appropriate response agencies in a coordinated and efficient manner. Once in the field, the coordination can continue as new data can be added and disseminated via wireless applications and internet or intranet connectivity. A lot of spatial information is now available on different web sites which can be valuable input for GIS applications. Examples of data concerning India are depicted in Fig. 17.1. Recently Google Earth and other web sites provide access to remote sensing data covering almost whole earth however the data strength and coverage may vary. Nevertheless, the scope of developing a dynamic GIS based on geographic data available from web sites and internet has increased tremendously. With such developments in GIS and internet technologies, a lot of applications and business opportunities are becoming possible.

GIS technology is considered to be a new research tool in academia today and one of the fastest growing high-tech careers for students as well. GIS technology provides powerful tools for geographic analysis for almost any academic discipline. A GIS allows students and researchers to ask and answer geographic questions by designing and analysing maps using user-selected criteria. It is widely used for administration, research, community services, libraries, community colleges and teaching. GIS is multi-facet and hence a wide range of applications.

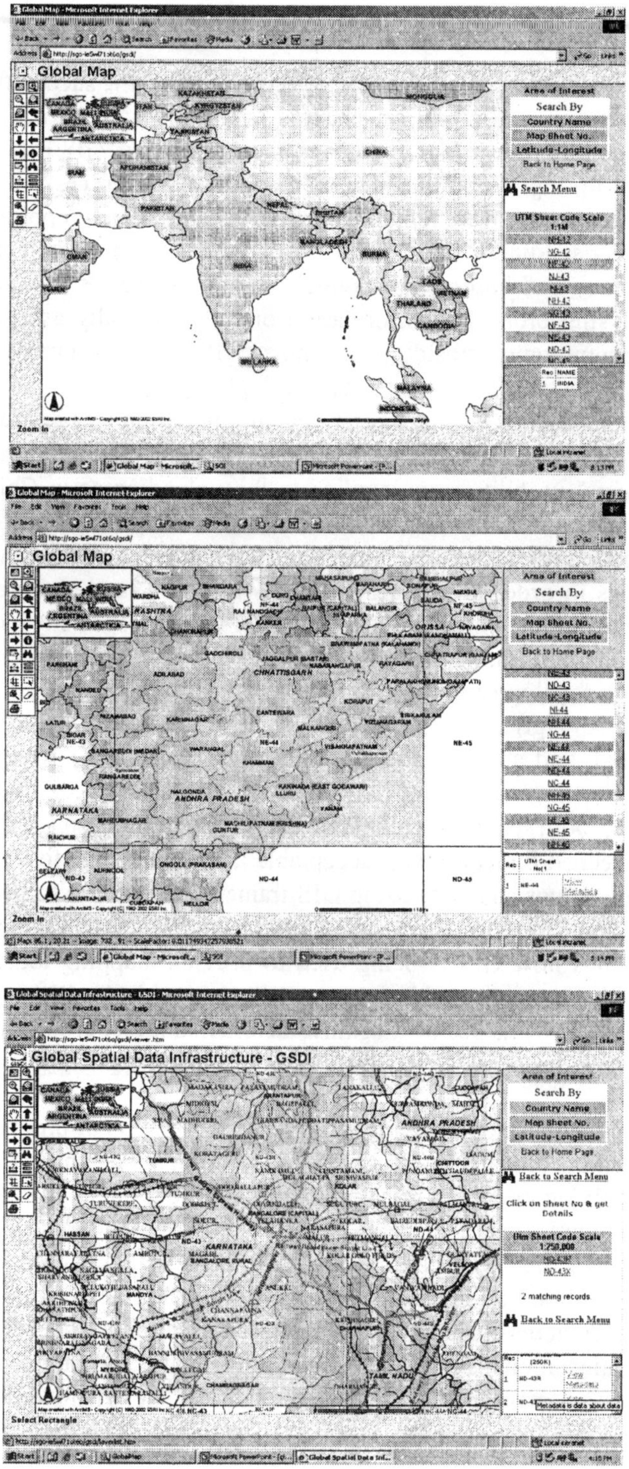

Fig. 17.1: Example of GIS data on web concerning India (*Source*: GSDI)

GIS for Agricultural Development

Even the GIS software multinationals like ESRI is committed to the successful implementation and benefits of geographic information systems in all key agricultural sectors including production, agri-business, research and in government activities. GIS has been proved to be useful for farmers, crop inputs and manufacturers, post harvest processors and research. The study of geographic features and the relationship between them can be applied to all agriculture sectors by better understanding of how features within the landscape interact, decision-makers can optimize efficiency and improve economic returns. Regardless of scale, whether at the subfield level analysing crop yield information or internationally assisting governmental organisations with commodity subsidy programmes, GIS is becoming commonplace in the information technologies domain, even in the developing countries.

In case of Thailand, the GIS and expert system Integration was carried out for land and agriculture development at sub-district level. The land development regional offices created a tool with more general usability to be applied to the upland areas in Chiangmai aiming to design a soil and water conservation and soil loss evaluation at pre- and post-project levels that included cost estimation for decision-making. Land development village was to be established to integrate activities in order to provide farmers with more land and sustainable land use alternatives. The utility of integrated GIS and the expert system, especially its use for supporting land development village planning, was displayed and discussed, with special attention to user interfacing (www.gisdevelopmenr.net/aars/acrs/1995/PS2/PS2003.asp).

US Department of Agriculture's Risk Management Agency (RMA) is suppose to monitor possible waste, fraud, and abuse in the crop insurance programme. The usefulness of GIS information for monitoring and flagging purposes was to be considered. It was to bring together multiple data sources, both spatial and relational database types. Reasons for the crop insurance claims was to be screened for legitimacy, acceptable farming practice for crop, and acceptable combination of crop, water and soils using GIS frame. RMA is also to provide a sound risk management programme to reduce agricultural producers' losses applying GIS in its business processes. It illustrates how GIS is being used to provide mapping tools and products for decision-makers and the process of integrating GIS data with the RMA policy database to visualize various aspects of its business against factors that affect crop insurance.

(www.esri.com/news/arcnews/winter/0102articles/many0usda.hmtl)

GIS is being widely used for the survey of crops, estimation of production and varieties. In India, the production of sugar cane, wheat and tea crops is estimated during pre-harvest stage. This helps in fixing minimum price protection for the farmers. In case of the developed countries, this technology is being used for a variety of crops. For the annual survey of sunflower in the Central Plains of the US, GIS is being used. Such survey includes the states of South Dakota, Minnesota, Nebraska, Kansas, Colorado, Missouri and Texas. Data is collected at one location for each 5,000 acres of sunflowers planted. Information is collected for various weed, disease and insect infestations, and bird damage to the plants. The sample points, survey data and interpolated contour maps are available on the Internet site. The GIS generated sensitivity maps are served on the Internet using ArcIMS 3.1 and internet information server, SP6, at http://134.129.78.15/website/sunflower/viewer.htm. Individuals can use a common web browser to access the site and interact with the survey data in a spatial format.

National Sunflower Association's annual Fall Sunflower Crop Survey is a research project coordinated by North Dakota State University Extension. Volunteers from the various levels of the sunflowers industry conduct the survey annually in September. The survey teams of three individuals estimate yields, insect, disease and bird damage and note management practices. Surveyors will use GPS units to mark the survey locations. The sunflower crop survey supplies information to university researchers and individuals in the commercial sunflower industry. Research and extension personnel use the survey results for justification when applying for Section 18 emergency labels. The commercial sunflower industry uses the information for a variety of supply, sales and marketing decisions. Further, in the North Dakota State University Agricultural and Biosystems Engineering Department the geospatial extension specialists coordinate development of GIS layers from the survey information and publishing the data on the internet using ArcIMS software. The ArcIMS web site allows users to view survey results for each state included in the survey. Users can also add or remove desired GIS layers for each survey factor and for the general soil type. Users can also query the survey information at each survey site. (www.gis.esri.com/library/userconf/proc02/pap1238/p1238.html)

For the preparation of the GIS layers for ArcIMS, the survey field data is assembled into a Microsoft Exel database file. The database file includes the latitude and longitude of each survey point. The dbf format data is added to ArcView as a table. The data is added to and ArcView view as an event theme. The event theme is interpolated as a grid using the ArcView Spatial Analysis extension and reclassified into four categories. The reclassified grids are converted to shape files and served as layers in ArcIMS. The web page is served as image map services to facilitate serving speed and to allow greater versatility of web browsers.

In a developing country like Bhutan, GIS is being used for land use planning. In 1992, A PC ARC/INFO based system was established within the Land Use Planning Section of the Ministry of Agriculture, Royal Government of Bhutan. The primary task was to produce a national coverage of maps in scales of 1:50,000, 1:100,000 and 1:250,000, based on data collected from satellite hard copy images, existing topographic maps and fieldwork. Later, the system has been used for the production of analytical and mapping tasks including local planning maps at 1:25,000 scale and terrain modelling. The choice of a scaleable low end, PC-based GIS, is regarded as crucial for the success of the implementation of the system, both in terms of human resources, institutional development and technically. (www.gis.esri.com/library/userconf/europroc97/4environment/E6/e6.html)

The Land Use Planning Project (LUPP) is funded by the Danish development assistance (DANIDA) and was to continue until July 1997. The ambitions, in terms of GIS, were fairly modest in the beginning. The need was a facility for efficient production of maps, by combination of information from a variety of data sources. Only later the expected wish for analytical capabilities came true. Further, the challenge of producing useful and accurate maps and a digital map database for the entire Bhutan, considering the limited number of people and time available, was overwhelming. It would have been impossible if it wasn't for the facilities provided by modern GIS-technology. The choice of a low-end system was considered crucial for the success of the project. The development of desktop GIS is mainly targeted for the data presentation and, to a lesser extent, analytical capabilities. In most developing countries, where data collection still has a major role to play, the need of well developed and supported low-end tools like PC ARC/INFO is important. (www.gis.esri.com/library/userconf/europroc97/4environment/E5/e5.html)

GIS for Cave and Karst Management and Research

Cave and Karst managers and researchers understand the importance of geography. This discipline indicates why and where features exist and how they relate to the surrounding environment. Management and research decisions are based on geography. Geographic analysis and modelling provides answers to a variety of questions, and helps make informed decisions. A GIS provides tools that will help in integrating a variety of data sources and types, maintain and manage inventories, visualize data and related information using dynamic maps, make decisions about resource management, and perform modelling and analysis.

Karst is an interesting feature of the Romanian landscape and where the mesozoic-age limestone outcrops can be found in many parts of the Carpathian Mountains. Such areas abound in caves, and while many have been mapped in detail, few such paper maps have so far been published. The case of Ratei, one of Romania's longest caves, with more than 7 km of passages and chambers, many of them flooded is interesting. Located in central part of the Romanian Carpathians, this cave was surveyed in the 1980s by two Romanian speleological clubs and undertook mapping at 1:2,000 scale. This, together with topographical and hydrographic information drawn from large scale maps (1:10,000 scale) and field observations, yielded data for a GIS database. This was constructed by Geosystems of Bucharest, utilising ArcView GIS 3.2 and ArcGIS 8.1 software for the purpose. A 3-D image of the Ratei karst area was also generated in ArcScene 8.1 3D visualisation software. This offers speleologists and other researchers some key answers regarding the connection between geomorphological and hydrological features of the main and sub-surface karst. It also opens up new opportunities for exploration and provides basic support for conservation efforts in the region.

(www.balwois.mpl.irdfr/balwois/papers/abstracts/balwois-abstract-book.pdf)

GIS for Libraries, Museums and Public Access

Making information accessible and easy to understand is the goal of libraries and museums everywhere. A GIS organizes and presents information in ways that spark greater understanding of places and events around the globe and close to home. It helps in administration, exhibition and displays, public access, research and teaching. Libraries and museums provide a gateway to all kinds of information. GIS manages census data, zoning and tax assessment maps, digital aerial photographs, and satellite images, providing public access to information that previously was difficult to use. GIS also provides tools for exploring information through maps and images, helping inquisitive minds to see things in a new way. Museum exhibits about topics such as natural history and geology come alive with dynamic maps to illustrate processes and events.

Regarding public access of information, there is an example of Monmouth Battlefield State Park. In 1770s, seven well-established farms occupied an area of hilly farmland and hedgerows of 2,000 acres of Manalapan and Freehold townships in New Jersey. This bustling farming community was occupying a future battlefield. Monmouth Battlefield State Park preserves this splendid 18th century landscape, which includes miles of hiking and riding trails, picnic areas, a restored revolutionary war farmhouse, and a visitor's centre. On June 28, 1778, one of the largest and hottest battles of the American revolution took place in Monmouth county, New Jersey. As the British Army marched from Philadelphia to New York City, General Washington and the Continental Line (Army) attacked the rear guard of the British Army. Even with a short review of the battle, the many intimate details that the public could be and is interested in learning more about are seen. This project utilized ArcView GIS to develop an interactive interface that the public can utilize at the park to gather the volumes of information collected by the state park historian. Integrating the volumes of information with pictures, maps, and explanations allows the visitor to experience the fullest the battlefield has to offer. This presentation was to review the goals of the project, the choices made, the lessons learned, and the plans for the future. A demonstration of the interface and the information used was to be the highlight of the presentation. (www.gis.esri.com/library/userconf/educ02/abstracts/a5127.html).

With information technology and electronic communications making inroads everywhere, the media is making use of GIS, particularly when it is dealing with aerial of geographic description of an event. GIS combines layers of information about a place to give a better understanding of that place. GIS software, such as ArcView, provides the data model and tools needed to store, analyse, and display information about places. Reporters can map everything from census data to crime statistics to traffic accidents quickly and accurately. With wireless and internet access, data can be retrieved at the scene or before the reporter even leaves the office. GIS provides journalists with more effective tools to report the news and put it in context for readers, and also manages the networks that facilitate media subscriptions and distribution.

There are several examples where GIS has been successfully used for academic purposes. The 3-D visualisations of the beautiful Parco del Corno alle Scale in Italy's Northern Apennines were generated by Faenza-based GEOgrafica for a multimedia teaching CD ROM. This aimed to show geomorphological features of the area, vegetation distribution and trail paths—all data being derived from the National Park's ArcView-based GIS or from GPS trail readings. The software employed to generate the imagery was 3D Nature's Visual Nature Studio (VNS) and in which vector isolines were gridded to produce 20m DEMs of the area. ArcView shapefiles were imported and gridded in a single-pass operation. The entire gridding process was extremely rapid, taking just five minutes to tile an area of 30×30 km into 64 DEMs of around 20m grid size. CORINE Landcover coverage was used to attach ecosystems to the terrain. Some fifteen different eco classes were prepared in VNS to fit the CORINE legend; then, while importing the CORINE shapefiles, they were automatically attached, through attributes, to the imported and georeferenced vectors. Further, each ecosystem is a combination of ground, grasses and trees that fit the description of CORINE landcover and includes real images of plants, 3-D building objects and simulation algorithms of natural distribution. In this way, it was possible to simulate a higher resolution than that of CORINE itself. Furthermore, TIFF with TFW scanned maps were used to drape the ground and display a more "technical" view of the terrain.

With the help of multiple cameras, several views of the area, from zenithal to birds-eye, were rendered. This was accomplished on a simple PC with 512MB RAM and running Windows 2000. These renders were inserted in the multimedia CD ROM and also used at visitor information points in the park. The public appreciated the views, especially for their "virtual reality" look, while the park management were delighted with this fusion of 3-D photorealistic visualisation with validated and georeferenced data. (www.geoplace.com/ge/2001/0111/0111VS-2.asp).

GIS in Real Estate

The real estate industry has always known geography matters since location is what GIS is all about. GIS-based solutions are useful in all segments of the real estate industry, from map-based contact management to sophisticated investment analysis in large real estate investment. GIS allows the real estate practitioner to integrate a wide variety of data into one common format, a map. The presentation of a wide variety of data affecting the desirability and value of a property on one or two maps can give a far more accurate picture of the property's suitability as a first time residence, acquisition for a portfolio, or site for a retail outlet than any number of generalized market studies, photographs, and marketing text.

In case of RealtiCrop a real estate company, the first six months were spent laying out groundwork to make GIS. After some time the growth and flexibility were factored into the design of the organization so that very little lead time is needed to open up offices in new areas when good opportunities are identified. The company makes extensive use of the internet utilizing frame relay and T1 connections to transmit information immediately to regional offices. The firm has operated primarily in the southeastern portion of the United States.

RealtiCorp is built on four interlocking components: (a) Research and Investment, (b) Sales and Marketing, (c) Technology Departments, and (d) regional offices. Sales and marketing staff works with past and potential clients to identify their requirements and impending needs. The Technology Department constantly develops new ways to utilize information technology to spot and manage opportunities. The Research and Investment Department uses ArcView GIS and ArcView Spatial Analyst to analyse data and determine areas of high suitability that can be referred to regional offices for further development. The company uses GIS to get know their customers better than the customers know themselves and for decision support. These robust applications analyse risk and returns on commercial real estate investments. It has chosen ArcView GIS because the company was looking for an enterprise wide solution that offered flexibility in terms of customization of the interface in using data in various formats from public and private sources. The lower start-up costs associated with a desktop application and the ability to make ArcView GIS work with Word were also attractive to the company. Analysts use ArcView GIS to track expanding MSUs and research their specific locational requirements. This information is overlaid with the company's base GIS to locate areas of high suitability.

A network of regional offices provides local savvy and locates promising deals on either a to-order basis or by seizing an opportunity they discover. Regional offices validate property characteristics for specific requirements and monitor the progress of ongoing deals. Lotus Notes ties the company's departments together. As areas of high suitability are located, that information gets forwarded to broker partners using Lotus Notes. These local representatives are given information on promising areas and the specific site requirements for clients. Brokers visit properties to qualify the physical characteristics of the property. RealtiCorp plans to bring the power of GIS out of the back room and make it more accessible to brokers through developing more intuitive applications and is exploring the use of MapObjects to supply low-tech users with real estate information. (www.gis.esri.com/news/arcusers/1098/regis.html)

GIS for Environmental Management

Geography is the key to helping government and commercial entities monitor the environment more efficiently and cost-effectively. It also is the key to helping organizations comply with the multilateral environmental regulations that are the result of growing global markets. The digital version of such expertise, in the form of GIS, has been useful for site remediation, natural resource management, waste management, groundwater modelling, environmental impact assessment, policy assessment, environmental compliance and permit tracking, water quality and vegetation mapping. Nevertheless, since the inception of the technology, GIS has been an integral component in environmental management. A GIS links geographic and attribute data, such as lab and field data, which allows you to do predictive modelling for planning and compliance purposes. GIS technology reduces the cost of doing environmental business while giving the organisation that uses it a competitive advantage in an ever-evolving market place.

An example of the application of AutoCAD Map in arboreal activities may be mentioned here. Arbortech Consulting Ltd. offers consulting services to assess, diagnose and manage clients' trees and shrubs. These services include preparing management plans and specifying pruning, fertilizing, and monitoring trees for insects or diseases, as well as providing specifications for planting, transplanting, and removing trees. The company is often called on to consult with clients on issues such as how existing trees and landscaping might affect building, land development, and related infrastructure installation and renovation.

Relying heavily on AutoCAD and AutoCAD Map software, Arbortech collected survey data about the trees and landscape, while associated companies collect information regarding storm, water, sanitary system data and the like. The resulting GIS is to help to take important decisions regarding proposed land use and community development. With approximately 1,000 acres of land and water resources and more than 8,000 trees to inventory, the team had quite a challenge. Moreover, the team needed to collect and store the data in a GIS format so that the future contractors, land developers, and university staff could use and build on.

As surveyors from Murray and Associates collected field data about storm, water, and sanitation sites, Arbortech created a Microsoft Excel database containing the size, species, location, and condition of all trees on the property and surrounding residential areas. This effort included using AutoLISP routines for plotting to scale the radii of tree drip lines. Each tree's drip line was represented by a circle, so one could easily visualize the radius of each tree. The data collected by both companies was then output to Excel and dBASE files for compilation.

As long time AutoCAD users, Murray and Associates turned to AutoCAD Map software to create the final maps of the area. AutoCAD Map simplifies the process of creating and maintaining maps and makes it easy to share that data with others. The team depended on

AutoCAD Map software's ability to integrate data from a variety of sources. AutoCAD Map gave the power to present a large amount of complicated data in a way that anyone could easily understand. Obviously, getting ideas across in layman's terms is crucial. The specific applications were (a) to select a specific piping and view the database to find its age, material, diameter, depths, and so on; (b) the mapping and inventory of trees for planning future development; and (c) preparation of plant operations schedule and budget treatments for pest control and to keep track of hazardous conditions like sick or dying trees. (www.autodesk.co.jp/adsk/servlet/item).

Another example of GIS application is for the development and use of an environmental information system (EIS) for power station atmospheric emissions. ArcInfo has been used to develop an EIS for atmospheric data. The system combines the output from computer models of atmospheric stack and cooling tower plume dispersion and deposition with monitored air quality data, and third-party datasets (such as SSSI boundaries), to provide a range of information about power station stack emissions, their dispersion and their potential environmental effects. This multi-user application includes custom grid and contour visualization, automatic context-sensitive legend generation and complete "loading" and "saving" of all displays. It uses AML, C, AWK and UNIX shell scripts for OpenWindows or Motif environments. (www.gis.esri.com/library/userconf/europroc96/PAPERS/PN18/PN18F.HTM)

National Power (NP) is the UK's largest electricity generator holding about 25 per cent of the market and an increasing investment in electricity generation worldwide. Effective management of the environment is vital to the commercial success of National Power. As well as managing the existing assets with the environment in mind, it is necessary to develop the business strategy to take account of the likely direction of future legislation and technology—and the views of everyone with an interest in the business. The national power environmental approach is guided by five principles:

(a) to integrate environmental factors into business decisions,
(b) to monitor compliance with environmental regulations and to perform better than they require, where appropriate,
(c) to improve environmental performance continuously,
(d) to review regularly at Board level, and to make public, the company's environmental performance, and
(e) to establish a reputation for effective environmental management.

The above principles were considered important to understand its impacts on the environment, manage them cost-effectively, and to communicate its performance and the progress.

The NP environmental performance is backed by a team of environmental specialists who provide the underlying scientific and technical expertise. Its Environment Unit uses modelled and measured information about the environmental effects of power station operation both for development of environmental strategy and for input to site-specific environmental effects evaluations and impact assessments. Historically these modelling studies were performed using separate systems and integrated only at a later stage. The continuing drive for increased productivity motivates the search for opportunities to integrate existing systems within a single framework. This integration has several clear benefits: (a) it reduces the number of separate software systems that need to be supported, (b) it provides many opportunities for synergies between the combined areas (sometimes unforeseen), (c) the ability to contextualize model output using other datasets adds considerably to effectiveness of use and ease of interpretation, and (d) it provides a unified approach to data management. The integrated Environmental Information System (EIS) used ArcInfo initially for atmospheric environmental studies.

(www.gis.esri.com/library/userconf/europroc96/PAPERS/PN26/PN26F.HTM)

There is a similar case in France where GIS was used for environmental concerns for power transmission. During the sixties, this country needed electricity everywhere. There was no understanding of the best way to transmit electricity from one point to another. Decisions were being based on economical considerations—the straight line was quite often the best one. At the end of the 1970s, when the Very High Voltage (in French Tres Haute Tension) network was being reinforced, the context changed: the global needs had been satisfied and ecological consciousness (following the events of May 1968 and other demonstrations like Woodstock) was rising. It was at that time that the French power establishment, EDF, began to take account of the environmental concerns. In 1982, it signed an agreement with state authorities to show its good will. Accordingly, in 1990 its strategic plan included environmental measures. EDF also concluded an agreement with the authorities to take steps to integrate the transmission network components into their surroundings. Later, EDF asked its regional divisions to plan the enhancement of the regional network considering the environmental constraints.

(www.gis.esri.com/library/userconf/europroc96/PAPERS/PN40/PN40F.HTM)

At the end of 1993, some years after the EDF's decision to take account of the environmental concern, Energie Aquitaine (an EDF Generation and Transmission division) decided to build Pro-SIGEA, a dedicated GIS to deal with its region's environmental data. It was planned to show interferences from the power transmission network on maps and to form these interferences into a hierarchy. In addition, the system might improve communication between different organisations dealing with long term planning problems. In order to achieve the goals, ArcView with Access was chosen as the main query tool. Now technical EDF staff use Pro-SIGEA for various purposes within their areas of operation. Today, such a system is strongly dependent on the data input into the computer. Though it is not difficult to form a hierarchy of classifications based on juridical aspects, it is not the same for the landscaped and ecological criteria, partly as a consequence of input data. Hence, the main conclusion is that the task of modelling interferences between the environment and human activities is difficult precisely because they interact together in space and time. Nevertheless, Pro-SIGEA can be considered as a first step towards that goal.

The database and ArcView based tool that form Pro-SIGEA, the EDF version of SIGEA, are useful. Energie Aquitaine, a regional division of EDF Generation and Transmission, uses it as an aid in the decision-making process in its technical field. The environmental database and its use through SIGEA is now the responsibility of Aquitaine DIREN (DIrections Régionales de l'Environnement). An agreement has been concluded between EDF and Aquitaine DIREN to protect it and its use by third-party members. SIGEA which allows the exchange of data between EDF and DIREN facilitates communication between these two entities.

The results that can be issued from Pro-SIGEA (as for SIGEA) depend strongly on the data input into the computer from both a quantity and quality point of view. Fortunately data is usually updated quickly. For the transmission network geo-databases the updating is permanently undertaken by EDF. Further, if the object of forming a hierarchy of environmental constraints has been achieved for the legal aspects, the results are not so good for the ecological and the landscaped aspects. This can be explained by the difficulty of measuring the constraints attached to those criteria and also by some limitations of the software and of the environmental database. Either the precision of data location is not sufficient or knowledge of the land's cover is missing making it difficult to obtain correct results.

Another interesting case related to environment and power is regarding oil spill trajectory and fate model using GIS. It is widely understood that the oil spill is a great threat to environment in general and mankind in particular. It is a challenge for petroleum industry as well. The conventional processes are much tedious, less accurate and time consuming. The solution lies in using GIS along with all the features of conventional processes with decision support system, analysis and visualization capabilities, developing spill predictive model, providing strategies and cleanup plan, charting optimum shipping roots and to support contingency plan for quick action.

The study area was Rawa oil field located along the east coast of India. The methodology adopted for this purpose is as follows:

- After pre/post-monsoon surveys, available data like physical and chemical parameters are processed using Geomedia professional.
- Identification of sensitive areas like mangrove, reserved forest, fishponds, paddy fields, terrestrial and aquatic ecosystems.
- Analysis in this fragile environment, by GIS to predict the extent and location of a spread.
- Use of mathematical model to calculate amount of oil evaporated, spread etc.
- Probable spill behaviour under local conditions.
- Creating maps to analyse different zones, for resultant directions and speeds of wind and current.
- To predict the extent of spread at specific time interval and also the spreading of spill towards the coast.
- Analysis of relationship of dispersion rate with wind speed.
- Calculation of the affected shoreline area.

The GIS system was considered for planning for the clean up and predicting how the oil will move, accessibility, preparedness in a situation of oil spill, quick analysis and follow up action. The application of GIS is meant to save valuable time which in turn can be used to save the nature and human life.

GIS for Banking and Insurance

USAA has developed a web based, graphical model that provides users flexibility in estimating the number of potential, active and remaining claims for any type of catastrophe. The model uses ESRI software applications with some custom development on the base GIS infrastructure. For this purpose, a CAT model is designed based on ArcIMS application to assist loss reserving and cat operations personnel in the viewing and analysis of homeowners, fire and renters policy in force data, as well as other catastrophe-specific data. There are four options with which to view the policy in force information:

- *Quick view:* An interactive tool which provides tools for simple viewing and querying of the U.S. policy in force data.
- *Java view:* An interactive tool similar to the *Quick View*, but with additional analysis capabilities.
- *Pre-processed policy in force maps:* This option shows policy in force maps at the U.S., state, county and ZIP code levels.
- *Policy in force (PIF) reports and geocoding hit rates:* This option shows policy in force counts (both total PIF and geocoded PIF) at the state, county and ZIP code levels. The geocoding hit rate is the percentage of PIF that were successfully geocoded to the address level, however other levels of geocoding are available in the application.

When a major catastrophe occurs, senior management will rightfully expect accurate, timely (daily) information that they can use to make critical decisions. Sending the wrong number of adjusters with the wrong skill sets to the wrong places can result in poor service to the membership when they need us most, and can result in unnecessarily high expenses. Inaccurate estimates of loss and additional living expenses (ALE) costs can lead to improper decision-making by senior management. Also, the push towards shorter financial closing schedules (eventually, to a virtual close) makes it imperative that robust catastrophe data be accessible on a daily basis to facilitate the accurate estimation of catastrophe liabilities. Waiting until delivery of correct month-end data to analyse catastrophes is not an option.

The benefits of such initiatives could be estimated at the time of crises. The case of the early benefits in San Diego County (Alpine) Fire can be mentioned here. Around 5:00 AM on Wednesday, January 3, 2001 a fire started about 30 miles east of San Diego. With the winds blowing at 65 mph towards the west the fire quickly got out control. It started threatening the moderately populated city of Alpine. By mid-morning, evacuations began to take place. Looking at a 3-hour visible satellite loop the smoke from the fire could be seen blowing across San Diego and then out over the Pacific.

By late morning the WESTRO (USAA Sacramento office) requested that Cat Operations plot a map showing the southwest portion of San Diego county. They asked that the map include the homeowner and fire policies in force (PIF), by risk location, so that they could assess the exposure. This was a standard request, but given that the fire was out of control and moving quickly WESTRO had need for immediate information. They expressed concern about how quickly that need could be met. Plotting a hard copy map in the home office generally takes about an hour, if the map is fairly simple. Then the map must be shipped over night to the region which means they wouldn't receive it until the next morning. With the new GIS web application in place, WESTRO did not have to wait. It only took about 5 minutes for them to pull up a web browser and plot their own map on the screen. This new application provides the same information (plus more) that is available in a hard copy plotted map and it is available immediately.

Fortunately, the fire did not continue to rage. By the next day, the winds were down to a mere 30 mph and then by Friday, the fire fighting teams were able to start controlling the fire. USAA only received 6-7 claims and only one of those actually had any fire damage. The rest were for smoke damage or Additional Living Expenses. But it could have easily been much worse. (www.gis.esri.com/library/userconf/proc01/professional/papers/pap1105/p1105.htm)

GIS as Serving Maps

Brooklyn, New York, has three million people and thousands of restaurants, featuring virtually any cuisine in the world. For the Brooklyn Chamber of Commerce, Community Cartography developed MapObjects-based internet applications to enable people to find restaurants and other information about their communities. Ease of use was essential to the design strategy. Focus was on creating pages that were simple for the average internet user to easily navigate, and we dealt with many issues of web design that become critical when GIS is served on the internet. The new web site, iBrooklyn.com, includes map pages for property data, transportation, census data and administrative districts as well as restaurants. It also provided Brooklyn Atlas on the web. The first map application we designed was for Brooklyn Eats, in preparation of the 4th Annual Food, Wine and Beer Tasting Festival on October 12, 2000.

(www.gis.esri.com/library/userconf/proc01/professional/papers/PAP729/pap729.htm)

The Business Directory contains the current members of the Brooklyn Chamber of Commerce and is searchable by company name, contact name, category, and subcategory. Additional information on each member includes a description, company name, address, phone number, email, and links to the company's homepage. The membership data on the map server is updated on a monthly basis. With over 200 business categories and subcategories to choose from, it was decided to implement two drop-down menus where the subcategory is dynamically refreshed when a category is chosen. This way we avoided the problem of the user having to scroll through one long drop-down menu or have them guess at the proper key words to type into a search text box.

The Brooklyn Atlas consists of four pages containing maps for property data, administrative districts, transportation networks, and census data. Unlike the Brooklyn Eats and Business Directory pages, these atlas maps have multiple data layers that can be identified. The user can change the active layer by clicking on its legend symbol which appears above the map. These legend graphics are form elements, so the page is resubmitted automatically when one is selected—a "refresh map" button is not necessary.

The Property Map page shows tax lots, parks, historic districts, business districts, and economic development zones. Users can query and identify tax lots to find property information including the owner, zoning, land use and building class. Properties may be found either by street address or by block and lot number. Currently, this is the only online resource to find all of this information on one easy-to-use page.

This project posed some interesting challenges in designing internet mapping applications. It was found that in seeking to simplify the site's design, one had to implement more complex technical solutions. The customizations we designed were not difficult to programme, but it took considerable planning and testing to arrive at. The feedback gained from many rounds of testing enabled to streamline a sequence of GIS functions to a great extent. A search produces the map that is most useful to the user with the least number of intermediate steps, and brings the results that they expect to see. Creating the "go there" button to highlight one record from a selected set helped non-GIS users to see the relationship between the map and the table.

GIS in Internet Mapping

Internet has evolved as a viable tool for public access to information. Governments, institutions, industries, corporations, municipalities and the like have found this web based technology to promote their agenda and business interests. Internet combined with GIS has created further wonders in making the spatial data more accessible to people at large. Web GIS or internet GIS, as they are popularly called, have wide range of map based solutions ranging from applications on finding directions to mapping biodiversity to creating the Geography Network. In case of India, the new National Map Policy announced in 2005, allows the users to publish maps on the web with or without GIS database. This liberalization has yet to make an impact in terms of internet mapping. However, internet has been accepted as a powerful tool for public access and transparency in the functioning of the government and its agencies because the information is easily available to all and reduces time for making inquiries through telephones or otherwise. Nevertheless, some examples from other countries are given below:

Table 17.1: Examples of internet GIS

Title	*Publisher and Place*	*Content summary*	*Purpose etc.*
Iowa Gap Analysis Stewardship Application	Iowa Gap Analysis ProgramAmes, Iowa	Data includes conservation and recreation boundaries from municipal, county, state and federal agencies.	For conservation managers, planners.
Sistema Informativo Territoriale—Istituto Promozione Industriale (SLI)	Istituto Promozione Industriale	Location of a new industrial plant using ArcView IMS and actually is only in Italian language	To provide public access to the geographic content.
Jefferson County, CO GIS	Jefferson County, CO GIS	Delivers address related property, political, and physical information to residents using HTML, Perl/CGI, SQL/Informix, and VB/MapObjects IMS components.	To provide public access to the geographic content.
Johnson County, KS—AIMS Interactive Maps	Johnson County, KS	Provides various interactive mapping solutions using property maps, aerial photos, planimetrics, and other pertinent county information.	To provide public access to the geographic content.
Ruhrtal à la Karte	Kommunalverband Ruhrgebiet (KVR)	Shows public points of interest, tourist sights, landmarks, and bicycle paths in German.	To plan bicycle tour.

(contd.)

(contd.)

Title	*Publisher and Place*	*Content summary*	*Purpose etc.*
LOJIC Home Page—Interactive Maps Section	LOJIC	Information on polling sites and contact information and flood plain, property, zoning, sewers and much more.	Public access to the geographic content.
Natural Resources, Physical Features	Louisiana Department of Natural Resources	Displaying and overlaying the mineral, energy, water, biological and other natural resources.	For navigational, engineering.
Real Estate Information System of Lucas County, Ohio	Lucas County Auditors Office	Parcel attributes, tax information, interactive digital base maps and aerial photos.	To provide public access geographic content.
Maine Office of GIS	TopoZone—USGS Topographic Maps; Maps a la carte, Inc. North Chelmsford, MA, USA	Interactive, seamless topographic map of the entire United State. Maps are available a 1:25,000, 1:50,000, 1:100,000 and 1:200,000 scales.	TopoZone is designed for all topographic map users, including engineers, planning professionals, outdoor recreation enthusiasts, genealogists, and others.
Meridian Live Internet Demos	Meridian Systems Oy	Online geocoding, automatic vehicle tracking demo, and MOSAIC land information service.	Public access to the geographic content.
Nashvill/Davidson County Online Property Maps	Metropolitan Nashville/Davidson County Planning	Viewing and querying of property, ownership, zoning, topographic, photographic and planimetric data	Access to public property and ownership records. This is not a
	Department, Nashville, TN	Print maps and property records. Search by owner, address, subdivision etc.	legal document.
Montana State Library Natural Resource Information System	State Library	To create detailed thematic maps based on user-selected criteria. "Topographic Map Finder" lets you browse USGS quadrangle images or zoom to the image of any feature or Township/Range/Section in the state.	Public access to the geographic content.
On-line Browser for Topographic Maps of Montana	Montana State Library	For viewing MrSID-compressed USGS 1:24000, 1:100000, and 1:250000 Topographic Maps. Search tool for any Public Land Survey Section, Latitude/Longitude coordinate, UTM Coordinate, or any feature name on a Topographic Map.	Finding topographic maps of any place in Montana.
Zoning and Platting Cases On-line	Municipality of Anchorage	To have access to information on case descriptions, staff reports, zoning	To give the public access to timely

(contd.)

(contd.)

Title	*Publisher and Place*	*Content summary*	*Purpose etc.*
		regulations and Municipal codes, hearing agendas, schedules, and meeting minutes and view many GIS data layers and print a map.	information and for public comment.
National Atlas of the United States Solution	National Atlas of the United States	It delivers easy to use maps of America's natural and socioeconomic landscapes. More than 20 Federal organizations have collaborated to produce reliable, integrated, and documented national maps. These are offered online in interactive mapping presentations that allow Americans to visualize and understand complex relationships between environments, places, and people.	This site is intended to promote geographic literacy and national self-awareness. The Atlas is designed to provide a reliable summary of national-scale geographical information. Partnered with ESRI to develop innovative and reliable products and services that characterize America through maps.
The National Geographic MapMachine	National Geographic Society	The MapMachine is an interactive mapping site that enables users to access a wide variety of geographic information covering the world. It offers dynamic maps from many of the world's leading data providers, atlas maps from the National Geographic 7th Edition Atlas. It is powered by ESRI Internet mapping technology and built on content from the Geography Network.	MapMachine is intended to support the education and conservation communities as well as others interested in geographic information about our world.
National Land Survey of Finland	National Land Survey of Finland	These web sites in English provide you with information on the maps and services in real estate, topography and the environment for the needs of citizens, other customers and the community at large and information on the NLS and its international activities.	It also promotes the shared use of geographic information.
NGS Survey Control Map	National Oceanic and Atmospheric Administration	It provides a map of the US, which allows you to find, and display survey control points stored in the database of the National Geodetic Survey and access the geodetic data associated with the points.	Provides detailed information on the NGS CORS sites located throughout the nation.
National Atlas of Canada	Natural Resources Canada	Using a combination of high-quality small-and large-scale digital	This site is intended to provide public access

(contd.)

(contd.)

Title	*Publisher and Place*	*Content summary*	*Purpose etc.*
		data, users can select custom maps and perform advanced spatial analysis. Users can zoom in on a particular area of a map, create, query, save and print customized maps, download base maps, and access educational resources.	to the geographic content described above.
Nelson and Wallery, Ltd./Nursing Home INFO	Nelson and Wallery	Find a nursing home based on location, religious affiliation, medical specialization, language skills, operator, and other database navigational aids. Create a map of your vicinity—then generate a list of nearby nursing homes from a list of over 18,000 nursing home in the US.	This site is intended to provide public access to the geographic content described above.
AfriView	SDS Mapping	An interactive mapping site of Africa and Madagascar based on a publicly funded watershed mapping exercise sponsored by the UN/FAO and WWF. By simulating Internet Map Serving (IMS) technology, AfriView allows "users": to dynamically browse, pan, and zoom through six layers of increasingly higher resolution map graphics of Africa.	It introduces users to NIMA and USGS that provide near global data coverage of the world. It is a preliminary map display interface that is part of a much larger and as yet complete data viewing tool entitled "NIMA in a Nutshell".
Regional Economic Development Information (REDI) System	San Diego Association of Governments	Assists industrial developers, economic development analysts, small business owners, community planners, and others to explore land development opportunities and visualize relationships among available employment land, existing land use, planned land use, average daily traffic volumes, and other geographic layers. A unique aspect of the REDI system is the ability to map on top of a digital ortho quad aerial photo base.	This site is intended to provide public access to the geographic content described above.
Inmarsat GPS Tracking and Display Solution	Sky Satellites System and ESRI France	This service can be used to display GPS position and status through Inmarsat satellite link.	This service can be used to display the position and status of a fleet of various vehicles (car, truck, boat, etc.).

(contd.)

(contd.)

Title	*Publisher and Place*	*Content summary*	*Purpose etc.*
eMAP	Systematics Technologies	Provides commercial services in Hebrew and English to many other Israeli sites and portals, including demographic data, generate driving directions between any two addresses or find all the gas stations, banks and health centers in Israel.	This site is intended to provide public access to the geographic content described above.
TBO.com	Hillsborough County Office, Tampa Police Department, and Pinellas County Office	The crime data is available at http://crimetracker.tbo.com/crimetracker/definitions.htm. Data fields include the date, location, and offense for each crime reported. Segmented the data is available into four separate Map Services: 2000 Crime Data for both Hillsborough and Pinellas Counties and 2001 Crime Data for both Hillsborough and Pinellas Counties. The street data was purchased from GDT.	Allows to find out about the reported crime data. Helps in taking decisions about personal safety, home protection, and relocation, and increases community awareness.
Afghanistan GIS and satellite image visualization service	Telemorphic, Inc., Berkeley, CA	It is a free web browser-based interactive GIS and satellite image visualization service powered by Telemorphic's Maplicity and ESRI's ArcIMS technology. Telemorphic built the service entirely with public domain data including low-resolution digital elevation models, medium-resolution Landsat Thematic Mapper (TM) satellite imagery, and small-scale roads, rivers, populated places, and political boundaries.	To provide a basic understanding of Afghanistan, along with geospatial data such as satellie imagery, digital terrain models, and basic map information related to place names, rivers, roads, and country boundaries.
Australian White Pages Directory	Telstra Corporation	It used SDE to build a White Pages directory that provides online access to 8,000,000 residential, business, and government listings. The directory is updated daily.	This site is intended to provide public access to the geographic information.
Thailand Environment Institute	Thailand Environment Institute	The institute provides Internet based access to various environmentally related spatial information maintained by them.	Provides public access to the geographic content.
Centre for GIS, Qatar	CGIS, Qatar	The Map Server application called "Explore Qatar" lets you locate street names and addresses, and landmarks such as hotels, post offices, banks, and schools. A spatial query feature is also available..	This site is intended to provide public access to the geographic content.

(contd.)

(contd.)

Title	*Publisher and Place*	*Content summary*	*Purpose etc.*
National Seismic Hazards Mapping Project	U.S. Geological Survey	USGS offers seismic hazards maps showing peak horizontal and horizontal spectral response accelerations on a national scale.	To provide public access to the geographic content.
Malaysia Internet Yellow Pages	Cybermall Sdn Bhd	Business and travel guide to Kuala Lumpur (Klang Valley), Malaysia. Powered by MO IMS, AV IMS and customized Java client.	To provide public access to the geographic content described above.
Flood Risk Mapping Solution	ESRI, Redlands, CA	Users can access FEMA Q3 Flood Data to help determine the relative flood risk of 1,200 counties across the US. The ESRI Flood Risk Maps are created using FEMA Q3 Flood Data and GDT Dynamap/2000 street data. The maps are intended as a tool to assist in screening property addresses or other locations to determine their flood prone status.	The Flood Risk Maps help identify a location falls in or near a Special Flood Hazard Area (SFHA). This flood area is the national standard on which the flood plain management and insurance requirements.
Base map for the Lithuania at Scale 1:50000, Vilnius map	Hnit-Baltic GeoInfoServisas	Digital Base map for the Lithuanian Territory at Scale 1:50000, was produced as National Basic Map in digital and paper format with the help of existing maps information, digital data bases together with the most recent panchromatic ortophoto material from SPOT satellite and with geometrically rectified multi spectral space images.	To be used for production of printing thematic maps. Ellipsoid GRS80, TM projection with the central meridian Lo =24°. Scale 1:50000. Coordinate system. ETRS-89 (ARC/INFO data format).
Korea-Japan 3D Tour	Infmedia Srl	The website is spectacular and interactive, for an impressive flight experience over the sites of the 2002 FIFA World Cup(TM). With 3DIMap technology, landscape meshes are dynamically built, and overlay images precisely positioned. On the Korea-Japan 3D Tour you can freely fly inside a virtual world with an extent of more than 3000 km!	To demonstrate the 3DIMap technology (3D Interactive Maps) as a powerful and spectacular way to deliver geographic content on the Web.

Conclusion

The above examples of GIS applications and business opportunities are only indicative. The case studies do not exactly fit in the earlier chapters, have been included here. But, as we keep on getting newer information through journals and conferences or even through internet, we find that there is no limit to such applications. The widening scope has opened new vistas for business opportunities for the GIS professionals. The integration of GIS with other technologies, such as web or internet, has further multiplied the scope of business in several areas. Such applications can be related to the identification or designing of recreational areas or resorts, siting of institutions and banks, e-commerce, geo-index for shoppers, business intelligence and the like.

REFERENCES

GIS Development (2000), Internet GIS software, *GIS Development,* Vol. IV, issue 7, pp. 28-32.

Gupta, Ravi (2000), GIS in the internet ear: What will India gain ?, *GIS Development,* Vol. IV, issue 7, pp. 37-40.

Pulsani, Preetha (2001), Internet GIS – One perspective, *Map India 2001 Conference Proceedings,* 7-9 February, pp. 19-22.

18

Spatial Data Infrastructures

In last few years a huge volume of spatial information has been available, stored and managed in the form of GIS for various applications which has opened a new vista for business opportunities. International Geo-sphere Biosphere research, digital earth, digital regions and several global and national projects related to digital information have given a boost to a new type of applications. These applications or projects are based on geo-spatial data and knowledge about such data-bases. The management of such data, information or knowledge depends on efficient and user-friendly GIS. The GIS technology now permits us to collect basic common sense knowledge linked with geo-spatial data, and also allows this knowledge to be closely associated with such data. The association of knowledge with spatial data makes the option of having feature-oriented objects in the GIS environment. The new applications look towards the data-base which can produce intelligent levels. Such a combination of data and knowledge has opened types of applications not only in the developed world but also in developing world. Hence, there has been a greater urgency to understand the spatial or geospatial data infrastructures and their various facets.

Geo-spatial data Infrastructure plays a crucial role in the development of the economy and promotion of industrialization at national and regional level. While being an expedient for economy, infrastructural support a highly complex and heterogeneous life system, having varied expectations and accountability depending upon its socio-economic structure. Infrastructure therefore cannot be merely visualized as physical infrastructure. These have to be planned as a life support system for the man as well as his environment with milestones indicating infringement limits and related course correction. The Spatial Data Infrastructure (SDI) concept continues to evolve as it becomes a core infrastructure supporting economic development, environmental management and social stability in developed and developing countries alike (Williamson, Rajabifard and Feeney, 2000 and 2003). Spatial data has been considered as an essential perquisite for any GIS initiative (*Geospatial Today,* 2004). SDI hierarchy can be organized hierarchically starting from a local to global data infrastructure (Fig. 8.1).

We live in information age which has become essential to solve problems of national development. Further, spatial information has proved to be an importance element underpinning decision-making for many disciplines because it deals with location. In the sophisticated digital environment, it has expanded to include databases, satellite positioning, communication networks including internet, web technology and wireless applications. It draws experiences from the geographic information systems, computer science, land administration, geography, surveying and mapping, legal and public administration disciplines. SDI initiatives in different

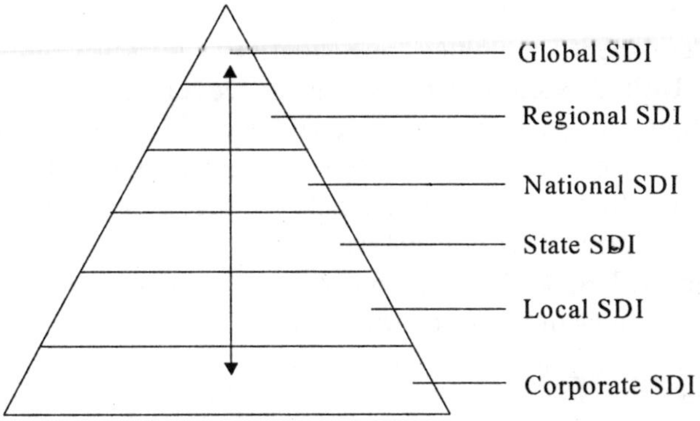

Fig. 18.1: SDI Hierarchy, After Williamson, Rajabifard and Feeney (2003)

countries have a common objective, i.e. to create an environment or framework where all stakeholders can cop-operate and interact with each other with relevant technology in order to meet objectives at different political and administrative levels. SDIs are meant to share data by avoiding duplication linked with generation, maintenance and integration of data and development of innovative applications related to business and national development. It appears that spatial information industry has evolved and will continue to grow. SDIs can become instrumental in having higher savings of all sorts. The expected structure of SDI is shown in Fig. 18.2.

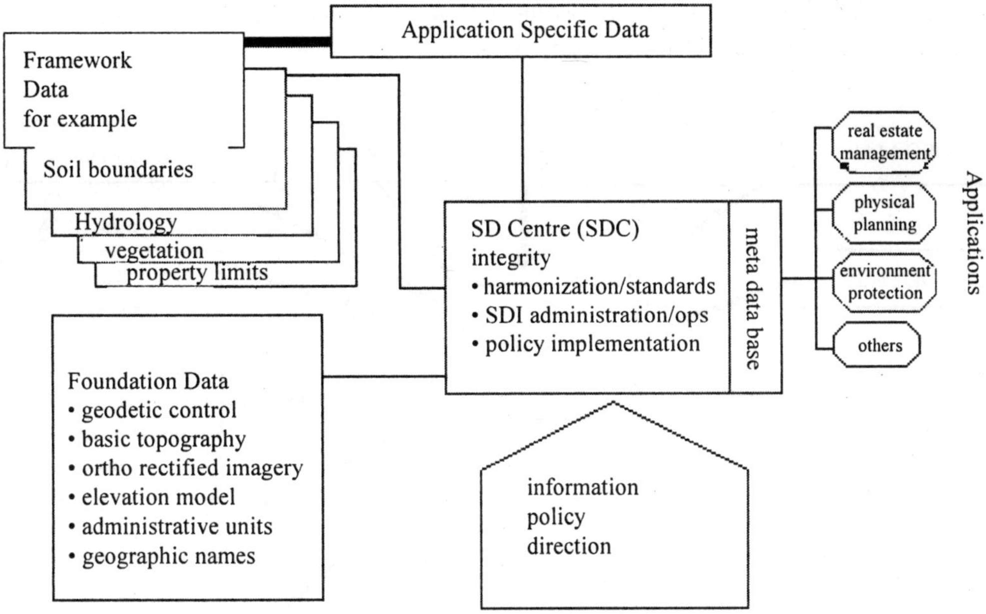

Fig. 18.2: Structure of SDI

At the global and international levels, SDI initiatives have been taken through different forums. The most important are being the Global Spatial Data Infrastructures (GSDI) conferences. UN sponsored Permanent Committee on GIS Infrastructures for Asia and the Pacific (PCGIAP), Permanent Committee for Spatial Data Infrastructure for Africa, Digital Earth Conferences, United Nations Committee on Development Information (CODI), International Cartographic Association, International Steering Committee for Global Mapping, International Society of Photogrammetry and Remote Sensing and the like are contributing to the understanding of the SDI concept. At the national level, the Executive order from the President of the United States Office of 1994 was a great boost. On one hand, complimentary initiatives followed in the US; while on the other, similar orders or initiatives were taken in other countries. The examples of the latter are Australia, Germany and Canada.

The Indian Situation

Technological advancements in the fields of computers, satellite sensors, global positioning systems, geographical information systems and digital photogrammetry combined with the increased user aspirations in a way to obtain, analyse and apply the spatial data, have forced even the developing nations to think of evolving mechanisms to make both spatial and non-spatial data accessible to users to a set standards. India has possibly the longest known tradition of systematically collected scientific spatial data. Survey of India having been doing so for the past 235 years. Survey of India (SOI) being the leader with 235 years of collection of topographical data followed by Indian Meteorological Department (IMD) and Geological Survey of India (GSI) each over 150 years. Similarly, other Departments such as Central Ground Water Board (CGWB), Forest Survey of India (FSI), National Bureau of Soil Survey and Land Use Planning (NBSSLUP) etc., have collected wealth of data over long periods of time (Fig. 18.3). In the past three decades, Department of Space (DOS) has mastered the space and

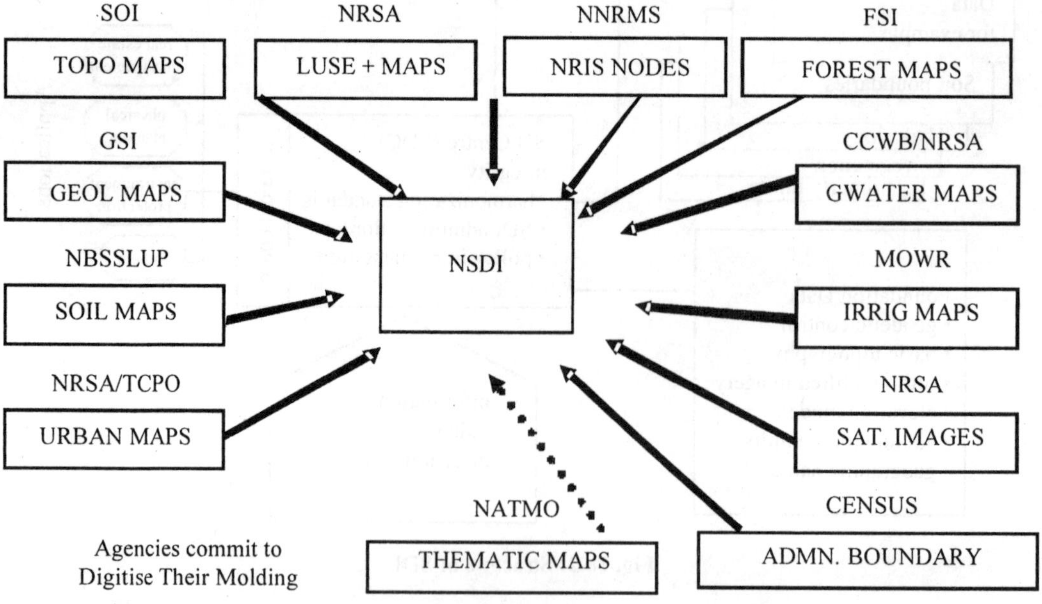

Fig. 18.3: Indian NSDI

produced a variety of data from various satellites. However, this data could be utilized for limited developmental activities. The basic reasons could be classified under three categories viz. (a) technical, (b) institutional and attitudinal, and (c) policy issues. Technical problems are mainly platform oriented. Fortunately, in India, most of the spatial data is based on Survey of India topographic maps on scale 1:50,000 or 1:250,000. However, the analogue data when converted into digital format is obtained in the machine formats of the platform on which it is generated. In spite of the best conversion software available, there is loss of data when converted from one format to another. SOI has taken the lead in preparing a national standard exchange format for Digital Vector Data (DVD). However, it does not cater to all requirements.

The institutions who have generated the data are highly possessive of the data generated and do not want to part with that. Even if they are willing to share, in the absence of metadata, it has become really difficult to know what is available. Lastly, the present policies of the government do not permit digital data of topographical nature being utilized freely and there is a specific ban on putting the data on network.

Considering these issues, a Task Force was constituted by Department of Science and Technology (DST) in November 2000 with Surveyor General of India as the Chairman. Members were drawn from various departments, organisations, users and NGOs providing spatial data. The Task Force has met several times and has set up Working Groups for metadata, data exchange formats, network protocols, communication strategy etc. Three national/international Workshops were held and the vision of NSDI was decided. Meanwhile, Department of Science and Technology obtained all approvals for bringing out a second series of maps which could be made available without any restrictions. The task force has brought all the data providers to a common platform and to think alike. Thus, there is a better understanding of each other view points and a consensus has emerged to evolve a National Spatial Data Infrastructure (NSDI). The data providers are engaged in the task of conversion of analogue data into digital domain and also to generate the metadata. The metadata server is ready and will be launched soon.

What is NSDI ?

India, has over the past years, produced a rich "base" of map information through systematic topographic surveys, geological surveys, soil surveys, cadastral surveys, various natural resources inventory programmes and the use of the remote sensing images. Further, with the availability of precision, high-resolution satellite images, data enabling the organisation of GIS, combined with the Global Positioning System (GPS), the accuracy and information content of these spatial datasets or maps is extremely high.

India has a National Spatial Data Infrastructure (NSDI) strategy and the *NSDI Strategy and Action Plan* Report has been adopted in 2001, through the coordinated efforts of the NSDI Task Force. The NSDI will be an important element for supporting economic and sustainable growth in the country. With the recognition that spatial information is a national resource and citizens, society, private enterprise and government have a right to access it, appropriately. Only through common conventions and technical agreements, standards, metadata definitions, network and access protocols will it be easily possible for the NSDI to come into existence. While government must provide the lead, private enterprise, NGOs and academia have a major role to play in making the NSDI reality. NSDI will require for coming together of various groups and harmonizing their efforts in making this national endeavour a success. The NSDI model for India has been proposed (Fig. 18.4).

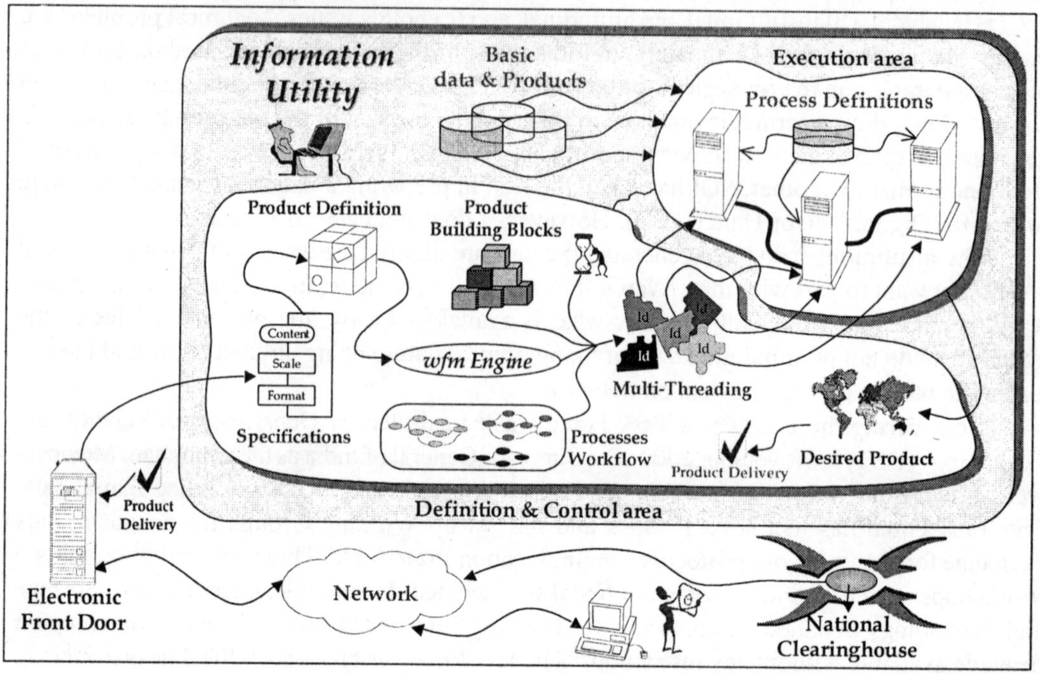

Fig. 18.4: Example of process model

Since the adoption of the NSDI Strategy in 2001, the NSDI Task Force has been working on developing a NSDI implementation mechanism through drafting an executive mechanism and also drafting a possible NSDI bill. The NSDI vision is:

> " National Infrastructure for the availability of and access to organized spatial data, use of the infrastructure at community, local, state, regional and national levels for sustained economic growth."

It has been decided that the National Spatial Data Infrastructure (NSDI) be established under a legislation that will set out essential operational features of the NSDI including its composition and administrative structure, decision-making powers, and the broad policy issues on which the NSDI will be entitled to make rules and regulations. The stated objective of the NSDI is the creation of a national infrastructure for the generation, availability of and access to organized spatial data to be used at community, local, state, regional and national levels to meet the requirements of spatial data for sustainable development. It is proposed that the NSDI be established as a statutory body, empowered to carry out various functions necessary in order to achieve the broad objectives stated above. The functions of this overarching framework will be as follows:

(a) Function as a repository for spatial data currently available with, as well as all future data created and maintained by, participating government agencies;

(b) Co-ordinate, control and manage the receipt of multiple sources of spatial data from participating government agencies and supervise the appropriate format conversion of this data to ensure uniformity of information and inter-operability of data;
(c) Develop the NSDI metadata of NSDI spatial information;
(d) Develop a physical infrastructure compromising a network of servers on which an intranet for the storage, exchange, control and access to NSDI data operates;
(e) Determine the requirement for particular categories of spatial data and require the creation or collection of spatial data, by any of the participating agencies, to fill such requirement;
(f) Co-ordinate and authorize the regular revision of existing spatial data in order to reflect changes in spatial features;
(g) Establish an appropriate standard (the NSDI standard) defining the content, scheme, design, process, quality, protocols and other standards necessary in order to ensure standardization of the data contained in the NSDI repository as well as appropriate formats for its exchange and interoperability;
(h) Function as a repository for spatial data meeting the NSDI standard, that have been generated by industry and private enterprise;
(i) Develop and co-ordinate appropriate quality control measures to ensure the integrity, standardization, accuracy and quality of NSDI data, including but not limited to the establishment of appropriate certification and accreditation procedures;
(j) Create mechanisms, procedures, tolls and applications (including appropriate search and access protocols and an NSDI user interface) to facilitate public access to NSDI spatial data;
(k) Develop common solutions for the discovery, access and use of spatial data in response to the needs of diverse user groups;
(l) Offer consultancy services on specialized subjects including but not limited to geodetic, geophysical, topographical surveys and digital database to government and non-government organisations;
(m) Co-ordinate multi-disciplinary mapping activities in collaboration with governmental and non-governmental organisations; generate a digital database of unrestricted Open Series Maps (OSM) as per National Map Policy; and
(n) Increase the awareness and understanding of the vision, concept and the NSDI.

In the event of establishing NSDI on the approval of the Government, the Union Department of Science and Technology will create the proposed body through the administrative act of appropriate order constituting the body. Similar action was taken by the US by Executive Order 12906, "Coordinating Geographic Data Acquisition and Access: The National Spatial Data Infrastructure" which was signed in April 1999 (http://www.fgdc.gov/fgdc.html).

Major Constraints for the Spatial Data Base

Many planning department and other national organizations such as Survey of India, National Remote Sensing Agency and National Atlas and Thematic Mapping Organisation have a long history of using topographical maps and remotely sensed imageries for preparing base maps for spatial planning. The methodologies of designs in such projects are invariably worked out on the basis of requirements projected by the resource users. The multidisciplinary approach to planning is often effected by non-availability of reliable, systematic and accurate data to all those involved in spatial planning. A number of planning organizations have developed need based GIS to assist in analysing and making suitable projections. Non-availability of accurate and consistent spatial data base is, however, proving to be the biggest constraint. There also appears to be hesitation in standardization the database content on account of subjectivity in the requirement of data. Major constraints for standardizations of data are as follows:

(i) Geodetic framework and projections of topographical maps (base map) are often not in conformity with that of the cadastral maps.
(ii) The information on the base map is generally outdated and does not contain the details required by the planners.
(iii) The scale of map does not conform to the requirement of spatial planning at village/town level.
(iv) Value added information on artifacts have to be collected from widely distributed sources and ground survey for items not depicted on map.
(v) Metadata, i.e. Definitions and description of parameters are not standardized.
(vi) The mechanism for public participation depends on the local legal provisions and practices.
(vii) Lack of technology upgradation.
(viii) With no commitment available, the data producers are hesitant to invest in creating and maintaining the infrastructure for spatial database.
(ix) Lack of legal commitment for dissemination of data to resource users.

The Department of Science and Technology and the Department of Space have come together and taken an initiative to establish a platform, viz. National Spatial Data Infrastructure (NSDI) for coordination and integration of activities between various data producing agencies to enable standardization, collection, compilation of data on a variety of related themes in a sustainable manner. Such a platform would also play a crucial role in formulating policies on dissemination of data and regulations for commercialization/privatization in various theatres of information technologies.

Major Functions of NSDI

After a series of discussions with the members of the Task Force, the major functions of the NSDI have been identified as follows:

(i) To define policy concerning dissemination of spatial data to the government and private organisations.
(ii) To identify and fix the standards of data to form fundamental data sets for topographical, cadastral and thematic applications.
(iii) To lay down standards to which digital technology should conform and remain valid for a fixed period.

(iv) Identify sponsors and formulate guidelines for funding for sustaining the infrastructure.
(v) Initiate standardisation of geo-referencing systems, plan for generating transformation parameters to support transportation of data in other popular geo-referencing systems and projections commensurate with the user requirements.
(vi) To identify potential users, their resources and data requirements, economics of data transfer and legal cover etc., for setting up clearing houses.

The current model of the Spatial Data Infrastructure in the country is shown in Fig. 18.5. The data users and the data providers have one to one association and such connections do not necessarily process the transaction through the NSDI gateway. Obviously, this model will change with the implementation of NSDI.

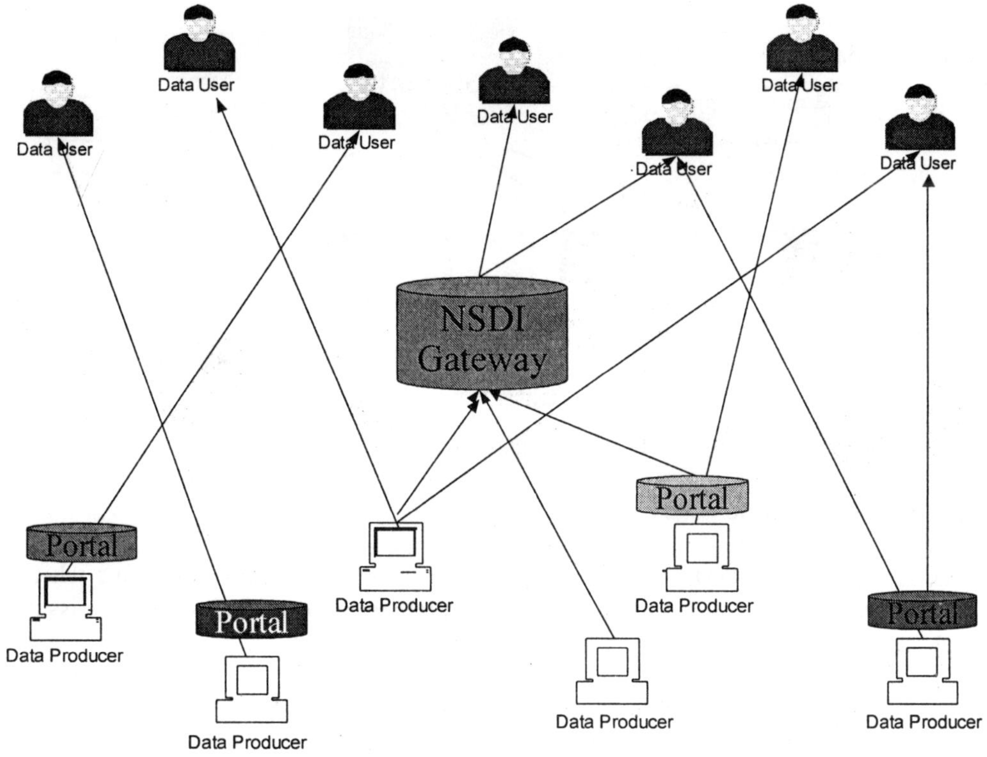

Fig. 18.5: Current model of NSDI

Expected Results from NSDI are:

(i) Single window clearance for all users of geospatial digital data.
(ii) Standardisation of geospatial digital databases including format, hardware, software, linkage, access and sharing and meta data.
(iii) Finalisation of geospatial accuracy requirements for different users.
(iv) More effective modelling for planning and development projects.

The future model of NSDI is indicated in Fig. 18.5 where portals of different data producing institutions will be connected with the NSDI portal. The queries can be directed to the NSDI portal as well.

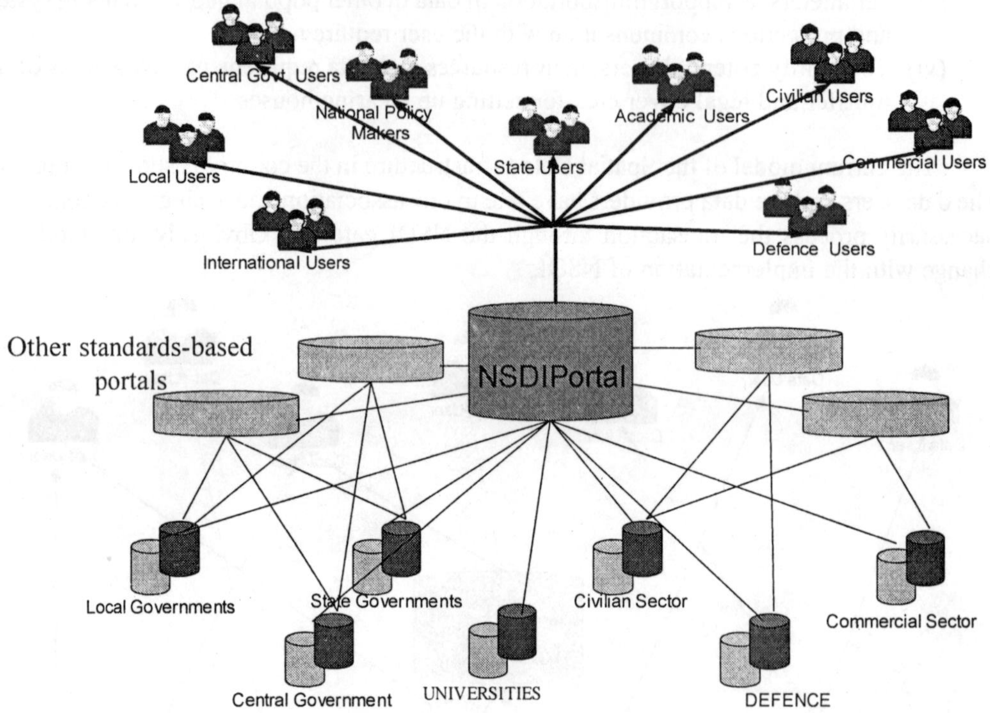

Fig. 18.6: Future model of NSDI

Major Issues Related to NSDI

The issues related to NSDI are being formulated through various sub-groups which are deliberating with user agencies to assess their requirement of data set. Some of the issues are as under:

(i) To lay down priorities and time frames for the development of infrastructure.
(ii) To lay down standards for metadata in the fundamental data sets comprising of topographical and thematic data.
(iii) To identify data generating organizations who can meet the standards of metadata and work on mechanism for transfer of standardized data to the fundamental data sets.
(iv) To set-up subcommittees to evaluate and scrutinize the on data transfer format, exchange format and other technical standards for the infrastructure.
(v) To identify the custodians of fundamental data sets, methodologies for its upkeep, updation, operational linkages and evaluation *vis-à-vis* meta data standards.

(iv) Initiate standardization of geo-referencing systems and projection commensurate with the user requirements.

(vii) To identify potential users, their resources and data requirements.

(viii) To identify data requirements of users and lay down the contents of user data sets inconsonance with the government policies on desensitization of spatial data. Policies on networking for access to data for browsing and value addition are defined.

These are many and varied stakeholders to the NSDI and would include government at local, state, national levels (as both users and data collectors/owners); NGOs, educational and scientific organisations, private and public agencies. As of now, the major players as contributing to NSDI are: Department of Space, Survey of India (SOI), India Metereological Department (IMD), National Atlas and Thematic Mapping Organisation (NATMO), Geological Survey of India (GSI), National Informatic Centre (NIC), Census Department and the like.

Progress so far

- Various Central Government Departments, DST, DOS, MoEnF, Mines, Urban Development, Rural Development, NIC, Census Department, DAC, DARE, DOD etc. have initiated demonstration programmes towards NSDI Agency activities.
- The finalization of the text of a NSDI Bill—which will now be processed by DST—first as an Executive Order and then the NSDI activities as an inter-agency programme could start.
- Drafting of seven NSDI Standards documents—through inter-agency committees. The following documents have been initiated and are in different stages of finalization:
  —NSDI Content Standard—being coordinated through a committee and based on inputs from different NSDI agencies.
  —NSDI Metadata Standard—being coordinated by a committee. Design and development of the Metadata Server also being taken up.
  —NSDI Exchange (NSDE) Standard—being coordinated through a committee. The NRIS and SOI DVD formats are being merged into a NSDE Standard.
  —NSDI Search and Access Protocol—being coordinated through the committee for NSDI Metadata.
  —NSDI Networking Framework—being coordinated through a committee.
  —NSDI Quality Standard. Just initiated as a draft based on NRIS Standard.
  —NSDI Access Rules—being worked out by ISRO/DST.
- The task force has developed a NSDI web-site and a NSDI Metadata Server—which, when launched, will be able to provide information on spatial data availability to users. The NSDI Metadata Server is a catalogue of all spatial information of the NSDI Agencies, as per a NSDI Metadata Standard. The design of the clearinghouse search for spatial data based on internet has already been proposed (Fig. 18.7).

Major Issues

The major issues still to be considered—and very crucial to the implementation of NSDI are:

Use of NSDI

The utility of NSDI and its possible support to sustainable and economic growth in the country and for supporting developmental activities is well accepted by all. During 2nd NSDI Workshop held at Ooty in July 2002, the business potential of NSDI was discussed and it has emerged that the value of data and data products could be of the order of a few billion rupees. There is a need to evolve a business plan for NSDI as well as the organisations participating in NSDI.

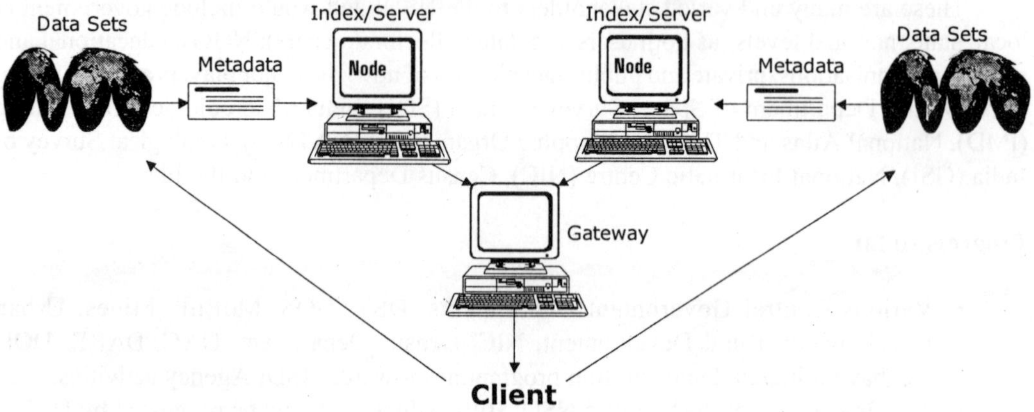

- Multiple catalogs searched on internet in parallel by gateway
- Results are collated for presentation to client
- A single session can search hundreds of collections and accesses current content!

Fig. 18.7: Clearing house search metaphor

Discussion and Debate on NSDI Standard Documents

The NSDI Standard documents prepared need to be discussed, debated and, if endorsed, adopted for implementation. Now they are available on the web.

www.nsdiindia.org/www.nsdindia.com/www.nsdiindia.net

Map Policy and a NSDI Policy

The Map Restriction Policy needs review and modification if NSDI has to be implemented. Actually the restriction on digital map data (topographical sheets and other maps; including datum and projection parameters restriction) is an important issue for NSDI. The other issue is making the data available on networks and for public domain. In terms Ooty Communique, DST has prepared a note for discussion on policy issues.

There is a need for a NSDI policy that takes of from a modified and tuned map policy and defines the policy scope of each NSDI agency (SOI for topographical data, GSI for geological data, FSI for forest data and so on). The NSDI policy must address issues like—access rules, costing, copyright and licences, value addition and so on. This issue needs to be debated in the NSDI workshop. In June 2006, NSDI has got the approval of the Government of India.

Institutionalization of NSDI

How should NSDI be implemented? As decided in Ooty, a draft government resolution is ready for discussion amongst ministries. Further, after three years of effort by the task force on NSDI, the following have emerged.

- A clear understanding on the next steps and the institutional framework – Government resolution with an ultimate aim to have the act.
- NSDI Task Force to continue its effort in the direction resolved at Ooty.
- NSDI Agency to commit specific agency programmes and resources for 10th Plan. Major thrust is to convert analogue data into digital and prepare metadata.
- Each agency designs its own business plan.
- Private sector to gear up for providing value added services and products to users.

Recognizing the need of accurate, consistent and updated geographical information, it is inescapable that the planners and decision-makers should adopt an integrated approach for holistic planning and aided with geo-informatics, arrive at rational solutions. Reliable and complete information on various themes will facilitate better understanding of complex issues on conflicting objectives and reduce the conflicts in decision-making, especially where the conflict has its origin in disagreement on 'facts' and specially 'spatial awareness'.

Keeping these aspects in mind, two major initiatives, viz. introduction of new series of topographical maps and establishment of NSDI taken by the Department of Science and Technology will go a long way to claim g-readiness of India to provide geographical information to resource users and industry for sustainable development of the nation.

Geospatial information, including maps and images, are vital to support decision-making at various level and implementation of action plans. With the availability of space borne imagery, Global Positioning System (GPS) data and Geographic Information System (GIS) technology, users are now able to process maps—both individually and along with tabular data and crunch them together to provide a new perception—the spatial visualization of information.

Space based images has the characteristic to cover vast areas and have larger information quality as well as multi-spectral and repeat observations ability, thus, it is suitable for regional and global environment monitoring. The application of satellite images not only used for mapping but also to monitor temporal changes in the environment.

The science of surveying and mapping encompasses a broad range of disciplines including Surveying and Mapping, Remote Sensing (RS), GIS and GPS. The technology of digital image processing, the global positioning system and GIS are being used to integrate, processed and analyzed the spatial data for sustainable development of the natural resources. The digital data (e.g. high resolution images, scanned aerial photographs, scanned maps and digital orthophotos) are currently available for this purpose. GIS is powerful system for decision-making tool is now in the hand of cartographers. Beginning with a computerized topographic map as its base, a GIS overlays and integrate graphic and textual information from separate databases. The end result is a customized and reliable tool that can support decision-making and problem solving and provides almost instantaneous answers to complex questions.

In view of the present day requirement, Survey of India (SOI) plans to provide digital topographical databases for entire country on 1:50,000 scale initially, in a years time. Topographical data/maps in digital and analogue form on WGS-84 (World Geodetic System-84) is made available for general public without any restriction (existing topographical maps of SOI are on Everest datum). The department uses the state-of-the-art technology in glossarial data acquisition, management and dissemination. Airborne Laser Terrain Mapping (ALTM), GPS, Electronic Distance Measuring (EDM) instruments, digital photogrammetry, modern printing technology and computer hardware/software are being used in map-making process. High-resolution satellite imagery, viz. SPOT, IKONOS data etc. being used for updating of topographical maps. All these data sets would be useful in developing Spatial Data Infrastructure that will better facilitate the availability and access to spatial data for all level including Government sectors, Commercial sectors, Academia and citizens in general. Survey Training Institute of Survey of India also provides short training course on the insight to NSDI which includes topics like standards, nodes and network, capacity building, role of service providers, legal framework and metadata (www.soisti.gov.in).

User Applications and Solutions with more Flexible Solutions

While technologies such as GPS and GIS have matured during the 1990s, increasing attention has been focused on the development of user applications and solutions. This technology segment can be defined as software/hardware solution bundles developed specifically to solve a geospatial information user's problem. The critical challenge in this area is for the technology supplier to gain an in-depth understanding of the user's business environment so that the solution is optimized to address the user's key business issues. Built on top of the core technologies, user applications/solutions are developed using combinations of database management, object-oriented programming and systems integration tools and techniques. Clearly, geospatial data and GIS were considered to be extremely important technologies for addressing solutions in a number of applications.

India, over the years, has generated a rich base of information through systematic data collection in the form of topographical surveys, geological surveys, soil surveys, cadastral surveys, various national resource innovative programmes and the use of remote sensing images. Further with the availability of digital topographic databases, high-resolution imageries and sophisticated techniques of data collection using GPS and tidal station, the accuracy of information content of these spatial data sets or map is very high. With the growing awareness for Information Technology among users, an initiative was taken by Department of Science and Technology in the form of National Spatial Data Infrastructure (NSDI). NSDI is the mechanism through which need for access to reliable, timely and spatial data are met. It is a first step in the direction of bringing together data providing agencies and to obtain commitment from them to make special data available and accessible to the user community, which conforms to accepted standards, The NSDI framework aims at decentralized approach to achieve the following objectives:

— Develop and maintain standard of collection of spatial data.
— Develop common solution for discovery, access and use of spatial data in response to the needs of diverse and new user groups.

— Relationship among participant's organizations to support the continuing development of NSDI.

— Increase the awareness of the vision, concepts and benefits of the NSDI.

This framework when fully implemented will be utilized for various developmental activities including preparedness, response and recovery purposes during disasters. The availability and use of spatial data shall have impacts on every aspect of society and will be made available to people who need them, when they need them and in the format at they can make decision with minimal pre-processing.

Open Geospatial Consortium (OGC)

For data standards at the international level, Open Geospatial Consortium or OGC has been established. It is a non-profit making international voluntary organisation. The core mission is to develop standards that enable interoperability and seamless integration of geospatial information, processing software and geospatial services. Geospatial information and processing encompass GIS, remote sensing, surveying and mapping, navigation, location based services, access to geospatial databases, sensor webs and other IT geospatial technologies and information sources. Further, it facilitates a consensus process to define, test and approve standards that promote the integration and use of geospatial data and services. In the forum provided by the OGC, participants from different information communities collaborate to achieve the vision of a world in which everyone is likely to benefit from geospatial information. Representatives of about 310 industries, academic and government organisations work together to make geospatial information and services accessible and useful with all kinds of applications.

In order to ensure that the OGC standards are harmonized with broader information technology standards, the OGC works closely with other standards bodies such as the International Organisation for Standards or ISO, the European Committee for Standardization or CEN, the World Wide Web Consortium or W3C and the like. The importance of OGC standards has been felt because the more one knows about the interoperability, the more he or she saves in technology lifecycle. The interoperability standards enable to go to users to increase their plug and play options. Developers and integrators go to market faster with more flexible solutions. Organisations join OGC not only to create the standards that enable such benefits, but also to be a part of the world's premier network of such users and providers of geospatial software, service and data. As a result, the technology developers submit interoperability requirements to meet anticipated needs of the market place.

The development of the standards does not require the members to give up any intellectual property or trade secrets. The use of open standards to connect components, applications and content—allowing a white box view on the components' functionality and interrelationships without revealing implementation details—fulfils the industry requirement for protection of intellectual property as well as the user requirement for transparency. Such transparency supports both interoperability and the credibility of the enterprise, or federated solution. Nevertheless, OGC offers its specifications free of charges to all and adheres to rigorous process to ensure that the specifications remain frees of royalties for use, a position taken by the World Wide Web Consortium as well. It promotes free and open web. For further details, log into the http://www.opengeospatial.org.

Conclusion

The characteristics of the geospatial data set are changing. First and foremost, in order to meet users demands effectively, the capacity for the real-time collection, synthesis and access must exist; data currency is essential. The data should be scale less, seamless, without artificial boundaries, and linked to a time component that has become critical to many applications, for example, traffic flow management, routing and delivery, and tidal and marine traffic. Moreover, as technologies become more advanced, geospatial information will be both more readily available and in greater demand.

There will also be a growing trend toward the collection and integration of non-traditional data using secondary reference systems like voting, culture and housing patterns, gender, sales and industry. Furthermore, as technologies and applications become more globally used, geospatial data will spread to and originate from non-traditional sources such as the voluntary sector, health councils, communities and peoples. However, regardless of what data are collected by whom, unless they are easily and readily accessible, their value diminishes; hence the importance of an exceptional geospatial data infrastructure. Furthermore, a well-developed national information infrastructure, enabling the dissemination and sharing of valuable, geographically referenced information, and with an ever-increasing audience of businesses, entrepreneurs, students and researchers, and communities, is widely accepted as an essential asset for any country to maintain and to advance its social and economic well-being. As such, geospatial data and the infrastructure in which they are organized can be considered to be a technology in its own right within the rubric of this technology.

NSDI is a reality and not just a dream of few visionaries. NSDI will be playing key role in the developmental activities of the country which has the requisite expertise, data, networking and other infrastructure. NSDI offers tremendous opportunities are there for the wealth of geospatial data to be for the benefit of all. The big picture is as follows (Fig. 18.8).

Fig. 18.8: The big picture of NSDI

Geospatial scientists must take advantage of the technology and the information available with it. The information will be available through net, web and other electronic media. The geographically referred information will be in the form of bits, bytes and pixels; and the volume of data in kilobytes, megabytes and gigabytes. This will also help in horizontal and vertical analysis of smallest areas such a building in a town or a parcel of land in the village. Further, the electronic flow of data will effect spatial applications in twenty first century by shifting (a) from so-called analogue theories to empirical models; (b) geographical enquiry from bottom up to bottom down; (c) developing indigenous models, and (d) revalidating established models, theorems and postulates.

REFERENCES

Adams, T. and S. Dworkin (1997), *WfMC Work or Handbook*, chapter Work of Interoperability Between Businesses, pages 211-21. John Wiley and Sons.

Alameh, N. (2001), *Scalable and Extensible Infrastructures for Distributing Interoperable Geographic Information Services on the Internet.* Ph.D. thesis, Massachusetts Institute of Technology (MIT), Massachussets, United States of America,.

Alfred Leick, Alfred (1995), *GPS Satellite Surveying*, John Wiley and Sons Inc.

Batty, M., *et al.* (1994a). Modelling inside GIS. Part 1: Model structures, exploratory spatial data analysis and aggregation, *International Journal of GIS*, Vol. 8, pp. 291-307.

Bernus, Peter and Laszlo Nemes (196), *Modelling and methodologies for enterprise integration.* In Proceedings of the International Federation for Information Processing (IFIP) Conference on Models and Methodologies for Enterprise Integration. Chapman and Hall.

Berry, J.K. (1987), Fundamental operations in computer-assisted map analysis, *International Journal of Geographical Information Systems*, Vol. 1, no. 2, pp. 119-136.

Bishr, Y., W. Kuhn, and M. Radwan (1999), *Interoperating Geographic Information Systems, chapter Probing the semantic content of information communities: a first step toward semantic interoperability*, Kluwer Academic Publishers, pp. 211-21.

Blakemore, Michael (2004), Globalisation, infrastructure and agenda, *GIM International*, Vol. 18, No. 1, pp. 11-3.

Buehler, Kurt (ed) (2003), *OpenGIS Reference Model.* Version: 0.1.2, No. OGC 03-040, Open GIS Consortium Inc.

Burrough, P.A. and McDonnell, R.A. (1998), *Principles of Geographical Information Systems: Spatial Information Systems and Geostatistics.* Oxford University Press.

Coordinates (205), NSDI in India:The reality behind the dream, *Coordinates*, December, pp. 34-38.

Dasgupta, A.R. (2003), Spatial data infrastructure: Concepts and efforts, *Geospatial Today*, Vol. 2, issue 4, pp. 16-20.

Forghani, Alan and Denise Gaughwin (2002), Spatial information technologies to aid archaeological site mapping, *GIS@Development*, Vol. VI, issue 6, pp. 31-35.

Gavin, Elizabeth (2003), Roadmap to SDI in Africa, *GIS@Development*, Vol. 7, issue 9, pp. 29-31.

Georgiadeou, Yola (2003), Back to future, *Geospatial Today*, Vol. 2, issue 4, pp.21-23.

Geospatial Today (2004), National SDI initiatives, *Geospatial Today*, Vol. 2, issue 5, pp. 25-46 (based on www.ngdf.org.uk).

Groot, Richard and John McLaughlin (2000), *Geospatial Data Infrastructure: Concepts, Cases and Good Practice.* Oxford University Press, Oxford.

Hutchinson, M.F. (1995). Interpolating mean rainfall using thin plate smoothing Spline. *International Journal of GIS*, Vol. 9, pp.385-404.

ISO/IEC (1996), *Information Technology—Open Distributed Processing—Reference Model: Foundations. Draft International Standard*, International Organization for Standardization, Geneva, Switzerland, No. 10746-2.

Konecny, Gottfried (2002), *Geoinformation: Remote Sensing, Photogrammetry and Geographic Information Systems*, Taylor and Francis, London.

Kottman, C. (1999), *Interoperating geographic Information Systems,* (chapter on *The OpenGIS Consurtium and Progress Toward Interoperability in GIS*). Kluwer Academic Publishers.

Lieberman, Josua (ed) (2003), OpenGIS Web Services Architecture. Discussion paper, Version: 0.3, No. OGC 03-025, Open GIS Consor-tium Inc.

Luisa Liliana, Alvarez C. (2003). *Geoinformation virtual enterprises design and process management.* Masters thesis, International Institute for Geo-Information Science and Earth Observation ITC, Enschede, The Netherlands.

Masser, I. (1998), *Governments and Geographic Information*, Taylor and Francis, London.

Mitasova, *et al.* (1995), Modelling spatially and temporally distributed phenomena: new methods and tools for GRASS GIS, *International Journal of GIS*, Vol. 9, pp. 433-46.

Moellering, H., H.J.G.L. Aalders and A. Crane (eds) (2003), *World Spatial Metadata Standards*, International Cartogaraphic Association (forthcoming).

Mohammadi, Hossein, Abbas Rajabifard, Andrew Binns and Ian P. Williamson (2006), Bridging SDI design gaps, *Coordinates*, Vol. 2, issue 9, pp. 26-29.

Morales, Javier and Mostafa Radwan (2002), Extending geoinformation services: A virtual architecture for spatial data infrastructures, in *Proceedings of the Joint International Symposium on GeoSpatial Theory, Processing and Applications* ISPRS Commission IV, Ottawa, Canada.

Nag, P. (2002), Maponomics – Map products and their commercialisation, *GIS@Development*, Vol. 6, issue 2, pp. 19-23.

NSDI (2002), *National Spatial Data Infrastructure: Strategy and Action Plan*, Department of Science and Technology, New Delhi.

NSDI (2003), *Data Exchange Format – NSDE Version 1.01*, National Spatial Data Infrastructure, New Delhi.

OGC (2002), *OGC Web Services*, DVD NTSC, Open GIS Consortium, Inc.

OGC (2006), *OGC Web Services*, OWS-3 Demonstration, Open Geospatial Consortium, Inc. DVD Rom/ Video.

Percivall, George (ed)(2002), The openGIS Abstract Specification Topic 12: Open GIS Service Architecture. Version 4.3, Open GIS Consortium, Inc.

Pulusani, Preetha (2003), Interoperability—Trend or reality ?, *GIS@Development*, Vol. 7, issue 9, pp. 16-19.

Rajabifard, A. and I.P. Williamson (2002), Spatial Data Infrastructures: An initiative to facilitate spatial data sharing, in R. Tateishi and D. Hasting (eds), *Global Environmental Databases—Present Situation and Future Directions*, Vol. 2, International Society for Photogrammetry and Remote Sensing (ISPRS-WG IV/8), GeoCarto International Centre, Hong Kong.

Rajabifard, A., M. Feeney and I.P. Williamson (2002), Future directions for SDI development, *International Journal of Applied Earth Observation and Geoinformation*, Vol. 4, pp. 11-22.

Rhind, D. (ed) (1997), *Framework for the World*, Geoinformation International, Cambridge.

Roy, P.S. and Samir Saran (2003), An interoperability model for Indian NSDI, *Geospatial Today*, Vol. 2, issue 4, pp. 36-8.

Satyanarayana, P. (1999), Open GIS consortium, *GIS India*, Vol. 8, April-June, pp. 36-37.

Sipkes, Jacques (2004), AfricaGIS 2003: How far have we come ?, *GIM International*, Vol. 18, No. 1, pp.55-59.

Siva Kumar, R. (2005), Standards and interoperability: For an Indian standard, *Geospatial Today*, Vol. 4, issue 2, pp. 54-56.

Williamson, Ian, Abbas Rajabifard and Mary-Ellen F. Feeney (eds)(2003), *Developing Spatial Data Infrastructures: From Concept to Reality*, Taylor and Francis, London and New York.

Williamson, Ian and Abbas Rajabifard (2004), Capacity building for SDI in geo information science, *Geospatial Today*, Vol. 3, issue 4, pp. 31-39.

19

Status of GIS in India

Since very beginning, cartography in India had been considered very secretive and protective and the map-making activities lie within the four walls of the obscure government offices in the country. Maps were considered to be confidential document and only to be consulted if there are no other option. However, a few industries did take up cartography and map printing, not always as their prime activity. Since independence and till the time of the advent of the computer-assisted cartography, the scenario remained almost same. The digital cartography got impetus with (a) introduction of the Geographical Information System (GIS) and (b) easy availability of remote sensing data, that too from the Indian remote sensing satellites.

In India, digital mapping and GIS related activities in the public sector and industry go in different directions. However, such activities are either sponsored by the government or are from abroad. The priorities, policies, nature of work and issues are entirely different. This was due to the so-called Map Restriction Policy which was in force till recently. A New Map Policy has been announced which is meant to promote, apart from other things, the GIS industry in the country (Appendix I). As per this policy, the new maps, known as Open Series Maps (OSM), will have WGS-84 datum, UTM projection, a different map numbering system and reduced content. This series in hard as well as soft copy will be made available to users without any restrictions. Hence, OSM is likely to boost GIS activities further in the country. Nevertheless, the priorities have shifted more towards detail maps and GIS applications, particularly for urban areas. Hence, digital data at a larger scale is to be generated.

Within the fold of government, there are institutions, which generate the basic data for mapping such as, Survey of India and the National Remote Sensing Agency under the Department of Space. Then there are institutions, which have major activity for preparing maps required for their specific use, particularly, mapping of natural resources. Their cartographic divisions have repository of relevant information. These institutions are Geological Survey of India and the Forest Survey of India. Then there are institutions like the office of the Registrar General and Census Commissioner which collect data on human resources and also prepare maps and atlases related to population. The case of NATMO is somewhat unique. In this case, the information is collected from every possible source and maps of relevant themes are prepared.

Survey of India (SOI) was established in 1767. It is the national survey and mapping organisation of the country. Survey of India has already accomplished various landmarks in the field of cartography including the mapping of the entire country on the scale of 1:50,000, provided high precision geodetic control survey etc. Besides its commitment of topographic mapping of the country, SOI provides infrastructure data for national development programmes, such as survey for road, rail, tunnel, alignment, irrigation, large scale river valley project, mining and excavation. With the advent of new era in computer and space technology and its rapid proliferation in every sphere of human activities, SOI has also stepped in and taped the immense potential of this modern technology. Initially the digital cartography was adopted as an R and D activity with the installation of Automap System in 1981. These R and D activity results were encouraging and it was decided to adopt the state-of-the-art technology at operational level. Thus during the years 1986-89, three centres, viz. Modern Cartographic Centre (MCC) a project with UNDP assistance and two digital mapping centres (DMCs) at Dehra Dun and Hyderabad were established in Survey of India. These centres of SOI were fully equipped with state-of-the-art technology of digital mapping and are fully operational to process the digital cartographic databases required for planning and developmental activities. The 1:250,000 and 1:50,000 series of maps have been digitized. At present, 1:25,000 series is being digitized. The availability of data at these three scales will provide much required basis for the promotion of GIS activities in the country. For this purpose, Survey of India has opened geospatial data centre for each state.

National Atlas and Thematic Mapping Organisation (NATMO) also under the Department of Science and Technology is a pioneer and nodal organisation for preparing atlases and thematic maps of the country. It was established in 1956. As per technological advancement in different field of cartography, NATMO has also adopted newer technologies in preparation of different maps on different themes, area and scales.

NATMO has a special division on Digital Mapping and Geographical Information System. This division is very efficiently preparing digital thematic maps. Digital Mapping System (DMS) has started the digitization of spatial, statistical and non-spatial databases available in the organisation. It has already generated data under the NRDMS programme for a part of Bankura district, West Bengal. Further, a map on *Infrastructure Profile of West Bengal* has been also produced by this division for the West Bengal Industrial Development Corporation. The division is also coordinated the scanning and digitization of SMART maps of Calcutta Municipal Corporation (CMC) (now Kolkata) with an industrial partner (IDS) and CMC, Kolkata. A project is also in progress for preparation of *Satellite Atlas of India* in collaboration of the Department of Space.

Geological Survey of India (GSI), one of the largest scientific organisations of the Government of India, was founded in 1851 and has been carrying out different geo-scientific activities of the entire land and recently also of offshore areas of India (covering EEZ) with aims and objectives enshrined in the charter of functions as laid down by the government. It has played a vital role for development of country's economy in the industries and mineral, power or energy, agriculture, irrigation and water management sectors and in environmental and natural hazard studies. Geological and other geo-scientific thematic maps on different scales are some of the major end products of GSI activities. Geological maps on 1:50,000 scale form the basic documents and are prepared primarily based on ground surveys. These are also supplemented at times by aerial photos, satellite imagery and various remote sensing techniques and aero-geophysical mapping. Huge amount of map data in analog forms have been collected through a long period of about 140 years by geo-scientific activities. Since geological data are collected mostly in analog format and not always quantifiable, conventional cartographic techniques are being followed for compilation of geological maps. However, it is intended to introduce digital cartography in phases so that the colossal information so far generated could be stored and retrieve efficiently thereby avoiding time gaps between the dates of surveys and the publication of the results. Airborne Mineral Surveys and Exploration (A.M.S.E) wing of the GSI is already processing and interpreting the airborne geophysical data in the form of maps by computer assisted methods in the Geophysical Mapping Centre at Bangalore. Application of GIS is also being attempted for storing and suitable retrieval of various geo-scientific and mineral data.

The Census of India is one of the largest producers of maps in South Asia. This has been achieved with intimate collaboration among geographers, statisticians and demographers. Census of India has been a discontinuous affair up to 1861. The restricted programme in 1941 due to Second World War restrained the cartographic activities. In 1951 Census, one of the major contributions was excellent *tahsil* or *taluk* maps published in *District Census Handbooks.* In 1961, *Census Atlas Project* was initiated for the first time while urban land use maps in state level town directory were introduced in 1971. It was also proposed to bring out administrative atlases for states and union territories, which may serve as a handy guide to the administrators and planners. In 1991, in view of the scientific and technical advancement, the Registrar General has initiated another plan project designated as 'Computer Assisted Cartography and Data Dissemination' to map the towns at ward or enumeration block (EB) level. The main objective of this plan scheme is to study the cities or towns through various mapping and cartographic techniques, fieldwork and with the introduction of computer aided cartography which aims at the generation of selected city or town maps at enumerative base (EB) level and the analysis of growth of urban centres through decades from different view points caused by religious, administrative and industrial stand points. The scheme also envisages preparing and updating selected town maps on large scale and create digital database for dissemination of census results as well as for uses in subsequent censuses and surveys etc. Final base maps of cities or towns were to be prepared on suitable scale indicating the ward. The base maps were to be utilized to generate thematic maps, utility maps, land use maps of city or town at micro level through computer. Similarly, the enumerative base maps will be digitized for further updation and use by various survey for conducting census and other vital statistics programme. Work has been initiated in a moderate way on a PC platform at Registrar General India (RGI) headquarters using ARC/INFO GIS software. Similar facilities have been extended in different states to generate cartographic database at town or ward level for further work on utility maps at micro level for area specific planning. In addition to pre-census and post-census cartographic activities, the census organisation also caters to the cartographic need of different departments and ministries.

In 1972, the successful launching of LANDSAT by the US provided a new source of data which has a lot of flexibility and with special processing, such data can be tuned for a variety of applications. Though the remote sensing establishments based on US data and later French data (SPOT) were established in late 1970s and early eighties, the successful launching of IRS 1A in 1988 provided a boost and a different orientation to GIS-related activities in the country. Space based observation system provide timely and accurate information for monitoring and optimum management of earth resources. The main advantages of space-based remote sensing are: (a) synoptic view-wide area can be covered by a single imagery;

(b) repetitivity—can collect the data of any area repeatedly; and (c) coverage of inaccessible areas like mountains, swampy areas, thick forests and deep sea. To day, the IRS system, comprising several Indian remote sensing satellites, is providing data for monitoring and managing our natural resources. The payloads on board the IRS satellites operate in different spectral bands and provide varying spatial and spectral resolutions.

The first operational remote sensing satellite, IRS-1A, was successfully launched on March 17, 1988. Even though designed for three years, IRS-1A served for almost eight years. An identical satellite, IRS-1B, launched in August 1991, is still in service. IRS-1B is placed in the sun-synchronous polar orbit of 904 km with equatorial crossing time of the descending node being at 10.25 a.m. The repetitivity cycle is 22 days. The payload on this satellite comprises two Linear Imaging Self-Scanning Sensors, LISS-I and LISS-II. While LISS-I provides a spatial resolution of 72.5m, the LISS-II sensors provide a spatial resolution of 36.25m. LISS-I has a swath of 148 km the two LISS cameras provide a composite swath of 145 km. The sensors operate in four spectral bands—0.45-0.52 micron; 0.52-0.59 micron; 0.62-0.68 micron and 0.77-0.86 micron. IRS-1C, launched in 1995, and IRS-1D was launched in 1997 incorporate sophisticated sensors with resolutions of around 23m in multi-spectral bands and 5.8m in the panchromatic band. They have stereo-viewing capability and on-board data recording facility (Roy, 2004). Cartosat-1 has been successfully launched in 2005. Its pixel size is 2.5m for panchromatic band. The data will be made available soon which should give a big boost to GIS activities. The Cartosat-2 with pixel size less than 1m is planned to be launched in a year or so.

Planning Commission recognized the importance of establishing National Natural Resources Management System (NNRMS) and within its umbrella, the Regional Remote Sensing Service Centres were opened for the satellite data processing for specific application of the concerned region in which they were located. Due to better launching facilities and higher resolution of data, the cartographic activities are likely to improve further.

An Integrated Mission for Sustainable Development (IMSD) was launched in 1992 which now covers 174 districts in the country. Under IMSD, locale-specific prescriptions are arrived at mainly using IRS data, merged with other collateral social economic data using GIS. The mission involves generation of thematic maps showing current land use or land cover, types of wastelands, forest cover or types, surface water resources, drainage pattern, potential ground water zones, land forms, geology (rock types, structural features, mineral occurrences), soil types, etc. Maps showing slope aspect is prepared using topographic contour information and the meteorological data is collected from existing data-bases. Using all these information map showing land capability, irrigability, soil erosion, run-off potential, priority watersheds needing immediate treatment, etc., are prepared. The action plans, for example, will include identification of sites for water harvesting through ponds and check dams, soil conservation through terracing and contour bunding, afforestation, agro-forestry and agro-horticulture, fuel wood and fodder development, sand-dune stabilization, mining and necessary conservation measures.

National Remote Sensing Agency (NRSA), an autonomous organisation under Department of Space was established in 1978. One of the primary responsibilities of NRSA is to provide remote sensing data not only from the Indian IRS series of satellites and from other satellites, such as Ikonoss or SPOT. The services provided by this agency include (a) generation and dissemination of standard and special satellite data products, (b) generation of large-scale line maps, (c) 3-D digital data generation for input GIS and generation of DTM or DEM,

(d) long-term and short-term training courses in the field of remote sensing, GIS and allied areas at IIRS, Dehara Dun and NRSA, Hyderabad, and (e) GIS as decision support system for natural resource management. All these activities have supported GIS activities in the country. Further, the world is being exposed in yet another aspect of the revolution of information technology the increasingly widespread acceptance and use of GIS. Owing to their strong analysis capabilities, GIS is playing an important role in the management of a country's natural resources. Efficient use of GIS is providing decision-makers, the information, for the proper utilization of these resources. Keeping this in view, NRSA has acquired several state-of-the-art GIS and applications are being developed using them.

Space Applications Centre (SAC) of the Indian Space Research Organisation (ISRO) at Ahmedabad is engaged in the research and development activities of remote sensing technology and GIS applications. The major activity of this centre are design and development of optical and infrared payloads for space and airborne platforms, design and development of microwave remote sensing payloads (both active and passive), image processing and data processing related software development and operationalization of applications and development of models for meteorological and oceanographic applications. SAC has been actively engaged in ISRO satellite programmes through contribution of payloads for IRS series of satellites, the processing of payload data for data products generation and their applications. SAC has been closely interacting with various user agencies and departments to developmental methodologies for remote sensing and GIS applications in the field of resources survey and environmental monitoring under the NRIS programme (Dasgupta, Rao and Gopalan, 2000). Some of the notable applications that have been carried out are the crop acreage and production inventory, coastal zone mapping, wetland mapping, monitoring of forests, geological applications for mineral targeting and the like.

Forest Survey of India (FSI) deals with forest mapping at national level by making use of GIS and remote sensing techniques. Its predecessor organisation, the Pre-investment Survey of Forest Resources (PISFR), took up aerial photo interpretation and mapping for forestry purpose for the first time in India in the year 1965 with the help of FAO experts. In 1968, FAO collaboration ended successfully and by that time the Indian counterpart acquired enough expertise in this field to carry out work independently. From 1st June, 1981, PISFR was converted to Forest Survey of India and reorganized with effect from 1st July, 1986, under the Ministry of Environment and Forests, Government of India. FSI prepares thematic maps on 1:50,000 scale by interpretation of existing latest aerial photographs on ten years cycle for the whole country and forest vegetation maps on 1:250,000 scale by interpreting satellite imagery in two years cycle. Maps under special projects are also prepared. Thematic mapping and vegetation mapping activities are described as under.

The conventional method of map production is tedious and time consuming. The advent of computer technology for handling spatial data sets has brought in powerful techniques for analysing and archiving map data. It increases flexibility in map making, as the digital cartographic databases are capable of carrying much more information that could be legibly presented on any one map. Digital mapping combined with other data sources can be used for variety of end uses such as forest cover monitoring and mapping selection and location of plantation sites, environmental change detection, assessment of damage done to forest, estimation of area under shifting cultivation etc. Data in digital form facilitate linkage between all other data sources—which are both physical (forest cover, soil capability classes, elevation, slope), and socio-economic (demographic distribution, marketing facilities, production and demand etc.). Digital cartographic system and GIS have been installed in FSI at Dehra Dun. The FSI has also taken up special projects for preparation of land use maps in Bhutan and Nepal. Further, this institute has developed a GIS model and village database for Uttari Kosi sub-watershed (Almora) for watershed management and wasteland development at micro-level planning during 1993. At present FSI is preparing a GIS project using remote sensing and GIS techniques for Sariska Tiger Reserve and its environ.

Natural Resources Data Management System (NRDMS)—a multi-disciplinary programme of the Department of Science and Technology, Government of India aims at developing a scientific data based approach for operationalising the concept of decentralized planning. Launched in 1982, the programme is a major initiative of the government to promote research and development-cum-demonstration in the field of GIS technology and its application to the problems of natural resources management. During the Eighth Plan, the focus has been on preparation of spatial district resource profiles, upgradation of the low cost GRAM GIS package, area specific application studies and training of users, academic institutions. Centre of Studies in Resource Engineering (CSRE), Indian Institute of Technology and R&D laboratories provided the technical back up. Priority areas have been identified for research and development. Over the last few years the areas of research studied were (a) land system classification, (b) watershed modelling, (c) energy planning, (d) groundwater budgeting, (e) large scale resource mapping, and (f) landslide studies. NRDMS is now poised towards further consolidating these efforts by upgrading the GRAM package to suit the data processing requirements of local level planning. Concerted efforts are being made to develop kits and tutorials for the training of potential users so that the methodology strikes roots in the resource management, practices at different levels of the planning hierarchy. UNDP supported NRDMS in this endeavour.

National Informatics Centre (NIC) was setup in March 1975 by the Government of India to play a promotional role in creating computer awareness and for developing and implementing computer-based information systems for decision support in the Ministries and Departments of the Central Government and the corporate sector. NIC has set up a satellite-based nation-wide computer-communication network, called NICNET, with over 1,000 nodes connecting the national capital, the state capitals and district headquarters to one and another. The services provided by NIC includes development of application software packages, systems software, database systems, distributed databases, networking, electronic mail, Electronic Data Interchange (EDI), access to international databases, computer-aided design, GIS, modelling and training.

The spatial information and GIS are fast becoming an important prerequisite in most government departments. With the development and availability of more than 2,000 data bases on NICNET consisting data of almost every activity of the government, it was felt to encompass the spatial dimension of data into various data bases. Applications ranging from demographic census to soil conservation, pollution control to location of a facility, all require one or other spatial information. After getting acquainted with the power and usefulness of computerized information, available through various databases on NICNET, there has been sufficient awareness and demand for spatial data component. Further the decreasing cost of hardware, versatile graphic peripherals and high speed network have made the prospects of developing spatial data bases at decentralized location a viable preposition.

NIC has developed an indigenous GIS software, GISNIC, with user-friendly interface and complex GIS functions built into an integrated package. GISNIC is designed to provide the users with a set of tools for retrieval, transformation and analysis of both spatial and non-spatial data which provides an environment to manipulate and manage locational and thematic data. The product comprises of modules DIGITIZE for digitization, MAPBASE for generating thematic maps with facility to query on attribute data base, PLOT for off-line generation of hard copy maps and SHOWMAP a callable routine for integrating maps with existing applications.

NIC has initiated the process of making many large spatial databases available on NICNET. GISNIC is being used for the development and retrieval of these databases. It is also being used as a presentation tool for preparing thematic maps for deriving the attribute information from the existing databases.

NIC has also setup the facilities for processing of remote sensing data with programmes to utilize data from IRS and other satellites. A major component of project under remote sensing programme envisages development of national level thematic database. Software design and development in this connection has been for processing raster data. Indigenous software on NICNET would provide a powerful tool for processing images, which would yield the thematic maps and strengthen GIS programme for spatial data analysis. NIC has been encouraging various pilot projects in the area of GIS and thematic mapping with a view to establish the procedures and technology. This technology can be horizontally transferred to various districts and states. NIC has been working in the area of natural resource informatics for sustainable development at grass root level.

In order to support the city administration NIC has also established a group, which looks after the mapping of utility lines. This group is fully equipped with digital photogrammetry equipment, GPS receivers and treated software. The group offers its services for establishment of utility mapping centres in ever-growing cities. The group at NIC is also involved in a facilities management project for entire National Capital Territory of Delhi. In this project an arbitrary grid has been established for Delhi. Same grid has been used for digital base mapping by NRSA. The base maps are prepared at 1:1,000 scale. These maps are then used for overlaying underground utility networks. The planned utilities include water, sewage, electricity and the road network. The objective is to support the city administration and the respective utility organisation. The project is also helping in the better management of city traffic and fire-fighting management. NIC has executed several projects on natural resources data management system for sustainable development at grassroots level through GIS based mapping.

Within the fold of the State Governments, there are two major sources of cartographic information: (a) cadastral map available with the land records and (b) detailed ward-wise maps with the Corporations and Municipal bodies. These maps are at a standard scale but they are old and almost in all cases, serious efforts are to be made for its updation and validation. Such maps are also important because due emphasis has been given on infrastructural development in urban and rural areas of the country. *Panchayati Ṛaj* system has been empowered with 73rd and 74th Amendments of the Constitution for taking up the developmental issues at the lowest level. Both types of maps have been used for GIS purposes as well.

Municipal GIS: Example of Kolkata

Several examples are now available where municipal GIS has been introduced in urban areas. Such cases are generally sponsored by the municipal bodies or by the government. Universities and the NGOs have taken up such projects as well. Further, in several cases, there have been joint efforts in this regard. The case of Kolkata Municipal Corporation (KMC) is a typical example of such initiative. About sixty per cent of its area is covered by Smart's maps which were prepared during the early part of the last century (1903-10). The example of Smart's map of a part of Calcutta, (now Kolkata), is included here (Fig. 19.1).

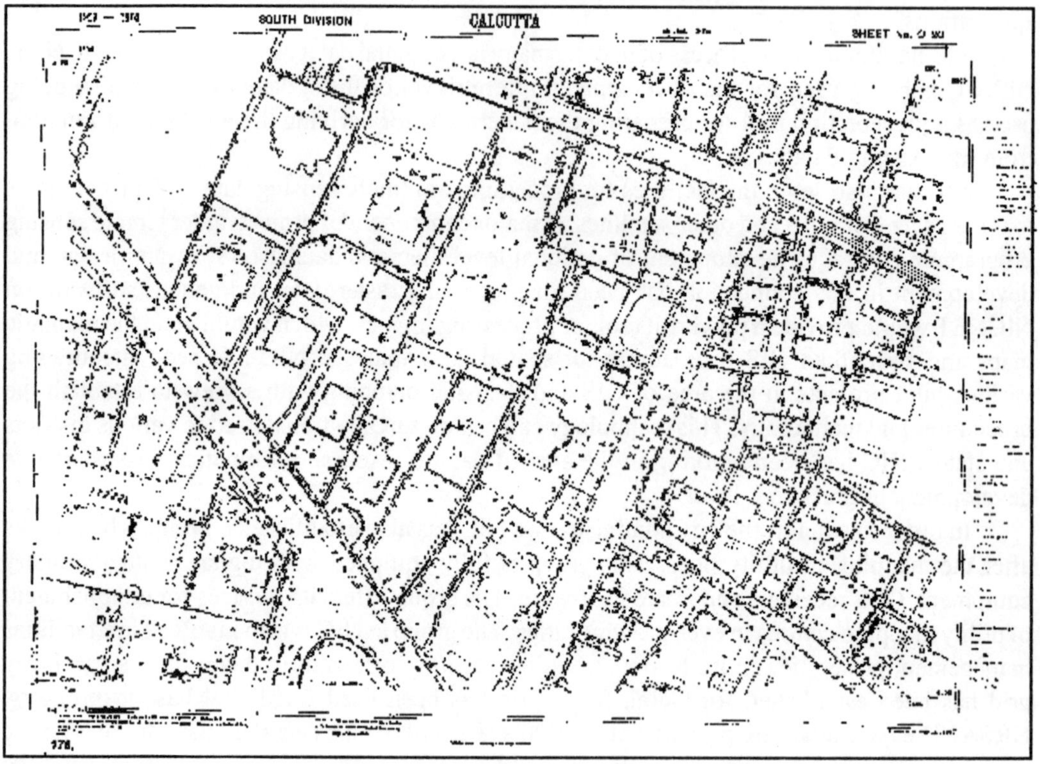

Fig. 19.1: Smarts map of a part of Kolkata (Ballygunge)

These maps contain some physical features and property boundaries and are considered as a legal document. Obviously, it has become necessary to update these maps along with other relevant information. Hence, a project was taken up by KMC in collaboration with the Kolkata based National Atlas and Thematic Mapping Organisation (NATMO). Integrated Data System (IDS) from the GIS industry was engaged in this project as well (Nag, 1999). The objectives were as follows:

(a) To scan the old smarts maps with out making changes.
(b) To keep film positives/negatives of the original maps as a backup.

(c) To incorporate the latest changes on a separate file.
(d) To attach data base required for record and tax administration.

It was felt that the digital data would be useful for following activities:

(a) Property records
(b) Development and planning initiatives
(c) Revenue collection
(d) Development of GIS
(e) Decentralization of development and taxation processes.

Recently, Kolkata Municipal Corporation or KMC has taken the implementation of Municipal GIS with the funding of UK's Department for International Development (DFID) and technical support of Tata Consultancy Services and the National Atlas and Thematic Mapping Organisation. KMC's vision is: "*KMC will be efficient, effective, equitable, citizen responsive, financially sustainable and transparent, delivering quality services to its citizens.*" The objective of the project is to develop a GIS based system to assist KMC in strategic planning and resource mobilization and utilization, streamlining municipal taxation (especially the property tax, but encompassing all the taxes and fee structures as envisaged under the Municipalities Act, 1980) and management functions, and planning and management of day-to-day operations in all sectors of activity. In addition GIS would provide a dynamic information base on poverty and poor settlements in the KMC to serve as a basic planning, monitoring and evaluation tool to improve governance of poverty alleviation programmes particularly the targeting and the delivery of services.

In KMC views computerization as a very important tool, which can be used to convert its vision into reality. Whether it is making KMC more efficient and effective in its operations or improving the financial health of KMC, or making KMC more responsive to citizens' needs and making its operation more transparent, the role of computerization cannot be overstated. With a view to resolve all the problems inherent in the current scenario, an integrated and comprehensive Municipal Administration System (MAS) is proposed to be developed by a separate group of consultants called MAS consultants. The proposed MAS would comprise of the following main components:

1. Municipal Transaction Processing System (MTPS)
2. Municipal Data Bank
3. Citizens Interface System
4. Municipal Geographic Information System (MGIS)
5. Municipal Management Information System (MMIS)

A suggested schematic diagram is given here (Fig. 19.2).

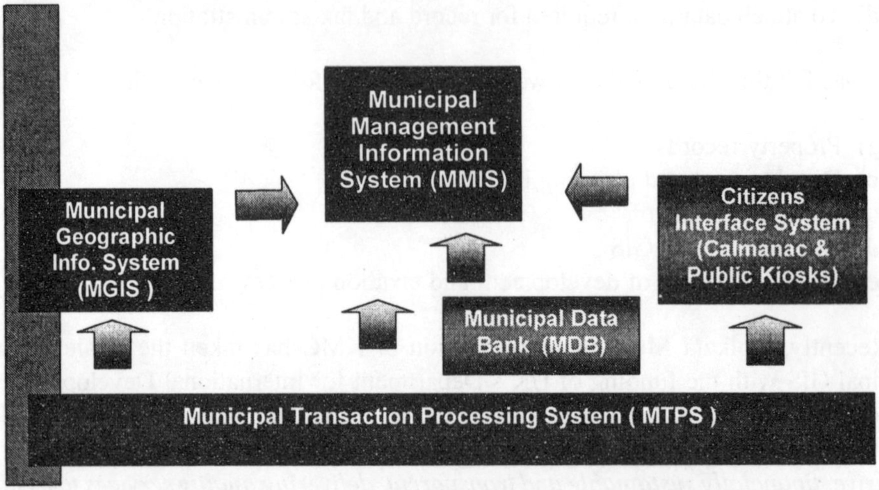

Fig. 19.2: Municipal Administration System (MAS) (*Source*: KMC)

The Municipal Transaction Processing System (MTPS) would provide the on-line transaction-processing interface, which would be used by operational users to key in all the transactions. This component would comprise of the various functional modules that would take care of the working of all the departments of KMC. The following functional modules are envisaged for TPS. This module would cover the following functions (Fig. 19.3):

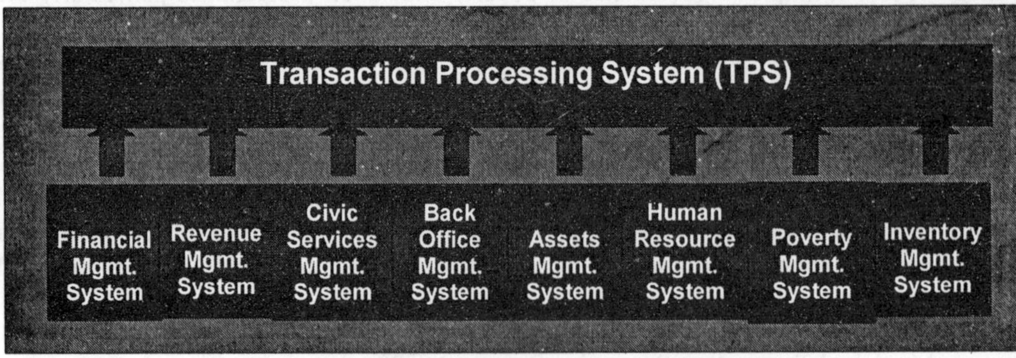

Fig. 19.3: Schematic diagram of TPS (*Source*: KMC)

- Finance
- Treasury
- Auditing
- Budgeting

Being the central module, this system would be linked to all the other management systems. Any financial transaction carried out in any department would automatically generate the required accounting entry in this module. Further, an important element in the process of strengthening planning and management capabilities is the development of an integrated Geographical Information System (GIS) for KMC. This will strengthen the strategic planning capacities and day-to-day management of core municipal functions by the local bodies. In this regard, it is proposed to implement a fully integrated GIS in KMC which has a population of roughly 4.6 million citizens and the city spread over an area of 187 sq. km.

This GIS would help the KMC in restructuring and rationalization of its revenue from various sources such as property tax, water supply, rental from municipal properties etc. and would provide a platform for effective planning, operation and maintenance of municipal utilities and infrastructure. It would also enable KMC to concentrate on improving access to municipal services in poorer areas of Kolkata. So the GIS would provide the different thematic layers superimposed on the base map of Kolkata. As in any GIS, the purpose here is to attach KMC's attributes data as different thematic layers to the spatial data. The air photograph based digital base map was prepared by the National Remote Sensing Agency (NRSA). Example of such map is as follows (Fig. 19.4).

The GIS proposed to be developed should have the following dynamic features:

(a) Interfacing the alphanumeric data from the Transaction Processing System of MAS, or any other deemed fit by KMC/Consultants.
(b) The updation of data (spatial or alphanumeric) should be dynamic in nature so that the GIS always produces up-to-date information of reality on ground.

The development of enterprise GIS map for this purpose has been shown in Fig. 19.5.

On the rural scene, all the states and union territories of the country have a directorate of land records. The directorate is on state level and there used to be subordinate offices known as Superintendent of Land Records (SLR) at district level and further as revenue inspectors or *kanoongo* at *tahsil* or *taluk* level. In these offices various kind of maps are prepared for different level. The directorate prepares and update the maps at state level that is when there use to be any change in administrative boundaries of state or districts or other units the map is redrawn and updated. Similarly, if any change at *tashil* or block level happens, then at the office of Superintendent of Land Records (SLR) the district and *tashil* level revenue maps are updated. At the *tahsil* office the cadastral or village maps are updated or redrawn. The cadastral map are updated or redrawn after the consolidations of landholding are made. This process is generally goes on periodical bases for one or the other village. Thus in the directorate of land records or department revenue, cartographic or map making is a permanent activity. From time to time the directorate of land records also publishes statistical maps and atlases based on available up-to-date statistics on the development of the states. Such maps are also published by the department of agriculture showing land use, crop distribution, crop production, soil quality, irrigation etc.

Now, many state governments' directorate of land records have introduced or adopted the technology of LIS or GIS and GPS for the management and maintenance of their land records and related matters. Government of Madhya Pradesh was the first state to adopt this modern technology. The other states like Andhra Pradesh and Karnataka, are following the same. It may be concluded that the map making and cartographic activity is widely done in almost in all government departments in one form or the other.

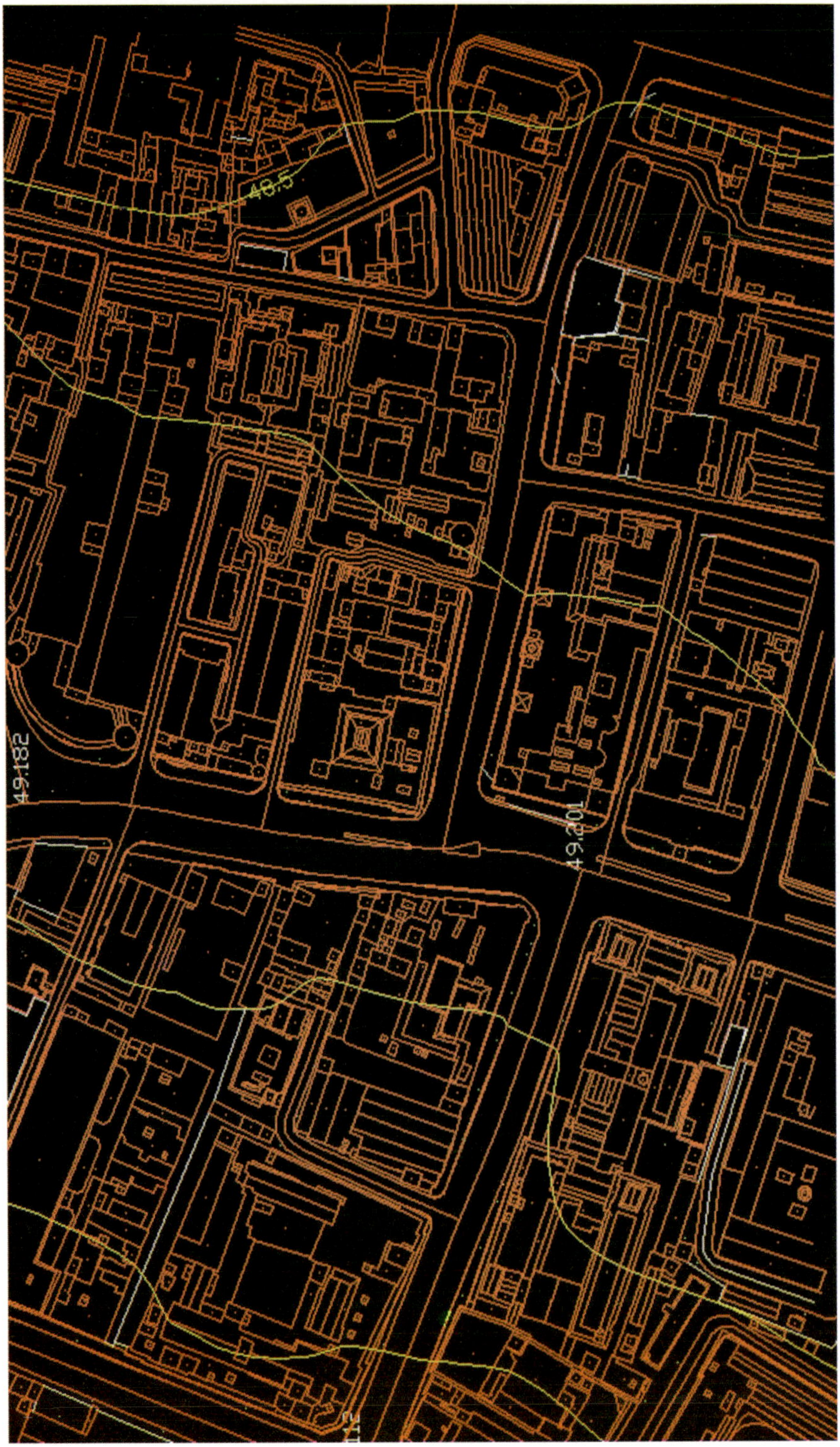

Fig. 19.4: Example of air photographs based digital map of a part of Kolkata City by NRSA

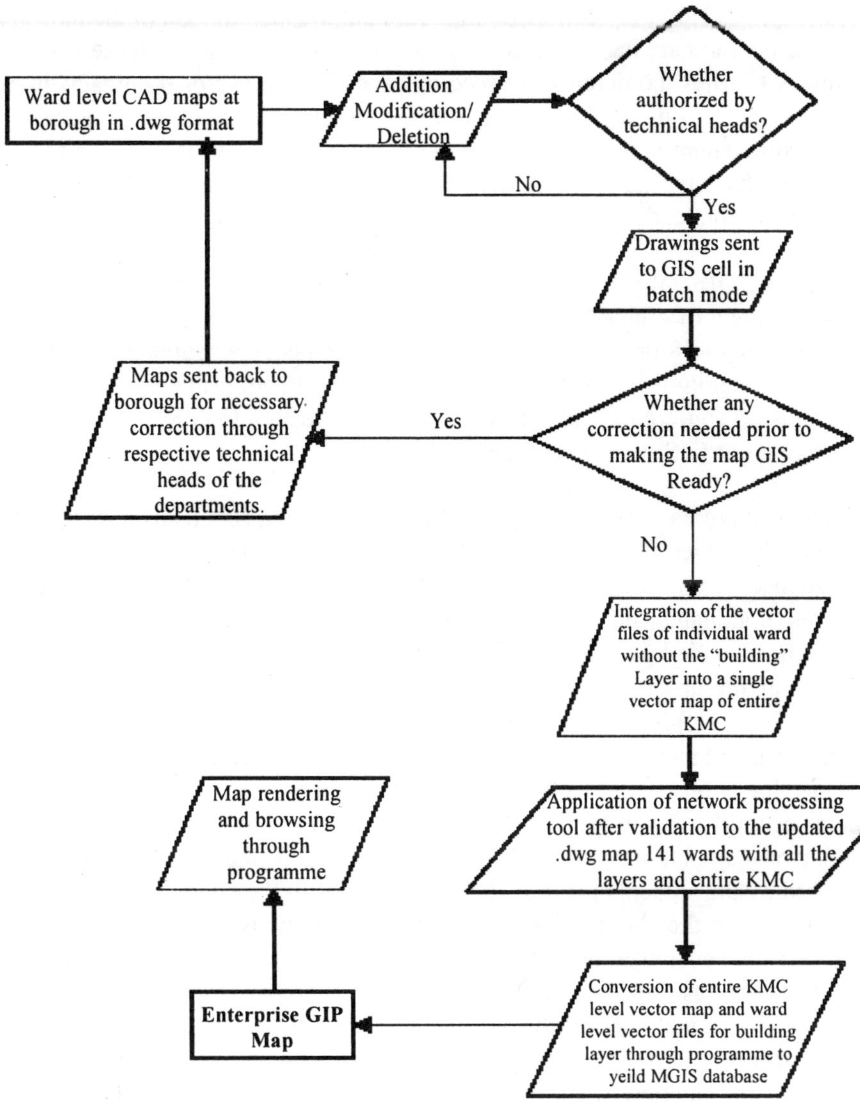

Fig. 19.5: Enterprise GIS map (*Source*: KMC)

GIS in Indian Industry

In the Indian Industry also the cartographic activities are fairly old as well. The making and production of maps for educational purposes and guide maps for tourists and pilgrimage were largely made by private publishing industries or companies. But now in the age of Information Technology (IT), industry is also involved in production of different kinds of maps and graphics. The GIS has revolutionised the field of cartography. In fact, GIS industry has taken over cartography outside the folds of the government. About hundred IT companies in the country are involved in GIS related activities. They are mostly involved in the application GIS packages in exploration, natural resource management and telecommunication, power, ports, roads, soil, water, forest, land, etc. However, it is difficult to find out the full strength of the industry since projects handled by them are hardly publicized. One has to go through their web sites, annual reports, company profiles, conference proceedings and through personal contacts in order to make some assessment. National Spatial Data Infrastructure is trying to involve them but not yet fully successful. There is no association or confederation of GIS industry in the country. Attempts made so far have not yielded any results. Based on the information available, the activities of some of the leading companies or industries in the area of GIS are as follows:

(a) Supplying GIS software
(b) Development of digital bases for GIS applications
(c) Undertaking GIS projects basically from government sponsored activities
(d) Undertaking digitization and GIS projects from abroad, primarily from the west
(e) Organising conferences, exhibitions and events related to GIS
(f) Bringing out GIS related journals

GIS related activities of some industry are described here. This list is not exhaustive but only indicative. Wherever, the Indian industry has taken up GIS projects, they are mostly government sponsored (Marutish, 2004). The examples are regarding enterprise system for electric utilities, command and control system for safety. Recently surveys of Indian GIS industries have been made by CSDMS and Spatial Network Pvt. Ltd. (www.spatialindia.com). They have been taking up projects on local problems as well. An example of Chennai based WTI Advanced Technologies (www.wtiatl.com) is given in Fig. 19.6.

Besides the above mentioned companies or industry there are many other industries who are engaged in preparation and production of maps. These are the text and other books publishers, wall and guide or tourist city maps producers. The newspaper publishing houses also do a lot of cartographic or map making jobs. Unpublished maps are available in large numbers in the theses submitted by the students for their different degrees in the educational institutes also.

As apparent form the above, the industries which are involved in cartographic work are of different sizes and specialization. There are purely cartographic institutions who have been dedicated to this work as their primary field of activity. Over time, they have changed their orientation due to digital methods and the quantum jump in technology. However, such activities in industry are of recent past. They got impetus due to the possible higher profitability in this area owing to the introduction of automated or computer assisted systems. Nevertheless, the units of this industry have not only kept the cartographic work as their sole objective. There is a set of back-up activities on which they can fall, in the event of lack of activities in cartography. Such back-up activities are related to sale of hardware, software, data product, specialized services, conversion activities and supply of qualified manpower. At the lower end, these companies do carry out data processing, DTP, project evaluation and related activities. Hence, their outfits can take shape as per the market forces. In fact, we could find out the units, which can raise highly train personnel and infrastructure as and when required. They are waiting for projects, which may exclusively or partially fall in the arena of cartography. Further, from the examples given above, some industries have been closed down, some have closed GIS activities while some have modified their vision and action.

Unfortunately, the origin of almost all the cartographic projects in the country is from government sources and this industry looks forward to central and state governments or their

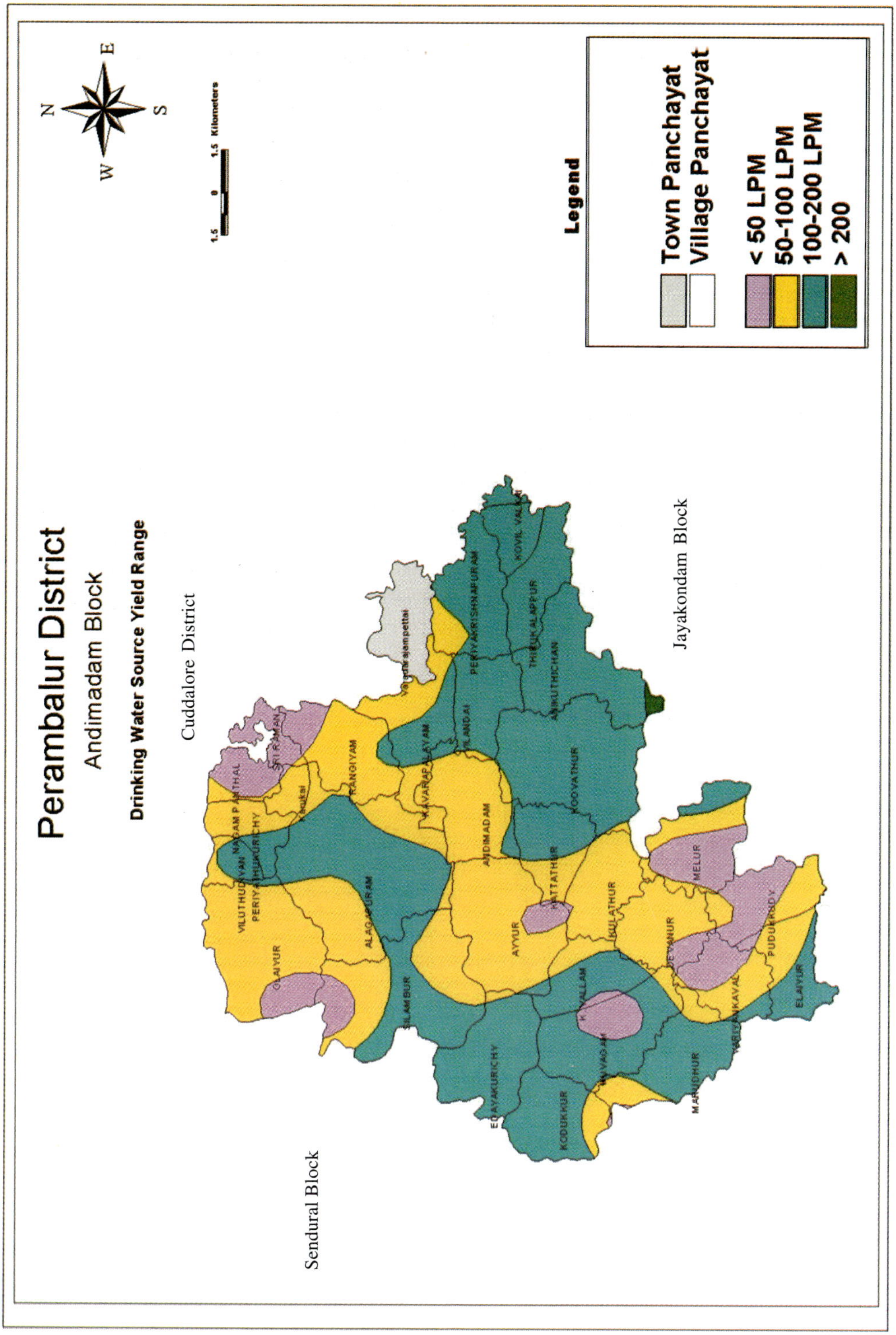

Fig. 19.6: Indian GIS industry (*Source*: WTI Advanced Technologies)

specialized agencies or to municipalities for different projects. There are some other companies in India who use GIS technology for their marketing and planning like HLL, ITC, Domino's Pizza, Tata Chemicals, Tata Power, Tata Teleservices, Lintas and the like. Since the cartographic services are specialized in nature, the comparison of the quality of the work at pre-project status is difficult to assess. Due to this reason, the current practices of awarding jobs through tender and contract system do not find acceptability among the captains of the industry. Their claim is regarding the professionalism, quality and the post-project technical support. Some of the government departments, like Survey of India and National Remote Sensing Agency did develop a method of short-listing the industries for out-sourcing. This system appears to be acceptable with the industry as well as the government. The identification of such industries also gives them recognition to take up jobs from other government departments as well.

With the liberalization of economy in the recent past, it has been seen that a lot of information in spatial format is required for NRI and multi-national investments. The information is to be processed so that the market strategies and location of industries can be worked out. Further, maps are also required for infrastructural development related to telecommunication, road development and urban development where the industry is directly involved. Hence, there is a healthy trend for more cartographic jobs to be done by one sector for another sector of the industry. But such development is likely to have a shorter life. Hence, the cartographic activities within the industry will have to find out new ways and options for its survival. Nevertheless, larger industry groups like Reliance and Infocom have initiated their own GIS projects for internal use. The same digital data is being used by other group of companies or sister concerns like Reliance Energy.

GIS Offshore Projects by Indian Industry

As mentioned earlier, GIS industry in India depends on the offshore projects for their survival. The bigger companies have their offices in major cities abroad. Projects are taken up by these outposts and transferred. The smaller GIS companies do get a share of such projects which are outsourced of partially off-loaded in India. In some cases, the smaller companies never come to know the original client in the countries in the US of Europe. Similar to the success of call centre type of out-sourcing, GIS projects are passed on to this country due to cheap technical manpower, knowledge of English and favourable time difference. But the so-called emerging GIS centres do not have the advantage of 24×7×365 situation. The projects come in phases and the flow is unpredictable. Nevertheless, Indian GIS industry is making its dent allover the world. It may not be too far when India becomes a hub for GIS activities on a global scale. The following Table 19.1 will indicate the pattern of offshore projects taken up by the GIS industry.

Table 19.1: Some examples of GIS projects undertaken by Indian industries from abroad

Project	*Country/Agency*	*Indian Company*
GIS based solutions and mapping	Tokyo Electric Power Corporation, Japan	Tata Infotech
GIS based solutions and mapping	Hokkaido Electric Power Corporation, Japan	Tata Infotech

(contd.)

Table 19.1 *(contd.)*

Project	*Country/Agency*	*Indian Company*
A geographical management information system (GMIS) of Sri Lanka for the World Bank and provided various geographic and attribute data from a data hub to certain agencies of Government of Sri Lanka. The GMIS prototype are designed for specialise application like: • Land use and thematic mapping. • Utility network fault-call and proximity analysis.	World Bank	Tata Consultancy Services, New Delhi
• Environment impact Assessment. • Monitoring of Environmental Quality. • Solid waste management.		
Developing a geographic database (GDB). Performing sector specific analysis and bringing out map and tabular outputs Regional Information System application development to help sector experts make *on-line* queries and *what-if* analysis.	The Japan International Co-operation Agency (JICA)-Sri Lanka.	Tata Infotech
Creation of data base for corridor in Madhya Pradesh, West Bengal, Orissa, Bihar, Northeast India, Assam and Himachal Pradesh.	Nynex, U.S.A and Reliance Telecommunication	IN-RIMT, Hyderabad
Creation of data base for Gujarat and Maharashtra.	AT&T, U.S.A. and Birla Communications	IN-RIMT, Hyderabad
Digitization of geophysical data.	Emron Oil and Gas, the largest oil companies in the world.	INDICOS, Mumbai
Digitization of several hundred land parcel maps and re-engineering of GIS software including digital land base files with the attribute information attached to them.	U.S.A. based company	INDICOS, Mumbai
Software re-engineering of GIS product to port it from BTOS platform to DOS and UNIX.	K and A Intl., U.S.A.	INDICOS, Mumbai
Digitization of maps and drawings for the C and P Bell Telephone, Florida Power, Seminote Electric, Con Edison, Burns and Roc. etc.	A Business Associate in U.S.A.	INDICOS, Mumbai
Digitised the drawings of railnet work and other engineering applications of New York City Transit Authority (NYCTA).	New York City Transit Authority (NYCTA)	INDICOS, Mumbai
Digitization of oil and gas exploration maps.	GEO-PROS, Venezuela.	INDICOS, Mumbai
Flood Rate Maps of USA digital conversion from paper maps with different flood levels, county boundary in AutoCAD with 0.0024" accuracy.	U.S.A. (prime contractor R.M.S India Pvt. Limited)	Caltech India, Kolkata
Digitization of drawings.	Imagine Technologies, USA.	DataVision, Pune
Digital Maps.	Map Info Corporation	RMSI, New Delhi
Digital Maps.	Motorola and ESRI.	RMSI, New Delhi

(contd.)

Table 19.1 *(contd.)*

Project	*Country/Agency*	*Indian Company*
Digitzation and GIS application.	M/s IMMCO INC., USA.	Pixel Infotek Pvt. Ltd., Bangalore
Digitization of Maps.	Canadian Government Highway Engineering Division.	DataVision, Pune
GIS based solutions and mapping.	Unisys, Canada.	Tata Infotech
CAD setup with GIS and digital mapping.	Voist Alpine Industrilagnabahu, Austria	RCE-I, New Delhi
GIS based application to maintain image, spatial and attribute data in a central ORACLE database.	British Telecommunication United Kingdom	Tata Infotech
Digital map needed for telecommunication net working planning.	Siemens, Germany	RMSI, New Delhi

With the above it is seen that Indian industries have done a number of GIS, digitization and mapping related projects for different overseas companies or governments. They have the potentialities to expand the GIS activities in future as well. Such activities can be carried out for the Indian projects. For this, the availability of base maps and date bases extremely essential. Indian industry would like to have access to such sources without any restriction. Since the present policy of government on map availability or production has already changed the GIS industry is likely to flourish with Indian projects. They may not have to look for offshore projects for their survival. However, they may be able to export GIS products and services in large volume and can work for more and more foreign countries or foreign companies in the field of GIS or cartography or mapping related work. The concerned industries are earning valuable foreign exchange by doing GIS related projects for foreign companies or countries.

Indian GIS Software

In India the standard GIS packages have been widely used. They are MapInfo (http://mapinfo.com), Idrisi (http://www.idrisi.clarku.edu), Geomedia (http://www.intergraph.com), or ArcInfo (http://www.ersiindia.com) (GIS Development Team, 2000). But there is always demand for software by which both, GIS and remote sensing data, can be handled. Hence, a software package which can equally manage raster and vector is preferred. One of the important reason is the provision of in-built compatibilities for using any type of digital data at a reasonable cost. For example, GRASS (http://wgbis.ces.lisc.ernet.in/grass) is preferred because it is a open source GIS with capabilities such as raster analysis, vector analysis, image processing, map algebra, and other functionalities which are comparable to any commercially available software (Ramachandra *et al.,* 2004).

The first attempt to develop a GIS software in India was made by the Indian Space Research Organisation (ISRO). This was developed for wider application of the remote sensing data available from the Indian satellites in GIS activities. This software is known as Isrovision. On the other hand, in the Department of Science and Technology through the Natural Resources Data Management System (NRDMS) with the financial assistance of the United Nations Development Programme (UNDP) and technical assistance from the Centre for Studies in

Resources Engineering (CSRE) of the Indian Institute of Technology, Bombay developed a window based GIS software called Geo-References Area Management++ or GRAM++ software package which has the following features (www.nrdms.nic.in):

—Data preparation tool (DPT)
—GRAM Viewer
—Raster analysis kit
—Network analysis
—TIN Generation
—Statistics
—Digital image processing
—Mapping control tool
—GIS tutor

Modules like GRAMNET and GEO_SQL have been developed to facilitate resource network analysis and spatial querying. Further, a user friendly module VEC Viewer has also been developed to support viewing of geo-referenced layers with collateral attribute information with query facility. A GIS tutor around GRAM++ has also been developed (Venkatachalam *et al.*, 2001; Prasad, 2002).

GRAM++ is a comprehensive suite of GIS tools primarily useful in preparing and implementing development plans by district level line departments, Panchyati Raj institutions (PRIs), communities and the non-governmental organisations (NGOs). Obviously, its potentials for the development of business geographic are yet to be explored. Indian industry is taking some time to accept it in a big way. With training it is making way in the administration at district level. Universities and colleges are accepting it due to attractive cost. IIT Bombay (CSRE) is now marketing this software. Recently, Government of Madhya Pradesh in collaboration with the Department of Science and Technology has developed its own version of this software, known as GRAM Dristi (NRDMS, 2003).

Further, an initiative was also taken up by the industry to develop a low cost GIS package which can equally handle remote sensing data. The target group was universities and collages where GIS and remote sensing courses have been introduced in the courses but these educational institutions were not having sufficient funds for having commercially available software. JT Technologies, a Noida (near New Delhi) based company, specializing in the domain of development of software products and solutions in the field of GIS has recently released a comprehensive GIS software called JT Maps (http://www.jtmaps.com). This package has a variety of features which does address almost the entire spectrum of GIS activities. The software supports following features:

—Image processing which supports a range of satellite data formats
—Data creation
—Analysis: Spatial analysis, Network analysis
—SQL support
—External database connectivity
—Export-import compatible with industry standard data formats
—Projection transformation

There are also attempts to develop or customize the available GIS packages for specific applications or for specific institutional use. The case of the Maharashtra Remote Sensing Applications Centre (MRSAC) is an example where the data base development was considered to be cumbersome task when done through the basic GIS operations and commands. As a result MRSAC developed data base structuring tools (DBS-TOOLS) which is a menu driven, user friendly customized software package built over ARC-Info software using Arc Macro language (Tamhane and Despande, 2000). Further, GRDASS, a open source GIS with capabilities such as raster analysis, vector analysis, image processing, map algebra and other functionalities has been developed which is claimed to be comparable to any commercial software available in the market (Ramachandra, 2004). Similar attempts are being made in the GIS related institutions and industries. Unfortunately very little is known about such initiatives.

Indian Journals on GIS

With all the problems being faced by the GIS community in India, the publications of journals is doing fairly well in India. The existing journals connected with geography, remote sensing, aerial photography or science in general were giving due attention to GIS issues. But it was felt that there is more potentiality to commercially explore this field. Industries, NGOs, universities and government sponsored institutions appear to converge on this aspect of GIS. Organisation of conferences, exhibitions and events, publishing of journals, launching web sites and the like have lead to a new brand of professionalism, i.e. *GIS journalism* or *Geojournalism.* New journals are being brought out and at least in one case the quarterly has become a monthly journal. Issues to meet different regional requirements are also being brought out. As a result every activity related to GIS or mapping is closely watched and becomes a news item. The GIS related journals are as follows:

— GIS Development: www.GISdevelopment.net
— GIS India: gisindia@hd2.vsnl.net.in
— Geospatial Today: www.geospatialtoday.com
— Geography and You: www.geographyandyou.com
— ISG Newsletter: bsahai@wilnetonline.net
— Coordinates: www.mycoordinates.org
— Photonirvachak: Journal of the Indian Society of remote Sensing: www.isrsindia.org
— Manchitrian: Journal of National Atlas and Thematic Mapping Organisation: www.natmo.gov.in
— Manchitrian Taknik: Manchitran@hotmail.com (in Hindi)

GIS Training in India

Two decades earlier, when GIS was getting introduced in the country, no was sure in which discipline it should be included (Saraf, 2005). Hence, there was a degree of uncertainty about its status under the aegis of the University Grants Commission or All India Council for Technical Education. But with the success of the Indian satellite programme, remote sensing was introduced in a big way which included GIS. Hence, GIS got a foothold in the country and developed as a multi-discipline basically for assessment and mapping of natural resources.

The nature of GIS education has been influenced by such development. As a result, there is less emphasis on business GIS or GIS related activities in the industry. Obviously, the course contents are more inclined towards the development of GIS for natural resources using remote sensing data. However, GIS is being taught in the civil engineering, geography or earth sciences curriculum. Further, the emphasis or combinations with other specializations vary from social aspects to geodesy to surveying and mapping. GIS courses have wholly or partially been integrated with Bachelor of Engineering or Technology, M.Sc. or M.Tech. or even doctoral level research programmes (Kulkarni, 2005). In the Indian universities and colleges, the theoretical foundation has already been laid and trained persons are also available. It is up to the academics to take further initiatives in promoting GIS (Parihar, 2001).

A survey was carried out by CSDMS team in 2000 about the educational qualifications and training of the GIS users through out the world. The trends reveal that the large number of users in India is post graduates. They are mostly involved as GIS analysts, mostly with 7-10 years experience and employed in private sector (GIS Development Team, 2000). The GIS training institutes in India are as follows:

- Indian Institute of Remote Sensing (IIRS), Dehra Dun
- National Remote Sensing Agency, Hyderabad
- Space Applications Centre, Ahmedabad
- Regional Remote Sensing Service Centres: Kharagpur, Dehra Dun, Nagpur, Jodhpur, Bangalore
- Centre for Studies in Resource Engineering, IIT Bombay, Mumbai
- Institute of Remote Sensing, Anna University, Chennai
- Geological Survey of India
- Indian Institute of Technology, Kanpur
- National Bureau of Soil Survey and Land Use Planning, Nagpur
- Survey Training Institute, Survey of India, Hyderabad (www.soisti.gov.in)
- National Atlas and Thematic Mapping Organisation, Kolkata
- Birla Institute of Technology, Meshra, Ranchi
- Jawaharlal Nehru Technological University, Hyderabad
- Jadavpur University, Kolkata
- The University of Calcutta, Kolkata
- GIS Institute, CSDMS, Noida
- Symbiosis Institute of Geoinformatics, Pune
- Bhaskaracharya Institute for Space Applications and Geoinformatics, Gandhinagar
- Visvesvarayya Technological University with Karnataka State Remote Sensing Centre, Bangalore

There is a need that the GIS course content which allows university programmes to remain abreast of market needs as well as providing students with a well-rounded education in the use of technology (Hall, 2005). "GIS is potentially much more than a technology. Education in and for GIS now needs to be much wider than has hitherto been normal, if we are to gain the full potential available. Good technology—and the skills to use it—form a necessary condition for success but not a sufficient condition" (Rhind, 2005; Roy, 2001; Venkatachalam *et al.,*

2001). GIS Education is not only limited to universities and specialized institutions, but it is being taught in schools as a tool to teach science (Phoenix, 2004).

Possible GIS Products and Applications in India

In order to have proper utilizations of GIS products, there is a need to identify potential market to commercialization of its data and products. The first step in commercialization process is to develop a consensus among the industry and users on what markets are critical to future growth. There could be ten types of uses, which are potential application of geospatial data. Obviously, the products and applications cannot be in isolation. There has been a convergence of technologies within GIS and related fields. As a result, different combinations of data, technologies, and delivery systems are being worked out. Now such integration of technologies is also identified under the realm of geomatics or geoinformatics. The possible GIS products and applications are:

Land and Natural Resources

The natural resources market for geospatial data is land, forestry, agriculture, fishing, mining etc. These are derived from the earth and its water. The principle of sustainability and bio-diversity protection are increasing demands for more and better data. High-resolution imagery, GPS and integration of GPS, GIS and remote sensing (RS) technologies can develop the required information. Attempts are made to model groundwater vulnerability using GIS technology, particularly pollution assessment applying Drastic model (Mohan and Gopi, 2004). The example of remote sensing based GIS application for preparing land use map a part of Dehra Dun town is shown below (Fig. 19.7).

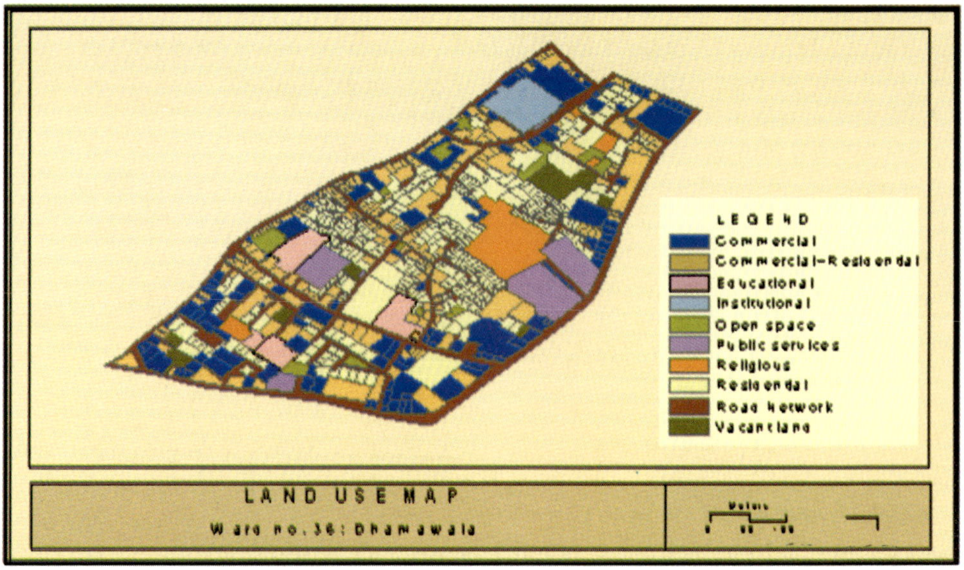

Fig. 19.7: Application of GIS and remote sensing for preparation of urban land use map of a part of Dehra Dun town (*Courtsey*: Suman Jain)

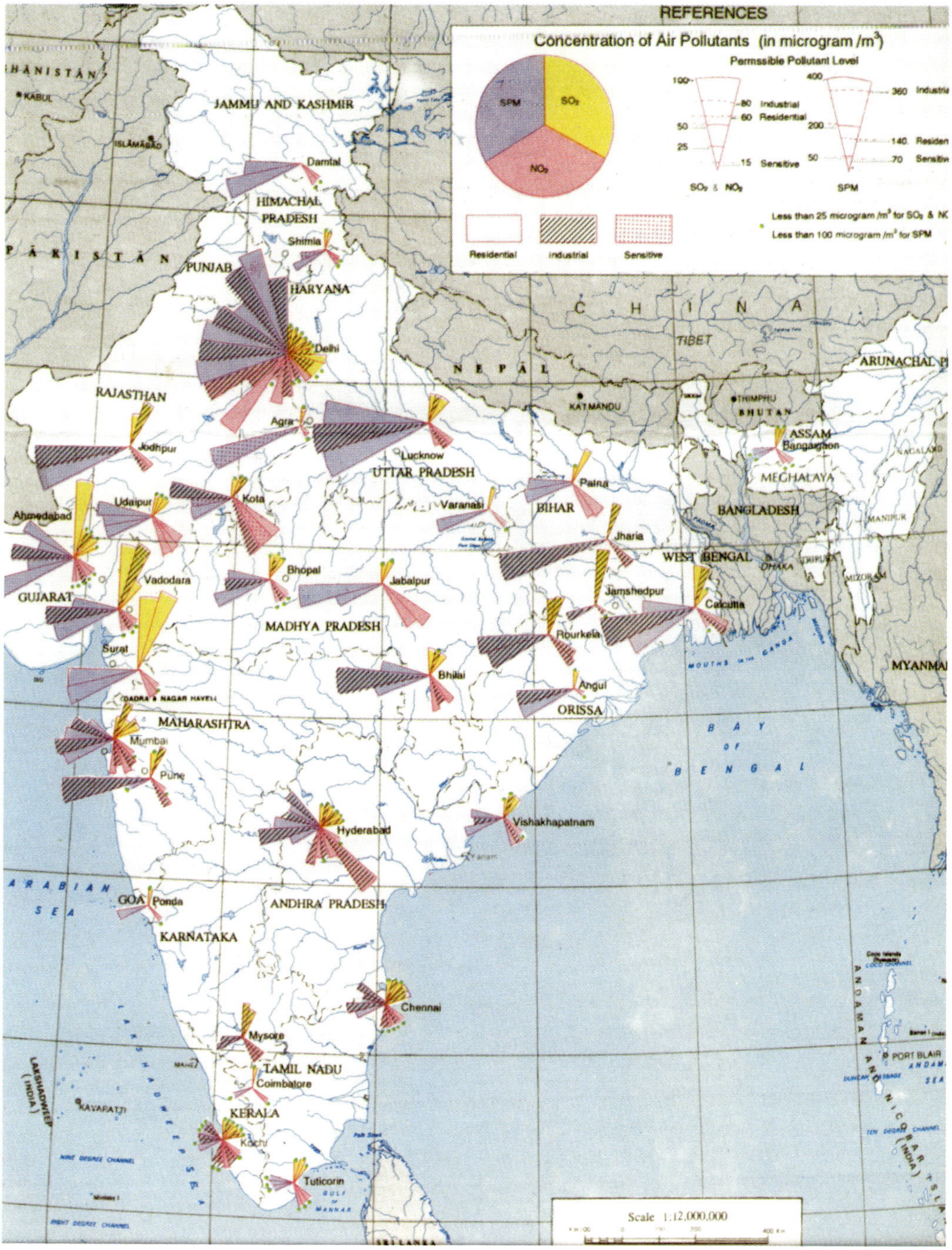

Fig. 19.8: Pollution over Indian cities (*Source*: NATMO)

Environment

The environmental market for GIS goods and services is related to a wide range like pollution prevention, environmental conservation, protection, remediation and enhancement technologies, processes, products and services. Environmental reporting is increasing because of the growing recognition of the importance of environmental management (Ramaswamy, *et al.*, 2002). Two examples from the National Atlas and Thematic Mapping Organisation (NATMO) has been taken to demonstrate the possible GIS applications for (a) monitoring air pollution over major Indian cities, and (b) mapping of flood effected areas in Maldah district of West Bengal. The former example is for small scale GIS application (Fig. 19.8), while the latter one is for large scale operation (Fig. 19.9) in the field of environment.

The complexity of environmental issues demands the collection, analysis and presentation of massive quantities of data, and the mathematical and surface modelling of multifarious processes, usually over time. GIS technologies are ideally suited to this area (Patil, Ravan and Kaushal, 2003). Global Position System (GPS), for example, offers an excellent tool to monitor society's impact on the environment. It can be used to track the movement of everything from icebergs to polar bears, and offers exciting opportunities in the area of weather forecasting. Remote sensing technologies and GIS, which are ideally suited for handling such data sets, and many other technologies, are regularly being employed in many environmental applications.

Property and Urban GIS

In terms of GIS goods and services, the property market includes the building and maintenance of property infrastructures, which involve systems for survey, registry and assessment. Other activities include surveying and mapping in support of land development and providing information in support of property-related transactions such as conveyancing, mortgaging, and valuing and environmental assessing. The business of providing information in support of property transactions is also changing from its more traditional ways. Traditional conveyancing and mortgaging work is diminishing for a variety of market, economic and technical reasons; however, opportunities to provide new services such as environmental assessment are gradually increasing. Unless the industry can successfully introduce new business models, it can expect very little growth in this area. In general, it will become increasingly important to be able to work in an on-line world, conversant both with the latest developments for accessing and analysing property-related information and with the world of electronic commerce (Upadhyaya, 2002). Bhoomi Project in Karnataka state is a good example of some degree of success in providing access to the farmers about their land records (Chawla and Bhatanagar, 2003; Sipkes, 2003). More important than technology *per se*, however, will be an in-depth understanding of areas such as cost/benefit and risk management and the capacity to quickly incorporate them into new GIS business models (Umashankar, 2003). The case study of municipal GIS for property management, and related issues for Hyderabad city has been depicted (Fig. 19.10).

The case study of Kolkata municipal GIS has been discussed in this chapter earlier. National Urban Renewal Mission (NURM) is funding several major cities of the country for their development, including preparation of spatial data base for 137 cities. These data bases are to be used for development of GIS for the respective cities and towns. In addition, international funding agencies like DFID (UK) and JICA (Japan) are also keen in providing resources and expertise for several cities. Municipal GIS is also has the potentiality to be used for urban services such as electricity, transportation, fire, flood and drainage. Further, there is also scope for using this data base for business opportunities as well. Mobile, location based, retail, taxi and several services can be based on such GIS data base. In India there is a big scope for such businesses and services. In near future a good number of GIS trained man power will be required for managing such GIS activities as well.

Infrastructure

The infrastructure market for GIS goods and services includes engineering, construction, transportation and utilities industries. This sector supports both business and non-business activities in a country through the design, engineering, construction and maintenance of capital projects and the provision of ground, air, water and space transport services, telecommunications, and water and energy supplies. The use of GIS technologies in this area not only facilitates the efficient, cost-effective and environmentally responsible development and operation of these activities but also may preclude the need for the more costly expansion of the infrastructures. As such, the GIS industry will play an increasingly important role in the infrastructure market.

The construction sector is responsible for building the facilities and supportive infrastructure necessary for a nation to produce wealth and shelter its citizens. Some GIS-related activities include land planning and development, design, location selection, impact assessments, surveying, mapping and other associated municipal activities such as urban planning and development. Utilities, hydro grids, pipelines, communications systems and so on must monitor, maintain, accurately locate and coordinate extensive amounts of expensive, geographically dispersed capital stocks.

GPS is another area that is having a major impact on the utilities market. The ability of GPS receivers to synchronize themselves to atomic clocks on satellites is proving to be a boon to the wireless communications industry.

There is currently in process a technology on electric power that will be of interest to the GIS industry. It will focus on four issues: asset optimization, intelligent power system delivery, end-use efficiency and convergence, and small-scale generation and renewable. An example of the use of GIS data base and GPS is included here for planning location of pylons for power transmission (Fig. 19.11). Here, the options for alignments of transmission lines have been worked out on the GIS data base and also on the digital elevation model. Location of pylons, their coordinates and expected sag between the pylons have been calibrated as well.

In combination with advanced computing and communication technologies, GIS technologies are dramatically improving the efficiency and environmental cleanliness of all modes of transportation and commodity exchange. Intelligent Transportation Systems (ITS), for example, are revolutionizing the way transportation-related issues such as congestion; safety and logistical optimization are being handled. ITS can be applied in areas ranging from travel

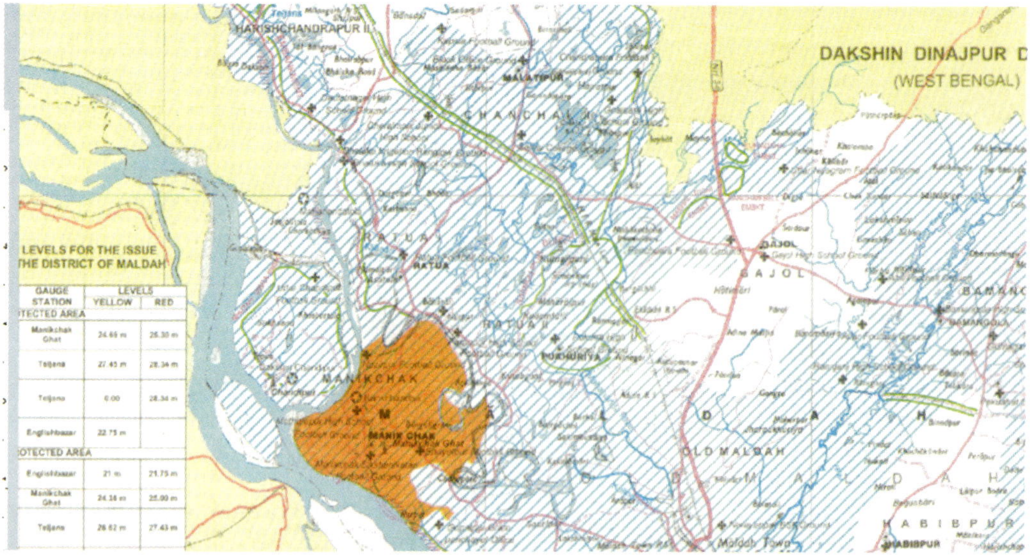

Fig. 19.9: Flood map of Maldah district, West Bengal (*Source*: NATMO)

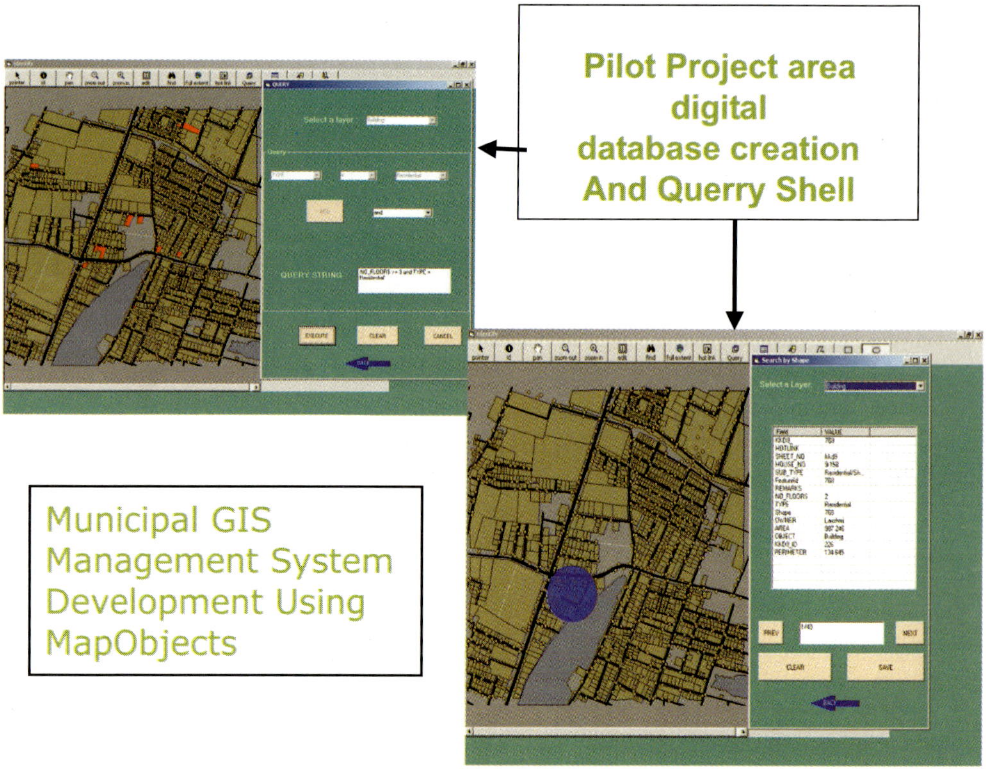

Fig. 19.10: Municipal system for property management, tax etc. (*Courtesy:* RSI)

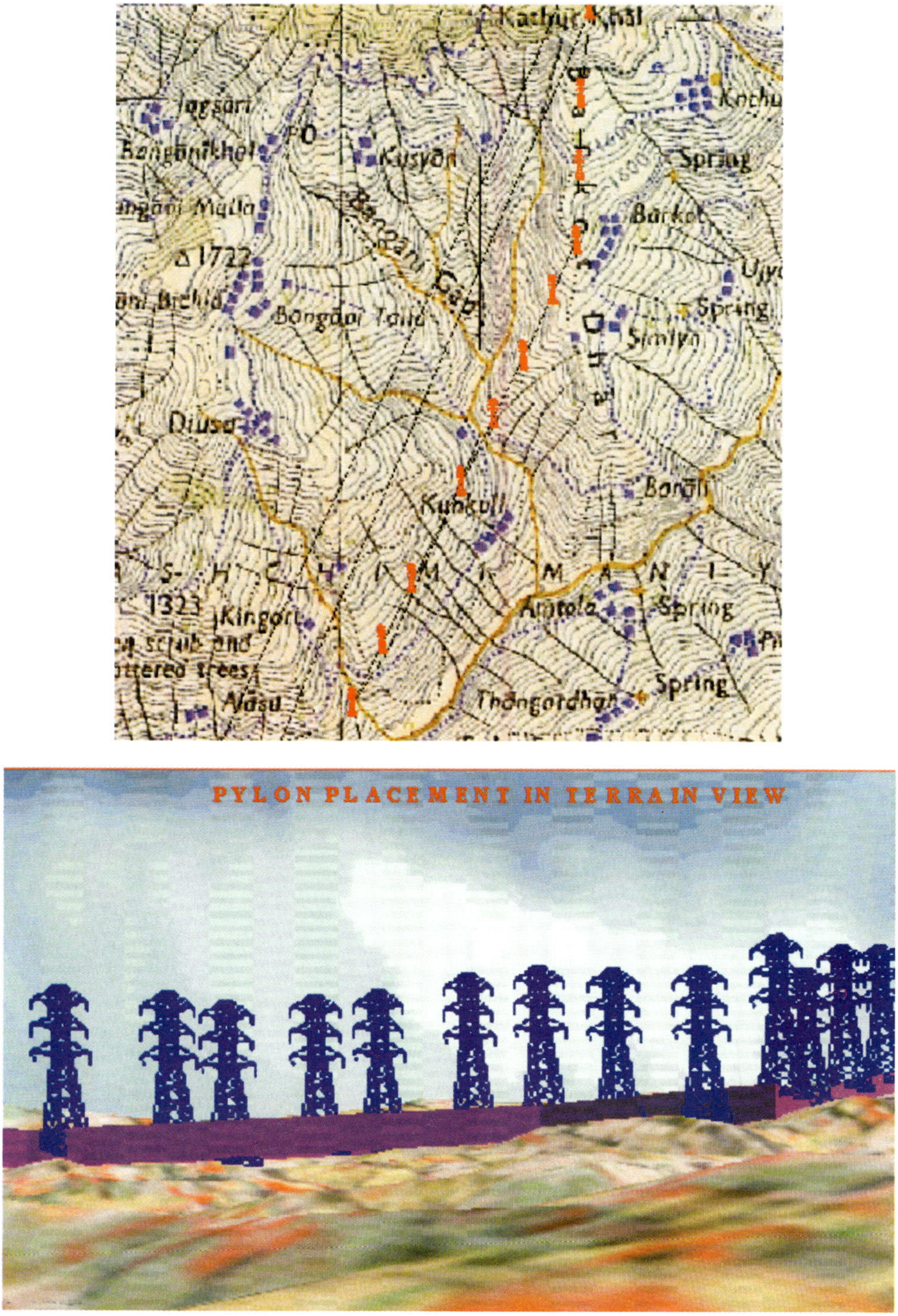

Fig. 19.11: Placement of pylons on GIS data base (*Courtesy*: M.V. Bhatt)

planning, adaptive traffic control, real-time traffic information, public transit operations, commercial vehicle operations, fleet management and emergency vehicle routing to, in the future, automated vehicle control systems.

Access to geospatial information will play an essential role in the development of these market opportunities. As we have seen, United States is currently the leader in the development and deployment of ITS and, as emerging industrial nations' transportation systems become more congested, their markets will become very important as well.

Health

As the GIS industry has evolved, there has been limited involvement between it and the health care sector. Where links have been made, they have been concerned with the location of facilities (i.e. clinics, ambulance stations and practitioners' offices) and the routing of emergency service vehicles. The growth of GIS in this area has been hampered by limitations on the data, technology constraints, lack of training and education, and the structure of the health care sector. These constraints are interrelated and, taken together, have resulted in limited use of geospatial data in the health care sector. However, both the GIS industry and the health care sector are undergoing major changes.

With GIS expertise residing in highly specialized units, there have been few opportunities for interaction between these experts and experts in the health care sector. As a result, awareness and appreciation of the potential usefulness of geospatial data in the health care sector have been limited. Furthermore, even where there was an awareness of the technology, it was sometimes discounted because of the presumed high costs associated with its use.

Geospatial data provide the health care sector with a new sub discipline for collecting, storing, retrieving and analyzing health care data. An example of Uttaranchal has been included here (Fig. 19.12). NATMO has not only prepared such a data base for a particular state, but an whole Atlas of Health and Diseases of India (Nag, 2000). To take advantage of this sub-discipline, the geospatial and health data must be coordinated across the organizations, institutions and individual providers in the community. A coordinated data infrastructure supports:

- Human resources planning
- Facilities planning and optimized service delivery
- Epidemiological studies
- Informed decision-making by users and providers
- Proactive and preventive health care
- Emergency preparedness
- Health imaging techniques development.

Application of GIS in social, health and welfare activities are generally government sponsored. It is considered as an extension of thematic mapping. NATMO has been contributing a lot in this regard (Massey, 2002). However, there is a lot of scope to develop GIS for the social sector.

Emergency Response and Disaster Relief

Both natural and human disasters such as floods, tornadoes, storms, oil spills and the release of poisonous gases require timely, accurate and supporting geographic information for relief operations, logistics of supply, national search and rescue capabilities, prediction modelling and environmental monitoring associated with these disasters (Sharma, 1999). Considering the topography, village boundaries and information regarding the floods, a GIS data base has been developed for identifying effected villages (Fig. 19.13). This will help in drawing immediate attention for relief and rescue.

Current and future GIS technologies used for emergency response and disaster relief efforts include integrated information systems, precision and mobile GPS, sophisticated planning and prediction modelling, wireless dissemination and multi-purpose operational support systems. As future disasters emerge and as relief efforts are mobilized, GIS technologies will be required more and more to accelerate the planning and response activities. Technologies that provide ease of access and integration will be important, as will online access to information sources through integrated infrastructure shared between the private and public sectors. Timely access and rapid dissemination are key variables to aid relief efforts. Information distribution must also support differentiated views of supporting geographic information, along with the tool kits to use wireless and remote communications to disseminate critical information to users in the field and in planning and coordination centers (Sarkar and Kanungo, 2001).

Military, Defence and Security

The military GIS market offers some of the greatest potential for new applications and integrated GIS technology and systems. The military GIS market has traditionally been one of the larger consumers of GIS technology, information and related services, although most production activity so far has been conducted in-house. As we move towards the millennium, the military market remains large and strong, with an emphasis on maintaining global capabilities that enable the mobilization of operations over land, sea and air (Singhal, 2000).

Overall, the military GIS market spans a range of different requirements. Military GIS systems and technology requirements are diffuse, vary greatly and tend to be user driven. The nature of the market has moved from customized large-scale developments to commercial-off-the-shelf (COTS) products. Similarly, the pace of technological development and changing operational environments have resulted in a change in requirements toward flexible components, interoperability, standardized tool kits and data, and integrated systems development.

Future market opportunities for GIS technologies in this sector are numerous. Key areas include advanced communications and secure networking to meet demands for on-line, just-in-time delivery of information and decision support capabilities. Development of experimental, soon-to-be-operational mobile warrior and integrated systems for in-field application continues to dominate research and early pilot phases. These and other technology areas such as advanced guidance, high-resolution imagery, operational picture and precise location determination remain critical areas for GIS technologies, often within larger operational command and control systems.

As ever, military systems require well-supported access to geographic information. These requirements fuel an ever-expanding demand for innovative geospatial information discovery,

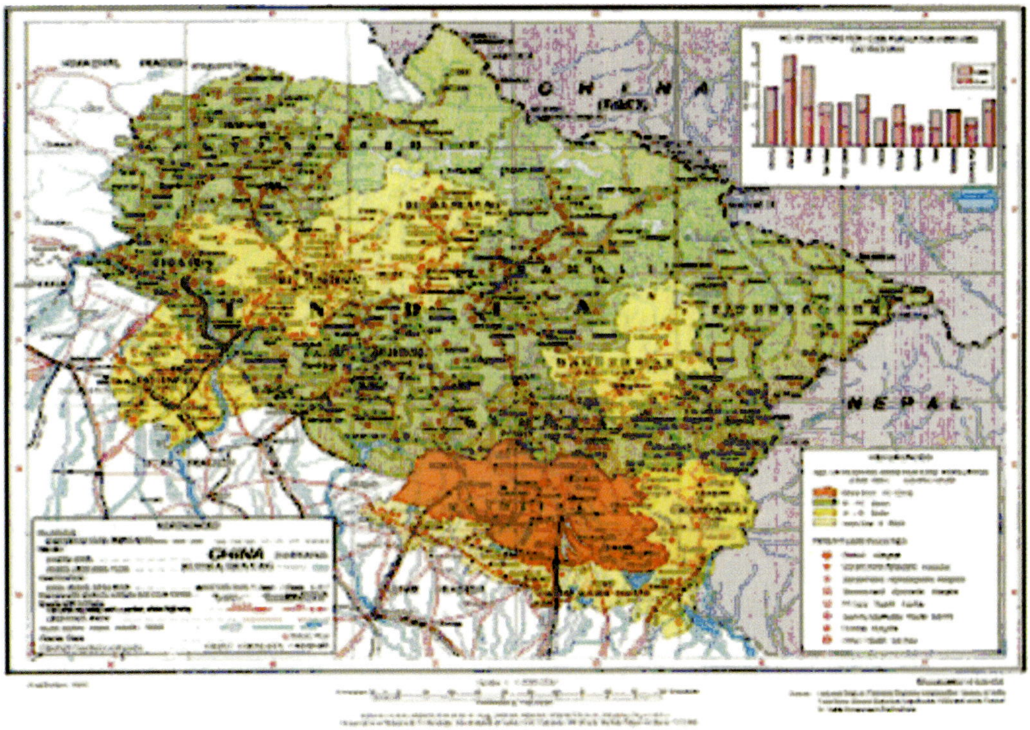

Fig. 19.12: Uttaranchal: Health Service (*Source*: NATMO)

Fig. 19.13: Low lying villages effected by floods (*Source*: NRDMS)

access, integration and management tools to support the provision of responsive, accurate and timely global geospatial information systems. Delivery of this information is moving from internally produced, discrete hard copy products toward interoperable, seamless, on-line delivery of services via national and global infrastructures, in essence, mobilizing technology to support the delivery, storage and accessibility of information from within and outside military sources.

Business Geographic

The business market includes applications in commercial sectors such as shipping, banking, retail sales, food services, real estate and insurance, and also in vehicle navigation and routing. There has been a clearly identified shift in the mapping or GIS business opportunities away from traditional surveying and mapping data capture to the outsourcing of data services. Large organizations with legacy databases are now asking the surveying and mapping providers to offer "value-added" services—where their legacy non-graphic databases are joined with their graphics, delivered by a data capture firm, to produce intelligent business analysis of their day-to-day operations. This trend will continue, and many organizations will find themselves moving from the existing traditional data capture model to a new and more lucrative business GIS model. In addition, the traditional business geographic market now has expanded to encompass this new and more lucrative market. An example of Hyderabad city is depicted here (Fig. 19.14). The GIS for this purpose was jointly developed by Survey of India and M/s Spinfo, Bangalore under the MapQue Project. Similar GIS data base for Bangalore is also available.

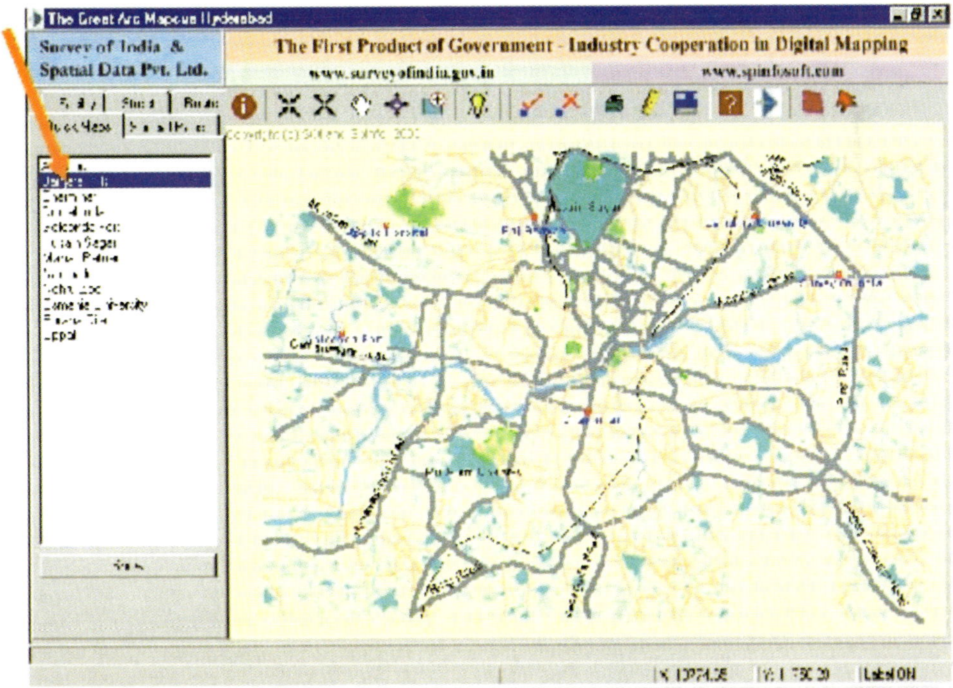

Fig. 19.14: Business geographic for Hyderabad City

The traditional business geographic market addresses the needs of the commercial business sector for business demographic and geographic market trend analysis, while the new business GIS market encompasses non-traditional GIS analysis in market segments such as health and welfare. Medical and social market analyses are but two of the new business market areas being addressed by GIS.

The market today includes any organization with large data sets that wishes to share those data on a corporate intranet. In addition, the market is also the family household, and their desktop personal computer, which offers both a marketing and sales opportunity for that large corporate database through the use of the Internet. Companies familiar with handling databases, both graphic and non-graphic, will have excellent opportunities to service these new market areas.

Currently, managing and manipulating digital map and non-graphic databases are now providing services for organizations not familiar with the graphic presentation of data in a geographic context. These new business opportunities are offering challenging and rewarding experiences for companies previously confined to the traditional mapping or GIS business model. The expression "a picture is worth a thousand words" is proving that the value of presenting previously proprietary non-graphic data in a public forum on a graphic backdrop is both effective and exciting. The analysis now available using the previously "graphically static" databases is very impressive.

Entirely new approaches are being used to distribute and display these data sets by marrying the legacy databases, graphic and non-graphic, with the companies, which produced them. The technology is intranet or internet software. The products necessary to "explode" this market include access to digital mapping, national, provincial, regional and local data sets, interoperable software and the associated non-graphic databases.

The convergence of technologies and applications in the mapping sciences has led to geographic information processing being used on the development of a much wider range of products. As mentioned, traditional GIS has concentrated on applications in environment, natural resources, infrastructure, cadastre and defence. More recently, business applications, vehicle navigation systems and health applications have increased in importance. Education and entertainment, especially the latter, are the most recent and promising markets for GIS. The entertainment segment of the "edutainment" market is probably larger but much less focused than the educational segment.

Consumer Services

This market currently consists of individuals whose primary use of geographical information is for such purposes as travel planning and navigation in both land and water vehicles and for outdoor recreational activities such as hiking, camping and hunting. These kinds of demands have traditionally been met through the use of paper map and atlas products and less frequently through the use of aerial photography or satellite image products. Advances in two key technologies, GPS and GIS, combined with other technological developments in data storage and communications (especially the internet) and computer visualization, make possible a broad range of new GIS applications. For example, consumers now have access to mass-produced, hand-held GPS receivers that are being used for a variety of recreational activities. Basic desktop mapping and GIS functionality are being bundled with off-the-shelf office management software, providing consumers with the capability to manage and use all kinds of information in a geographic context.

M/s Eicher is involved in providing GIS based services in several cities. This company has also taken up joint projects with the Survey of India. Example of GIS application in telecom and cellular services has been included above (Fig. 19.15). From their GIS data base, M/s Eicher has produced details maps of cities like Delhi, Mumbai and Chennai.

Within the next five years, other applications of GIS technology for the consumer market will include Yellow Pages searches by geographic parameters, and travel and play in three-dimensional virtual reality based on geography. Given the scope of daily activities that are location-dependent, applications for the integration of geography with other types of information are limited only by the imagination. Constraints on the rate of development of the society or the consumer market segment include the limited amount of up-to-date geographical information publicly available, restrictions on its access and the small size of the present-day consumer market.

Summary

Like elsewhere in the world, GIS in India too has its mating dance with the information technology mainstream, as the two players circle around each other, cautious about surrendering autonomy but conscious of the certainty of the conclusion (Goodchild, 2003). In the federal and state governments, a lot of GIS activities are being sponsored and promoted by their ministries and departments of information technology. Further, we would like to have the benefit of the IT mainstream without losing the special nature of GIS. With a very strong presence of IT community in the country, the threat of losing uniqueness is greater. Perhaps the best way out is to develop GIS in India as a very special application area of information technology.

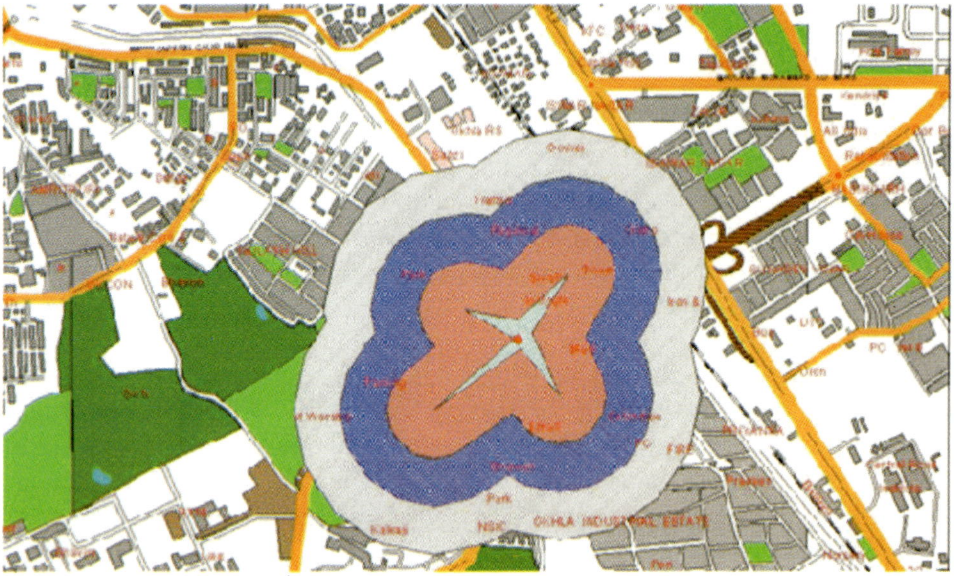

Fig. 19.15: Telcom and cellular services (*Source*: Eicher)

Note: Cellular Services Industry: Figure depicts usage of base digital map data for range analysis of cellular (mobile phone) coverage by superimposing clutter data as a layer.

REFERENCES

Agarwal, Pragya (2000), GIS in cultural resource management in historic urban centres, *GIS Development,* Vol. 4, Issue 1 1, pp. 32-34.

Ahuja, J.S. (2000), Legal, political, technical and commercial issues of land information system, Pre-conference *Proceedings of the National Conference on Land Information System,* 18-19 October , New Delhi, pp. 72-75.

Ahmad, Jamil (2003), Step-by-step approach to build a municipal GIS, *GIS Development*, Vol. 7, Issue 2, pp. 42-46.

Baijal, R., M.K. Arora and S.K. Singh (2002), A GIS assisted knowledge-based approach for military operations, *GIS Development*, Vol. 6, Issue 10, pp. 21-24.

Chawla, Rajiv and Subash Bhatnagar (2003), Bhoomi: A boom for farmers, *Geospatial Today*, Vol.1, Issue 5, pp. 13-17.

Cooper, Paul (2003), GIS gets cool!, *GIM International*, Vol. 17, No. 7, pp. 28-33.

Dangermound, Jack (2005), GIS: Helping manage our world, *GIS Development*, Vol. 9, Issue 12, pp. 26-27.

Dasgupta, A.R., Mukund Rao and A.K.S. Gopalan (2000), The National (Natural) Resources Information System, *GIS Development*, Vol. 4, Issue 12, pp. 18-23 and 47.

Dent, Borden (1999), *Cartography: Thematic Map Design*, WCB McGraw Hill.

Dorling, Daniel and David Fairbairn (1997), *Mapping: Ways of Representing the World*, Addison-Wesley.

Georqiadou, Yola (2003), Is Survey of India ready for the NSDI ?, *GIS Development*, Vol. 7, Issue 2, pp. 38-39.

Gibbons, Scott (1999), Mirzapur: A CIS that works, *GIS Development*, January-February, pp. 39-41.

Goel, R.K. (u.d.), National (Natural) Resources Information System (NRIS) – Design standards, unpublished paper from Space Applications Centre, Indian Space Research Organisation, Ahmedabad.

Goodchild, Michael F. (2003), GIS—Finding the mainstream, *GIS Development*, Vol. 7, Issue 1, pp. 8-15.

Gopal Rao, M. (2000), LIS in India : Perspective and Retrospective, Pre-conference. *Proceedings of the National Conference on Land Information System,* 18-19 October, New Delhi, pp. 1-6.

GIS Development Team (2000), GIS survey. An analysis, *GIS Development*, Vol. 4, Issue 10, pp. 25-28.

GIS Development Team (2000), GIS software at a glance, *GIS Development*, Vol. 4, Issue 11, pp. 20-21.

Gupta, V.K. (2001), Monitoring crime, *GIS Development*, Vol. 5, Issue 4, pp. 23-25.

Hadley, Clare and John Hammond (2003), Survey of India – A view from abroad, *GIS Development*, Vol. 7, Issue 2, pp. 29-31.

Halari, Rakesh (2000), Computerization of urban land records – Case study for city survey land records of Ahmedabad district – Gujarat state, Pre-conference. *Proceedings of the National Conference on Land Information System,* 18-19 October, New Delhi, pp. 46-49.

Hall, G. Brent (2005), Curriculum development in university-level GIS education, *GIS Development,* Middle East, Vol. 1, Issue 1, pp. 18-21.

Harmsen, Karl, Yogesh Kant and Shilpi Singh (2004), Assessment of the use of RS and GIS in India, *GIS Development*, Vol. 8, Issue 4, pp. 36-41.

Jain, Suman (2005), Vulnerability identification of low income settlements, *GIS Development*, Vol. 9, issue 10, pp. 26-28.

Joao, Elsa Maria (1998), *Causes and Consequences of Map Generalisation*, Taylor and Francis.

JT Map (2004), *JT Maps 2005 User Manual,* Jlet Technologies Pvt. Ltd. *(info@JTMaps.com).*

Khare, Manoj, Ashok Kaushal, Sandeep K. Srivastava and Krishna Kant (2005), Disaster management: Get real, pick sites for flood forecast, *Geospatial Today*, September-October, pp. 49-51.

Krishna, Bal (2003), Breaking inertia: Survey of India embraces a new era, *GIS Development*, Vol. 7, Issue 2, pp. 40-42.

Kulkarni, Madhav N. (2005), GIS-GPS academics in India: Present and future, *GIS Development*, Vol. 9, Issue 1, pp. 32-33.

Lemmens, Mathias (2003), Geospatial intelligence, *GIM International*, Vol. 17, No. 7, pp. 50-51.

Longley, Paul and Graham Clarke (2000), *GIS for Business and Service Planning*, Bell and Bain, Glasgow.

Massey, Jack (2002), A tribute to social GIS in India, *Geospatial Today*, Vol. 1, Issue 4, pp. 75-77.

Marutish, V. (2004), An enterprise GIS system for electric utilities: new approach to profitability, *Geospatial Today*, Vol. 3, issue 3, pp. 30-33.

Mishra, Prabhakar (2000), Cadastral surveys in India – A critique, Pre-conference. *Proceedings of the National Conference on Land Information System,* 18-19 October, New Delhi, pp. 15-17.

Mishra, Prabhakar (2004), Thoughts on LIS in India: Policies and reorganization, *GIS Development*, Vol. 8, Issue 4, pp. 20-24.

Mohan, S. and Kethari Gopi (2004), Modelling groundwater vulnerability using GIS, *Geospatial Today*, Vol. 3, issue 1, pp. 46-56.

Munshi, M.K. (2004), Integrated command and control system: dail your way to safety, *Geospatial Today*, Vol. 3, issue 3, pp. 34-36.

Nandan, Vaishali (2001), ...And Gorakhpur follows the suit, *GIS Development*, Vol. 5, Issue 5, pp. 29-32.

Nag, P. (1987), A proposed base for a geographical information system for India, *International Journal for Geographical Information System*, Vol. 1, No. 2, pp. 181-87.

Nag, P. (1999), Getting smarter with Smart maps, *GIS Development*, January-February, pp. 20-24.

Nag, P. (2000), Health and Disease Atlas of India, *GIS Development,* Vol. IV, No.5, pp. 26-27.

Nori, Chandrasekhar (1999), Towards effective utility management – A GIS perspective, *GIS India*, Vol. 8, April-June, pp. 41-44.

NRDMS (2003), *Natural Resources Data Management System: Vision Document*, Department of Science and Technology, New Delhi.

Pandey, Piyush Kumar and Ruma Shukla (2005), Web GIS based asset life cycle management, *GIS Development*, Vol. 9, Issue 10, pp. 42-43.

Parihar, Seema M. (2001), GIS in Indian colleges, *GIS Development*, Vol. 5, issue 11, pp. 36-41.

Parker, V.N. and D. Sampat Kumar (1999), A facelift to Mumbai: Courtesy, *GIS Development,* January-February, pp. 16-19.

Patil, Uday, Shirish Ravan and Ashok Kaushal (2003), GIS based air pollution surface modelling, *GIS Development*, Vol. 7, Issue 8, pp. 45-47.

Phoenix, Michael (2004), Global GIS education, *GIM International*, Vol. 18, No. 2, pp. 30-33.

Prasad, Manish (2002), Product development, *GIS Development*, Vol. 6, issue 8, p. 19.

Raghavendran, S. (2003), GIS in water resource management, *GIM International*, Vol. 17, No. 7, pp. 41-43.

Raja Rao, T. and P. Raj Sekhar (2000), LIS-Issues, technologies and standardization, *Pre-conference Proceedings of the National Conference on Land Information System*, 18-19 October, New Delhi, pp. 62-65.

Ramaswamy, S.M., C.J. Kumanan and K. Palanivel (2002), GIS based solutions for waste disposals, *GIS Development*, Vol. 6, Issue 12, pp. 40-42.

Ramachandra, T.V., Uttam Kumar, Vaishnav and S.N. Pradad (2004), Geographic resources decision support system – an open source GIS, *Geospatial Today*, Vol. 3, issue 3, pp. 52-59.

Rhind, David (2005), GIS education for a changed world, *GIS Development Middle East*, Vol. 1, Issue 1, pp. 14-17.

Robinson, Arthur, Joel Morrison, Philip Muehreke, A. Jon Kimerling and Stephen Guptill (1995), *Elements of Cartography*, John Wiley and Sons.

Roy, Aniruddha (2001), Web mapping solution, *GIS Development*, Vol. 5, Issue 5, pp. 33-36.

Roy, P.S. (2004), A qualitative journey into the technique, *Geospatial Today*, Vol. 3, issue 3, pp. 19-27.

Roy, P.S. and P.K. Joshi (2000), Making of GIS: 2000, *GIS India*, Vol. 9. No. 3, pp. 9-11.

Roy, P.S. (2001), Remote sensing training and education needs in India, *GIS Development*, Vol. 5, Issue 1, pp. 20-23.

Saraf, Arun K. (2005), GIS education in India: Issues and challenges, *GIS Development*, Vol. 9, Issue 1, pp. 30-31.

Sarkar, S. and D.P. Kanungo (2001), Landslide identification and mapping using remote sensing and GIS, *GIS Development*, Vol. 5, Issue 10, pp. 42-44.

Satyanarayana, P. (1999), Open GIS consortium (OGC), *GIS India*, Vol. 8, April-June, pp. 36-37.

Sharma, Vinod (1999), Use of GIS related technologies for managing disasters in India – An overview, *GIS Development*, Vol. 3, Issue 3, pp. 26-30.

Singh, Bhoop (2005), Landslide mitigation, a rugged tract, *Geospatial Today*, September-October, pp. 30-34.

Singhal, Alka (2000), In search of military GIS, *GIS Development*, Vol. 4, Issue 10, pp. 29-31.

Slocum, Terry (1999), *Thematic Cartography and Visualization*, Pretince Hall.

Sipkes, Jacques (2003, The Bhoomi Project: Computerisation of land records, *GIM International*, Vol. 17, No. 6, pp. 51-53.

Sreeshaila, G. and Satish Selvalumar (1999), The garden city blooms with GIS projects, *GIS Development*, January-February, pp. 33-38.

Subudhi, A.P., B.S. Sokhi and P.S. Roy (2000), Land information systems for municipal planning possibilities, problems and perspectives, Pre-conference. *Proceedings of the National Conference on Land Information System*, 18-19 October, New Delhi, pp. 35-40.

Tamhane, Rajen M. and Ajay S. Despande (2000), Customised GIS software package, *GIS Development*, Vol. 4, Issue 11, pp. 18-19 and 22.

Tarafdar, Ayon (2002), Community GIS vs GIS community, *GIS Development*, Vol. 6, Issue 9, pp. 31-33.

Tarafdar, Ayon (2005), Understanding the Indian map policy, *GIS Development*, Vol. 9, issue 7, pp. 18-26.

TIFAC (1999), *Report of the Special Core Group on Government-Industry in Cartography*, Technology Information, Forecasting and Assessment Council, Department of Science and Technology, New Delhi.

Umashankar, C. (2003), Land record systems revisited, *GIS Development*, Vol. 7, Issue 10, pp. 30-36.

Upadhyaya, Alok (2002), State-wide digital database and GIS of cadastral maps, *GIS Development*, Vol. 6, Issue 12, pp. 28-29.

Uttarwar, P.S. (2000), Transparency in public dealing departments through GIS, *GIS Development*, Vol. 4, Issue 11, p. 28-31.

Ventatachalam, P., B. Krishna Mohan, J.K. Suri, Aarthi T. Chandrasekar and Vikas Mishra (2001), Teaching GIS principles through multimedia based GIS tutor, *GIS Development*, Vol. 5, Issue 1, pp. 24-27.

Verbyla, David (2002), *Practical GIS Analysis*, Taylor and Francis, London.

20

Conclusion

Cartography has now become a repository of enormous quantity spatial digital data acquired, processed, stored, retrieved and displayed. The instant communication to the user, in the form, mode, speed and variety desired by user is possible with electronic media. The technology and the information age have transformed cartography from the art, science and technology of making maps in to an "information transfer process". Cartography can be visualized as an information science performing a service to a wide range of other disciplines. Disciplines that are linked in spatial relationships are history, geography, geology, sociology, meteorology, astronomy, ecology, demography etc. There are many users and owners of different data involved in the building of Geographical Information System (GIS). It is the cartography who has always coordinating all sources of inputs and responsible for the concept and design of the product. It is also entrusted with the design and management of GIS and other sub-sets of the information system such as Land Information System (LIS) cadastral information system etc. Digital Cartographic databases have increased the capability and flexibility, which allow investigation of potential methods for information display to meet a wide variety of requirements. Perspective visual simulation is one of the most effective display methods to create natural appearing image, allowing associating geographic data with the real world. There are many applications of digital data such as tool for planners, Digital Elevation Model (DEM) which could be used to generate inter-visibility maps, 3D oblique view of terrain is useful for the military planner in assessing approaches and relative size and shape of key terrain features. DEM is especially useful in air-borne operation planning by providing a pilot's eyeview of the terrain that will be encountered in an actual mission. Spatial data in an information system and representation technology of cartography continue to make the scope of applications almost limit less, far outstretched from the old tasks of map making and data handling.

In the past, it sometimes took years to provide map product containing the new information. The delays in the cartographic procedures made such new information to obsolete ones. The map making tools have undergone revolutionary advancement such as high resolution imaging, Global Position System (GPS), Airborne Laser Terrain Mapping (ALTM) etc. which has the capability of rapid collection of data and has reduced the response time, which is becoming increasingly critical as a user requirement. In this context the aspect of 'Time' has become an additional dimension. A typical data that is collected with time as an element is that of meteorology. There are many other types of data that lend themselves for arrangement with time as the main element, like population censuses. Such data sets with time enable modelling of the processes to provide forecasts of future patterns. This is a field of "dynamic cartography", which will have a greater role with the technological advancement in the field of electronic hardware and software.

Cartography and GIS

Cartography and GIS are strongly related to each other, yet they remain fairly independent in their central objectives. Cartography is a much older field than GIS and some of its areas of interest are not well suited for GIS like visual impacts of different mapping techniques, map theory etc. In other areas cartography is strongly supported by GIS like data storage and manipulation, spatial inter-relational studies, experimentation with different mapping techniques. Therefore, GIS operations do not have to lead to the production of maps as final products. Although, it helps in producing more user input-friendly spatial products. Updating and changing of data and maps using GIS tools is easier than using manual method. GIS in their nature are strongly connected with fields like cartography, geography, computer science, databases, computer graphic, land shape architecture, spatial statistic, management planning and so on.

Cartography with GIS could be used in various fields, *viz.* national topographic mapping, navigation, property and administrative boundary, geology, geography, surveying, other social and natural sciences, defense, emergency services, marketing and other commercial, population census, public administration and others.

GIS has come long way since manual GIS systems evolved from the discipline of cartography, where architects or site designers needed to visually compare the building plan with the site survey. With more elaborate urban planning in the 19th and early 20th centuries, *map overlays* became popular. These were translucent paper (and later, plastic) sheets upon which maps were drawn or printed. For example, a map of site drainage could be overlaid on a topographic map of the site, or on a street plan. This could help the urban planner or site designer determine whether or not to locate houses or other buildings at a given place (e.g., close to a creek or river floodplain). The same overlay process can be used to superimpose structural data on an aerial photograph. This technique was often used by site designers or architects when presenting designs to a prospective client.

In the 1960s, computers available to domestic government agencies and university laboratories had few graphics capabilities, generally operated in batch mode, and were quite expensive and unreliable by today's standards. Although little cost analysis information is available, it appears as though the automated methods were not yet competitive with manual map production, due largely to the high cost of procuring and maintaining computer hardware.

A more fortunate circumstance occurred with the establishment of the Environmental Science Research Institute by R. Dangermond in 1969, who eventually produced the ARC/INFO package that has become something of a standard in GIS. Intergraph, ComputerVision, and Synercom were also major players in the 1970s, and most of these approached GIS from the CAD/CAM area. But by 1976, there were at least 285 computer software packages that handled spatial data and were developed outside USGS, which number rose to over 500 in 1980. Due to lack of contact between software developers, much duplication resulted. Large states in the US developed their own GIS software, some using CGIS, others adapting SYMAP, and still others generating proprietary, in-house products. Tomlinson describes the 1970s as a period of "lateral diffusion" rather than innovation. In the defence community, the CIA developed a world data bank that subsequently was made available in the public domain. No comprehensive history exists of these local approaches, but the Bureau of the Census and USGS have made representative progress, namely linking cover from the USGS 1:100,000 topographic maps with tracts for the 1990 Census, which has motivated further GIS developments in the US.

In the 1990s, GIS has matured somewhat, with research directed away from the basic issues of map production and encoding, which have been solved after a fashion. The current issues of importance are more related to engineering concerns, such as map accuracy, precise co-registration of maps and imagery, and the vexing problem of combining datasets of diverse formats. In the latter case, much data has accumulated from earlier vector-format GIS, which is difficult to accurately integrate with current GIS systems.

For some time now GIS has been expected to turn a corner and enter into the mainstream of computing, and may be society. Chances are there that this will happen through a more 'location aware' society supported by services built on positioning, distributed geospatial databases and connected with ubiquitous telecommunication coverage. Will geographic information or GI science lead the way, or will GI technology drive developments ? Due to the potential of 'location' to serve as the critical link connecting the virtual and real worlds it is argued that the geospatial domain plays a critical role in establishing the business foundations of any information society.

GIS is not a pre-compiled map. Rather, it enables users to construct maps dynamically by displaying the selective information in the way in which they choose to view it. GIS software helps in developing the maps in chosen formats for better visualization of the geo-referenced data. Obviously GIS makes it possible to link or integrate data which is otherwise difficult. Since all most all the information vital to the government and industry are generally in one way or other location-based, GIS allows better understanding and clear visualization of the action and decisions. This aspect of GIS has a vital role in applications which we have seen in the earlier chapters.

There is definitely better business opportunity with the help of GIS. Linking location to information is beneficial for decision-making in business. As we have seen that GIS has helped the industry or business houses in choosing a site, targeting a market segment, planning a distribution network or to meet an emergency situation. Where are the customers or potential customers? In which neighbourhoods or zip code areas or pin code areas are important for a particular business? Anyone trying to take a right decision, the intelligent digital maps made in GIS environment is required. The location-based data allows government and the business managers to better understand and visualize the impacts of their decisions. For an emergency situation, not only the government but also the industry benefits from the intelligent use of GIS capabilities.

GI for Business

When we discuss current trends and a vision where we expect to be standing several years from now, we first have to establish a clear frame of reference. GIS as software architectures links technology and methods to application and business domains. Geographic Information (GI) science researches the conceptual background and methodology frameworks implemented in GI systems and guiding applications. GI manifested in geospatial data collections and data streams represents the real world, thus describing application domains and making them accessible for digital workflows. GI technology reaches beyond the software in GI systems to include sensor systems and dedicated interaction devices from handhelds to vision aids. For a discussion on trends and visions, we will have to pursue an integrated perspective with a focus on what we will be able to do many years from now. The science behind that, its translation into technology and systems, and the availability of multiple and detailed digital representations of past, current and future realities are being integrated towards future capabilities. All GI combined will serve the purpose of providing an explicitly 'spatial user interface' for our world across all its different scales, facets, themes and user perspectives.

Before attempting to look into future perspectives, we first need to review recent trends as an indicator of where substantial change is currently occurring. Of course, any such listing is highly subjective, but most experts will likely agree on the majority of items mentioned here:

- More or less isolated desktop systems have become part of distributed architectures. Recent and upcoming releases of GI software clearly demonstrate that they work best against a backdrop of services feeding into the workflow of the locally-based analyst.
- Developments initiated through the Open Geospatial Consortium (OGC) have crystallized the interoperability concept and contributed to a unified web of services evolving into our GI infrastructures.
- The past 'data bottleneck' where projects were stifled by the enormous efforts needed to supply data has by and large given way to a 'flood of data'. Data mining and 'intelligent' extraction of relevant information are critical to effective operations.
- Part of this 'flood' is provided by a multitude of sensors above and at the earth's surface. These sensors supply geo-referenced streams of measurements, effectively tying the virtual world of geographic models into the real world around our lives.
- Largely due to these sensors and ubiquitous connectivity, capabilities for near real-time monitoring have evolved. For a large part of GIS' past, we were essentially processing historical data, only recently we are moving towards 'current affairs'.
- This development has been made feasible by the now pervasive presence of positioning devices. Our societies are increasingly mobile in every respect: individual, environmental and business 'objects' change position frequently, without tracking our model of the real world would not be current.

Framework for GIS

All too often we are focusing on 'what can be done with technology' and 'how we can leverage progress in GI science'. Preferably we should look into 'use cases' and analyse what needs in society are being and need to be supported by progress in the GI fields from technology to science. GIS is still burdened (like many ICT fields) by the 'automation trap' of computers being fast and good at clearly defined repetitive tasks. Therefore, long established jobs are more effectively done, which led to file cabinets moving to databases, typing to word processing, drafting to CAD and of course mapping to GIS. The quest for the 'map looking like the one we've always had' is still on in many people's minds.

For real progress, GIS clearly needs to focus not on doing a better job at established tasks, but rather to see how tasks can best be organized to reach our societal, environmental and business objectives. Analysing how to reach these objectives, breaking down workflows into tasks and then researching any missing concepts or methods through GI science will lead to newly defined 'jobs' for GIS instead! So, what are GIS good at now? Documentation through static representations, change detection and visualization probably rank among the most frequent applications. Spatial analysis helps create new insights by generating information from data through advanced methods. Dynamic process models and scenario techniques support spatial decision support applications. All of these will be needed and provided in the future, but is there more to come?

Certainly, we will be moving from analysing and presenting manifest data sets towards working with streams of mostly georeferenced data. Without the 'where?' any data on 'how much?', 'what?', 'how?', 'when?' are missing much of their value, hence sensors deliver a fixed or moving location integrated into their reports. Through data networks, these reports are delivered right away. This means that instead of analysing past patterns, we are looking at the current picture. Insight into the past gives way to real time applications. Instead of learning from what did not work well to do it better the next time around, we can react right away. GIS, therefore, is moving towards real time control. Cartography develops into a human computer interface discipline for spatial views supporting interaction with current processes or for deep, recursive analysis of spatial data. In essence, the long feedback loop is shortened substantially: what used to be an elaborate cycle of data acquisition, analysis, visualization and decision-making is for many applications being compressed into real time monitoring and control.

GIS Trends

Observing the above outlined current trends and changes to business logics involving geospatial information, we now might want to assess where we expect to stand several years from now. Distinguishing technology developments from GI science quests will be helpful while acknowledging that both areas are of course tightly linked in a symmetric stimulus—response relationship. First, looking at expected technological progress, the main drivers clearly are broad, multiple use developments demonstrating that GIS is increasingly becoming an integral part of mainstream ICT:

- Sensors, based on the sensor-web framework are penetrating our world widely. This means that the state of our environment, business assets and society's infrastructure is thoroughly monitored and available for diagnostic and controlling action.
- Positioning technologies are being integrated into sensor frameworks, and positioning/navigation applications will become independent from specific technologies like GPS, GSM, Wifi or others. Transparently roaming across positioning technologies across a ubiquitous positioning infrastructure is a key enabler for novel applications.
- Due to pervasive networking and 'the network being the system' (a) local (geo-) processing capabilities are much less critical for GIS applications than access to the network.

- Interaction with digital information is becoming much less abstract. Working directly with spatial views (Google and Microsoft Virtual Earth currently leading the way) ties the 'online domain' directly into daily individual experience and perception. This likely will turn out to be a major driver towards the development of spatial data infrastructure services.
- Finally, access and interface devices will be defined by workflows, not by GIS. We need a big desktop screen to look at a map? Well, do we actually need to look at a map or can we relate the required information otherwise? A majority of geospatial applications will not be tied to map-like user interfaces, but rather provide 'location intelligence' in the background.

Interacting with technology trends like these, GI science will set priorities continuing from current research as well as explore new areas. Again, there are of course many legitimate perspectives assessing future emphasis which are:

- First and foremost, although this might come as a surprise to many, privacy and preservation of civil rights clearly top the list. Location is such a powerful key for integrating and accessing data collections that this issue has a huge potential for a backlash against leveraging spatial information to benefit society and our lives. How to protect individuals' locations and tracks, how to design applications without generating a 'big brother' effect will have to be high on our research agenda.
- As a related theme, geospatial enabling will be accepted as a means towards more public participation, supporting the empowerment of citizens. Citizens watch (instead of being watched!) and participate in planning and decisions affecting their lives and interests. Geospatial applications as a local communication platform require a high level of spatial literacy, this has to be confronted as an educational task.
- With increasing flow of data, methodology research is being focused on approaches to spatial pattern recognition in a wide sense. Filtering data streams to extract meaningful information is needed to environmental monitoring, security, infrastructure operations and transportation applications.
- Visual interaction (and other human sensory interfacing) ties the real world into our virtual models. Designing spatially enabled Human-Computer Interaction (HCI) is a key theme supporting penetration of geospatial applications throughout personal and professional applications, providing assistance for mundane tasks as well as 'performance support'.
- towards 'intelligent' infrastructures integrates many currently leading topics in GI science. Infrastructures serve a purpose, they should do so efficiently and effectively and without causing undue harm. Research in semantic web, including cognitive aspects, emphasizing sustainability and minimum impact all come together in this meta-topic with a long overlooked strong GI science dimension.

Education Perspective

Education towards professional qualifications has frequently been named as a key factor towards extending the reach of the geospatial industry, with the idea that qualified specialists ranging from IT-focused via methodology-centered analyst to application experts. While this observation is certainly true and needs to be supported through the development of suitable curricula and implementation of programmes across traditional boundaries between disciplines, another aspect might be even more decisive. Traditionally, one aspect of schooling is the establishment of literacy as a foundation for critical thinking, creativity and participation. The spatial angle has been covered by the building of map reading skills supporting education towards spatial awareness and understanding many facets of our world. Well, map reading skills are changing substantially and are applied in the computing domain. 'Maps' now move, change perspective, can be actively navigated and are much more an orientation device and spatial user interface.

Most surveys and visionary perspectives claim that the geospatial industry is moving from a technology-oriented niche into the mainstream of society. GI only has limited growth potential in the business-to-business sectors of corporate decision support, workflow optimization and day-to-day operations. Rather, a customer-focused evolution towards spatially enabled and locally centred services is considered as a major growth opportunity. If we want customers to use a search engine's spatial portal as their default web page, bookmark their community's map interface, work out an itinerary from a spatial perspective and shop, eat and interact with friends based on ICT-facilitated proximity, we need support from general education.

Spatial literacy is not about teaching GIS in schools. Rather, creating spatial awareness, navigational ICT skills and an understanding that all or most of today's problems have a significant spatial aspect to them need to become objectives throughout schools curricula. Only spatially aware citizens with the requisite ICT skills will be able to critically participate and contribute to an information society.

Will our lives be different?

Of course, the answer will be yes or no, depending on what we consider the defining aspects. Will our future lives be different due to progress in geoinformatics? For all of us, and many others, yes. In what ways will life be different? This will depend largely on lifestyles, professions and interests, and of course, our geographical and societal positions on the globe. Geospatial information will more and more affect us on different scales. The 'global village' and the much quoted links between local action and global thinking will become less philosophical and more practical. Beyond the obvious assistance with travel and transportation, geography affects everything around us. The environment, sustainability and infrastructures are all inherently geographical issues. As they are being represented through geoinformation to enable direct, location aware interaction, digital geospatial interfaces essentially become our dashboards as we navigate life.

On a practical level, the trend towards networked services means that local availability of computing and processing power becomes much less critical than the availability of bandwidth. If ample bandwidth is provided differences in education, opportunities and participation will decrease. Bandwidth as an indicator for access to information and interaction, therefore, is the key differentiating factor in future geographies.

Towards the Geo-Information Society

The European Union's 'Information Society' concept claims that the spatial aspect is pertinent to a majority of information aspects and it also provides interesting perspectives to toy with the term *'Geo-Information Society'.* We believe that, in a way, we are moving into this direction. Let us briefly look at information from an abstract meta-perspective: information objects or features can be located in an 'information cube' which has defined geographical research since the early days of the (then) 'quantitative revolution'. Objects have attributes characterising their position in subject, temporal and spatial domains. A cube with the dimensions of subject (what?), time (when?) and place (where?) can therefore be used to place information objects (Fig. 20.1)

This view provides support for the argument that GIS is not a distinct area along the lines of the 'spatial-is-special' phrase, but geography is rather a view, often a very important view on our world through the lens of information representations. In simple terms, the spatial dimension is a key component of the 'context' of objects, and which affects our actions. Providing full context for decisions and actions is essential, as we know from our many 'profiles' we use to make software and services more focused on our preferences and intentions. Context is a key enabler of semantic-rich interaction (semantic web, see reference), and context is rarely complete without location.

So, the concept of a geo-information society as a more context-sensitive, less abstract concept might not be so far-fetched after all! Technologically speaking, it is built on three pillars: (Fig. 20.2) geospatial data as digital representations of much of our world; positioning services putting people, assets and other 'objects' into this context; and mobile telecommunication connecting everything and providing access for users.

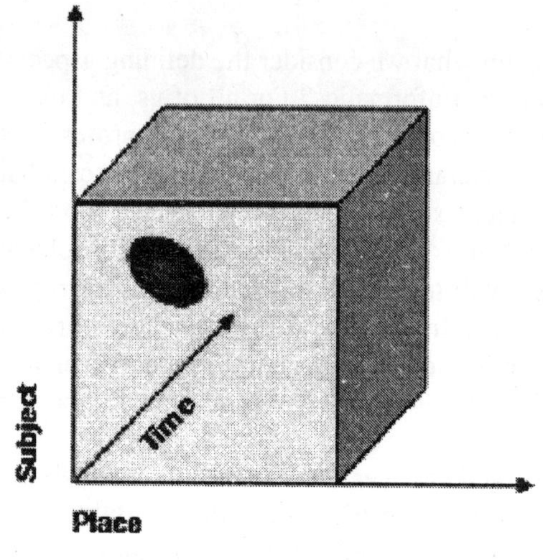

Fig. 20.1: Attributes to place information objects

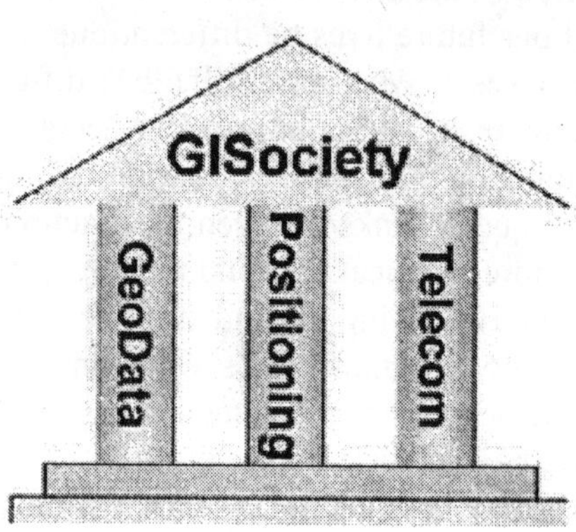

Fig. 20.2: The Geo-Information Society

GI Science and the Information Economy

In 1995 and 1997, the magazine *Economist* declared the 'death of distance' caused by information infrastructures making location largely irrelevant. Telecommunication, telecommuting and globalization appeared as clear indicators. Interpreting Tobler's 'First Law of Geography' (stating 'Everything is related to everything else, but near things are more related than distant things') this should mean that geography had to be declared defunct as well. Further, in March 2003, though, the same publication titled an article 'The revenge of geography', starting with the observation: "In the early days of the internet boom, there was much talk of the death of distance. The emergence of a global digital network, it seemed, would put an end to mundane physical or geographical constraints. There was some truth in this, but ..." Clearly, the authors had recognized that location provides the best general means of connecting virtual and real worlds. This connect in turn is required to provide business models for the information economy—Geography and GI Science are well placed to provide some of them!

Just as there are 'macro' and 'micro' aspects to economy, we recognize this as a valuable distinction in geography and now GI science. Up to now we have mostly worked in the frame of macro-geography. Now due to positioning services, detailed geospatial data, and powerful communication GIS moves more towards micro-geography. Actually, there would be no macro-economy without the micro-level of businesses. We wonder what it means that GIS has come the other way. Due to the availability of better and finer data and also the technology for processing the same, the concept of "digital geography" is getting currency. It is becoming increasingly possible to have a bottom-up approach in developing geographical (also spatial) analysis, models and postulated. A new set of indigenous concepts in geography is likely to evolve with the help of computer processed huge data sets. Earlier, models and concepts will be subject to revision, modification and revalidation.

Similar to above, in the business and industry, a fresh approach is likely to evolve for providing better services, functionally and economically, to the clients and people at large. In order to achieve this goal, GIS cannot be effective in isolation. It has to integrate with on one hand with other mapping technology like remote sensing of GPS; while on the other, with other emerging technologies such as information and communications technologies. In short, due to the expanding horizon of the digital technologies, integration is the core issue in any business solution involving GIS as well. In the previous chapters we have seen the GIS is perhaps a small part of a business solution, but effective and essential. The developing countries do become a virtual laboratory for the application of GIS as in these places the vast population reside in less organized way. For any attempt for economic or social development, application of GIS cannot be ignored.

Increasingly, state and local economic development activities are coming to rely as much on GIS technology as are environmental and natural resource concerns. These problems would be challenging enough, but they are complicated by the inefficiencies of traditional government planning processes. As authorities admit, "traditional planning is fractured, takes decades to implement, and often spawns lawsuits between developers and environmentalists." With population growing at skyrocketing rates, these were conditions different counties could no longer tolerate. Increasingly, national, state, regional and municipal, social and economic development activities are coming to rely on GIS technology. In order to have an economically viable approach, GIS capabilities are also being deployed in an effort to improve health, enhance emergency preparedness, and save lives. GIS has also been applied to reduce the incidence and cost of major diseases like diabetes, cardiovascular disease, and cancer, improve health outcomes through prevention and earlier detection, and increase the health of residents through fitness and nutrition education programmes. There are several instances where the government officials have used GIS technologies to accurately gauge the size of the affected population, the optimum sites for distributing medicines, the impact of weather conditions at the time of the event, the relative capacities of nearby hospitals, clinics, and pharmacies in the area, and the travel patterns of affected individuals in a given time before and after the event. Further, as governments have deployed GIS capabilities more widely, many have discovered that greater usage of GIS data may be the quickest route to more efficient government. The civic authorities can browse through a variety of general information, like census data, housing profiles, home values, and so on. San Francisco GIS Manager observes a few of the ways in which the city and county have already implemented GIS capabilities:

—The city assessor and real estate staff rely on GIS information to improve their ability to manage city property.

—Police officers can review Internet-based geographical data directly from their squad cars in order to help speed their response to emergencies.

—The police department later this year will begin using GIS-based crime data as the basis for reassigning task groups to neighbourhoods with the highest crime rates.

—A new application called SFViewer enables users to input an address or block number in order to view parcel information, aerial photographs, or street maps.

—As in Dallas, city residents can enter their address into the city's web site in order to locate a range of neighbourhood-related data, from the name of their legislative representative to times when their streets will be cleaned.

—And businesses can use an online application called SF Prospector to help them to determine the best location for a new business, based on office rental vacancies, neighbourhood demographics, and so on.

The key to the success of these efforts, GIS manager believes, has been an unprecedented degree of executive level support, including the Mayor of the City. This approach has enabled several departments to employ GIS successfully with minimal investment. Further, most government agencies deploy GIS technologies to manage such geospatial data as land parcels, roads, and facilities. By linking this GIS data with the supporting documents, such as engineering drawings, scanned deeds, or road construction contracts, employees are given quick and easy access to the data required to make immediate and informed decisions. All that changed with the adoption of a fully integrated electronic GIS system. For example, finding related information all of the documents pertaining to a certain project will involve in locating someone who knows which file cabinet to search. What typically happened therefore was that a request for information would be sent to the records section and it would take days for staff to collect the required information and respond to the request. Residents and other members of the public looking to the municipality for legal contracts, payment of taxes, financial records, engineering documents, geographical maps, satellite maps, and land parcel ownership records and correspondence hence often found it difficult if not impossible to obtain access to records that corresponded with information in the GIS.

As we have seen, with the newly integrated GIS system, documents can be associated with projects though text-based searches as well as geographically based searches. By linking geographic documents with corresponding map features in this way, staff will spend less time searching for records, thus increasing their productivity. Instead of spending time looking for records that might not be found, the staff can now email requests directly to the appropriate parties and receive a prompt reply, sometimes in a matter of minutes. This whole activity can be out-sourced as well. There is a potential business opportunity in such activities as well. Obviously, integrated GIS solution has become a very powerful strategic enterprise tool for managing costs, complying with government regulations, and improving customer service as well.

There are other instances also which require immediate response, such as firefighters and emergency medical services where the personnel work around the clock to put out fires and to respond to the health emergencies of individual citizens; or engaged in daily operations to identify, track, and mitigate life-threatening diseases; or respond to accidents, keep roads safe and analyse accident patterns to develop strategies that reduce injury and death. All of these operations have two things in common: (a) they are responsible for saving lives each day, and (b) they have to depend upon the information that has a spatial or geographic context, such as digital aerial photography, accurate street and parcel maps, topographic maps, critical infrastructure maps, and maps of hazardous materials. These information sources when combined with geospatial and other complementary technologies as GIS, computer-aided dispatch systems, routing software, the GGPS, automated vehicle location, remote sensing, and others contribute to the foundation for a modern public safety information infrastructure. During major emergencies such as terrorist attacks, flood, fire, earthquakes, or hurricanes, the data can immediately be used to support the efforts of first responders. They also can support hazard mitigation operations such as the tracking of potential terrorists and environmental monitoring to prevent emergencies from happening in the first place.

Certain life saving operations, like *tsunami* or earthquakes, extend beyond the borders of local jurisdictions including state and regional boundaries, however international and national emergency management operations must use these same data so that support and response efforts can be quickly and easily integrated with efforts at national and global levels. It is, therefore, essential that spatial data be built to comprehensive, consistent, and nationally and internationally agreed upon standards. Global Spatial Data Infrastructure (GSDI), Permanent Committee for GIS Infrastructure for Asia and Pacific (PCGIAP) and similar regional bodies, International Steering Committee for Global Mapping (ICGM: www.iscgm.org) and the National Spatial Data Infrastructure (NSDI) and similar national bodies have been working on joint initiatives for developing common data bases. Spatial data must be created through national, regional and global initiatives to ensure that they are available to all who require their use. In case of India, following the approval of the Cabinet, the state governments are also taking steps to form the state level spatial data infrastructure (SDI). This process will further filter down to district or even at sub-district levels.

Creation and deployment of the public safety components of the NSDI or state SDI will have many additional benefits. Local and state governments can use the same data to provide a base for countless non-emergency operations and applications as well, such as services, formulation of business strategies, e-government initiatives, economic development and planning, waste removal, street cleaning, code enforcement, environmental protection, construction, inspections, capital construction and so on. These applications of spatial data are known to increase workforce productivity, streamline business processes, reduction in costs and improve services delivered to the public. To take advantage of the technology, a GIS based on reliable data should already be available with every government agency deployed in extending the services to the people. The public safety components of the NSDI need to be developed immediately particularly for a country like India. Necessary steps should be taken well before the measures are required to respond to a catastrophe.

Finally......

Geographic Information Systems (GIS) deal with the storage, management, retrieval, conversion, analysis, modelling and display of spatially related data in a systematic way. GIS software is potentially applicable in nearly any situation that calls for decisions involving a spatial component. In other words, GIS can be a useful tool in situations as diverse as area to determining the environmental impact. It is impossible to list all the possible applications of GIS, as they are practically limitless. As computer power grows, enabling the use of more complex models and the more efficient incorporation of a time component in analysis, the more this will be true. Advances in computer software and hardware, increased familiarity with the power and applicability of GIS and its related technologies, and a greater breadth of accessible data have driven the growth in demand for GIS and will continue to do so in the future. Throughout the knowledge-based economy, both software and hardware have become much more powerful and simple to use while becoming less expensive. Clients in all markets are now comfortable with computers and have access to the necessary equipment to run even complex GIS packages. Furthermore, not only has a greater range of geospatial data become more readily available but also non-traditional data such as those found in the social sciences.

In a developing country like India, having huge population base with democracy, public access to information and freedom of expression, GIS has several additional function to play. It is consider being a tool for getting in touch with the government, lodging protests, and having interface with people. In addition, there is a wide scope for business enterprise as well. GIS has a definite role to play in the following sectors:

(i) Highways survey and mapping.
(ii) Natural disaster management.
(iii) Telecommunication services.
(iv) Power projects.
(v) Ports projects.
(vi) Urban land use planning.
(vii) Rural land management or agricultural land management.

(viii) Waste land reclamation projects.
(ix) River basins or watersheds management.
(x) Imagination and canal developments.
(xi) Consolidation of land holdings and land records.
(xii) Digitization of maps on various scales.
(xiii) Updating and digitization of wards and services maps of different cities and towns for their betterment both in terms of economy or finance and environments.
(xiv) Mapping of places of tourist interest or pilgrimages both in hard copy and soft copy forms.
(xv) National natural resource management.
(xvi) Monitoring of environment.
(xvii) Terrain and height information or positions for frequency requirement for the purpose of All India Radio, Door Darshan and telecommunication.
(xviii) For better and fool-proof security with up-to-date and accurate maps in the form of digitized or special software both for military and civilian purposes.
(xix) Development of accurate navigational maps both at ocean or water and land for better traffic-flow and least risk of accident.
(xx) Production of good quality maps and atlases for the use of students and others.
(xxi) Maps and GIS for government officials and the common people to understand the real positions and facts in their day-to-day jobs or life.

Although a number of national GIS initiatives are underway, many of the most interesting developments in the use of GIS technology are taking place at the state and local level. The use of geographic information over the past few years has increased dramatically, with a focus particularly on public safety and economic development. In some countries in the west, an extensive crime-mapping system is being implemented that will provide real-time information to police authorities regarding criminal activities. In terms of economic development, the city authorities in order to attract investments have sponsored interactive mapping web site that allows the public to review and analyse economic and demographic data for business site selection. The sense of achievement lies in getting support from the leadership to enterprise GIS effort, particularly freeing agencies from having to individually procure and develop data and systems on their own. This approach has enabled several departments to employ GIS successfully with minimal investment. The success is due to the collaborative efforts of other GIS professionals as well. Without support and initiative from all concerned, the GIS programme would be much more difficult to implement. Nevertheless, the main constraint has been the inability, at times, to get relevant agencies to collaborate toward a collective goal of using, sharing, and disseminating standardized geospatial data, the issue which is very significant for the implementation of NSDI in India as well. From spatial data technology perspective, implementing of a protocol that preserves data ownership, while still emphasizing the sharing and use of common GIS datasets, is a challenge. The solution is a hybrid approach where the agencies still retained ownership of their information, but in which enterprise GIS servers are able to dynamically connect to their own servers among themselves or with NSDI hub for data access and retrieval. The key to success is collaboration.

While the technology hurdles are considerable, organizational and political concerns have tended to be much more challenging. Resolving issues such as data distribution to the public, or privacy concerns related to public data, have proven to be more complex than implementing the technology. Some solutions have been worked for India by establishing NDSI, officially in June 2006. It has committees at two levels, first the National Spatial Data Committee (NSDC) for polices, strategies and programmes; and the second the NSDI Executive Committee for implementation and executive functions.

Funding and marketing have been the challenge. From a business perspective, policy makers, legislators, and administrators do not buy what they do not understand. The potential application of geography, digital geospatial data, and the ability to manipulate it in GIS environment are to be marketed well enough to obtain the necessary funding from government officials. The success lies in public/private partnership aimed at modernizing the land records and other development and public service related data in a standardized manner. Champions of GIS face the same challenges as those seeking to implement other forms of IT, i.e. how to demonstrate value to the business side of government institutions. This issue has become more vital due to government's emphasis on revenue generation. Each institution should have a revenue model as well. However, such initiatives have to take into account of (a) heavy initial costs, (b) data integration expenses, and (c) long payback periods. It will soon become necessary to quantify the financial benefits of GIS implementation in terms of public good, operational efficiency and also political returns. The latter involves public relation, public safety, economic development and increased participation in government sponsored activities. Obviously, GIS is an integrating technology; its value is realized when multiple data sources reveal spatial relationships between geography and a collection of information on economic, environmental, health, social and infrastructure. GIS implementations often begin, not at an enterprise level, but from organizational initiatives that have recognized the value of GIS as a part of their mission. GIS startup carries substantial costs for thee procurement of data, digital imaging, map digitization, data quality assurance, data quality control and software and hardware purchase and implementation. Solid project management and supportive governance are essential to a successful GIS implementation.

It is seen that the Indian industry is fully capable and able to take the challenge of the demand for mapping or cartographic work. The concerned Indian industry is fully prepared to take cartographic work in collaboration with the government departments or agencies. They are doing a number of cartographic or mapping jobs for and of the government. They are also doing Geographical Information System (GIS) or Land Information System (LIS), digitization of maps or documents etc. for different organisations and agencies. It is also found that there is a great demand for thematic maps on large and small scales on 1:1,000, 1:10,000, 1:50,000 and 1:1,000,000 on different themes for projects in the field of telecommunication, business, irrigation, power, transport and tourism, resource management, town and urban planning, etc. It has been assessed that there is a huge potential for export of cartographic products and services, especially in the field of GIS or land information system, scanning and digitization, remote sensing technology and developing different products, services and applications not only for India but other countries as well. Many Indian industries have done several mapping and digitization work for foreign companies or countries and are doing for many others. Many

Indian industries are ready to collaborate in the field of cartographic or GIS with global partnership. Further, India has become a hub for in sourcing IT projects from abroad. Mini-Bangalores have cropped not only in the fringes of the bigger Indian cities, but the medium size cities have not been spared as well. With the successful experience of IT, trained man-power available with fairly good knowledge of English and capacity to pick up other foreign languages, India is likely to become hub for global GIS or geospatial industry as well. GIS or geoinformatic institutes for teaching, training and research are coming up in every city. The time is not too far when India will become a favourite destination for GIS industry as well. Geotechnology is the new challenge, a new determination and a new dimension to the world of science.

REFERENCES

Ramchandra, T.V., Uttam Kumar, B. Vaishnav and S.N. Prasad (2004), Geographic resources decision support system—an open source GIS, *Geospatial Today,* Vol. 3, issue 3, pp. 52-59.

Strobl, Josef (2005), GI Science and Technology—where next?, *GIS@Development,* August (josef.strobl@sbg.ac.at).

APPENDIX I

National Map Policy-2005

1. Preamble

All socio-economic developmental activities, conservation of natural resources, planning of disaster mitigation and infrastructure development require high quality spitial data. All advancements in digital technologies have now made it possible to use diverse spatial databases in an integrated manner. The responsibility for producing, maintaining and disseminating the topographic map database of the whole country, which is the foundation of all spatial data vests with the Survey of India (SOI). Recently, SOI has been mandated to take a leadership role in liberalizing access of spatial data to user groups without jeopardizing national security. To perform this role, the policy on dissemination of maps and spatial data needs to be clearly stated.

2. Objectives

- To provide, maintain and allow access and make available the National Topographic Database (NTDB) of the SOI conforming to national standards.
- To promote the use of geospatial knowledge and intelligence through partnerships and other mechanisms by all sections of the society and work towards a knowledge-based society.

3. Two Series of Maps

To ensure that in that furtherance of this policy, national security objectives are fully safeguarded, it has been decided that there will be two series of maps, namely:

(a) *Defence Series Maps (DSMs)*—These will be the topographical maps (on Everest/ WGS-84 Datum and Polyconic/UTM Projection) on various scales (with heights, contours and full content without dilution of accuracy). These will mainly cater for defence and national security requirements.
The series of maps (in analogue or digital forms) for the entire country will be classified, as appropriate, and the guidelines regarding their use will be formulated by the Ministry of Defence.

(b) *Open Series Maps (OSMs)*—These will be brought out exclusively by SOI, primarily for supporting development activities in the country. OSMs shall bear different map sheet numbers and will be in UTM Projection on WGS-84 datum. Each of these OSMs (in both hard copy and digital form) will become "Unrestricted" after obtaining a one-time clearance of the Ministry of Defence. The content of the OSMs will be as given in Annexure 'B'. SOI will ensure that no civil and military Vulnerable Areas and Vulnerable Points (VA's/VP's) are shown on OSMs.

The SOI will issue from time to time detailed guidelines regarding all aspects of the OSMs like procedure for access by user agencies, further dissemination/sharing of OSMs amongst user agencies with or without value additions, ways and means of protecting business and commercial interests of SOI in the data and other incidental matters. Users will be allowed to publish maps on hard copy and web with or without GIS database. However, if the international boundary is depicted on the map, certification by SOI will be necessary.

In addition, the SOI is currently preparing City Maps. These City Maps will be on large scales in WGS-84 datum and in public domain. The contents of such maps will be decided by the SOI in consultation with Ministry of Defence.

4. National Topographical Database (NTDB)

SOI will continue to create, develop and maintain the National Topographical Database (NTDB) in analogue and digital forms consisting of following data sets:

(a) National Spatial Reference Frame,
(b) National Digitial Elevation Model,
(c) National Topographical Template,
(d) Administrative Boundaries, and
(e) Toponomy (place names).

Both the DSMs and OSMs will be derived from the NTDB.

5. Map Dissemination and Usage

- Open Series Maps of scales larger than 1:1 million either in analogue or digital formats can be disseminated by SOI by sale or through an agreement to any agency for specific end use. This transaction will be registered in the Registration database with details of the receiving agency, end use etc.
- Thrugh the agreement, SOI will allow a user to add value to the maps obtained (either in analogue or digital formats) and prepare his own value-added maps.
- The user should be able to share these maps with others—the information of all such sharing will also require to be logged in the Map Transaction Registry.

6. Applicability of Previous Instructions

The Ministry of Defence has from time to time issued detailed guidelines on various aspects of map access and use. These instructions shall continue to hold good but for the modifications cited herein.

GUIDELINES FOR IMPLEMENTING NATIONAL MAP POLICY

1. Background

The Central Government announced the National Map Policy (NMP) on 19th May 2005. The NMP document authorizes Survey of India (SOI) to issue detailed guidelines on the implementation of the NMP. These guidelines are issued in the light of the above cited authorization. The guidelines are clarificatory in nature and does not create any new rights nor abridge any existing rights which are enforceable in courts of law.

2. Information of Maps and Guidance of Indentors

(a) All up-to-date information on SOI maps shall be available at the various offices, map sales counters and Geospatial Data Centres (GDC) under the SOI. The information shall also be available on Survey of India website www.surveyofindia.gov.in created and maintained by the SOI. The information provided *inter alia* includes all relevant details of maps such as scale, information content, date of data capture, price, mode of data dissemination whether MOD clearance available for issue as Open Series Maps (OSMs) etc. The website also has a search engine by which the exact sheet number of a given locality can be traced by giving elementary details like name of district or any other prominent feature. SOI offices may be contacted for further information on OSMs.

(b) The lay out of OSM shees on UTM projection and WGS-84 datum is shown in Annexure 'A'. The annexure also shows the incidence of the existing topographical series on the OSM series.

(c) Data content: Approved data content of OSMs is given in Annexure 'B'.

(d) Export of all maps/digital data in 1:250K and larger scales through any means in prohibited *vide* Ministry of Finance (Department of Revenue) Notification No. 118-Cus./F.No. 21/5/62-Cus. I/VIII dt. 4th May 1963. Digital Topographical data will be licensed to only Indian individuals, organisations, firms or companies.

3. Sales/Data Dissemination

(a) *Analogue/Paper Maps*: of all scales shall be made available from Survey of India (SOI) offices and Map Sales Counters on payment at prices as may be fixed by SOI from time to time. The list of such outlets, their addresses with telephone numbers shall be made available in the website www.surveyofindia.gov.in. These maps can also be sold by any retailer. Digitisation of SOI analogue/paper maps is strictly forbidden.

(b) *Map Transaction Registry* (MTR): As stipulated in the National Map Policy, Survey of India will establish an on-line MTR for recording of all transactions relating to digital maps. Each user will be allotted a unique user ID and each transation with a unique transaction ID. For all future correspondence, this user ID and Transaction ID should be referred.

(c) *Digital Maps*: include both Raster and Vector forms. The ownership of all digital data vests solely with SOI and will be given only under licence against indent and on payment. Unauthorised copying and distribution of SOI digital data are strictly prohibited. All licences will be issued through the Map Transaction Registry (MTR). The format of the licences is available in SOI web site www.surveyofindia.gov.in. The indents may be made in the prescribed proforma.

4. Licensing of Digital Maps

Digital data will be available in single/multiple/commercial licensing for general use, value addition and marketing. All digital maps will be provided with encryptions/mechanisms which may corrupt the data while copying unauthorisedly or while attempting the same. Every such attempt shall attract criminal and civil liability from the user without prejudice to the corruption of data or software/hardware for which the SOI will not be liable.

SOI digital data will be licensed based on usage. Following are the categories:

- Digital Licence
- Publishing Licence
- Internet Licence
- Media Licence
- Value addition Licence

Terms and conditions governing each of the licence is available in SOI web site www.surveyofindia.gov.in.

5. Procedure for Value Added Products

Inasmuch as the National Map Policy is aimed at encouraging a flourishing geospatial industry and related knowledge services, there is much expectation from the industry to do value additions to the products of SOI. SOI encourages individuals/industry to do value addition. Intending value adders will submit a business model indicating the products to be generated and marketing strategy and also execute a Memorandum of Agreement (MOA) with SOI.

6. Pricing of Various Products

(i) The SOI shall notify from time to time the price list of its standard products.
(ii) The prices are subject to revision from time to time.
(iii) An indentor has to pay the amount at the rate prevailing on the data of supply unless the indenter has already paid at the earlier rates. In other words, the indenter will not be eligible for supply on payment on the plea that the rates prevailing on the date of indent were lower.
(iv) The SOI reserves its right for differential pricing, such as concessions to Universities, Research organizations, deserving NGOs on the basis of predetermined policies, different prices between single and multiple user licences etc.

7. e-Payment

Digital maps, on following the procedure, will be made available on variety of media like Compact Disks (CDs), DVD etc. Maps on 1:1M and smaller scale on compressed JPEG format will be available for download *against* online submission of licence agreement and on e-payment. The self-guide provided in www.surveyofindia.gov.in shall enable an indentor to make e-payment and download the requisite maps.

8. Settlement of Disputes

In case any dispute arises on the applicability or interpretation of these guidelines between the SOI and any other person, the matter shall be referred to the Secretary, Department of Science and Technology, Government of India, whose decision shall be binding on both the parties.

9. Applicability of Previous Instructions

The Ministry of Defence has from time to time issued detailed guidelines on various aspects of map access and use. These instructions shall continue to hold good but for the modifications cited in the NMP.

10. General

(i) Copyrights of all SOI maps (both digital and analogue) vest exclusively with the SOI. Any person resorting to unauthorized copying or use or attempts to do so shall make him liable to criminal and civil liability under existing laws.

(ii) While every effort will be made to ensure availability of maps of all areas containing all necessary details and with highest accuracy, SOI will not be responsible for any loss caused to any person on account of failure on any of the above factors or any other reason.

(iii) The SOI reserves the right to add, delete, modify amend any or all of these guidelines without notice and without assigning any reason.

Annexure 'A'

अनुबंध 'क'
Layout of Sheets
शीटों का विन्यास
Datum: World Geodetic System (WGS)-84
आधार : विश्व ज्योडीय पद्धति-84
Projection: Universal Transverse Mercator (UTM)
प्रक्षेप : यूनिवर्सल ट्रान्सवर्स मर्केटर

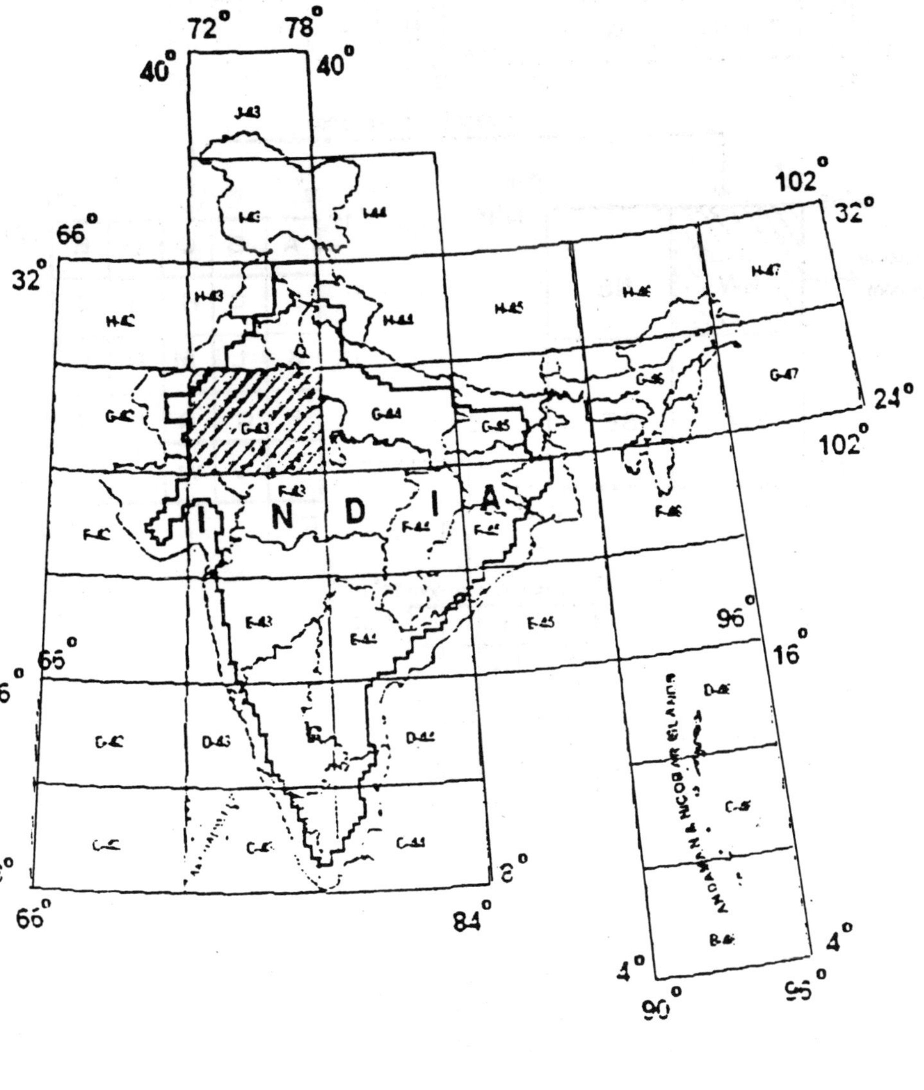

Annexure 'A' (contd.)

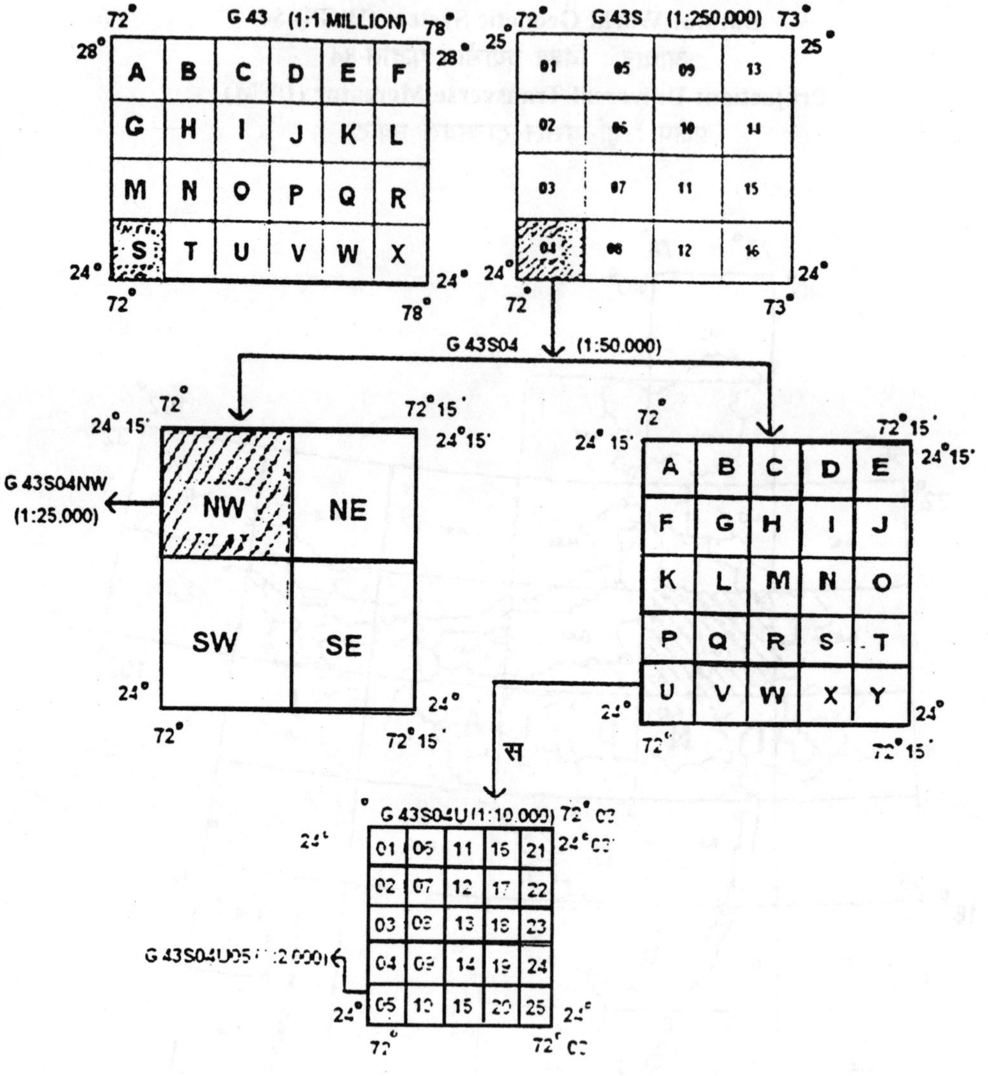

Annexure 'B'

Contents of OSM for Small Scale Maps

Sl.No.	*Category/Layer*		*Sub-details*
1.	General		Latitude/Longitude Name of State/District/Administrative index Topo sheet Number/Year of Survey/Edition No. Index to topo sheets Magnetic variation from true North direction Map reference Bar scale/Representative Factor
2.	Administrative Boundaries	Names Boundary Boundary Pillars	Administrative/Locality or tribal International to village, Forest, all boundary pillers, village trijunctions
3.	Communications/Roads	Roads Tracks Railways Embankments Other Lines	All Roads All Tracks, pass, footpath All gauges with stations, tunnels Light railways or tramway, all embankments, road/rail/tank
4.	Hydrology	Stream/Canals Dams Rivers and Banks Wells, Water Features	All streams/canals All earthwork dams All rivers with details, banks, islands All wells/tube wells/springs All Tanks (excluding overhead tanks), Lightship buoys, anchorages
5.	SI I II MI Ni/Cultural Details	Towns Villages Offices Settlements	Village inhabited, deserted and forts Huts, Tower, Antiquities Religious places, tombs/grave All post/telegraphic/Police stations hut All Bungalows
6.	Transmission Line		
7.	Relief/Hypsography	Contours Sand Features Ice Forms Heights Benchmarks	Contours with sub-features All sand features Ice forms (all features) Spot height, Approximate height Benchmarks (Geodetic tertiary, canal)
8.	Vegetation	Plantations, Trees	All trees, Vine on Trellis, grass, scrub.
9.	Forest		Reserved/Protected

* Contours and heights will not be available in restricted zones as per MOD's instructions.

Annexure 'B' II

Contents for Large Scale Maps

1. Boundary (Polygon) • Administrative Division State District Tehsil Village Ward Municipal/Corporation Locality Hill Range Tribal Area Slum Area	• Boundary (Line) State Municipal/Corporation District Tehsil Village City Colony/Locality with Pincode Block/Sector/Sub Colony Limits
2. Built Up (A) Commercial Bank Money Changers ATM Cinema Theatres/Hall District Centre General Business Milk Booth Hotel Motels Hostels Dharamshalās Guest Houses (Commercial) Hotel with Star Categorisation Restaurants	Non-Hierarchy Commercial Centre Central/State and Other Trade Centres—Shopping Area Wholesale/Warehousing Depots
(B) Government Building Courts Supreme Court High Court Magistrate/Sessions Court Munsif Court Civil Court Family Court Customs and Excise Appellate Tribunal Motor Accidents Claims Tribunal University and School Tribunal Consumer Courts Jail Office Post and Telegraph Office Rail Reservation Centre Air Line Office Check Post Octroi Posts Toll Points	• Manufacturing Brick Kiln Group Industry Service Centre • Public/Semi Public Cremation and Burial Electric Crematoriums Cemeteries Graveyards Tower of Silence Education and Research School College University Institution Music/Dance/Drama/Martial Art School Professional/Financial/Commercial/ Philanthropic and Other Institutions

(contd.)

(contd.)

Facility Centre
    Electricity Complaint Office
    Sewer Complaint Office
    Water Complaint Office
    Birth/Death Registration Office
    Marriage Registration Office
    Travel Agencies
    Fire Station
Medical and Health
    Hospital
    Dispensary
    Nursing Home
    Blood Bank
    Accident and Trauma Centre
    Ambulance
	Post HQ and Police Lines
	    All Police Including State, Railway, Home Guards and Civil Defence, Traffice
	Police Chowki
	
	Socio Cultural Institutions
	    Old Age Homes
	    Art and Cultural Institutions
	    Art Galleries
	    Marriage/Banquet Halls

Recreational
    Amusement or Water Parks
        Places of Tourist Interest
        Film City
    Auditorium
    City Forest
    Club—Libraries—Govt./Public
    Distrcit Parks
    Golf Course or Race Course
    Historical Monuments
        Forts
        Place of Trourist Inerest
        View Points
        Statues
        Tombs
	Major Parks—Gardens and Parks
	Play Ground
	Regional Parks
	Religious Places
	    Temple
	    Gopuram
	    Mosques
	    Idgah
	    Gurudwara
	    Church
	    Synagogue
	    Fire Temple
	National Park/Sanctuary
	    Sports Centre
	    Sports Complex
	    Stadium (Indoor/Open)
	    Fountain

Residential
    Foreign Mission
        Embassies with their Office
    Mixed Residential
        Important Building
        Apartments/Flats/Quarters
        Govt. Colonies
        Bungalows of Chief Minister and All Ministers/Mayor
	Residential
	    Build up Area
	    House Number Information
	Gate
	Wall—Fort Boundaries

3. Land Cover
• Agriculture
    Cultivation (Irrigated/Non-irregated)
Nursery
    Farm Houses
    Farms
	• Road Circulation
	    Cart Track
	    Divider
	    Express Way—Arterial Road
	    Flyover

(contd.)

(contd.)

Vegetated Area Bound Tree Cover
- Vegetation
  - Forest
  - Reserved/Protected Forest
  - Forested Area/Wooded Area
  - Orchard
  - Plantation
- Govt. Land
  - Rented
  - Leased
  - Open
- Green Belt
  - Fallow Area
  - Green Belt
  - Poultry and Dairy Farming
- Parking and Other Open Space
  - Other Open Space
  - Parking
- Land
  - Rural Area
  - Barren Land
  - Marshy Land
  - Mud/Mangrove Swamp
- Water ways
  - Harbour
  - Jetty
- Airport
  - Domestic and International (Boundary wall and internal details will not be shown)
- Railway Line
  - Railway Line
    - Broad Gauge
  - Electrified/Single/Double/Multiple Line
    - Meter Gauge
    - Narrow Gauge
    - Metro Railway
- Railway Station
  - Railway terminal
  - Integrated Pass Terminal
  - Metro Pan Terminal
  - Goods Terminal
- Sewerage
  - Dust Bin Sanitary Landfill
- Solid Waste—Dumping ground
- Water
  - Overhead Tank
  - Water Swimming Pool

Bridges/Culverts
Level Crossing Rail/Road
Ford/Ferry Points (with periods)
Foot Path
Temporary Roads/
Bridges with Periods
Foot Bridges
Mountaineer Chock
- Route Tentative

National Highway
- With Number

State Highway—With Number
Other Roads
- Unmetalled Roads
- Lanes/Bylanes
- Subways
- One Way Road

Traffice Island/Light/Lamp Post
Traffic Restriction with Timing
- Bus Depot
  - Inter State Bus Terminus (ISBT)
  - Other Bus Depot—Bus Shops/Bus Terminals

Texi/Auto Stand
- Other Auto Stand
- Prepaid
- Truck Terminal—
  - Integrated Freight Complex—Parking
- Communication
  - Signal Point
- Drain
  - Drain—
    - Single and Double Line Drain
- Electricity
  - Lamp Post
  - Tower Pylons
- Fuel
  - CNG Station
  - LPG Distribution Centres
  - Petrol Pump—Petrol and Diesel/
    - CNG with Owner Information
  - Gas Pipeline Pole/
    - Service Station
- Water Bodies

River—Flow Direction Arrows For River
- River Island
  - Water Falls with Heights
  - Sand

Tank/Well/Springs/Reservoirs/Salt Pan

(contd.)

(contd.)

Other Tower Chimney	Water Limit Lakes Ponds Built Up Tanks Canals (Single and Double Line)
	Creek
	High Water Line Low Water Line Topographical Features Surveyed Tree

Note: Large Scale Maps of towns and cities falling in the restricted zone near the International boarder, Line of Control and Actual Line of control will not be available in the public domain.
Authority: Ministry of Defence order no. 2(7)/2003-D(GS-III)/Vol. V dt. 14/21st March 2006.

APPENDIX II

National Spatial Data Infrastructure : Data Format

Table 1: Description and Format of First Two Records Common to All Files

Rec. No.	*Field*	*Contents*	*Type (Fortran)*	*Format*	*Starting Byte*	*Ending Byte*	*Remarks*
1	1	Map Sheet Code	Character	A40	1	50	Top Sheet Number as per Scale e.g. 56G/12
		Filler	—	—	51	52	2 Spaces
	2	NSDE Version	Character	A6	53	58	Version of the exchange format in which data is organized e.g. 1.0.1
		Filler	—	—	59	60	2 Spaces
	3	NSDI Layer	Integer*2	I4	61	64	NSDI Main Layer, e.g. Topography-001, Geology/Mining-002 etc.
	4	NSDI Sub-Layer	Integer*2	I4	65	68	Sub-layer Classification e.g. Coastal Geomorphology etc.
	5	Agency Code	Integer*2	I4	69	72	Unique Code of Data Generating Agency
2	1	File Name	Character	A7	1	7	7-Digit file name as per figure 1.
	2	Filler	—	—	8	9	2 Spaces
	3	Volume Code	Integer*4	16	10	15	6-Gidit Code, Unique for Data Set. Provided for future use.
	4	Data Type	Integer*2	12	16	17	Indicates type of Spatial Data 1 if data is base categorywise (Vector Format) 2 if data is Layerwise (Vector Format) 3 if raster multiband image (band-wise) 4 if Single band raster image values 5 if Single band coded raster image
		Filler	—	—	18	18	1 Space
	5	Serial Number of Media	Integer*2	I6	19	24	For data on Multiple Media Only. Otherwise this value is zero
	6	Total Number of Records in the file	Integer*4	I12	25	36	
	7	Base Category/ Layer serial No.	Integer*2	I6	37	42	Corresponds to field 1 of Record number 4 to 3+p of VOLDIR. Applicable only for files DATACAT and DATAFIL. FOR other files this value is zero.
		Filler	—	—	43	4	2 Spaces
	8	Data Classification	Integer*2	I2	45	46	0 Topological Non-structred 1 Topologically Structured
		Filler	—	—	47	72	26 Spaces

Table 2: Volume Directory (Voldir) Records

Rec. No.	*Field*	*Contents*	*Type (Fortran)*	*Format*	*Starting Byte*	*Ending Byte*	*Remarks*
3	1	Availability of Data as One Single file or Multiple Files Or No. of Layers Or No. of Bands (p)	Integer*2	I6	1	6	This value is zero for single file and non-zero (p) for multiple files, where p is the total number of base categories/Layers mentioned in the volumes. Record no. 4 onwards will appear only if this value is non-zero. Also the next field of this record is meaningful only if this value is non-zero. In case of Raster data, each file will contain data for a particular band. In case of Layer data, the file will contain data pertaining to a layer.
	2	Whether data is base category-wise/layerwise/ Image bandwise	Integer*2	I6	7	12	This value is 1 if data is base categorywise (Vector Format). 2 if data is Layerwise (Vector Format) 3 if raster multiband Image (band-wise) 4 if Single band raster image values 5 if Single band coded raster image
	3	Whether File TOPOINFO exists	Integer*2	I2	13	14	Relevant only for Topographic data 1 for yes 0 for no
	4	Whether File QUALINFO exists	Integer*2	I2	15	16	Flag for existence of file containing Quality processes and parameters 1 for yes 0 for no
		Filler	—	—	17	72	46 Spaces
4 to 3+p	1	Base Category/ Layer/Band Serial Number	Integer*2	I6	1	6	Order of listing of Base Categories, Layers, Bands
	—	Filler	—	—	7	7	1 Space
	2	Whether Attribute table Exist	Integer*2	I2	8	9	0 Indicates no attribute data attached in DATAFIL, ELSE number of attribute tables associated
	—	Filler	—	—	10	10	1 Space
	3	Description of Base Categories/ Layers/Bands	Character	A50	11	60	As per Section 3. Description of Base Categories, in case of vector topographic data Description of Layers, in case of user domain Layers in vector or raster format Description of Spectral Bands in case of multiple band image data (raster).

(contd.)

(contd.)

Rec. No.	*Field*	*Contents*	*Type (Fortran)*	*Format*	*Starting Byte*	*Ending Byte*	*Remarks*
	4	Whether Base Category/Layer/ Band included in the volume	Integer*2	I6	61	66	1, if includes Otherwise 0
	5	Serial Number of Media in which the Base Category/ Layer/Band is contained	Integer*2	I6	67	72	Applicable for Data on Media only

Table 3: General Information (GENINFO) Records

Rec. No.	*Field*	*Contents*	*Type (Fortran)*	*Format*	*Starting Byte*	*Ending Byte*	*Remarks*
3	1	Agency Acronym	Character	A10	1	10	Allotted by NSDI (ref NSDI metadata document 1.1.1.1)
	2	Layer Name	Character	A30	11	40	Allotted by NSDI (ref NSDI metadata document 1.1.1.2)
	3	Layer Type	Character	A20	41	60	Allotted by NSDI (ref NSDI metadata document 1.1.1.2)
		Filler	—	—	61	61	1 Space
	2	Number of States/ UT = (m)	Integer*2	12	62	63	As per Central Heading. Mostly applicable for SOI topographic data
	3	Source Type	Integer*2	19	64	65	0—Original 1—Compiled or Aggregated... *(Add new table)*
		Filler	—	—	66	72	7 Spaces
4	1	Scale	Integer*4	19	1	9	Scale of the map
		Filler	—	—	10	11	2 Space
	2	Edition Number	Integer*2	14	12	16	Edition of the source material
		Filler	—	—	17	18	2 Space
	3	Edition Year	Integer*2	18	19	26	Edition year e.g. 2001
		Filler	—	—	27	72	46 Spaces
5 to 5+m	1	StateName	Character	A68	1	64	One state per record (ref NSDI metadata document 1.1.1.4)
	2	Number of Districts (n)	Integer*2	I4	65	68	No. of Districts in the State
	3	Number of Location points(o)	Integer*2	I4	69	72	No. of unique reference locations
5+m to 5+m+n	1	DistName	Character	A72	1	72	Districts will be listed One District per record (ref NSDI metadata document 1.1.1.5)
5+m+n to 5+m +n+0	1	LocPtName	Character	A72	1	72	One Location name per record (ref NSDI metadata document 1.1.1.6)

(contd.)

(contd.)

Rec. No.	*Field*	*Contents*	*Type (Fortran)*	*Format*	*Starting Byte*	*Ending Byte*	*Remarks*
6+m+ n+o	1	Surveyed Year/ Data of acquisition	Character	A60	1	60	Year(s) of Survey of the source material e.g. 1968-69 Date of Acquisition in case of Satellite raster image
		Filler	—	—	61	68	8 Spaces
	2	Security classification of dataset	Integer*2	14	69	72	0 Not Classified 1 Classified
7+m+ n+o	1	Data Type Code	Integer*2	I2	1	2	1 if data is base categorywise (Topographic Vector) 2 if data is Layerwise (Vector Format) 3 if raster multiband image (band-wise) 4 if Single band raster image values 5 if Single band coded raster image
	—	Filler	—	—	3	8	6 Spaces
	2	MapProjection	Character	A4	9	12	POLY—Polyconic, UTM LAMB—Lambert conical orthomorphic SOM—Space Oblique Mercater Any Other
		Filler	—	—	13	14	2 Spaces
	3	Units	Character	A6	15	20	Units of measure for ground planimetric co-ordinates. Its Value is METERS
		Filler	—	—	21	22	2 Spaces
	4	Grid Size (Resolution)	REAL*4	F8.2	23	30	The true ground distance in meters
	5	Number of Control (Tick) Points (q)	Integer*2	I6	31	36	Usually 4 corners of the sheet for Topographic data Could be more fr User Domain data
		Filler	—	—	37	72	36 Spaces
8+m+ n+o	1	CentralLongitude	Real*4	F12.5	1	12	Projection Parameters
	2	CentralLatitude	Real*4	F12.5	13	24	Projection Parameters
	3	StandardParallel 1	Real*4	F12.5	25	36	Projection Parameters
	4	StandardParallel 2	Real*4	F12.5	37	48	Projection Parameters
	5	ScaleFactor	Real*4	F6.4	49	54	Projection Parameters
		Filler	—	—	55	56	2 Spaces
	6	Spheroid	Character	A8	57	64	EVEREST, WGS84, WGS72 etc.
		Filler	—	—	65	72	8 Spaces
9+m+ n+0	1	OriginLatitude	Real*4	F12.5	1	12	Projection Parameters
	2	OriginLongitude	Real*4	F12.5	13	24	Projection Parameters
	3	OriginEasting	Real*4	F12.2	25	36	Projection Parameters
	4	OriginNorthing	Real*4	F12.2	37	48	Projection Parameters
		Filler	—	—	49	72	24 Spaces

(contd.)

(contd.)

Rec. No.	*Field*	*Contents*	*Type (Fortran)*	*Format*	*Starting Byte*	*Ending Byte*	*Remarks*
10+m+n+o to 10+m+n+o+q	1	Control (Tick) point Label	Integer*2	I4	1	4	1 = SW/LL, 2 = SE/LR, 3 = NE/UR, 4 = NW/UL, 999 = Reserved for Identifying the origin of Raster Image (For Raster data) 5 to 998 can be used for identifying other control (Tick) points within the sheet/dataset
	2	Raster Image Orientation	Character	A2	5	6	Relevant for Control (Tick) Point Label = 999 (i.e. raster data) BL = Indicates raster image origin at Bottom left and rasterization is bottom up (row/scan lines) and Left-right (Columns/Pixels) TL = Indicates raster image origin at Top left and rasterization is Top-down (row/scan lines) and Left-right (Column/Pixels)
	3	Latitude	Real*4	F12.6	7	18	In Degrees of Tick Points
	4	Longitude	Real*4	F12.6	19	30	In Degrees of Tick Points
	—	Filler	—	—	31	36	6 Spaces
	5	X Co-ordinate	Real*4	F12.6	37	48	Projected in ground terms in meters
	6	Y Co-ordinate	Real*4	F12.6	49	60	Projected in ground terms in meters
	7	User Tick Label	Character	A10	61	70	User defined label for Tick point
	—	Filler	—	—	71	72	2 Spaces
11+m+n+o+q	1	O.S. Name	Character	A18	1	18	Specifies name of source Operating System
	2	GIS Package	Character	A18	19	36	Specifies name of source GIS Software
	3	DBMS Name	Character	A18	37	54	Specifies name of source Database
	—	Filler	—	—	55	72	18 Spaces

Table 4: Quality Information (QUALINFO) Records

Rec. No.	*Field*	*Contents*	*Type (Fortran)*	*Format*	*Starting Byte*	*Ending Byte*	*Remarks*
3	1	Process Citation	Character	A69	1	69	General statement on data generation processes
	2	No. of Process stages (m)	Integer*2	13	70	72	Number of distinct process stages
4 to 4+2*m	1	Process Description	Character	A62	1	62	Processes description
		Filler	—	—	63	64	2 Spaces
	2	Process Year	Integer*4	18	65	72	Process/Source Data Year
	1	PAccHorX	Real	F12.2	1	12	Positional accuracy in X direction
	2	PAccHorY	Real	F12.2	13	24	Positional accuracy in Y direction
	3	PAccVer	Real	F12.2	25	36	Positional accuracy in Z direction
	4	TAcc	Real	F12.2	37	48	Thematic accuracy in %
	5	TAccConf	Real	F12.2	49	60	Confidence level for thematic accuracy
	6	TAccArea	Real	F12.2	61	72	Thematic Area accuracy in %
5+2*m	1	PACCLayer* (horizontal)	Real4	F12.2	1	12	Cumulative RMS Horizontal Positional accuracy (ref. NSDI metadata document 1.1.1.9)
	2	PACCLayer* (vertical)	Real	F12.2	13	24	Cumulative RMS vertical Positional accuracy (ref. NSDI metadata document 1.1.1.9)
	3	TACCLayer*	Real4	F12.2	25	36	Final % thematic accuracy (ref. NSDI metadata document 1.1.1.9)
	4	TACCLayer*(area)	Real4	F12.2	37	48	Final thematic area accuracy
		Filler	—	—	49	72	24 spaces

Table 5: General Topographic Information (TOPOINFO) Records

Rec. No.	*Field*	*Contents*	*Type (Fortran)*	*Format*	*Starting Byte*	*Ending Byte*	*Remarks*
3	1	Whether COMPIND Data Exists	Integer*2	I2	1	2	1 for yes, 0 for No
	2	Whether ADMIND Exists	Integer*2	I2	3	4	1 for yes, 0 for No
	3	Number of SPECIAL FOOTNOTES lines (m)	Integer*2	I2	5	6	
	4	Magnetic Variation E or W of True North	Character	A2	7	8	E = East of True North W = West of True North
	5	Annual Change in Magnetic Variation with sign	Integer*2	I4	9	12	+ for increasing, – for decreasing
		Filler	—	—	13	14	2 Spaces

(contd.)

(contd.)

Rec. No.	*Field*	*Contents*	*Type (Fortran)*	*Format*	*Starting Byte*	*Ending Byte*	*Remarks*
	6	Year of Magnetic Variation	Integer*2	I6	15	20	
	7	Copyright	Character	A24	21	44	Example: Govt. of India Copyright
	8	Year of Copyright	Integer*2	I6	45	50	Example: 1975
	9	Contour Interval	Integer*2	I6	51	56	Contour interval in meters
	8	Mean Grid North Alignment	Integer*2	2(13)	57	62	In degrees and minutes. For un-gridded sheets this field and the next are blank
		Filler	—	—	63	64	2 Spaces
	9	East and West of True North	Character	A1	65	65	E = East of True North W = West of True North
		Filler	—	—	66	67	2 Spaces
	10	Name of Grid	Character	A5	68	72	User Defined
4	1	Authority	Character	A64	1	64	Name of Authority under direction the data is published/generated
	2	Admn. Boundary Verification Remark	Integer*2	I2	64	66	1 = Adm. Boundaries are verfied 0 = Adm. Boundaries are not verified from appropriate Authority
	—	Filler	—	—	67	68	2 Spaces
	3	Triangulated heights and contours adjusted with spirit-leveled heights	Character	A1	69	69	Y = Adjusted N = Not Adjusted
	—	Fillers	—	—	70	71	2 Spaces
	4	Territorial water upto 12 Nautical miles from the coastline	Character	A1	72	72	Y in case of data having coastline; N in case of data not having coastline.
5to5+m	1	SPECIAL FOOTNOTES RECORES	Character	A72	1	72	

Table 6: Data Category (DATACAT) Records

Rec. No.	*Field*	*Contents*	*Type (Fortran)*	*Format*	*Starting Byte*	*Ending Byte*	*Remarks*
3	1	Base Category/ Layer/BAnd Serial Number	Integer*2	I6	1	6	Order of listing of Base Categories, Layers, Bands (Corresponds to field 1 of records 4 to 3+p ob Table 2)
	2	Number of major categories/Layer Categories (m)	Integer*2	I6	7	12	• Number of Base categories in case of Topographic Data • Number of Layer Categories in case of user domain thematic data in vector or raster form • Not relevant in case of Image data

(contd.)

(contd.)

Rec. No.	*Field*	*Contents*	*Type (Fortran)*	*Format*	*Starting Byte*	*Ending Byte*	*Remarks*
	3	Number of tables(nt)	Integer*2	I6	13	18	• Number of attribute tables attached to the layer
	—	Filler	—	—	19	72	54 Spaces
4 to 3 +m	1	Category Code	Integer*2	I2	1	2	Major Category/Layer Category Code
	—	Filler	—	—	3	8	6 Spaces
	2	Category Name	Character	A64	9	72	Full category name as per Appendix A Major Category/Layer category Name/Bandwidth
4+m	1	Number of Nodes (nn)	Integer*4	I8	1	8	Set to Zero for Raster Data
	2	No. of Lines (nl)	Integer*4	I8	9	16	Set to Zero for Raster Data
	3	Number of Areas (na)	Integer*4	I8	17	24	Set to Zero for Raster Data
	4	No. of Texts (nx)	Integer*4	I8	25	32	Set to Zero for Raster Data
	5	No. of Rows (nR)	Integer*4	I6	33	38	Set to Zero for Raster Data
	6	No. of Columns (nC)	Integer*4	I6	39	44	Set to Zero for Raster Data
	7	No. of Pixel Bytes (nBytes)	Integer*2	I2	45	46	Set to Zero for Vector Data, relevant to Image, Grid data 1. Single byte, binary for images 2. Two byte, binary for images/noded grids 4. Four byte, binary for DEM
	8	Byte Order (depending upon architecture)	Character	A2	47	48	II/MM for more than one Byte meta II indicates Intel, Big Endian architecture MM indicates Motorola, Little Endian Architecture
	9	Record Type	Character	A6	49	54	RASTER or VECTOR
	10	Format of Value Records	Character	A10	55	64	Content would be LOGICAL*1 for Single byte, binary images INTEGER*2 for two byte, binary images/coded grids REAL*4 for four byte, binary DEM data (as per the details in field 7 above)
		Filler			65	72	8 spaces

Table 7: Raster (Grid) Data Records

Rec. No.	*Field*	*Contents*	*Type (Fortran)*	*Format*	*Starting Byte*	*Ending Byte*	*Remarks*
Value Records 3 to (nR*nC* nBytes)/ 72 + 3	1	Raster Values • Reflectance Values for image data • Elevation Values for DEM • Coded Values for Coded Raster Layer	Logical*1 or Integer*2 or Real*4 (As per record 4+m, field 7 in table 6)	Binary	1	NR*Nc *nByte	Order of Grid cell locations for Values Row 1: Col 1, Col 2,...... Col-nC Row 2:... Row-nr:....................................

Table 8: Node Records

Rec. No.	*Field*	*Contents*	*Type (Fortran)*	*Format*	*Starting Byte*	*Ending Byte*	*Remarks*
3	1	Record Type	Character	A4	1	4	NODE
	2	Element Internal Id No.	Integer*4	18	5	12	This is Unique value for each Node
	3	Co-ordinates of Node Point	Real*4	2F12.2	13	36	X, Y co-ordinate in ground meters
	4	Angle	Real*4	F12.4	37	48	Orientation of point feature with X-axis in degrees (Relevant for Topographic Data)
	—	Filler	—	—	49	72	24 Space

Table 9: Line Records

Rec. No.	*Field*	*Contents*	*Type (Fortran)*	*Format*	*Starting Byte*	*Ending Byte*	*Remarks*
3	1	Record Type	Character	A4	1	4	LINE
	2	Element Internal Id	Integer*4	I8	5	12	This is Unique value for each Line
	3	Starting Node	Integer*4	I8	13	20	Internal ID Number This refers to data element 2 of the Node records
	4	Ending Node	Integer*4	I8	21	28	-do-
	5	Number of points in the line (p)	Integer*4	I8	29	36	Number of x-y Co-ordinate pairs (excluding the end points)
	6	Height	Real*4	F8.2	37	44	Height in Meters
	7	Number of Feaure Codes or Levels (q)	Integer*2	I4	45	48	
	8	Object ID	Integer*4	I8	49	56	
	—	Filler	—	—	57	72	16 Spaces
Co-ordinate records	1 to 2p	x, y co-ordinate string	Real*4	3	(2F 12.2)1	.....	Ending byte number is as per the number of points in the line
Code records	1 to 2q	Major and Minor List or	Integer*4	6(216)	1	....	• Major and Minor Codes for topographic data • Will extend to next line as per the number of feature codes
		Layer Category Code List	Integer*4	2(136)	1	.....	• Layer category Codes for User Domain Thematic data • Will extend to next line as per the number of feature codes

Note: All POINT features included in LINE type records. A POINT feature is distinguished from a LINE feature as follows:

- A POINT feature has the same start and end node ID (Data elements 3 and 4) and
- The number of points (Data element 5) for a POINT feature is zero (0)

The key for linking with the text in the file can also be used for linking with non-graphic data outside the file.

Table 10: Area Records (Topologically Structured)

Rec. No.	*Field*	*Contents*	*Type (Fortran)*	*Format*	*Starting Byte*	*Ending Byte*	*Remarks*
3	1	Record Type	Character	A4	1	4	Area
	2	Element Internal ID No.	Integer*4	I8	5	12	This is Unique value within all areas (polygons)
	3	Co-ordinates of Area Point	Real*4	2F(12.2)	13	36	The area point (polygon label) is always within the pologon it represents
	5	Number of lines bounding the area (b)	Integer*2	I4	37	40	Number of x-y Co-ordinate pairs
	6	Number of Feature Codes (r)	Integer*2	I4	41	44	
	7	Object ID	Integer*4	I8	45	52	The key for linking with the text in the file, can also be used for linking with non-graphic data outside the file
		Filler	—	—	53	72	20 Spaces
Line ID records	1 to b	List of Line ID's	Integer*4	9(18)	1	.....	Internal ID number of lines bounding the area. Ending byte number is as per the number of lines in the area (polygon)
Code records	1 to 2r	Major and Minor Code List	Integer*4	216	1	12	• Major and Minor Codes in pairs for topographic data • Will extend to next line as per the number of feature codes
		Filler	—	—	13	18	6 spaces
		Feature Name	Character	A54	19	72	
		OR					
		Layer category Code List	Integer*4	2(136)	1	.....	• Layer category Codes for User Domain Thematic data • Will extend to next line as per the number of feature codes

Table 11: Area Records (Topologically Non-Structured)

Rec. No.	*Field*	*Contents*	*Type (Fortran)*	*Format*	*Starting Byte*	*Ending Byte*	*Remarks*
3	1	Record Type	Character	A4	1	4	Area
	2	Major Code	Integer*4	I6	5	10	Major Code of the Feature
	3	Minor Code	Integer*4	I6	11	16	Minor Code of the Feature
	6	Feature Type	Character	A52	17	68	
	5	Number of vertices (p) excluding end points	Integer*4	I4	69	72	Number of vertices
Co-ordinate records	1 to 2p	x, y coordinates of Area Points	Real*4	3(F12.2)	1	......	Ending bytes as per the number of vertices in the area feature

Table 12: Text Records

Rec. No.	*Field*	*Contents*	*Type (Fortran)*	*Format*	*Starting Byte*	*Ending Byte*	*Remarks*
3	1	Record Type	Character	A4	1	4	Text
	2	Element Internal Id	Integer*4	I8	5	12	This is Unique value for all text records
	3	Text Size	Integer*2	I4	13	16	In one hundredth of an inch
	4	Text Justification	Integer*2	I4	17	20	0 for left, 1 for right, 2 for centre justification
	5	Angle	Real*4	F12.4	21	32	Angle of Text in Degrees
	6	x, y Co-ordinates of reference Point	Real*4	2F(12.2)	33	56	
	7	Number (r) of attribute codes	Integer*2	I4	57	60	Number of x-y Co-ordinate pairs
	8	Number Text Character strings	Integer*2	I4	61	64	
	9	Object ID	Integer*4	I8	65	72	The key for linking with the object in the LINE or Area Records
	—	Filler	—	—	53	72	20 Spaces
Second	1	Text Characters	Character	A72	1	72	This record contains the text data describing Object ID
Third	1-2r	Major and Minor List	Integer*4	2I6	1	12	• Major and Minor Codes for topographic data • Will extend to next line as per the number of feature codes
		Filler	—	—	13	18	6 spaces
		Feature Name OR	Character	A54	19	72	
		Layer category Code List	Integer*4	2(136)	1	.....	• Layer category Codes for User Domain Thematic data • Will extend to next line as per the number of feature codes

Table 13: Attribute Records

Rec. No.	*Field*	*Contents*	*Type (Fortran)*	*Format*	*Starting Byte*	*Ending Byte*	*Remarks*
3	1	Type	Character	A4	1	4	Indicates attribute table type (ATTR)
	2	Attribute Table ID.	Integer	I6	5	10	Attribute table Id.
	3	Name of Attribute Table	Character	AI2	1	24	Indicates name of the attribute table as per source specification.
	4	No. of fields (af)	Integer	I6	11	16	Indicates number of columns in attribute table
	5	No. of records (ar)	Integer	I6	17	22	Indicates number of records in attribute table
	—	Filler	—	—	23	72	50 Spaces
4 to 4+ af	1	Field name	Character	A12	1	12	Represents Field Item name
	2	Field type 1	Character	A12	13	24	Definition of Field giving Width, Output, Type, number of decimal places separated by comma.
	3	Field type 2	Character	A8	25	32	Represents ANSI standard format
	4	Key-Field	Character	I2	33	34	Indicates 1 if the field is Key-indexed, ELSE 0
	—	—	—	—	35	36	2 Spaces
	5	Remarks	Character	A36	37	72	Description/Remarks of the Field
Data/ Records 5+ af to 5+af+ar	1-af	List of field values	Character	2(A36)	1	....	• List of field data values • Will extend to next line as per the number of filled values in attribute records

Note: Records numbered 3 onwards... will be repeated as many times as the number of attribute tables indicated by number nt in datacat file.

APPENDIX III

List of GIS Vendors and Related Software Developers

- ACE Software *from PCI—USA*
- ARGUS Technologies (http://www.argusmap.com/)
- Autodesk *AutoCAD Map* (http://www3.autodesk.com/adsk/section/0,,693177-123112,00.html)
- Avenza (http://www.avenza.com/)
- Bentley Systems (http://dn1.bentley.com/products/geproducts) *Microstation Geographics*
- Blue Marble Geographics (http://www.bluemarblegeo.com/) *GeoView and other tools*
- Caliper (http://www.caliper.com/) *Maptitude, TransCAD, GIS+*
- DeLorme (http://www.delorme.com/)
- Digital Land Systems Research (http://www.dlsr.com.au/welcome.html) *SAGE*
- EarthResource Mapping (http://www.ermapper.com/) *ER Mapper*
- EPPL7 (http://www.mnplan.state.mn.us/EPPL7/)
- ESRI (http://www.esri.com/) *Arc/Info, ArcView, ArcCAD, SDE, MapObjects*
- ESRI (http://www.esri.ca/) *Canadian Distributor*
- Etak (http://www.etak.com/) *Digital Mapping*
- Genasys (http://www.genasys.com/)
- Geodyssey Limited of Calgary (http://www.cuug.ab.ca/~russellj) *Hipparchus*
- GEOsurv Inc.—10 Years (http://infoweb.magi.com/~geosurv/)
- Geosystems (http://www.geosys.com/) *GeoLocate*
- GRASS (http://www.cecer.army.mil/grass/GRASS.main.html)
- IBM's GIS (http://giswww.pok.ibm.com/gishome.html) *Several new software products*
- IDRISI (http://www.idrisi.clarku.edu/)
- Innogistic Software plc. (http://www.innogistic.co.uk/)
- INTERGRAPH (http://www.intergraph.com/) *MGE, FRAMME*
- INTERMAP (http://www.intermaptechnologies.com)
- MacGIS (http://shiva.uoregon.edu/macGIS/macGISHomePage.html)
- Mapcom (http://www.mapcom.com/) *M4*
- MapInfo (http://www.mapinfo.com/)
- MicroImages (http://tnt.microimages.com/) *TNT Familiy of products*
- Northwoood Geoscience LTD (http://www.northwoodgeo.com/)
- PCI (http://www.pci.on.ca/) *Remote Sensing Technologies*
- Planet One Software (http://www.planetonegis.com/)
- Ramas (http://www.ramas.com/) *Ecological Models*
- Scan/US (http://www.scanus.com/)
- Sedona Geo-Services (http://www.sedonageo.com/)

- SHL System House (http://www.gis.shl.com/) *Vision Solutions*
- Siemens-Nixdorf (http://www.mch.sni.de/public/sicad/sicad_us/sicad.htm) *SICAD, GDB-X, WinCAT*
- Smallworld (http://www.smallworld.co.uk/)
- SST Development Group (http://www.sstdevgroup.com/) *Consultants and Farm Managers*
- Star Informatic (http://www.star.be/) *Star Carto, Star Viewer*
- Tactician Corporation (http://www.tactician.com/)
- TEXO GIS (http://www.popnet.it/TEXO/texo_eng.html?)
- Thinkspace (http://www.thinkspace.com/) *Map Factory*
- TYDAC *SPANS* (http://www.tydac.com/)
- Universal Systems (http://www.universal.ca/) *CARIS*
- Vision International (http://www.gis.shl.com/) SQS
- Wolleysoft (http://www.woolleysoft.co.uk/) landscape visualization software

APPENDIX IV

GIS Software Applications

At GeoPlan Center there was an attempt to make an effort to show people what their GIS software choices are. This Appendix will list any GIS software information which the Center has come across. They used quite a number of them at the Center, but they were interested in discovering what else is out there. If anybody know of any that aren't listed here, please write about them (e-mail us). According to the GeoPlan Center:

- We tried to list only the general-use GIS software here. There are lots of GIS software that are made for certain professions, such as hydrologic scientists, petroleum industry specialists, etc. that aren't listed here.
- Note that most of the software descriptions were written by the Software Makers/ Distributors.
- You may be able to get some of this software for less than indicated by the prices listed here. In all cases we listed the commercial price, not the educational/student, or similar discounted prices.

ArcCad
ESRI
www.esri.com
PLATFORM: Windows (95, NT)
"ESRI's ArcCAD software gives you full geographic information system (GIS) functionality within AutoCAD. ArcCAD provides powerful mapping, data management, and display tools that work directly with AutoCAD software's design and drafting tools. ArcCAD users can create, access, and exchange ARC/INFO data, the *de facto* standard for local, state, and national GIS installations."

ArcExplorer
ESRI
www.esri.com
PLATFORM: Windows
"ArcExplorer software is a lightweight GIS data explorer. ArcExplorer can work on its own with local data sets or as a client to Internet data and map servers. ArcExplorer is built using MapObjects, ESRI's powerful collection of GIS and mapping components for application developers. As a stand-alone application, ArcExplorer is a complete data explorer, allowing users to display and query a wide variety of standard data sources."

ArcFM
ESRI
www.esri.com
PLATFORM: Windows (NT)
"Arc Facilities Manager (ArcFM) is a powerful ARC/INFO-based application for the editing, maintenance, modelling, and data management of utility information. ArcFM is all-relational utility application software developed using ARC/INFO software's Open Development Environment and object-oriented programming."

Arc/Info
ESRI
www.esri.com
PLATFORM: Most Unix platforms, Windows (NT)
"Professional GIS offers advanced modelling and analysis capabilities"

ArcView Business Analyst
ESRI
www.esri.com
PLATFORM: Windows (3.51, 95, 98, NT)
"ArcView Business Analyst allows you to focus on getting the data analysis you need, rather than worrying about the details of the underlying technology. A wizard interface assists you with tasks ranging from basic, such as mapping customers, to complex, such as performing analyses or creating what-if scenarios."

ArcView GIS
ESRI
www.esri.com
PLATFORM: Most Unix platforms, Windows (95, 98, NT)
"The leading mapping and GIS software for the desktop."

Atlas GIS
ESRI
www.esri.com
PLATFORM: Windows (95, 98, NT)
"Business professionals worldwide rely on Atlas GIS to help them make decisions that improve the efficiency and profitability of their organizations. Atlas GIS is both powerful enough to perform sophisticated desktop mapping and intuitive enough that nontechnical professionals feel comfortable using it."

AutoCAD Map 2000
Autodesk
www.autodesk.com
PLATFORM: Windows (95, 98, NT)
"AutoCAD Map 2000 is the Autodesk software optimized for precision mapping and geographic

information system (GIS) analysis in the AutoCAD® environment. It provides the GIS analysis tools and productivity improvements you need to easily create, maintain, and produce maps and geographic information."

Autodesk MapGuide
Autodesk
www.autodesk.com
PLATFORM: Windows (95, NT)
"Autodesk MapGuide software delivers live maps and linked databases directly to your desktop, via your intranet or the Internet—anytime, anywhere."
Autodesk MapGuide Viewer
(A PC and a Java Edition)

Autodesk
www.autodesk.com
PLATFORM: Macintosh, Sun Solaris, Windows PC
"Works with your web browser and sites running Autodesk MapGuide"

Autodesk World
Autodesk
www.autodesk.com
PLATFORM: Windows
"Autodesk World 2.01 is a powerful Microsoft desktop application to integrate, analyse and display spatial data. With Autodesk World we are bringing GIS technology down to earth. It is a flexible, powerful spatial data information system that allows users to manage their data, their way. Autodesk World has an industry-standard interface that removes the complexity of GIS, while providing robust functionality."

BusinessMAP PRO
ESRI
www.esri.com
PLATFORM: Windows (3.1x, 95, NT)
"BusinessMAP PRO is an affordable mapping application that transforms information from databases, contact managers, spreadsheets, and electronic phone books to pushpin and color-filled maps. It will turn static data into meaningful images to help you understand your business better and help with sales automation and mission-critical decision making processes."

Cartalinx
Clark Labs, Clark University
www.clarklabs.org
PLATFORM: Windows (95, 98, NT)
Clark Labs' "first dedicated digitizing, database development and topological editing software product. CartaLinx is intended as a companion to a variety of popular Geographic Information System (GIS) and Desktop Mapping software products such as IDRISI, Arc/Info, ArcView, MapInfo, and so on.

CartaLinx is used to create coverages, consisting of the spatial definitions of features in vector format and associated attribute values files. These data are then typically exported to a GIS either as entire coverages or as a series of map layers. For example, CartaLinx might be used to develop a spatial frame of property parcel boundaries along with an attribute database recording the address, owner, area, property valuation, area and primary land use of each parcel. One might then choose to export land use and property valuation map layers to a GIS. Alternatively, one might wish to display and analyse the contents of a coverage using CartaLinx directly."

Data Automation Kit
ESRI
www.esri.com
PLATFORM: Windows (3.x, 95, NT)
"This software package perfectly complements ArcView GIS software and other desktop mapping software by allowing you to create fully intelligent (topological) data sets by digitizing or converting data from many formats, with functionality that previously existed only in a full ARC/INFO system."

Didger
Golden Software
www.golden.com
PLATFORM: Windows (3.51, 95, 98, NT)
"Didger is a highly accurate digitizing program that will be an invaluable addition to your software library. In seconds, Didger precisely transforms points, lines, or areas from your paper maps, graphs, aerial photos, or other plotted information to a versatile digital format you can use with your other software. You name it and Didger can handle it quickly, accurately, and usefully. And don't be fooled by the low $119 price. With Didger's multitude of features and ease-of-use, this is an unbelievable value considering the time and effort savings you will realize! You will soon wonder how you have done your job without this indispensible tool."

DLGV32
USGS
DLGv32 Page
PLATFORM: Windows 95/NT
"Windows 95/NT software for viewing a variety of digital cartographic data. It was originally developed in support of USGS Digital Line Graph (DLG) production activities. This software allows preview and evaluation of USGS data. It contains no editing capabilities, and is not a substitute for commercial Geographic Information System (GIS) software."

EPPL7
Minnesota Planning
EPPL7 Page
PLATFORM: Windows 95
"EPPL7 is a fast but powerful raster based GIS with a well stocked toolbox. EPPL performs a variety of complex map algebra functions as well as Boolean operations. There are powerful

filtering algorithms in EPPL7 as well as commands which perform interpolation, viewshed, slope, and aspect analysis. Users can view, query, and print data."

ERDAS IMAGINE Advantage
ERDAS
www.erdas.com
PLATFORM: Windows (95, NT), Most Unix platforms
A subset of ERDAS IMAGINE Professional
"Analyse data from your imagery via image mosaicking, surface interpolation, and advanced image interpretation and orthorectification tools."

ERDAS IMAGINE Essentials
ERDAS
www.erdas.com
PLATFORM: Windows (95, NT), Most Unix platforms
A subset of ERDAS IMAGINE Advantage and ERDAS IMAGINE Professional
"Our basic map production system enables you to produce quality maps through image visualization and analysis, information integration and geocorrection."

ERDAS IMAGINE Professional
ERDAS
www.erdas.com
PLATFORM: Windows (95, NT), Most Unix platforms
"Quick image testing and modelling allows you to automatically and quickly generate broad land cover maps. Perform graphical spatial data modelling and integrate radar imagery."

ERDAS MapSheets
ERDAS
www.erdas.com
PLATFORM: Windows
"All that work you put into gathering and analysing geographic data... without a quick and simple way to visually present the information to your boss, your client or your prospective customer.
Because its fully Microsoft Office Compatible, MapSheets gives you the same look and feel of other Microsoft applications, as well as the power to create a convincing PowerPoint presentation, an effective Word report or a compelling Web page."

ERDAS MapSheets Express
ERDAS
www.erdas.com
PLATFORM: Windows
"All that work you put into gathering and analysing geographic data... without a quick and simple way to visually present the information to your boss, your client or your prospective customer.

Because its fully Microsoft Office Compatible, MapSheets gives you the same look and feel of other Microsoft applications, as well as the power to create a convincing PowerPoint presentation, an effective Word report or a compelling Web page."

ER Mapper
Earth Resource Mapping
www.ermapper.com
PLATFORM: Windows, Unix
"Used in more than 142 countries worldwide and supported by 514 ER Mapper resellers around the globe, ER Mapper is the world's most popular desktop integrated image processing software."

ER Viewer
Earth Resource Mapping
www.ermapper.com
PLATFORM: Windows, Unix
"ER Viewer is an easy to use image viewer featuring interactive roaming and zooming with very large image files."

Geographic Calculator
Blue Marble Geographics
www.bluemarblegeo.com
PLATFORM: Windows
"The Geographic Calculator converts individual coordinates, point databases, and map files from virtually any coordinate system, datum, and map projection to any other."

Geographic Explorer
Blue Marble Geographics
www.bluemarblegeo.com
PLATFORM: Windows (95, NT)
"The Geographic Explorer extends the Windows 95 and NT Explorer to view, translate, and find map files. The Geographic Explorer provides a number of powerful utilities. You can explore map files directly on your desktop with the Geographic Explorer."

Geographic Tracker
Blue Marble Geographics
www.bluemarblegeo.com
PLATFORM: Windows
"The Geographic Tracker delivers all of the power and utility of GPS to the Windows environment by allowing you to link a GPS receiver to other Microsoft Windows applications through standard Windows DDE messaging.
This standard "plug-and-play" GPS interface component supports most of the leading GPS receivers on the market. Simply plug your GPS receiver into the serial port of a computer running Windows and the Geographic Tracker will GPS-enable your applications."

Geographic Transformer
Blue Marble Geographics
www.bluemarblegeo.com

GEMI Store
www.gemistore.com
PLATFORM: Windows
$799
upgrade: $299
"The Geographic Transformer allows you to easily establish an "image-to-world" relationship between image and map coordinates and reproject an image into a georeferenced image map."

Geographic Translator
Blue Marble Geographics
www.bluemarblegeo.com
PLATFORM: Windows
"The Geographic Translator supports comprehensive data and coordinate system translation of map files in the AutoCAD DWG & DXF, Microstation DGN, MapInfo MIF & TAB, and ESRI Shape formats. "Hands free" batch mode operation is just one compelling reason to add this new Blue Marble tool to your GIS toolbox."

Geomatica FreeView
PCI Geomatics
http://www.pcigeomatics.com
PLATFORM: Windows 95, 98, NT, 2000
"The Geomatica[tm] FreeView includes tools for fast image roaming and magnification, enhancements, numeric value display, and attributes table display."

Geomedia
Intergraph
http://www.intergraph.com
PLATFORM: Windows
"GeoMedia is your information integrator, serving as a visualization and analysis tool and a platform for custom GIS solutions. A universal GIS client, GeoMedia's live data connections allow you to maintain GIS data in native repositories, and ensure that your data is always up-to-date. You can integrate data automatically using on-the-fly coordinate transformation and feature definition. Perform simple and complex spatial analysis, such as incident and proximity analysis, to support decision-making activities."

Geomedia Viewer
Intergraph
http://www.intergraph.com
PLATFORM: Windows

"Create Thematic Maps... View and anlyze data in either Microsoft Access or Arc/View Shape format... Load a satellite raster image and easily manipulate your map with zoom-in, zoom-out capability... it's FREE!!!"

GRASS (Geographic Resources Analysis Support System)
U.S. Army Construction Engineering
Research Laboratories (USACERL)
www.cecer.army.mil

Baylor University (Official
Headquarters for GRASS support,
research and development-get GRASS here)
www.baylor.edu/~grass/
PLATFORM: Linux, Sun Solaris, Silicon Graphics Irix, HP-UX, DEC-Alpha and Windows 95/98/NT
GRASS is a raster-based GIS, vector GIS, image processing system, graphics production system, data management system, and spatial modelling system.
"The Geographic Resources Analysis Support System (GRASS) is a public domain image processing and Geographic Information System (GIS) that was initially developed by the US Army Construction Engineering Research Laboratories (USACERL) during the late 1980s and early 1990s. USACERL stopped developing GRASS, the last full USACERL release was in 1993. Many organizations have taken the public domain GRASS software code and enhanced its functionality beyond what was taken the public domain GRASS software code and enhanced its functionality beyond what was possible with federal dollars. Baylor University, Blackland Research Institute, GPZ Technologies, Logiciels et Applications Scientifiques, (LAS) Inc., and the National Oceanic and Atmospheric (NOAA) Geophysical Data Center have all released various public domain and commercial versions of GRASS on a variety of computer platforms."

Idrisi
Clark Labs, Clark University
www.clarklabs.org
PLATFORM: Windows
"IDRISI is the industry leader in raster analytical functionality covering the full spectrum of GIS and Remote Sensing needs from database query, to spatial modelling, to image enhancement and classification. Special facilities are included for environmental monitoring and natural resource management, including change and time series analysis, multi-criteria and multi-objective decision support, uncertainty analysis (including Bayesian and Fuzzy Set analysis) and simulation modelling (including force modelling and anisotropic friction analysis). Yet, despite the highly sophisticated nature of these capabilities, the system is as easy to use as it is inexpensive to buy."

Label-EZ
MapText, Inc.
www.maptext.com
PLATFORM: Windows 95, 98 and NT, UNIX
"A new productivity tool from MapText, Inc., Label-EZ places text automatically, without any human intervention, eliminating the need for manual text placement.The current version of the software works seamlessly with map data from: ArcInfo, ArcView, and MapInfo."

LandForm C3
GEMI Store
www.gemistore.com
PLATFORM: Windows
"LandForm C3 allows users to fly through terrain data in real-time. It runs on Windows based desktop and laptop PCs, and reads many file types including DEM, DTED, TIFF, GeoTIFF, and others. LandForm has a unique capability to load DTED files automatically, based upon viewpoint and create a seamless land surface, at user-defined resolution. Aerial photographs, satellite images, and digital maps can be overlayed on the 3-D terrain. The viewpoint can be dynamically controlled by driving with the mouse, or the viewpoint can follow a flight track or ground track. At $995, this is the first product in its price range which allows users to combine topographic data with maps, satellite, aerial imagery, GPS and flight track data. Further, the program allows users to fuse these data into 3-D VRML models, which can be posted on Internet or intranet Web sites and viewed by Web browsers. The program was designed to be easy to use."

Manifold System
manifold.net
http://www.manifold.net
PLATFORM: Windows
"Manifold is a software package that combines database and mapping functionality. Manifold also includes a variety of statistics, logic, mathematics, and networking capabilities for use on maps, on databases, or all together at the same time."

MapInfo Professional
MapInfo
www.mapinfo.com/
PLATFORM: Windows, Macintosh, Sun, HP
"MapInfo Professional is the world's leading desktop mapping software that helps you see patterns and trends in your data. Trends that may otherwise be impossible to identify."

MapInfo ProViewer
MapInfo
www.mapinfo.com/
PLATFORM: Windows
"MapInfo's easy viewing tool designed for MapInfo Professional users. With ProViewer, you

can share your MapInfo Professional maps and analyses with clients and co-workers, who may easily view and manipulate your creations even if they do not have MapInfo Professional."

Maplex
ESRI
www.esri.com
PLATFORM: Windows (NT)
"Maplex is a fully automated cartographic name placement software. Name placement can be one of the most time-consuming aspects of map production, often requiring considerable manual intervention to ensure that names do not overlap and are clearly associated with the features they annotate. Maplex provides the means to overcome this bottleneck in map production."

MapMarker and
MapMarker Plus
MapInfo
www.mapinfo.com/
PLATFORM: Windows
They both utilize "street maps to allow you to visually display all your geocoded data. MapMarker ships with MapMarker Streets—a five layer street display product based on TIGER data from U.S. Census Bureau. MapMarker Plus ships with MapMarker Plus Streets—a five-layer street display product based on data from GDT, Inc."

Maptitude
GEMI Store
www.gemistore.com
"MAPTITUDE Geographic Information System combines sophisticated geographic analysis capabilities, a powerful DBMS engine, and a tremendous amount of data all for $395. Maptitude lets you edit and update your maps, import maps from other products, or create your own maps from scratch. The MapWizard utility lets you create thematic maps with a single click of the mouse. Anyone can make beautiful maps with Maptitude. It includes two CD ROMs full of detailed maps and demographic data including every street in the U.S. It also has maps and data for the rest of the world with somewhat less detail. MAPTITUDE allows you to display and edit satellite images and digital orthophotos."

MAPublisher
Avenza
www.avenza.com
PLATFORM: MacOS PowerPC, Windows (95, 98, NT)
"MAPublisher 3.5 is a suite of Xtras for Macromedia FreeHand and Plug-ins for Adobe Illustrator 8.0.1 that bridges Geographic Information System (GIS) technology with high-end graphics software for high resolution printing and electronic publishing technology."

MAPublisher GPS
Avenza
www.avenza.com
PLATFORM: Windows (95, 98, NT)
"MAPublisher GPS allows you to publish your data quickly, easily and intuitively while speaking to a GPS receiver through a satellite/radio network from the Illustrator graphics environment. This data can efficiently be made available on the internet/intranet by using Avenza's pdfPLUS technology."

MAPublisher LT
Avenza
www.avenza.com
PLATFORM:MacOS PowerPC, Windows (95, 98, NT)
"MAPublisher Lt for Adobe Illustrator and Macromedia FreeHand imports raw GIS data files from Mac, PC or UNIX systems directly into the high end graphics environment with all attributes intact. MAPublisher Lt lets you import vector data (shapes, boundaries, routes, elevations, etc.) from MapInfo, ARC/INFO, ArcView, AutoCAD and others with all of the drawings data attributes intact. MAPublisher Lt incorporates GIS object management tools such as feature selection and query logic tools for easy data selection. Use MAPublisher Lt to GEOenable your websites or for simply importing your GIS/CAD files into high end graphics programs."

Map Viewer
Golden Software
www.golden.com
PLATFORM: Windows (95, 98, NT)
"MapViewer is a thematic mapping package that creates full-color maps by linking worksheet data to areas or points on a map. MapViewer includes numerous boundaries and data files for the United States and World to get you started, or you can create your own boundaries for use with MapViewer. By placing your data in the full-featured MapViewer worksheet, you can easily link your data to areas or points on the map, and generate instructive, publication quality maps in virtually no time. With the numerous map types, overlays, and drawing tools you will create the most informative maps possible."

Microdem
Bureau of Land Management
BLM ftp download site
"A program to manipulate gridded digital elevation models (DEMs), satellite imagery, and vector map data like DLG and TIGER."

MicroStation GeoGraphics
Bentley Systems, Incorporated
www.bentley.com
PLATFORM: Windows (3.1x, 95, NT), DOS. Releases are planned for Intergraph CLIX,

Hewlett Packard HP/UX, Sun Solaris, DEC Alpha NT, Silicon Graphics, IBM RS6000, and Macintosh Power

"MicroStation GeoGraphics is a comprehensive mapping toolbox that bridges the gap between CAD and GIS. MicroStation GeoGraphics combines MicroStation's industry-standard data capture and editing tools with a comprehensive database interface and a powerful spatial analysis engine. MicroStation GeoGraphics integrates seamlessly with MicroStation 95, and provides tools to input, validate, manage, analyse, and visualize geographical information. MicroStation GeoGraphics also has the ability to manage a diverse combination of vector data, raster data, and non-spatial attributes."

MrSID Stand Alone Viewer
LizardTech
www.lizardtech.com/
PLATFORM: Win95/NT, Macintosh & Unix(beta)
"View MrSID images on your personal computer."

NCAR Graphics
University Corporation for Atmospheric Research
ngwww.ucar.edu
PLATFORM: UNIX
"NCAR Graphics software continues to evolve its integrated environment for data processing and scientific visualization. Many tools come bundled with the distribution to perform such functions as viewing and editing metafiles, creating animations, converting between raster formats, resizing and compressing raster images, and zooming on images."

OziExplorer
ESRI
www.powerup.com.au/~lornew/oziexp.html
PLATFORM: WIN 95/98/NT4/NT5/WIN2000
"OziExplorer GPS Mapping Software will work with Magellan, Garmin, Lowrance and Eagle GPS receivers for the upload/download of data and most brand of GPS receivers for on screen tracking of position (Moving Map)."

PC Arc/Info
ESRI
www.esri.com
PLATFORM: DOS, Windows (3.x, 95, 98, NT)
"PC ARC/INFO is a series of six integrated software modules that combine fundamental geographic information system (GIS) tools and utilities for cartographic design and query, data entry and editing, data translation, polygon overlay and buffering, and network analysis and modelling."

pdfPLUS
Avenza
www.avenza.com
PLATFORM: MacOS PowerPC, Windows (95, 98, NT)
"pdfPLUS (formerly known as JAMBuddy) is an Adobe Acrobat 4.0 plug-in that allows you to read important database information embedded within a special PDF file. Save your MAPublisher creations and keep the data attributes alive and accessible for use on an intranet or the internet. Point and click on anything in the PDF and discover the hidden information. Query your data and watch pdfPLUS find the qualifiers."

Philcarto
Philippe WANIEZ
http://perso.club-internet.fr/philgeo/
PLATFORM: Macintosh or Power Macintosh
"Philcarto is a classical mapping software for spatialy agregated data (like census data). The input map is an Adobe Illustrator® file. The output map is an Adobe Illustrator® file too, with the same scale than the input map. Points and areas (polygons) are handled by the program. The data is a text-tabulated file containing an identifier, a name, and columns as variables. The different types of maps are:

- choropleth map
- proportional circles map
- proportional circles map coloured by a second variable
- proportional half-circles map
- choropleth associated with proportional circles map
- dot map
- links map."

Philexplo
Philippe WANIEZ
http://perso.club-internet.fr/philgeo/
PLATFORM: Macintosh or Power Macintosh
"Philexplo is a new software for cartographic exploratory data analysis. The purpose is to explore data and map using visualization tools. Data and map files are the same as with Philcarto. You can explore the statistical displays (histograms, box plots, bivariate plots and triangular diagrams) using tools like arrow, knife, lasso, etc.... The result (a selection of spatial units) can be stored in pict file or exported into a texte file compatible with Philcarto (to obtain a better cartographic output)."

Quad Companion
GEMI Store
www.gemistore.com
"Quad Companion was designed specifically to solve the problem of clipping collars and seaming DRGs together. Quad Companion clips off collars and seams DRGs together on-the-fly, just specify the region you are interested in and let the software do the rest."

QUADConvert
Applied Global Technologies, Inc.
www.appliedglobal.com/frames.htm
PLATFORM: DOS, Windows (3.1x, 95, NT), UNIX
Conversion Software for LTWG QUAD Format Satellite Scenes

R2V
Able Software Corp.
www.ablesw.com
PLATFORM: Windows (95/NT)
"The most advanced raster to vector conversion software for automated map digitizing and GIS data capture applications. R2V for Windows 95 & NT supports fully automatic vectorization and interactive line tracing from scanned maps and drawings. R2V inputs bi-level, grayscale and colour images in TIFF, GeoTIFF or BMP formats, exports vector data to ArcView Shapefile, Arc/Info Generate, MapInfo MIF/MID, AutoCAD DXF and MapGuide SDL formats. R2V supports complete on-screen image and vector editing, labelling, geo-referencing using user specified control points, image rubber sheeting, colour image classification and separation, text symbol detection and recognition (OCR), automatic polygon layer creation, 3D DEM and Grid creation, 3D display and image draping."

SAGE GIS for DOS Suite
DLSR (Digital Land Systems Research)
www.dlsr.com.au
PLATFORM: DOS
"The SAGE GIS for DOS system is a suite of three DOS programs which make up a functional and powerful GIS system. The system is comprised of: SAGE GIS (raster-based), SAGE Capture and SAGE Cartographer."

SAGE GIS for Windows
DLSR (Digital Land Systems Research)
www.dlsr.com.au
PLATFORM: Windows (95/NT)
"SAGE GIS support the analytics found in the SAGE GIS for DOS system, but with greatly improved speed and ease of use—perfect for training, teaching and consulting work."

Surfer
Golden Software
www.golden.com
PLATFORM: Windows (3.1, 95, 98, NT)
upgrade: $139
"Surfer is a contouring and 3D surface plotting program that runs under Microsoft Windows. Surfer quickly and easily converts your data into outstanding contour maps and surface plots. Virtually all aspects of your maps can be customized to produce exactly the presentation you want. Producing publication quality maps has never been quicker or easier. Both the 32-bit and 16-bit versions are included."

Terrain Professional
Maptech

GEMI Store
www.gemistore.com
PLATFORM: Windows (95, 98, NT)
"High-quality USGS topographic maps on CD-ROM. Upto 300 exportable maps per CD, at 160 dpi image resolution. Advanced features let you export and reproject full, georeferenced maps for use in GIS applications. Annotate and print customized maps. NGS (National Geodetic Survey) Data Sheets included."

Terrain Tools Freeware Edition
Softree
www.softree.com/betat e s t/t t ools.htm
PLATFORM: Windows 3.1, 95, 98 or NT
" Terrain ToolsTM—Freeware Edition is a quick and easy mapping package. Coordinates can be entered using the mouse, typed in, or imported from external files. Scanned topographic maps, air photos and other bitmap images can also be incorporated."

TNTatlas
MicroImages, Inc.
www.microimages.com
PLATFORM: Every Platform
"a free product that lets you publish and distribute your spatial materials on CD-ROM. TNTatlas CD's contain multiple versions of the software so that a single CD can be used on any popular computing platform. TNTatlas stacks are prepared with the HyperIndex feature in TNTmips."

TNTmips
MicroImages, Inc.
www.microimages.com
PLATFORM: Windows (3.1, 95, NT), Macintosh, UNIX
"TNTmips is the Map and Image Processing System, advanced software from MicroImages, Inc. for geospatial analysis. TNTmips is used in more than 70 nations around the world for geographic information analysis (GIS), advanced image processing, CAD, desktop cartography, electronic atlas preparation, and other spatial database management and visualization applications. TNTmips offers a level of data and process integration that is unmatched in any other professional system. And you can use TNTmips on PC, workstation, and Macintosh computers with exactly the same set of features and user interface on every platform. TNTmips is the most technically advanced and easy to use professional system on the market."

TNTview
MicroImages, Inc.
www.microimages.com
PLATFORM: Every Platform

"TNTview gives you the same powerful visualization and interpretation tools contained in MicroImages' industry-leading TNTmips product. TNTview is easy to learn and provides low cost visualization of your organization's project materials at sites where the users' scope of work does not include the preparation, processing, or final production of your spatial data."

WinCATS
Applied Global Technologies, Inc.
www.appliedglobal.com/frames.htm
PLATFORM: Windows
"WinCATS is the cutting edge intel tool for displaying database information over a vast host of geographic backgrounds. Advanced enough for the most complex cartographic demands, yet simple enough for the everyday software user, WinCATS is a serious mapping tool that remains user friendly."

Tcl/TK extensions
TK Geomap
www.tkgeomap.org
PLATFORM: Unix/Linux X11
Free set of extensions to Tcl/TK for map drawing.

WinCATS Viewer
Applied Global Technologies, Inc.
www.appliedglobal.com/frames.htm
PLATFORM: Windows (3.1x, 95)
The freeware version of WinCATS.

Source: www.geoplan.ufl.edu/software.html

APPENDIX V

Free GIS Software

1. **Free GIS** and CAD **Software**: geocomm.com—42k
2. **Free software**, data, and resources from ESRI ; www.esri.com/company/**free**.html—12k
3. FreeGIS.org: **freegis**.org—14k
4. **GIS Software**: www.**gis**development.net/downloads/**gi**
5. **Free GIS** mapping program **software** and resources: www.mapcruzin.com/**free_gis**.htm—22k
6. ArcExplorer — **GIS** Data Viewer: www.esri.com/**software**/arcexplorer—10k
7. EPIC 2001: **Free GIS software** for Minnesota Government: www.lmic.state.mn.us/EPPL7/EPIC—32k
8. **Free Software GIS**—Resources—PROCIG: www.procig.org/ing/resources-soft.htm—54k
9. **Free GIS Software**—GisWiki: www.**gis**.org.nz/index.php/**Free_GIS_Software**—21k
10. GRASS **GIS**—The World Leading **Free Software GIS**: grass.baylor.edu—9k
11. **Free** Maps and **GIS** Data: www.manifold.net/download/**free**maps.html—28k
12. **Free GIS Software**: www.**gis**.com/implementing_**gis/software/free_software**.html—11k
13. **Free GIS** Stuff: www.**gis**lounge.com/ll/**freegis**.shtml—23k
14. **Free GIS** Data: data.geocomm.com
15. **GIS** at CIP: www.cipotato.org/**gis**—6k
16. Christine **GIS** system: www.christine-**gis**.com—10k
17. ESRI Press: **GIS** for Everyone, Third Edition: **gis**.esri.com/esripress/display/index.cfm?...&websiteID=43&moduleID=1—15k
18. DIVA-**GIS** Homepage: www.diva-**gis**.org—8k
19. **GIS** Associations: www.reach.net/~maps/clearmaps/**free**.html—16k
20. **Free GIS** and Image Processing **Software!**: www.cof.orst.edu/cof/teach/for421/**Software**.html—6k
21. **Free** Tech Stuff: www.tenlinks.com/**free**.htm—21k
22. TNTlite for Archaeology: www.microimages.com/archaeol.htm—14k
23. SpatialNews, the leading **GIS**, LBS, and Geospatial technology news resource "spatialnews.geocomm.com/features/viewers2002 – 42k
24. **GIS** Data, Jobs, **Software**, News, and Fun—GeoCommunity: www.geocomm.com/channel/gps/**software**—49k
25. **free** downlaod **software**: www.spatialhydrology.com/**free**download.html—61k
26. GPS **Software**: www.**gis**development.net/downloads/gps—11k

27. [gislist] **Free GIS Software**—GISLIST: lists.directionsmag.com/discussion/read.php?f=20&i=34020&t=34020—78k
28. GISuser.com **GIS**, GPS, LBS, CAD, mapping news, jobs, **software**, data, community: RSS: View as XML: www.**gis**user.com
29. AGIS for Windows—simple mapping and **GIS** shareware: www.a**gis**map.com—12k
30. **GIS Software** and Resources: www.geoplan.ufl.edu/**software**.html—55k
31. **GIS Software**: home.earthlink.net/~**gis**webmaster/id14.html—20k
32. **Free Software**: www.hoslink.com/**free**stuff/**freesoftware**.htm—62k
33. **GIS** Dictionary: www.geo.ed.ac.uk/agidict/welcome.html—4k
34. GeoCommunity: The Premier Portal For Geospatial Technology Professionals: www.geocomm.com
35. USGS (U.S. Geological Survey) EROS, Sioux Falls, SD: edc.usgs.gov/geodata/public.html—8k
36. **Free GIS Software**: www.cmtinc.com/downloads/ftppc**gis**32.html—4k
37. **Free GIS Software**: www.cmtinc.com/downloads/pc**gis**.html—5k
38. **software**: www.yale.edu/**gis**/soft.htm—12k
39. Companies—Specializing—**Free Gis Software Software**: www.soft411.org/companies/**free-gis-software**.html—37k
40. Official GRASS **GIS** Homepage—The World Leading **Free Software GIS**: www.geog.uni-hannover.de/grass/index.php—9k
41. **Free GIS** Data Viewers: www.**gis**lounge.com/ll/**free**dataviewers.shtml—25k
42. Geographic information system—Wikipedia, the **free** encyclopedia: en.wikipedia.org/wiki/Geographic_information_system—66k
43. Open Directory—Science: Social Sciences: Geography: Geographic Information Systems : **Software**: dmoz.org/Science/Social_Sciences/Geography/**...**/**Software**—24k
44. **GIS** and Spatial Data Links: docs.lib.duke.edu/maps/guides/**gis**.html—28k
45. **GIS software**,arcgis,gwn-cogo,gwn-dtm,gwn-surf,mapviewer—Environmental **Software**: sci**software**.com/environmental_**software**/index.php?cPath=26_66&.
46. **Free** & Open Source **Software** Portal:**Software**/Productivity Tools/**GIS**—Geographical Information System: unesco.org/cgi-bin/webworld/portal_freesoftware/cgi/page.cgi?...&d=1
47. GInfoServer: **GIS-Software** + -Shareware: www.geo.uni-bonn.de/members/haack/**gis-software**.html—14k
48. Commercial **GIS Software**: www.**gis**.com/implementing_**gis**/**software**/commercial_**software**.html—11k
49. Download PC **Software** from NGS: www.ngs.noaa.gov/PC_PROD/pc_prod.shtml—29k
50. Main Page – GisWiki: www.**gis**.org.nz/index.php/Main_Page—9k
51. ArcLessons—Your Resource for **GIS** Educational Exercises: **gis**.esri.com/industries/education/arclessons/arclessons.cfm
52. Google Earth: www.keyhole.com/body.php?h=downloadKeyhole—13k
53. ESRI Canada-**Free**!: www.esricanada.com/english/bottom/**free**.asp—23k
54. **free** statistical **software**: gsociology.icaap.org/methods/soft.html—31k
55. Land Management Information Center Home Page: www.lmic.state.mn.us—33k

56. Jenness Enterprises—ArcView Extensions: www.jennessent.com/arcview/arcview_extensions.htm—50k
57. MicroImages: www.microimages.com—12k
58. **GIS** Data, Shaded Relief Images, DEM, Satellite Imagery, Geology of Madagascar, Global Mapping, Minerals, Fossils: www.gospatial.com—32k
59. Desktop **GIS** Web Publishing **software** products: www.collins**software**.com—8k
60. **Software** (RDMS/UD): www.rdms.udel.edu/rdms/main/**software**.html—31k
61. **Software** Ahoy!: www.ucalgary.ca/~clarko/mapsoft—24k
62. Ecological **Software** Solutions: www.ecostats.com/**software**/**software**.htm—11k
63. Yahoo! Groups: **free-gis**: groups.yahoo.com/group/**free-gis**
64. **GIS Software**: www.ies.wisc.edu/ltc/nap/**gis**vend.html—10k
65. **GIS** and Remote Sensing **Software**: www.indiana.edu/~dms/**gis**/**software**—9k
66. ESRI Conservation Program: supporting conservation geography, activism and multicultural social change: www.conservation**gis**.org/conservationb.html—26k
67. **Software** Resources—My Community, Our Earth: www.aag.org/sustainable/resources/**software**.html—10k
68. GISuser.com **GIS**, GPS, LBS, CAD, mapping news, jobs, **software**, data, community: www.**gis**user.com/content/view/3143/1—91k
69. **GIS** Internet Resources Body: www.tec.army.mil/**gis**/**gis**int_body1.html
70. **GIS software software** downloads: www.**free**-download-soft.com/**Gis**+**software** – 30k
71. Map Maker—Desktop Mapping: www.mapmaker.com—14k
72. GISuser.com **GIS**, GPS, LBS, CAD, mapping news, jobs, **software**, data, community: www.**gis**user.com/content/view/4199/1—75k
73. SWEGIS, Consultant and **GIS Software** Provider—**Free** stuff: www.swe**gis**.com/english/download/**free**.asp—37k
74. WG VI.1/**Free software**: www.photogrammetry.ethz.ch/general/persons/jana/isprs/ed_**free**sw.html—21k
75. **Software**: **GIS**, remote sensing, hydrology and GPS **software**,**free** downlaod: spatial hydrology: www.spatialhydrology.com/**software**_**gis**rsgps.html—36k
76. **GIS** Internet Resources Body: www.tec.army.mil/**gis**/**gis**int_body1.html—47k
77. **Software** Resources—My Community, Our Earth: www.aag.org/sustainable/resources/**software**.cfm—10k
78. Central Region **GIS Software** and Utilities: cpg.cr.usgs.gov/pub/**software**.html—30k
79. ESRI **GIS Software**: www.esriindia.com/Products/ArcPad.htm—13k
80. Surveying **Software**: www.sonic.net/~trollhei/survsoft.html—4k
81. The Impoverished Social Scientist's Guide to **Free** Statistical **Software** and Resources: www.hmdc.harvard.edu/micah_altman/socsci.shtml—23k
82. **GIS** Links: www.acadweb.wwu.edu/**gis**/links/**gis**.htm—25k
83. PTTC Rockies—**Software** Fair: www.mines.edu/research/PTTC/**software**fair/**gis**.html—20k
84. Open Source **GIS**: www.opensource**gis**.org—118k
85. ALOV Map/TMJava. **Free** Java **GIS**: alov.org—10k
86. RockWare- earth science & **GIS software**: **free** catalog and trialware CD-ROM: www.rockware.com/catalog.html—18k

87. Forestry **GIS** (fGIS)—**Free GIS** Viewer and Shapefile Editor: www.digitalgrove.net/fgis.htm—43k
88. Manifold System—The ultimate **GIS** and Mapping Product!: www.manifold.net—102k
89. GRASS Geographic Information System Information: www.cecer.army.mil/grass/GRASS.main.html—2k
90. TatukGIS: www.tatuk**gis**.com—15k
91. Companies—Specializing—**Gis** Mapping **Software**: www.soft411.org/companies/**gis**-mapping.html—35k
92. **Free GIS** Data Viewers: **gis**.geog.ttu.edu/texastech_**gis**/viewers.htm—6k
93. **GIS** in Schools: Education Programme: Ministry of Education, Youth, and Culture – Jamaica: www.moec.gov.jm/projects/**gis**sep/techsupport.htm—11k
94. Quantitative Decisions: www.quantdec.com—16k
95. **Free Software**: www.geo.tu-freiberg.de/~apelm/miscsoft.htm—14k
96. Geographic information system—Wikipedia, the **free** encyclopedia: en.wikipedia.org/wiki/**Gis**
97. **Software**: www.explorationist.com/**Software**.htm—28k
98. PTTC Rockies: www.mines.edu/Research/PTTC/**Software** Fair/**gis**.shtml—22k
99. ESRI Data: maps.unomaha.edu/Workshops/Career/ESRI/index4.html—11k
100. The NITLE News Volume 1, Number 1, spring 2002: newsletter.nitle.org/v1_n1_spring2002/**gis**.php—4k
101. CIPE | **Software** and Images: www.evisual.org/www/**Software**/Soft.html—11k
102. http://www.unchs.org/guonet/documents/urban_observer/Urban_Observer_2-2003.txt: unchs.org/guonet/documents/urban_observer/Urban_Observer_2-2003.txt—7k
103. **Free** Desktop **Software** Viewers: chesapeake.towson.edu/mapping/**free**viewers.asp - 9k
104. Product Information on **GIS Software**: vnu.bitpipe.com/plist/**GIS-Software**.html—31k
105. ESRI INDIA Your Partner in **GIS**: www.esriindia.com/index/ESRIINDIA.htm—12k
106. **GIS** File Library—Directions Magazine: www.directionsmag.com/files—47k
107. [GRASSLIST:3906] Linux-based **GIS software**: grass.itc.it/pipermail/grassuser/2002-June/006754.html—7k
108. United States—**Free GIS** Data—**GIS** Data Depot: data.geocomm.com/catalog/US—28k
109. GISCafe—Downloads: www10.**gis**cafe.com/link/display_categories.php—24k
110. NCSU Sci-Link: www.ncsu.edu/scilink/workshops_2003.html—9k
111. Companies—Specializing—**Free** Gps **Software Software**: www.soft411.org/companies/**free**-gps-**software**.html—39k
112. **Free** Statistical **Software**: members.aol.com/johnp71/javasta2.html—118k
113. ArchNet: Map Libraries, **GIS**, and Mapping **Software**: archnet.asu.edu/archnet/topical/maps/maps.html—23k
114. FGDL **Software**: www.fgdl.org/**software**.html—26k
115. The Soft FAQ: **Software**, freeware and FAQ on **GIS** and spatial statistics: www.ai-geostats.org/**software**/**GIS**_Softfaq.htm—20k
116. **GIS** Data from the Washington Natural Heritage Program: www.dnr.wa.gov/nhp/refdesk/**gis**/wnhp**gis**.html—27k

117. **Software**: www.mn**gis**lis.org/oldsite/education/**software**.htm—7k
118. India Map—India Maps, India **Free** Map, India **GIS**, India **GIS** Data: www.maptell.com—19k
119. **GIS** Internet Links, UT Dallas: www.utdallas.edu/~briggs/other_**gis**.html—132k
120. http://ils.unc.edu/~ruvam/INLS110_111/fall03/slides/Lesson_2_rev.ppt (MICROSOFT POWERPOINT): ils.unc.edu/~ruvam/INLS110_111/fall03/slides/Lesson_2_rev.ppt—621k
121. **GIS**/GPS Link and **Free Software**: www.jttaylor.net/contents/3-**GIS**-GPS—975
122. The Soft Earth: www.df.lth.se/~micke/s_earth.html—71k
123. About this CD-ROM: www.csc.noaa.gov/products/sccoasts/html/about.htm—14k
124. **GIS** Resources: **Software**: www.geog.umn.edu/courses/**GIS**/**software**.htm
125. **Free Software** – Christian Singer, Solutions for Geographic Information: www.icsinger.de/**free**ware.htm—5k
126. ArcScripts Home—ESRI Support: arcscripts.esri.com—10k
127. **GIS**, CAD, and mapping daily news provided by Spatial News: spatialnews.geocomm.com/dailynews/2001/aug/20/news4.html—29k
128. @SoftTop—top text messaging **software**, download **free** text messaging **software**: www.**software**-top.com/softdesc.asp?LinkID=11450—24k
129. Clearinghouse Search: **GIS**: www.csiss.org/cgi- bin/texis/webinator/clearsearch/+BwwFqmHflxzmwwww—7k
130. Commercial **GIS** Providers (**Software** and Consultants): www.inweh.unu.edu/unuinweh/**gis**categories/commercial.htm—20k
131. Cultural Heritage **GIS** Resources: www-personal.umich.edu/~roberta/cult**GIS**.htm—13k
132. **GIS Software** Vendor Directory: www.bitpipe.com/olist/**GIS-Software**.html—30k
133. Infotiger search results: **gis** linux **free software**: www.infotiger.com search?qs=**gis**+linux+**free**+**software**&OK=++search+—25k
134. City of Monterey **GIS**—**Software** Requiremnts: **gis**web.ci.monterey.ca.us/pub/html/requirements.htm—7k
135. Geographical Information Systems (**GIS**) WWW Resource List: www.geo.ed.ac.uk/home/**gis**www.html—97k
136. STOMP **GIS Software** Reading: www.tec.army.mil/research/**software**/Stomp/swread.htm—16k
137. INSIDE Tutorial: **Software**: inside.uidaho.edu/tutorial/**software**/**software**.htm—12k
138. **GIS Software**: www.bakersfieldcity.us/**GIS**/notes/tech/**software**.htm—12k
139. CAD **GIS** Computer Aided Design--**Free** Guide: www.cad-directory.com—22k
140. ESRI—The **GIS Software** Leader: www.clas.ufl.edu/users/puhalla/ESRI
141. Geographic Information Service: volusia.org/**gis**—11k
142. Extract Systems: Home Page: www.icomap.com—10k
143. Components of a **GIS**: www.rst2.edu/ties/GENTOOLS/comp_**gis**.html—5k
144. **GIS Software** and Training for K-12 Schools—December 2004: www.nd.gov/**gis**/news/20041207.html—7k
145. User:Tschirl—Meta: meta.wikimedia.org/wiki/User:Tschirl—17k
146. Teachers Resources for **GIS**: www.columbia.edu/~ja681/**GIS**/rs3.html—11k

147. Welcome to **GIS**\Solutions — makers of **GIS**\Key, the premier **GIS software** for Windows: www.**gis**key.com/core.htm—21k
148. PASCO and myworld **GIS software**: www.pasco.com/myworld—41k
149. A series of four different labs across the three-day annual meeting presented by Charlie Fitzpatrick of ESRI (PDF): ncge.org/events/meetings/birmingham/**...**/PP-ComputerLabs.pdf—86k
150. d/freegis: lwn.net/1999/1104/a/**freegis**.html—8k

Other details

- Global Mapper—Affordable **GIS** At Last Global Mapper is an easy to use package that allows you to view, edit, merge, rectify, convert, and analyse a huge variety of map data formats. Download a **free** demo today.: www.globalmapper.com
- **GIS**/Mapping **Software** Business **software** information guide. See how it can help your business: www.techbusiness.ws
- **GIS** Information Guide A Directory On **GIS** Services Mapping **Software** and **GIS Software**:
  www.advancedwebsearch.com
- **Free** 30-Day Trial of Autodesk Map 3D Increase geospatial design efficiency 25%. For **GIS**, engineering, mapping, survey professionals who want to create, manage, share spatial, design data. Integrate your CAD, **GIS** data: www.autodesk.com
- Mapping/**GIS Software** RockWare, Inc. supplies mapping **software** solutions including coordinate conversion, **GIS**, topo maps and more. We are licensed resellers of Arcview, Topo, Tralaine and more. **Free** demos: www.rockware.com
- **GIS** Information Guide: www.advancedwebsearch.com
- **Free** 30-Day Trial of Autodesk Map 3D: www.autodesk.com
- Mapping/**GIS Software**: www.rockware.com
- Find **Gis Software** Solutions: www.knowledgestorm.com
- **GIS** Information Guide A Directory On **GIS** Services Mapping **Software** and **GIS Software**: www.advancedwebsearch.com
- **Free** 30-Day Trial of Autodesk Map 3D Increase geospatial design efficiency 25%. For **GIS**, engineering, mapping, survey professionals who want to create, manage, share spatial, design data. Integrate your CAD, **GIS** data: www.autodesk.com
- Mapping/**GIS Software** RockWare, Inc. supplies mapping **software** solutions including coordinate conversion, **GIS**, topo maps and more. We are licensed resellers of Arcview®, Topo, Tralaine and more. **Free** demos: www.rockware.com

APPENDIX VI

2004 GIS Software Survey Notes

Program Name	Autodesk Raster Design 2005	Autodesk Survey 2005	Autodesk Land Desktop 2005	Autodesk Civil Design 2005	GEOPAK Survey	Power Survey	InRoads Survey	MicroStation GeoGraphics
Manufacturer/Distributor	Autodesk Inc.	Autodesk Inc.	Autodesk Inc.	Autodesk Inc.	Bentley Systems Inc.	Bentley Systems	Bentley Systems Inc.	Bentley Systems Inc.
Manufacturer's Phone Number	603/621-3270	603/621-3270	603/621-3270	603/621-3270	610/458-5000	610/458-5000	610/458-5000	610/458-5000
Cost Per Seat	Contact reseller	Contact reseller	Contact reseller	Contact reseller	Contact vendor	$1499	Contact vendor	Contact vendor
Date of First Release	1996	1998	1998	1998	1996	2003	2000	1995
SOFTWARE FEATURES								
Operating System/Network Support								
—Network Client-Server Support	No	Win NT/2000/XP	Win NT/2000/XP	Win NT/2000/XP	Yes	Yes	Yes	Yes
—Server Operating System	Win 2000/XP	No	No	No	Win NT 4.0/2000/XP/98/ME	Win NT 4.0/2000/XP/98/ME	Win NT 4.0/2000/XP/98/ME	WinXP/2000/NT 4.0/98
—Client Operating System	None	No	No	No	Win NT 4.0/2000/XP/98/ME	Win NT 4.0/2000/XP/98/ME	Win NT 4.0/2000/XP/98/ME	WinXP/2000/NT 4.0/98
—Internet Server Enabled	N/A	No	No	No	N/A	N/A	N/A	Via Bentley Publisher
Compatible Applications								
—Third Party Applications Designed to Run on Software	N/A	Yes, by vendor	[3]	Yes, by vendor	Yes	Yes	Yes	Yes
—Total Solution Packages Available	N/A	Yes	Yes	Yes	Yes	Yes	Yes	Yes
GIS Data Administration								
—Multi-user Edit Locking	Yes	Yes	Yes	Yes	N/A	N/A	Yes	[8]
—Versioning	No	No	No	No	N/A	N/A	Yes	[9]
—Metadata Maintenance	N/A	No	No	No	N/A	N/A	N/A	[10]
Database Management								
—Proprietary DBMS	No	No	No	No	No	No	No	No
—Relational Database Management System	No	No	No	No	Oracle, MS Access, SQL Server, Sybase, Informix	Oracle, MS Access, SQL Server, Sybase, Informix	Oracle, MS Access, SQL Server, Sybase, Informix	[4]
—RDBMS Spatial Data Warehouse	No	No	No	No				Oracle 8i/9i Spatial and Oracle 9i Locator
Native Graphic Data Structure and Format								
—Vector-Spaghetti	No	No	No	No	N/A	N/A	N/A	DGN, DWG, DXF
—Vector-Topologic	No	No	No	No	DGN, DWG, DXF	DGN, DWG, DXF	DGN, DWG	Oracle Spatial/DGN post processing
—Parametric	No	Yes	Yes	Yes	N/A	N/A	N/A	Yes
—3-D	No	Yes	Yes	Yes	Yes	Yes	Yes	Yes
—TIN	No	Yes	Yes	Yes	Yes	Yes	Yes	Yes, via Bentley GEOPAK Site module
—Grid	No	Yes	Yes	Yes	Yes	Yes	Yes	N/A
—Raster Image	Yes							
GIS Data Import/Export Utilities		Yes	Yes	Yes	Yes	Yes	Yes	[2]
—Direct Import Formats	N/A	[1]	[1]	[1]	[2]	[2]	[2]	DXF, DWG, MID/MIF, SHP, E00
—Direct Export Formats	N/A	[1]	[1]	[1]	N/A	N/A	Yes	DXF, DWG, MID/MIF, SHP, E00
Readable Formats That Do Not Require Translation					N/A	N/A	Yes	Oracle Spatial, DGN, DWG
—Import	[6]	DWG, LandXML, ASCII, FBK	DWG, LandXML, ASCII	DWG, LandXML, ASCII	[6]	[6]	Yes	Oracle Spatial, DGN, DWG
—Export	Image Formats, CALS, TIFF, PNG, BMP, JPG	DWG, LandXML, ASCII, FBK	DWG, LandXML, ASCII	DWG, LandXML, ASCII	[6]	[6]	Yes	Yes
GIS Data Entry and Editing								
—Board Digitizing	Yes	Yes	Yes	Yes	Yes	Yes	Yes	Via Bentley GEOPAK Site
—Coordinate Geometry/Precision Entry	1-8 Decimal Places	0-8 decimal places	0-8 decimal places	0-8 decimal places	Yes	Yes	Yes	Via Bentley GEOPAK Site
—Electronic Survey Data Import	N/A	Yes	Yes, ASCII Coordinate files	Yes, with Survey Add-in	Yes	Yes	Yes	
—Heads-up Digitizing	Yes	Yes	Yes	Yes	Yes	Yes	Yes	Yes
—Vectorization	Yes	Yes	Yes	No	N/A	N/A	N/A	Via Bentley Descartes
—Map Rectification	Yes	Yes	No	No	N/A	N/A	Yes	Yes
—Graphic Error Check/Correction	Yes	Yes	No	No	Yes	Yes	Yes	Yes
—Field Data Entry	N/A	Yes	Yes	Yes	Yes	Yes	Yes	Yes
Map Design and Composition								
—Interactive Map Composition	Yes	Yes	Yes	Yes	Yes	Yes	Yes	Yes
—Annotation from Attributes	Yes	Yes	Yes	Yes	Yes	Yes	Yes	Yes
—Global Symbol Change	No	Yes	Yes	Yes	Yes	Yes	Yes	Yes
—Thematic Mapping	Yes	Yes	Yes	No	N/A	N/A	Yes	Yes
Geographic Query and Analysis Functions								
—Attribute Query and Selection	N/A	Yes	Yes	Yes	Yes	Yes	Yes	Yes
—Map Measurements	Yes	Yes	Yes	Yes	Yes	Yes	Yes	Yes
—Address Matching	N/A	No	No	No	N/A	N/A	N/A	Yes
—Buffer Generation	N/A	Yes	Yes	Yes	N/A	N/A	N/A	Yes
—Point/Line-in-Polygon Analysis	N/A	Yes	Yes	Yes	N/A	N/A	Yes	Yes
—Polygon Overlay	N/A	Yes	Yes	Yes	N/A	N/A	Yes	Yes
—Network Analysis	N/A	Yes	Yes	Yes	N/A	N/A	Yes	Yes
—Raster Document Query and Access	Yes	With add-on	With add-on	With add-on	N/A	N/A	N/A	Yes
—Direct Access to Other GIS Format	N/A	Yes	Yes	Yes	N/A	N/A	N/A	SHP, MID, MIF, E00
Terrain Data Processing and Analysis								
—DEM Generation	No	No	No	No	Yes	Yes	N/A	Via Bentley GEOPAK Site
—Contour Map Generation	Yes	Yes	Yes	Yes	Yes	Yes	Yes	Via Bentley GEOPAK Site
—3-D Display/Profile Generation	No	Yes	Yes	Yes	Yes	Yes	Yes	Yes
—Map Draping	No	No	No	No	N/A	N/A	Yes	Via Bentley Descartes
—Slope/Aspect Analysis	Yes	Yes	Yes	Yes	Yes			
Raster Image Capabilities						Yes	Yes	Via Bentley GEOPAK Site
—Geometric Rectification	Yes	With add-on	With add-on	With add-on	N/A	N/A	Yes	Via Bentley Descartes
—Orthoimage Generation	Yes	With add-on	With add-on	With add-on	N/A	N/A	N/A	N/A
—Image Enhancement	Yes	With add-on	With add-on	With add-on	N/A	N/A	N/A	Via Bentley Descartes
—Spectral Classification	Yes	With add-on	With add-on	With add-on	N/A	N/A	N/A	N/A
Application Development Language								
—Proprietary Application Development Language	No	No	No	No	MDL, JMDL, Basic	MDL, JMDL, Basic	MDL, JMDL, Basic, C++	MDL, JMDL, MicroStation Basic
—Industry Standard Programming Environment	Yes, ActiveX	No	ActiveX, VBA, C++, Lisp	ActiveX, VBA, C++	C, C++, Visual Basic, JAVA	C, C++, Visual Basic, JAVA	C, C++, Visual Basic, JAVA	C, C++, Visual Basic, JAVA
Extra Notes	[7]	[2]	[4]	[8]		[7]		[11]
Reader Service Number	1	2	3	4	5	6	7	8

N/A = Not Applicable (*Source*: www.pobonline.com)

2004 GIS Software Survey Notes

Program Name	Bentley Water	Bentley WasteWater	Bentley Publisher	Bentley Descartes	Bentley I/RAS B	Bentley PowerField	Bentley WorkPrint	Bentley PowerMap
Manufacturer/Distributor	Bentley Systems Inc.	Bentley Systems Inc.	Bentley Systems Inc.	Bentley Systems Inc.	Bentley Systems Inc.	Bentley Systems Inc.	Bentley Systems Inc.	Bentley Systems Inc.
Manufacturer's Phone Number	610/458-5000	610/458-5000	610/458-5000	610/458-5000	610/458-5000	610/458-5000	610/458-5000	610/458-5000
Cost Per Seat	Contact vendor	Contact vendor	Contact vendor	Contact vendor	Contact vendor	Contact vendor	Contact vendor	Contact vendor
Date of First Release	1999	1999	1996	1993	1992	2004	2004	2003
SOFTWARE FEATURES								
Operating System/Network Support								
—Network Client-Server Support	Yes	Yes	Yes	Yes	Yes	Yes	Yes	Yes
—Server Operating System	Win XP/2000/NT 4.0/98	Win XP/2000/NT 4.0/98	Win 2000/NT	Win XP/2000/NT 4.0/98	Win XP/2000/NT 4.0/98	Win XP/2000/NT 4.0/98/Me	Win XP/2000/NT 4.0/98/Me	Win XP/2000/NT 4.0/98/Me
—Client Operating System	Win XP/2000/NT 4.0/98	Win XP/2000/NT 4.0/98	Any with a WEB Browser	Win XP/2000/NT 4.0/98	Win XP/2000/NT 4.0/98	Win XP/2000/NT 4.0/98/Me	Win XP/2000/NT 4.0/98/Me	Win XP/2000/NT 4.0/98/Me
—Internet Server Enabled	Via Bentley Publisher	Via Bentley Publisher	Yes	Via Bentley Publisher	Via Bentley Publisher	Via Bentley Publisher	Via Bentley Publisher	Via Bentley Publisher
Compatible Applications								
—Third Party Applications Designed to Run on Software	N/A	N/A	Yes	Yes	Yes	Yes	Yes	Yes
—Total Solution Packages Available	Yes	Yes	Yes	Yes	Yes	Yes	Yes	Yes
GIS Data Administration								
—Multi-user Edit Locking	[8]	[8]	N/A	N/A	N/A	N/A	N/A	N/A
—Versioning	No	No	N/A	Yes, via MicroStation	N/A	N/A	N/A	N/A
—Metadata Maintenance	Yes	Yes	Yes	N/A	N/A	N/A	N/A	N/A
Database Management								
—Proprietary DBMS	No	No	No	No	No	No	No	No
—Relational Database Management System	[4]	[4]	[4]	No	N/A	[4]	[4]	[4]
—RDBMS Spatial Data Warehouse	Oracle 8i/9i Spatial and Oracle 9i Locator	Oracle 8i/9i Spatial and Oracle 9i Locator	Oracle 8i/9i Spatial and Oracle 9i Locator	No	N/A	[1]	Oracle 8i/9i, Spatial and Locator	Oracle 8i/9i, Spatial and Locator
Native Graphic Data Structure and Format								
—Vector-Spaghetti	DGN, DWG, DXF	DGN, DWG, DXF	DGN	DGN, DWG, DXF	DGN, DWG, DXF	DGN, DWG DXF, MXD, SHP, LYR	DGN, DWG, DXF	DGN, DWG, DXF
—Vector-Topologic	Oracle Spatial/DGN post processing	Oracle Spatial/DGN post processing	Oracle 8i/9i Spatial and Oracle 9i Locator	N/A	N/A	Oracle Spatial (query-only)	[5]	[5]
—Parametric	Yes	Yes	Yes	Yes	Yes	Yes	Yes	Yes
—3-D	Yes	Yes	Yes	Yes	Yes	View-only	Yes	Yes
—TIN	N/A	N/A	N/A	Yes, via Bentley GEOPAK Site module	N/A	N/A	N/A	N/A
—Grid	N/A	N/A	N/A	N/A	N/A	N/A	N/A	N/A
—Raster Image	[2]	[2]	[14]	[14]	[17]	[2]	[2]	[2]
GIS Data Import/Export Utilities								
—Direct Import Formats	DXF, DWG, MID/MIF, SHP, E00	DXF, DWG, MID/MIF, SHP, E00	DXF/DWG	[14]	[17]	DGN, DWG, DXF, MXD, SHP, LYR	DGN, DWG, DXF	DGN, DWG, DXF, MXD, SHP, LYR
—Direct Export Formats	DXF, DWG, MID/MIF, SHP, E00	DXF, DWG, MID/MIF, SHP, E00	DXF/DWG	[14]	[17]	Read-only platform	DGN, DWG, DXF	DGN, DWG, DXF, SHP
Readable Formats That Do Not Require Translation								
—Import	Oracle Spatial, DGN, DWG	Oracle Spatial, DGN, DWG	Oracle Spatial, DGN, DWG	[14]	[17]	[3]	DGN, DWG, DXF	DGN, DWG, DXF, MXD, SHP, LYR
—Export	Oracle Spatial, DGN, DWG	Oracle Spatial, DGN, DWG	CGM, SVG, JPG	[14]	[17]	Read-only platform	DGN, DWG, DXF	DGN, DWG, DXF
GIS Data Entry and Editing								
—Board Digitizing	Yes	Yes	N/A	Yes	Yes	Yes	Yes	Yes
—Coordinate Geometry/Precision Entry	N/A	N/A	N/A	Yes	N/A	N/A	N/A	N/A
—Electronic Survey Data Import	N/A	N/A	N/A	N/A	N/A	N/A	N/A	N/A
—Heads-up Digitizing	Yes	Yes	N/A	Yes	Yes	N/A	Yes	Yes
—Vectorization	N/A	N/A	N/A	Yes	Yes	N/A	N/A	N/A
—Map Rectification	N/A	N/A	N/A	Yes	Yes	Yes	Yes	Yes
—Graphic Error Check/Correction	N/A	N/A	N/A	N/A	N/A	N/A	N/A	Yes
—Field Data Entry	N/A	N/A	N/A	N/A	N/A	Yes	Yes	Yes
Map Design and Composition								
—Interactive Map Composition	N/A	N/A	N/A	Yes	Yes	Yes	Yes	Yes
—Annotation from Attributes	Yes	Yes	Yes	No	Yes	Yes	Yes	Yes
—Global Symbol Change	Yes	Yes	Yes	Yes	N/A	Yes	Yes	Yes
—Thematic Mapping	Yes	Yes	Yes	Yes	Yes	Yes	Yes	Yes
Geographic Query and Analysis Functions								
—Attribute Query and Selection	Yes	Yes	Yes	Yes	Yes	Yes	Yes	Yes
—Map Measurements	Yes	Yes	Yes	Yes	Yes	Yes	Yes	Yes
—Address Matching	N/A	N/A	Available via additional services	N/A	N/A	N/A	Yes	Yes
—Buffer Generation	Yes	Yes	Yes	N/A	N/A	Yes	Yes	Yes
—Point/Line-in-Polygon Analysis	Yes	Yes	Yes	N/A	N/A	Yes	Yes	Yes
—Polygon Overlay	Yes	Yes	Yes	N/A	N/A	Yes	Yes	Yes
—Network Analysis	Yes	Yes	Yes	N/A	N/A	Yes	Yes	Yes
—Raster Document Query and Access	Yes	Yes	Yes	N/A	N/A	Yes	Yes	Yes
—Direct Access to Other GIS Format	SHP, MID, MIF, E00	SHP, MID, MIF, E00	N/A	N/A	N/A	Yes	Yes	Yes
Terrain Data Processing and Analysis								
—DEM Generation	N/A	N/A	N/A	N/A	N/A	N/A	N/A	N/A
—Contour Map Generation	N/A	N/A	N/A	Yes	N/A	N/A	N/A	Yes
—3-D Display/Profile Generation	N/A	N/A	N/A	Yes	N/A	Yes	Yes	Yes
—Map Draping	N/A	N/A	N/A	Yes	N/A	N/A	N/A	N/A
—Slope/Aspect Analysis	N/A	N/A	N/A	N/A	N/A	N/A	N/A	N/A
Raster Image Capabilities								
—Geometric Rectification	N/A	N/A	N/A	Yes	Yes	N/A	N/A	N/A
—Orthoimage Generation	N/A	N/A	N/A	N/A	N/A	N/A	N/A	N/A
—Image Enhancement	N/A	N/A	N/A	Yes	N/A	Yes	Yes	Yes
—Spectral Classification	N/A	N/A	N/A	N/A	N/A	N/A	N/A	N/A
Application Development Language								
—Proprietary Application Development Language	MDL, JMDL, Basic	MDL, JMDL, Basic	MDL, JMDL, Basic	MDL, JMDL, Basic	MDL	MDL, MicroStation Basic	MDL, MicroStation Basic	MDL, MicroStation Basic
—Industry Standard Programming Environment	C, C++, Visual Basic, JAVA, VBA	C, C++, Visual Basic, JAVA, VBA	ASP, JAVA Script, JAVA, Visual Basic, C, C++,	C, C++, Visual Basic, JAVA	Visual Basic	C, C++, Visual Basic, JAVA	C, C++, Visual Basic, JAVA	C, C++, Visual Basic, JAVA
Extra Notes	[12]	[13]	[15]	[16]	[18]			
Reader Service Number	9	10	11	12	13	14	15	16

N/A = Not Applicable

2004 GIS Software Survey Notes

Program Name	Geographic Transformer 5.0	Geographic Calculator 6.0	TransCAD 4.7	Maptitude 4.6	CARIS GIS	CARIS Carta	CEDRA AVseries [1]	ArcEditor	ArcView	ArcPad
Manufacturer/Distributor	Blue Marble Geographics	Blue Marble Geographics	Caliper Corporation	Caliper Corporation	CARIS	CARIS	The CEDRA Corporation	ESRI	ESRI	ESRI
Manufacturer's Phone Number	800/616-2725	800/616-2725	617/527-4700	617/527-4700	506/458-8533	506/458-8533	585 232-6998	800/447-9778	800/447-9778	800/447-9778
Cost Per Seat	$899	$499	$2995 (Base), $9995 (Standard)	$495	Contact vendor	Contact vendor	Contact vendor	Call ESRI	$1500	$495
Date of First Release	1996	1993	1988	1995	1983	2004	1985	2001	1992	2000
SOFTWARE FEATURES										
Operating System/Network Support										
—Network Client-Server Support	Yes	Yes	Win 98/NT and above	Win 98/NT and above	Yes	Yes	Yes	Yes	Yes	No
—Server Operating System	Win 98/NT and above	Win 98/NT and above	Win 98/NT and above	Win 98/NT and above	Win 95/98/NT/2000/XP/UNIX	Win NT/2000/XP	Win 95/98/NT/2000/XP/UNIX	Windows NT 4.0/2000/XP	Windows	No
—Client Operating System	Win 98/NT and above	Win 98/NT and above	Win 98/NT and above	Win 98/NT and above	Win 95/98/NT/2000/XP/UNIX	Win NT/2000/XP	Win 95/98/NT/2000/XP/UNIX	Windows NT 4.0/2000/XP	Windows	Windows CE
—Internet Server Enabled			With TransCAD for the Web	With Maptitude for the Web	No	No	Yes	Via ArcIMS	Via ArcIMS	Via ArcIMS
Compatible Applications										
—Third Party Applications Designed to Run on Software							ArcView or ArcGIS	www.esri.com/partners	www.esri.com/partners	www.esri.com/partners
—Total Solution Packages Available			2 [1]	3 [2]			12	Numerous	Numerous	Numerous
GIS Data Administration										
—Multi-user Edit Locking	Yes	Yes	Yes	Yes	No	No	Yes	Yes	No	No
—Versioning			No	No	No	No	Yes	Yes	No	No
—Metadata Maintenance			Yes	Yes	No	No	Yes	Yes	Yes	No
Database Management										
—Proprietary DBMS			No	No	No	No	No	No	No	No
—Relational Database Management System			Yes	Yes	Yes	Yes	Yes	Yes	Yes	Yes
—RDBMS Spatial Data Warehouse			Yes	Yes	No	Yes	Yes	[1]	[1]	No
Native Graphic Data Structure and Format										
—Vector-Spaghetti			Yes, can handle	Yes, can handle	Yes	Yes	Yes	Shapefile	Shapefile	Shapefile
—Vector-Topologic			Yes	Yes	Yes	Yes	Yes	Coverage, Geodatabase	Coverage, Geodatabase	No
—Parametric			No	No	Yes	No	Yes	Geodatabase	Geodatabase	No
—3-D			Yes (surfaces)	Yes (surfaces)	Yes	Yes	Yes	Shapefile, Geodatabase	Shapefile, Geodatabase	No
—TIN			Yes	Yes	Yes	Yes	Yes	ESRI TIN	ESRI TIN	No
—Grid	Yes	Yes	Yes	Yes	Yes	No	Yes	ESRI Grid	ESRI Grid	No
—Raster Image	Yes	Yes	Yes	Yes	Yes	Yes	Yes	[2]	[2]	Yes
GIS Data Import/Export Utilities										
—Direct Import Formats	[1]	[3]	16, including TIGER/Line, DWG, DGN	16, including TIGER/Line, DWG, DGN	Yes	Yes	Yes	Over 40 standard formats	Over 40 standard formats	N/A
—Direct Export Formats	[2]	[3]	8, including Shapefile, MID/MIF	8, including Shapefile, MID/MIF	Yes	Yes	Yes	Over 40 standard formats	Over 40 standard formats	N/A
Readable Formats That Do Not Require Translation										
—Import			Oracle Spatial, ESRI Shapefile, MapInfo TAB	Oracle Spatial, ESRI Shapefile, MapInfo TAB	Yes	Yes		Over 40 standard formats	Over 40 standard formats	Shapefiles, Mr SID/BMP, JPG
—Export			Oracle Spatial	Oracle Spatial	Yes	Yes		Over 40 standard formats	Over 40 standard formats	Shapefiles, Mr SID/BMP, JPG
GIS Data Entry and Editing										
—Board Digitizing			Yes	Yes	Yes	No	Yes	Yes	Yes	No
—Coordinate Geometry/Precision Entry	Yes	Yes	No	No	Yes	Yes	Yes	Yes	Yes	No
—Electronic Survey Data Import	Yes	Yes	GPS points	GPS points	Yes	No	Yes	Yes	Yes	No
—Heads-up Digitizing			Yes	Yes	Yes	No	Yes	Yes	Yes	Yes
—Vectorization			No	No	Yes	No	Yes	Yes	Yes	No
—Map Rectification	Yes	Yes	Yes	Yes	Yes	Yes	Yes	Yes	Yes	No
—Graphic Error Check/Correction	Yes	Yes	Yes	Yes	Yes	Yes	Yes	Yes	Yes	No
—Field Data Entry	Yes	Yes	Yes	Yes	Yes	Yes	Yes	Yes	Yes	Yes
Map Design and Composition										
—Interactive Map Composition			Yes	Yes	Yes	Yes	Yes	Yes	Yes	No
—Annotation from Attributes			Yes	Yes	Yes	Yes	Yes	Yes	Yes	No
—Global Symbol Change			Yes	Yes	Yes	Yes	Yes	Yes	Yes	Yes
—Thematic Mapping			Yes	Yes	Yes	Yes	Yes	Yes	Yes	No
Geographic Query and Analysis Functions										
—Attribute Query and Selection			Yes	Yes	Yes	Yes	Yes	Yes	Yes	Yes
—Map Measurements		Yes	Yes	Yes	Yes	Yes	Yes	Yes	Yes	Yes
—Address Matching			Yes	Yes	No	No	Yes	Yes	Yes	Yes
—Buffer Generation			Yes	Yes	Yes	Yes	Yes	Yes	Yes	No
—Point/Line-in-Polygon Analysis			Yes	Yes	Yes	No	Yes	Yes	Yes	No
—Polygon Overlay			Yes	Yes	Yes	No	Yes	Yes	Yes	No
—Network Analysis			Yes	Yes (best route)	Yes	No	Yes	Yes	Yes	No
—Raster Document Query and Access	Yes		No	No	No	No	Yes	Yes	Yes	No
—Direct Access to Other GIS Format	Yes	Yes	Yes	Yes	No	Yes	Yes	Yes	Yes	No
Terrain Data Processing and Analysis										
—DEM Generation			Yes	Yes	Yes	Yes	Yes	Yes	Yes	No
—Contour Map Generation			Yes	Yes	Yes	Yes	Yes	Yes	Yes	No
—3-D Display/Profile Generation			Yes	Yes	Yes	No	Yes	Yes	Yes	No
—Map Draping			Yes	Yes	Yes	No	Yes	Yes	Yes	No
—Slope/Aspect Analysis			Yes	Yes	Yes	No	Yes	Yes	Yes	No
Raster Image Capabilities										
—Geometric Rectification	Yes		No	No	Yes	Yes	Yes	Yes	Yes	No
—Orthoimage Generation			No	No	Yes	No	Yes	Yes	Yes	No
—Image Enhancement	Yes		No	No	No	No	Yes	Yes	Yes	No
—Spectral Classification			No	No	Yes	No	Yes	Yes	Yes	No
Application Development Language										
—Proprietary Application Development Language	Yes	Yes	Yes (Caliper Script)	Yes (Caliper Script)	Yes	No		No	No	No
—Industry Standard Programming Environment			Yes (.NET)	Yes (.NET)	VB	No		[3]	[3]	VBScript C/C++
Extra Notes			[37]	[40]				www.esri.com/arceditor	www.esri.com/arcview	[4]
Reader Service Number	17	18	19	20	21	22	23	24	25	2[illegible]

N/A = Not Applicable

2004 GIS Software Survey Notes

Program Name	ArcGIS Survey Analyst	Surfer 8	Didger 3	MapViewer 5	GeoMedia/IntelliWhere [1]	ERDAS IMAGINE	MapInfo Professional	MapX Mobile
Manufacturer/Distributor	ESRI	Golden Software	Golden Software	Golden Software	Intergraph Mapping & Geospatial Solutions	Leica Geosystems GIS & Mapping	MapInfo Corporation	MapInfo Corporation
Manufacturer's Phone Number	800/447-9778	303/279-1021	303/279-1021	303/279-1021	256/730-7191 [2]	404/248-9000	800/327-8627	800/327-8627
Cost Per Seat	Call ESRI	$599	$329	$249	Call for pricing information	$2200+	Contact vendor	Contact vendor
Date of First Release	2003	2002	2001	2002	1997	1990	1985	2000
SOFTWARE FEATURES								
Operating System/Network Support								
—Network Client-Server Support	Yes			Yes	Yes	Yes	Yes	No
—Server Operating System	Win NT 4.0/2000/XP			Windows, Novell	[3]	Win XP/2000, Sun Solaris	MS Windows logo certified	No
—Client Operating System	Win NT 4.0/2000/XP	Win 98/Me/2000/XP	Win 98/Me/2000/XP	Win 98/Me 2000/XP	Win Me/XP/2000	Win XP/2000, Sun Solaris	MS Windows logo certified	[17]
—Internet Server Enabled	Via ArcIMS				Yes	No	Yes, via MapInfo Discovery	No
Compatible Applications								
—Third Party Applications Designed to Run on Software	www.esri.com/partners				Several	[7]	Numerous	Numerous
—Total Solution Packages Available	Numerous				Yes	Yes	Yes	
GIS Data Administration								
—Multi-user Edit Locking	Yes, with core ArcGIS				Yes	No	Yes, via database	No
—Versioning	Yes, with core ArcGIS				Yes	No	Yes, via database	No
—Metadata Maintenance	Yes, with core ArcGIS				Yes	Yes	Yes	No
Database Management								
—Proprietary DBMS	No			Yes	No	CellArray	[1]	No
—Relational Database Management System	Yes			Yes	[11]	[1]	Yes	Yes [6]
—RDBMS Spatial Data Warehouse	[1]			Yes	[11]	ArcSDE	[2]	No
Native Graphic Data Structure and Format								
—Vector-Spaghetti	Shapefile	Yes	Yes	Yes		Yes	Yes	Yes
—Vector-Topologic	Coverage, Geodatabase				[4]	Yes	No	No
—Parametric	Geodatabase				Yes	Yes	No	No
—3-D	Shapefile, Geodatabase	Yes			Yes	Yes	Yes [3]	No
—TIN	ESRI TIN				Yes	Yes	Yes, via Vertical Mapper	No
—Grid	ESRI Grid	Yes			Yes	Yes	Yes [16]	No
—Raster Image	[2]	Yes	Yes	Yes	Yes	Yes	Yes [16]	Yes
GIS Data Import/Export Utilities								
—Direct Import Formats	Over 40 standard formats	[1]	[3]	[5]	[5]	[2]	[16]	No
—Direct Export Formats	Over 40 standard formats	[2]	[4]	[6]	[6]	[3]	[16]	No
Readable Formats That Do Not Require Translation								
—Import	Over 40 standard formats	SRF, GSI, GSB, BLN	PJT, GSI, GSB, BLN	GSM, GSI, GSB, BLN	[7]	[4]	[16]	Shape, ASCII
—Export	Over 40 standard formats	SRF, GSI, GSB, BLN	PJT, GSI, GSB, BLN	GSM, GSI, GSB, BLN	[8]	[5]	[16]	No
GIS Data Entry and Editing								
—Board Digitizing	Yes		Yes		Yes	Yes	Yes	Yes
—Coordinate Geometry/Precision Entry	Yes				Yes	No	Yes	Yes, with customization
—Electronic Survey Data Import	Yes		Yes		Yes	No	Yes	Yes
—Heads-up Digitizing	Yes, with ArcScan		Yes	Yes	Yes	Yes	Yes	Yes
—Vectorization	Yes, with ArcScan				Yes	Yes	Yes, via Vertical Mapper	No
—Map Rectification	Yes		Yes		Yes	Yes	No	No
—Graphic Error Check/Correction	Yes				Yes	Yes	Yes	No
—Field Data Entry	Yes				Yes	Yes	Yes	Yes
Map Design and Composition								
—Interactive Map Composition	Yes	Yes	Yes	Yes	Yes	Yes	Yes	Yes
—Annotation from Attributes	Yes		Yes	Yes	Yes	Yes	Yes	Yes
—Global Symbol Change	Yes	Yes	Yes	Yes	Yes	Yes	Yes	Yes
—Thematic Mapping	Yes			Yes	Yes	Yes	Yes	Yes
Geographic Query and Analysis Functions								
—Attribute Query and Selection	Yes, with core ArcGIS			Yes	Yes	Yes	Yes	Yes
—Map Measurements	Yes, with core ArcGIS		Yes	Yes	Yes	Yes	Yes	Yes
—Address Matching	Yes, with core ArcGIS				Yes	No [6]	Yes	No
—Buffer Generation	Yes, with core ArcGIS			Yes	Yes	Yes	Yes	No
—Point/Line-in-Polygon Analysis	Yes, with core ArcGIS			Yes	Yes	Yes	Yes	Yes
—Polygon Overlay	Yes, with core ArcGIS			Yes	Yes	Yes	Yes	No
—Network Analysis	Yes, with core ArcGIS				Yes	No	No	No
—Raster Document Query and Access	Yes, with core ArcGIS				Yes	Yes	Yes	No
—Direct Access to Other GIS Format	Yes, with core ArcGIS			Yes	Yes	Yes	Yes	Shape
Terrain Data Processing and Analysis								
—DEM Generation	Yes, with core ArcGIS				Yes	Yes	Yes	No
—Contour Map Generation	[5]	Yes			Yes	Yes	Yes, via Vertical Mapper	No
—3-D Display/Profile Generation	Yes, with ArcGIS 3D Analyst	Yes			Yes	Yes	Yes, via Vertical Mapper	No
—Map Draping	Yes, with core ArcGIS	Yes			Yes	Yes	Yes	No
—Slope/Aspect Analysis	Yes, with ArcGIS Spatial Analyst	Yes			Yes	Yes	Yes, via Vertical Mapper	No
Raster Image Capabilities								
—Geometric Rectification	Yes, with core ArcGIS		Yes		Yes	Yes	[4]	No
—Orthoimage Generation	Yes, with core ArcGIS				Yes	Yes	No	No
—Image Enhancement	Yes, with core ArcGIS		Yes	Yes	Yes	Yes	Yes	No
—Spectral Classification	Yes, with core ArcGIS				Yes	Yes	Yes, via Vertical Mapper	No
Application Development Language								
—Proprietary Application Development Language	No	Scripter		Scripter	[9]	IMAGINE Developers' Toolkit	Yes, MapBasic	No
—Industry Standard Programming Environment	[3]	C++, VB, Perl		C++, VB, Perl	Yes	C/C++ with some COM	[5]	[7]
Extra Notes	[6]				[10]			
Reader Service Number	27	28	29	30	31	32	33	34

N/A = Not Applicable

2004 GIS Software Survey Notes

Program Name	MapXtreme Java	MapXtreme 2004	Vulcan 4.5	Parcel Editor 8	ELT/5500	IMap	SOLO	TerraSync Professional Edition	GPS Pathfinder Office
Manufacturer/Distributor	MapInfo Corporation	MapInfo Corporation	Maptek Pty Ltd	NovaLIS Technologies	Paragon Imaging Inc.	SOKKIA	Tripod Data Systems	Trimble	Trimble
Manufacturer's Phone Number	800/327-8627	800/327-8627	303/763-4919	902/423-1800	800/937-4544	913/255-3913	541/753-9322	408/481-8000	408/481-800[illegible]
Cost Per Seat	Contact vendor	Contact vendor	Contact vendor	$2495 [1]	$3995	$1000 list	$1000	Contact vendor	Contact vendor
Date of First Release	1998	1998	2003	1996	2001	2000	1998	2000	1995
SOFTWARE FEATURES									
Operating System/Network Support									
—Network Client-Server Support	Yes	Yes	No	Terminal server	Yes, FlexLM	No	No	N/A	Yes
—Server Operating System	[8]	Win 2000/XP/2003	N/A	Unix, W2K, NT Server	Microsoft Windows	N/A		N/A	Win XP/2000/NT/Me/98/95
—Client Operating System	[13]	[13]	Win NT/2000/XP	Win XP/NT/2000	Microsoft Windows	Win CE/NT/98/2000/XP	Windows CE/Mobile	[1]	Win XP/2000/NT/Me/98/95
—Internet Server Enabled	Yes	Yes	No	N/A		No	No	Yes	Yes
Compatible Applications									
—Third Party Applications Designed to Run on Software	Numerous	Numerous	No	ArcGIS		No	RTI —RealTime Forest Inventory	N/A	N/A
—Total Solution Packages Available	Yes	Yes	Yes	Full Cadastral office suite		Yes	Yes	N/A	N/A
GIS Data Administration									
—Multi-user Edit Locking	Yes, via database	Yes, via database	Yes	Yes		No	N/A	No	No
—Versioning	Yes, via database	Yes, via database	No	Yes		No	No	No	No
—Metadata Maintenance	Yes	Yes	No	No		No	No	No	No
Database Management									
—Proprietary DBMS	[1]	[14]	Isis	Geodatabase		No	No	No	No
—Relational Database Management System	[9]	[9]	No	Geodatabase		No		No	No
—RDBMS Spatial Data Warehouse	[10]	[10]	Via ODBC	Geodatabase		No	No	No	No
Native Graphic Data Structure and Format									
—Vector-Spaghetti	Yes	Yes	Yes	Yes		Yes	Yes	Yes	Yes
—Vector-Topologic	No	No	No	Yes		Yes	Yes	No	No
—Parametric	Yes	Yes	No	No		No	No	No	No
—3-D	Yes, read-only	Yes, read-only from database	Yes	No		Yes	Yes	Yes	Yes
—TIN	No	No	Yes + solid triangulations	No		No	No	No	No
—Grid	Yes	Yes [16]	Yes	No		Yes	Yes	No	No
—Raster Image	Yes	Yes [16]	Yes	No	Native raster image support	Yes	Yes	Yes	Yes
GIS Data Import/Export Utilities									
—Direct Import Formats	[16]	[16]	[1]	[2]	[1]	[1]	[1]	ArcView Shapefile, Trimble SSF	[3]
—Direct Export Formats	[16]	[16]	[2]	[2]	NITF, GeoTiff and more	[2]	SHP, DXF, ASCII, CR5	ArcView Shapefile, Trimble SSF	[4]
Readable Formats That Do Not Require Translation									
—Import	[16]	[16]	[3]	[2]	[1]	[3]	UDF, SHP, MIF, DXF, DGN	Trimble SSF	Trimble SSF
—Export	[16]	[16]	[3]	[2]	[1]	[4]	UDF, CR5, SHP, DXF, ASCII	Trimble SSF	Trimble SSF
GIS Data Entry and Editing									
—Board Digitizing	No	Yes	Yes	Yes		No	No	No	No
—Coordinate Geometry/Precision Entry	No	Yes	Yes	Yes	[2]	Yes, no adjustments	Yes, no adjustments	N/A	Yes
—Electronic Survey Data Import	No	Yes	Yes	Yes		Yes	Yes	N/A	Yes
—Heads-up Digitizing	No	Yes	Yes	Yes		Yes	Yes	Yes	Yes
—Vectorization	No	No	No	Yes		No	No	No	No
—Map Rectification	No	No	Yes	Yes		No	No	No	No
—Graphic Error Check/Correction	No	No	No	Yes		No	No	No	No
—Field Data Entry	Yes	Yes	No	Yes		Primary function	Primary function	Yes	No
Map Design and Composition									
—Interactive Map Composition	Yes	Yes	Yes	Yes		Yes	Yes	Yes	Yes
—Annotation from Attributes	Yes	Yes	Yes	Yes		Yes	Yes	Yes	No
—Global Symbol Change	Yes	Yes	Yes	Yes		Yes	Yes	Yes	Yes
—Thematic Mapping	Yes	Yes	No	No		Yes	Yes	No	No
Geographic Query and Analysis Functions									
—Attribute Query and Selection	Yes	Yes	No	Yes		Yes	Yes	Yes	Yes
—Map Measurements	Yes	Yes	No	Yes	Image and map measurements	Yes	Yes	Yes	Yes
—Address Matching	Yes [11]	Yes	No	Yes		No	No	No	No
—Buffer Generation	Yes	Yes	No	No		No	Yes	No	No
—Point/Line-in-Polygon Analysis	Yes	Yes	No	Yes		No	No	No	No
—Polygon Overlay	No	Yes	No	Yes	Full polygon overlays	Yes	Yes	No	No
—Network Analysis	Yes [12]	No	Yes	Yes		No	No	No	No
—Raster Document Query and Access	Yes	No	No	No		No	No	No	No
—Direct Access to Other GIS Format	Yes	Yes (Oracle Spatial, Shape)	No	Yes		Yes	Yes	No	No
Terrain Data Processing and Analysis									
—DEM Generation	No	No	Yes	No		No	No	No	No
—Contour Map Generation	No	No	Yes	No		No	No	No	No
—3-D Display/Profile Generation	No	No	Yes + solid triangulations	No		No	No	No	No
—Map Draping	No	No	Yes	No		No	No	No	No
—Slope/Aspect Analysis	No	No	Yes	No		No	No	No	No
Raster Image Capabilities									
—Geometric Rectification	No	No	Yes	No	Yes	No	No	No	No
—Orthoimage Generation	No	No	No	No		No	No	No	No
—Image Enhancement	Yes	Yes	No	No	[3]	Yes	Yes	No	No
—Spectral Classification	No	No	No	No		No	No	No	No
Application Development Language									
—Proprietary Application Development Language	No	No	Lava	VB	Full API support with ELT/5500	Yes	Yes	No	No
—Industry Standard Programming Environment	Java and .Net languages	[15]	Perl	VB		Yes	Yes	No	No
Extra Notes							[2]	[2]	[5]
Reader Service Number	35	36	37	38	39	40	41	42	43

N/A = Not Applicable

2004 GIS Software Survey Notes

Summary of GIS Software Features

Software Feature Category	*Explanation*
Geographic Data Managemen	Database administration tools for managing data access by users, locking of data during edit and maintenance of metadata.
Tabular Attribute Data Management	Software environment and capabilities for storing and managing database attributes linked to map features in the GIS database. May involve use of a vendor-proprietary system for attribute storage or a commercial relational database management package.
GIS Data Import/ Export Utilities	Utility programs bundled with the GIS package for translation of GIS or CAD data to or from another format, including common industry-standard formats like DXF, SIF, DLG or SDTS.
GIS Data Entry and Editing	A range of interactive and batch processing functions for entry of map data through such means as board digitizing, coordinate geometry entry (COGO), scanning and heads-up digitizing, along with capabilities for editing GIS data, performing error checking and resolution, map rectification and transformation of coordinate systems and map projections.
Map Design and Composition	Interactive capabilities for the design of map plots and displays, automatic creation of thematic maps and legends, and modifying map symbology and annotation for custom map displays.
Basic Geographic Query and Analysis Functions	Basic tools for performing attribute or map-based queries and displays, basic distance and area measurements, query and access to scanned documents, buffer generation, polygon overlay operations and other query and analysis functions.
Network Analysis	Spatial analysis operations based on linear networks (e.g., road or pipeline systems), including such operations as shortest path tracing and region allocation. Network analysis capabilities in GIS packages often allow users to design network models based on attributes of network segments.
Terrain and 3-D Data Processing and Analysis	Capabilities for storing three-dimensional data normally in a grid or triangular integrated network (TIN) format with functions for 3-D analysis such as contour mapping, 3-D display, draping of map features over a 3-D display, slope and aspect analysis, etc.
Raster Image Processing Capabilities	Capabilities for the manipulation and processing of raster images (e.g., digital aerial photos or orthophotography, satellite images), including functions for the import and rectification of raw imagery, digital image enhancement and automated classification of multi-spectral imagery.
Application Development Languages	Programming environment for customizing applications accessing software functions provided by the package, including proprietary languages included with the GIS software package or industrystandard tools (e.g., C++, Visual Basic, Delphi) that may be used for application development.

Autodesk Inc.

1. ASCII, EOO, MID, MIF, ArcInfo coverage, ArcInfo Shape, DWG, LandXML, DGN, numerous data collectors.
2. Autodesk Survey is an add-in to Autodesk Land Desktop 2005. Survey adds bi-directional data collector support, traverse analysis, figures, basic coordinate geometry functions and annotation to Autodesk Land Desktop 2005.
3. Autodesk Raster Design, Autodesk Survey, Autodesk Civil Design.
4. Autodesk Land Desktop 2005 streamlines each phase of land development projects and provides a central location for project data that everbody on a team can access.
5. Autodesk Civil Design is an add-on for Autodesk Land Desktop 2005. The package includes tools to design roadways, grading plans, do hydrologic analysis, pond design and generate plan profile sheets.
6. Image Formats, CALS, TIFF, GeoTIFF, SID, ECW, PNG, BMP, JPG, GeoSPOT, JPG2000, DOQ, DEM, satellite imagery.
7. Autodesk Raster Design 2005e is an add-on application to AutoCAD 2005, Autodesk Map 2005 and Autodesk Land Desktop 2004 that enables customers to easily edit, convert, and integrate scanned drawings and raster files into the digital design process.
8. Autodesk Civil Design is an add-on for Autodesk Land Desktop 2005. The package includes tools to design roadways, grading plans, do hydrologic analysis, pond design and generate plan/profile sheets.

Bentley Systems Inc.

1. Oracle (query only): any ODBD compliant database (read/write).
2. Yes, GeoTIFF, iTIFF, JPEG, TIFF, HMR, RGB, RLE, TGA, PNG, BMP, CAL, IMB, CIT, COT, TG4, PCX, RS, MrSID, ECW, JPG2000.
3. Oracle Spatial, DNG, DWG, SHP, MXD.
4. Oracle, any ODBC compliant database.
5. Oracle Spatial/DGN post processing.
6. ASCII XYZ, TDS, Sokkia, Leica, Topcon, SMI, Geodimeter, C&C.
7. Power Survey is a Standalone application with CADD included. Power Survey includes: DTM Editing, Volume Calculations,expertation of Machine Control data formats.
8. Yes, when Oracle 8i/9i Spatial Locator is used.
9. Yes, with limitations (utilizing MicroStation Design History).
10. Yes, e.g. through links between geometry and a SQL database.
11. In the same way that MicroStation extends computer-aided design into an integrated engineering platform, MicroStation GeoGraphics transforms MicroStation into a comprehensive geospatial environment with premier mapping, data collection, editing, cleanup and analysis tools. MicroStation GeoGraphics allows users to integrate engineering and mapping applications on a single platform and extends the data capture and editing tools in MicroStation with a solid database interface and spatial analysis functionality. MicroStation GeoGraphics is a key engineering product that forms the foundation for geospatial applications. It significantly increases user productivity and offers an easy-to-use environment for geospatial users. With this discipline-specific platform, independent software developers can take advantage of highlevel functionality to increase application performance and reduce development time.
12. Bentley Water is a comprehensive, high-productivity engineering automation solution for water network design and management. All operations of a typical water supply network —from mapping and design to administrative functions such as maintenance and record-keeping—are addressed through an integrated environment. Municipal engineers can create a detailed system of water lines, valves, consumption points and other user-defined system components, while maintaining crucial network connectivity. Functions are provided for the placement, editing and attribution of facility graphics. Dynamic, database-driven facility annotation enables users to change existing annotation via a simple update function.
13. Bentley WasteWater is a comprehensive, highproductivity engineering solution for wastewater network design and management. All functions of a typical wastewater maintenance operation—from mapping and design to maintenance and record keeping—are addressed through an integrated environment. Municipal engineers can create a detailed model of storm water, sanitary and combined systems. The placement and edit functions create and maintain network connectivity, as well as distance, depth and connectivity of pipes and manholes.
14. GeoTIFF, iTIFF, JPEG, TIFF, HMR, RGB, RLE, TGA, PNG, BMP, CAL, IMG, CIT, COT, TG4, PCX, RS, MrSID, E.
15. Bentley Publisher is a powerful server-based application and a one-stop platform that allows users and implementers to publish their engineering information over the Internet. Bentley Publisher allows dynamic viewing and redlining of DGN, DWG and MicroStation GeoGraphics projects and images from Web browsers in raster, vector or hybrid raster vector format. It provides instant and universal access to valuable data, without requiring any change to existing workflows. Bentley Publisher offers multiple publishing capabilities, from simple queries to enhanced navigation using serverside processing, to full features using the Bentley Publisher Viewer.
16. Bentley Descartes is a comprehensive imaging solution for image mapping, visualization and raster-to-vector conversion projects. Designed for both production and visualization, Bentley Descartes is used worldwide by engineers and planners in many disciplines, such as civil engineering, plant design, mapping and architectural design. Part of the Geospatial product line, Bentley Descartes is used by governments and utilities for mapping, imaging and spatial analysis. Bentley Descartes adds high-performance raster operations to MicroStation, allowing warp, reference, resample, mosaic creation and 3D terrain draping to produce outstanding image scenes. Additionally, Bentley Descartes provides tools such as an integrated Optical Character Recognition (OCR) and Line Following engines, plus several tools to enhance and repair legacy drawings. Because Bentley Descartes is tightly integrated into MicroStation, it can be customized to integrate in workflows where automation and interface modification is required. MicroStation Descartes adds high-performance raster operations to allow warp, reference, resample, create mosaics and drape and produce outstanding image scenes. Additionally, MicroStation Descartes provides tools such as an integrated OCR and Line Following engines, plus several tools to enhance and repair legacy drawings. Offering an API of over 1,700 synchronous keyins and a fully customizable interface, MicroStation Descartes can be customized to integrate in workflows where automation and interface customization is required.
17. RLE, CIT, TG4, GP4, RT4, TIF, CAL, CRL, T29, MPF, RLC, LRD, HMR, iTiff.
18. Bentley I/RAS B is a powerful tool for organizations that work with scanned raster data. A proven raster editor, it offers a complete set of tools to meet the needs of engineers, architects, facilities managers and other professionals. Bentley I/RAS B is an essential element in any MicroStation workflow, as it is the ideal product to digitally archive, clean up or convert scanned drawings. With Bentley I/RAS B tools, you can restore and revise legacy drawings in raster format. Bentley I/RAS B includes a full suite of engineering tools to help in the conversion from raster to vector. A powerful Optical Character Recognition engine allows seamless conversion of raster text-to-text elements and automated conversion for lines and symbols. Bentley I/RAS B is ProjectWise compliant and fully integrated into MicroStation.

Blue Marble Geographics

1. JPG, JP2, BMP, PNG, TIFF, GeoTIFF, MrSID, ECW, BIL, BSQ, BSB, ADRG, CADRG.
2. JPG, JP2, BMP, PNG, TIFF, GeoTIFF, ECW.
3. DXF, DWG thru R2000, SHP, TAB, MIF, ASCII, SEG P1, dBase, XLS, LOTUS.

Caliper Corporation

1. TransCAD for the Web, TransCAD for Great Britain.
2. Maptitude for the Web, Maptitude for Redistricting, Maptitude for Great Britain.

CEDRA Corporation

1. Cedra AVseries includes AVland, AVsand, AVparcel, AVwater and AVcogo.

ESRI

1. With ArcSDE: Oracle, MS SQL Server, IBM DB2, and Informix.
2. ESRI Grid, ESRI SDE raster, ESRI Raster Catalogs.
3. VBA (included) and COM compliant languages.
4. ArcPad is a field data collection technology for use on Windows CE handheld devices. For more information visit www.esri.com/arcpad.
5. Yes, with ArcGIS Geostatistical Analyst & Spatial Analyst.
6. Since our extensions have to run on a version of our core GIS product you receive a huge array of functionality. ArcInfo, ArcEditor, or ArcView can be used in combination with any extension to increase your productivity no matter what industry you are in.

Golden Software

1. EMF, E00, DXF, DLG, LGO, LGS, BNA, CLP, WMF, SHP, MIF, BMP, JPG, PNG, TIF, TGA, PCX, DCX, WPG, PCT, GIF, and USGS SDTS TVP (DDF), XLS, DAT, SLK, WKx, WRx, CSV, TXT, BNA, GRD, DEM, HDR, DDF, DT0, DT1, DT2.
2. EMF, MIF, DXF, SHP, WMF, CLP, CGM, TIF, BMP, JPG, TGA, PCX, DCX, WPG, PNG, BNA, GIF, PCT, XLS, SLK, CSV, TXT, DAT.
3. GeoTiFF, TFW, RSF, EMF, BNA, DLG, LGO, LGS, DXF, PLT, CLP, WMF, DDF, SHP, MIF, E00, TIF, BMP, TGA, PCX, GIF, WPG, DCX, EPS, JPG, PCT, PNG, XLS, SLK, DAT, CSV, TXT, BNA, WKx, WRx.
4. LAS, EMF, DAT, DXF, SHP, BNA, GIF, CGM, MIF, WMF, CLP, TIF, BMP, TGA, PCX, WPG, PNG, JPG, PCT and DCX.
5. GeoTIFF, SHP, MIF, DDF, E00, EMF, WPS, PNG, DLG, LGS, WPG, DXF, LGO, BNA, WMF, CLP, TIF, BMP, JPG, GIF, TGA, PCX, DCX, PCT, XLS, SLK, WK1, WK3, WK4, WKS, CSV, WRK, WR1, DBF, MDB, DB, TXT, DAT.
6. SHP, MIF, BLN, EMF, PNG, DXF, WMF, CLP, CGM, TIF, BMP, JPG, GIF, TGA, PCX, WPG, PCT, DCX, BNA, XLS, CSV, TXT, DAT, SLK.

Intergraph Mapping and Geospatial Solutions

1. GeoMedia Suite of Products and IntelliWhere Suite of Products.
2. US toll free phone number: 800/791-3357.
3. Win NTS 4.0/NTW 4.0/2000 Server/2000 Professional/XP Professional.
4. Yes, pseudo in core products. Complete with Dynnamo or 3rd party—radius topology.
5. OGC WMS, WFS, GML, ESRI shape, ArcInfo binary, MapInfo binary, Microstation DGN V7, Microstation DGN V8, AutoCAD DWG, DXF, MGE, MGDM, Oracle Object, Framme, SQL Server, SmartStore, Text/ASCII , Access.
6. Microstation DGN, MapInfo MIF, Oracle Object, SQL Server, AutoCAD DXF, AutoCAD DWG, GML, ESRI shape.
7. OGC WMS, WFS, GML, ESRI shape, ArcInfo binary, MapInfo binary, Microstation DGN V7, Microstation DGN V8, AutoCAD DWG, DXF, MGE, MGDM, Oracle Object, Framme, SQL Server, SmartStore, Text/ASCII . Access, all industry standard raster formats including Intergraph raster formats, JFIF, PCX, TIFF, GeoTIFF, MrSID, ECW, ESRI bip and bil, bitmap, CALS 24, Hitachi, IGS, USGS DOQ,DTED., USGS DTM.
8. Can export from any supported format.
9. No, can use any OLE/COM compatible language.
10. GeoMedia and IntelliWhere Products: GeoMedia, GeoMedia Grid, GeoMedia Image, GeoMedia Image Light Table, GeoMedia Objects, GeoMedia WebMap, GeoMedia WebMap Professional, GeoMedia WebMap Publisher, GeoMedia Parcel Manager, GeoMedia Professional, GeoMedia PublicWorks Manager, GeoMedia SMMS, GeoMedia Terrain, GeoMedia Transaction Manager, GeoMedia Transportation Analyst, GeoMedia Transportation Manager, GeoMedia Viewer (FREE), GeoMedia-Hansen Interface, SMMS, SMMS GeoConnect, IntelliWhere OnDemand, IntelliWhere LocationServer, IntelliWhere TrackForce.
11. Oracle, SQL Server, IBMDB2, Access.

Leica Geosystems GIS and Mapping

1. Only through customization with Developers' Toolkit.
2. 100+ standard raster and vector formats.
3. 30+ standard raster and vector formats.
4. 50+ standard raster and vector formats.
5. ERDAS IMAGINE .img, TIFF, GeoTIFF, ESRI GRID, Shapefile, Arc/Info coverage, Simple Personal Geodatabase.
6. Can be done using Developers' Toolkit.
7. ESRI ArcGIS, VLS Feature Analyst.

MapInfo Corporation

1. Spatial Data support for MS SQL server and IBM Informix.
2. MS Access, MS SQL Server, Oracle, IBM Informix.
3. Read-only via database.
4. Yes, via Vertical Mapper and Geographic Transformer.
5. Integrated mapping via Visual Basic.
6. Pocket PC MS SQL support.
7. Embedded Visual C and embedded Visual Basic.
8. Win 2000, 2003, Unix, Linux, or any OS with Java VM jdk1.4 or higher.
9. Oracle, Microsoft SQL Server, IBM Informix.
10. Oracle Spatial, SpatialWare for SQL Server and IBM Informix.
11. Via MapInfo MapMarker.
12. Via MapInfo Routing J Server.
13. Any system with a web browser.
14. SpatialWare for SQL Server and IBM Informix.
15. All Microsoft .NET languages are supported.
16. Numerous formats supported.
17. Microsoft Pocket PC PDA systems.

Maptek

1. ESRI shape, TIN and GRID; DGN; DXF; ASCII: Datamine strings, triangulations and databases; Micromine strings and databases; Surpac strings and triangulations; Easimine triangulations; Geocomp strings; Phodat; Minex strings and triangulations; Petrosys DIG; Whittle 4X.
2. ESRI shape, TIN and GRID; DGN; DXF; ASCII: Datamine strings, triangulations; Micromine strings; Easimine triangulations; Pitboss; Eclipse.
3. Vulcan databases (.DGD & Isis) and ASCII.

NovaLIS Technologies

1. Subsequent seats $1495.
2. DXF, DWG, ShapeFile, GDB, Coverage.

Paragon Imaging Inc.

1. NITF, GeoTiff, Shape, JPEG and more.
2. Accurate coordinate/precision entry.
3. Advanced image enhancement functions.

SOKKIA

1. UDF, SHP, MIF, DXF, DGN, SIF, TIF, JPQ, DOQ.
2. DXF, ASCII, CR5, SHP, MIF.
3. UDF, SHP, MIF, DXF, DGN.
4. UDF, CR5, SHP, DXF, ASCII.

Tripod Data Systems

1. ESRI Shape, MapInfo MIF, AutoCAD DXF, Intergraph DGN.
2. RTI Application available through LandMark Systems.

Trimble

1. Windows XP/2000/NT/Me/98/95/CEv2.11/2.12/3.0/4.0.
2. The TerraSync software is a flexible GIS field data collection tool that can be used on a variety of field computers running Windows or Windows CE. The software is designed to integrate seamlessly with industry-standard GIS systems, and Trimble's GPS Pathfinder receiver range.
3. ArcView Shapefile, AutoCAD DXF, dBase, MapInfo MIF and Microsoft Access MDB.
4. ARC/INFO Generate, ArcView Shapefile, AutoCAD DXF, ASCII, MIF, MDB, DGN.
5. Trimble's GPS Pathfinder Office software supports all aspects of GIS data collection and data maintenance for Trimble's GPS Pathfinder and GeoExplorer systems.

ANNEXURE VII

Useful Web Addresses and Linkages

A

- **AGIS mapping/GIS shareware for Windows 3.x/95/NT** (http://www.agismap.com/) Free evaluation version of Windows mapping/GIS shareware. World map and demonstration examples provided.
- American Association of Geographers (AAG) **GIS Specialty Group** (http://www.cla.sc.edu/gis/aaggis)—online copies of their newsletter.
- **American Society for Photogrammetry and Remote Sensing** (http://www.asprs.org/asprs/) is to advance knowledge and improve understanding of mapping sciences to promote the responsible application of photogrammetry, remote sensing, geographic information systems and supporting technologies.
- **AM/FM Services for Electric Utilities** (http://srimap.com/) SRI utilizes the latest in GPS, Laser Rangefinders, and Pen Based Computers to complete a comprehensive inventory on power systems throughout North America.
- **AM/FM,GIS—Automated Digital Mapping** (http://www.4amfmgis.com/) Sentinel USA, Inc. specializes in AM/FM,GIS services, providing major utilities with Automated Digital Distribution Mapping, Field Inventory, and Digital Conversion Services.
- **AnGIS Software Home Page** (http://www.angis.com/) AnGIS Software is the developer of the AnGIS GIS system, as well as the Datamorpher engine. We support data conversions including ESRI SHPfiles as well as Landmark Argus GIS transfers.
- **ARGUS Technologies Inc.** (http://www.argusmap.com/) produces a Census Mapping GIS for Windows. Product information and downloadable data are available.
- **Array Systems** (http://www.array.ca/) Remote Sensing Image Processing and Archiving Systems.
- The **Association for Geographic Information** (http://www.geo.ed.ac.uk/agi/agi.html) The national coordinating body for GIS activities in the United Kingdom.
- **Atlas of Estonia** (http://atlas.ibs.ee/) Map server of Estonia with search by location or name.
- **Australian Centre of the Asian Spatial Information and Analysis Network (ACASIAN)** (http://www.asian.gu.edu.au/) is an academic and applied research institution specialising in Geographical Information System (GIS) databases for Asia and the former Soviet Union.
- **Australian Geodynamics on-line GIS** (http://www.ned.dem.csiro.au/AGCRC/4dgm/grasslinks/) Interactive GIS including information about Australian geology and geophysics.

- **Australian National University—Bioinformatics.** (http://life.anu.edu.au/) Biodiversity, Bio-information, Complex systems, and other material on complex systems research.
- **AutoDesk GIS Software** (http://www.autodesk.com/solution/gis/gis.htm) Descriptions of GIS solutions at AutoDesk, AutoCAD Map 2.0, AutoDesk World, and AutoDesk MapGuide.

B

- **Baltic Sea Drainage Basin GIS, Map and Statistical Database** (http://www.grida.no/baltic/) Europe's most comprehensive trans-boundary and multi-thematic GIS database. GIS data provided in PCARC/INFO and IDRISI formats. Thematic focus upon land cover and population. Links to related information.
- **Baltic Universities GIS (BUGIS) Project** (http://www.hydroconsult.se/UGIS/UGIS.html). Information resources and Data for the Baltic Countries.
- Bentley Systems **MicroStation Server** (http://www.bentley.com/) (see also Intergraph)
- The **Department of Geography** (http://www.geog.buffalo.edu/) at the State University of New York at Buffalo, and their **Geographic Information and Analysis Laboratory (GIAL)**. (http://www.geog.buffalo.edu/GIAL/welcome.html)
- **British Historical Geographical Information System Program** (http://www.qmw.ac.uk/~gbhgis) This major research programme is constructing a historical GIS covering the changing administrative units of Great Britain over the last 200 years combining computerised boundary information with a wide range of statistics.
- **BOSS International** (http://www.bossintl.com/) A recognized world leader in the field of civil and environmental engineering software.
- **Business Geographics Online** (http://www.geoweb.co.uk/) Comprehensive list of Census, geodemographic and lifestyle datasets. Locale on-line profiling system for generation of 1991 Census and Super Profiles local area reports. Description of Business Geographics Limited's activities, software and services. On-line click able maps and forms for finding UK property. An Internet GIS.

C

- **Caliper Corporation** (http://www.caliper.com/default.htm) Caliper Corporation performs GIS consulting and custom software development, and offers three Microsoft Windows-based GISs: Maptitude, TransCAD and GIS+.
- **CARIS GIS** (http://www.universal.ca/) Universal Systems Ltd.—home of CARIS GIS, CARIS Marine Information Systems and CARIS++ Spatial Objects, our new Application Development Toolkit.
- **Cartographic Technologies, Inc.** (http://www.sover.net/~cti) Cartographic Technologies, Inc. (CTI) provides a wide range of Geographic Information System (GIS), mapping, and information technology services to help clients gather and evaluate the information needed to make sound decisions.

- **Center for Advanced Spatial Technologies** (http://www.cast.uark.edu/) GIS Research, Professional Development Courses, Digital Data Catalog On-line Interactive Mapper, Other GIS links.
- **Southern California Traffic Report** (http://www.scubed.com/caltrans/transnet.html). This case study provides real-time mapping of the current traffic situation for commuters.
- Cartographic Communication research
- **CIEG—Federal University of Parana, Brazil** (http://www.cieg.ufpr.br/) CIEG (GIS center) it's one of the most important center for research, education and projects in South Brazil. Links to Brazilian GIS sites, Projects Going on, News.
- **Community 2020 (tm) GIS for Community Organizations** (http://www.hud.gov/cpd/2020soft.html)
- **GIS data for Connecticut, USA** (http://magic.lib.uconn.edu/) MAGIC, the Map and Geographic Information Centre at the University of Connecticut, is the geodata library for the State of Connecticut.
- **Consortium for International Earth Science Information Network (CIESIN)** (http://www.ciesin.org/) are a gateway to data and other information.
- **Consulting company for GIS, remote sensing and informatics—GISL** (http://www.gisl.co.uk/) Company profile, description of services and expertise, summaries of previous projects, links to other GIS sites and much more.
- **Convergent Group** (http://www.convergentgroup.com/) GIS systems integration and consulting, including Smallworld, ESRI, Intergraph, MapInfo, Genasys. Related technologies including Outage Management, CIS, Field Data, Document Imaging, Data Conversion, hardware and database management vendors.
- **Institute of Geography, University of Copenhagen** (http://www.geogr.ku.dk/) Research, educational and staff information at the Institute of Geography, University of Copenhagen. Also the homepage of WinCHIPS (a remote sensing software package developed at the institute) and NOAA satellite images.

D

- **Data-on-the-Map 3.0—Desktop GIS/mapping software** (http://www.chartwrite.se/) An innovative, cost-effective and user-friendly GIS/mapping software based on object oriented programming.
- **DDS Digital Data Services GmbH** (http://www.dds.ptv.de/) Geographic and Demographic data for Germany and Europe. Offers a single data source for geomarkting applications primarily for Germany, but also for many other countries.
- **Department of Geology and Geography at Hunter College, NY** *(http://www.geo.hunter.cuny.edu/)*
- **Deutches Klimarcchenzentrum** (http://www.dkrz.de/index-eng.html)(German Climatology Computer Centre). Climatic data and simulations.
- **Drake University Department of Geography and Geology** (http://www.geo.drake.edu/) Info about Drake's GIS Lab, courses, student projects and department of Geography and Geology (Des Moines, Iowa, USA).

E

- **EarthData International** (http://www.earthdata.com/) The Photo Science group specializes in the development of spatial data technologies to support the engineering, environmental and land management applications of its clients.
- **EarthWatch Incorporated** (http://www.digitalglobe.com/) is launching the world's first commercial, high-resolution satellite. Browse and order worldwide satellite imagery and geographic data from the Digital Globe database. The site also contains 1m and 3m satellite imagery.
- **Edinburgh University, Geography Department.** (http://www.geo.ed.ac.uk/home/gishome.html) The home of this page and other GIS Resources such as the AGI Dictionary, the Digital Elevation Data Catalogue and research projects including the first Arc/Info <-> WWW interface and the World-Wide Earthquake Locator.
- Edinburgh University, **Data Library,** (http://datalib.ed.ac.uk/) spatial meta-data research, UKBORDERS digital data and ESRC Research Publications Database (incl. some GIS research).
- **Environmental Data Center** (http://www.edc.uri.edu/rreapage/edc/)—GPS data and Rhode Island Geographic Information System Data. The site contains information about the Environmental Data Center, Rhode Island GIS data, and GPS Base Station data.
- **Environmental Management Earth Observation Group** (http://www.unn.ac.uk/~evz9/eog/eog.htm) at University of Northumbria, UK.
- **Environmental Systems Research Institute (ESRI)** (http://www.esri.com/). The vendors of Arc/Info and related GIS software and data. Includes Arc View V1.0 for free.
- **ERDAS, Inc.** (http://www.erdas.com/) Visualization and analysis tools to help you make more informed business decisions involving the geography that surrounds you.
- **ER Mapper** (http://www.ctv.es/ermapper) site for Southern Europe and North Africa. ER Mapper is a powerful yet easy to use geographical data viewer and processor.
- **ESRI Canada Ltd.** (http://www.esri.ca/) Official Canadian Distributor of ESRI, world leader in Geographic Information System technology. Resources include GIS products, consulting, training, support, partner program, us
- **ESRI Canada—Schools and Libraries Program** (http://www.esri.ca/k-12/) ESRI Canada is dedicated to promoting GIS awareness in schools!
- **ETAK—GIS Software and Digital Mapping Information** (http://www.etak.com/) A number of online demos for using maps on the internet, geocoding records over the internet, product descriptions.
- **Eurimage** (www.eurimage.com.) a major commercial course of remote sensing imagery
- **EUROGI** (http://www.eurogi.org/) the European Umbrella Organisation for GIS
- European Science Foundation, **GISDATA Programme** (http://www.shef.ac.uk/uni/academic/D-H/gis/gisdata.html) includes specialist meetings, GISDATA directory, news/outputs.

F

- **GIS in Production Agriculture** (http://www.farmsoft.com/) Farmer's Software Association (FSA) supports information tools for mapping and management of site-specific farming data in a GIS framework. Trainings are also held to educate agriculturalists in the application of GIS technology to farming.

- **Fenstermaker Online** (http://www.fenstermaker.com/) is your access gateway for interacting with our company. Learn about our GIS (Geographical Information Systems), Smart Maps, GPS (Global Positioning System) Surveying services, Engineering capabilities, comprehensive Environmental services, and some of our history.
- National Land Survey of Finland (the Finnish National Mapping Agency), **Geographic Information Centre** (http://www.nls.fi/index_e.html) includes details of their products and standards.
- **Institut Geographique National** (IGN) (http://www.ign.fr/)—the French national mapping agency.
- **Friends of the Earth** (UK) (http://www.foe.co.uk/cri/html/postcode.html) Describes their use of GIS with examples. Of particular interest is their Chemical Release Inventory.

G

- **Genasys II Inc.** (http://www.genasys.com/)—Information about their products.
- **Geodetic Survey of Canada** (http://www.geod.emr.ca/) Information about the organisation and the data products on offer.
- **GEODYNAMIC SOLUTIONS, INC.** (http://www.compassnet.com/gsi) Our company specializes in GIS applications and software solutions for the petroleum industry and for market, demographic and retail site selection analysis.
- **GEO/Graphics** (http://www.geograph.com/docs/html/homestart.html) GIS Consulting and Mapping
- **Geographical Institute (Slovenia)** (http://www.zrc-sazu.si/gi/) Geographic Institute has been established in 1948. It has been divided into 5 organisational units since 1993: Department for Geoecology, Department for Regional Geography, Department for Natural Disasters, Department for Geographical Information System and Department for Thematic Cartography. Activities have been organised into four research projects (basic, evolutional, applicational and aiming).
- **Geological and Geophysical Databases, Information and GIS Map Server, Cornell University** (http://atlas.geo.cornell.edu/) This page is designed to provide information, data, and maps for scientists, students and for those who are interested in the geological sciences.
- **Geomatics International Inc Web** (http://www.geomatics.com/) providing consulting services to private sector institutions, non-governmental organisations and all levels of government using leading edge GIS and web technology.
- **GeoNorth** (http://www.geonorth.com/) GeoNorth is a company owned by professionals dedicated to the development and support of GIS, database, and CAD systems. GeoNorth offers a full range of services including custom GIS and database application programming, GIS mapping and data conversion, and computer network administration. GeoNorth also offers a full line of the world's most popular software products including packages from ESRI, Sybase, and AutoDesk. We are also known for the development and marketing of out own, CD-ROM based products, and for developing software products for third parties.

- **GeoSoft Ltd.** (http://www.geosoft.co.uk/) are the U.K. based suppliers of Map Server and Map Server, which allow turnkey systems to be produced in Visual Basic.
- **GeoSolutions Consulting Inc.** (http://www.geosolutions.com/) GeoSolutions Consulting Inc. provides consulting and technical services in the area of Geographic Information System (GIS).
- **GeoSolutions, GIS Consultants and Application Developers** (http://www.geosolutions.ie/) GeoSolutions provide GIS services internationally. We focus on solutions and services specific to user requirements. GeoSolutions an ESRI Business Partner.
- **GeoSearch** (http://www.geosearch.com/) are a US-based GIS employment agency.
- **GEO/SQL** (http://www.geosearch.com/) Geo/SQL Corporation is an international company committed to providing a high level of consulting expertise in the design, implementation and maintenance of GIS systems.
- **GeoWeb—On-Line Resources for GIS/GPS/Remote Sensing** (http://www.ggrweb.com/) WWW Online Resources For Geographic Information System (GIS), Global Positioning System (GPS), & Remote Sensing (RS) Industries.
- GIS-L **Frequently Asked Questions (FAQ)** (http://www.census.gov/geo/www/faq-index.html) Page, which provides answers to many basic GIS questions.
- **GIS Internet Resources List** (http://www.lib.berkeley.edu/UCBGIS/gisnet.html) Extensive, annotated list of pointers, broken down into subject areas. Also has information on GIS education, research and facilities at UC Berkeley. This is part of the UCBGIS web site at: http://www.lib.berkeley.edu/UCBGIS/
- **GISLinx—complete GIS listing** (http://www.gislinx.com/) GISLinx is an Internet site devoted to listing any/all GIS related web sites. GISLinx contains links to GIS related software and hardware vendors, GeoMatics, GIS Events, and GIS Data.
- **GISL (UK) Ltd.—GIS, Remote Sensing and Informatics Consultants** (http://www.gisl.co.uk/gisl.htm) GISL is a UK firm specialising in the applications of GIS and remote sensing. Working worldwide on muti-disciplinary projects involving natural resources, rural development, environmental and urban planning. GISL provides support and training in most GIS and image processing software packages. We also have a register for freelance consultants with experience in GIS and/or remote sensing.
- **GISplan research group at Aalborg University, Denmark** (http://www.i4.auc.dk/gisplan/) A brief introduction to GISplan and the idea behind our research. There is also useful links to projects and partners around the world. We also have links to our own homepages. Expect to see updates every 3 month. We are committed to use the web for communicating our research and other information.
- **GIS World** (http://www.gisworld.com/), publishers of GIS World magazine and GIS Asia Pacific. Includes GIS conferences, books and other information.
- **Global Land Information System (GLIS)** (http://edcwww.cr.usgs.gov/glis/glis.html) Metadata and samples for many data products, such as AVHRR, TM, DEMs, Aerial Photographs, SLAR, World Ecosystems, and World Soil Series.
- **Global Positioning System Software (GPSS)** (http://www.gpss.co.uk/) Free GPS Software for Windows called GPSS—a mobile speech GIS. Information on what GPSS is, when it will next be on TV, and how to get a free copy, e.g. Personal Computer World CD-ROM Now available for whole World. Displays map and speak.

- **GRID-Geneva (Global Resource Information Database)** (http://www.grid.unep.ch/) GRID-Geneva is part of UNEP (United Nations Environment Programme). It provides GIS service and distributes free geo-referenced environmental data sets.
- **GTI (GIS Technology, Inc.)** (http://www.gistech.com/) All the GIS you can stand! Description of our products and services, articles, white papers, and other GIS links.
- **Geo InSight International, Inc.** (http://www.geoin.com/) GIS Training consulting and software development for private sector, state, federal govt, and environmental orgs.

H

- **Harlow Report Geographic Information Systems** (http://www.geoint.com/) Monthly newsletter on GIS since 1978. Free trail subscription available.
- **University of Hong Kong, GIS/LIS Research Centre** (http://www.hku.hk/cupem/hkugis)

I

- The **IDRISI Project** (http://www.idrisi.clarku.edu/) home page.
- An **IDRISI WWW Tutorial** (http://www.sbg.ac.at/geo/idrisi/wwwtutor/tuthome.htm) from the at the University of Salzburg.
- **ImageNet** (http://www.coresw.com/) from Core Software Technology—Search and Preview geospatial data (sat. imagery, air photography, vector data). There is a European node at **Eurimage** (http://www.eurimage.com/) in Italy.
- **Infomine Maps and GIS** (http://lib-www.ucr.edu/mapsinfo.html) University of California index to internet resource collections—access by subject, title, and keyword. Strong focus on California and the western United States.
- **Informed Solutions GIS and Spatial Data Consultants** (http://www.informed.co.uk/) A Group of Consultants specializing in the development of all leading GIS and RDBMS products. The leading independent consultancy in the country.
- **Institute of Advanced Technologies (Kyiv, Ukraine)** (http://www.iat-ant.com/) is the only organization in Ukraine, which is designed to develop geographical maps and atlases with help of exceptional GIS-technologies.
- **Intergraph Corporation** (http://www.intergraph.com), GIS product applications and Industry Solutions for Business and Governments.
- Iowa Department of Natural Resources, **Geological Survey Bureau** (http://samuel.igsb.uiowa.edu/). Information on geological and water resources in Iowa, including their Natural Resources GIS (with >400 documented databases) and publications.
- **International Institute for Aerospace Survey and Earth Sciences** (http://www.itc.nl/) The International Institute for Aerospace Survey and Earth Sciences (ITC) is the largest institute for international higher education in the Netherlands. The main objective of ITC is to assist developing countries in human resources development in aerospace surveys, remote sensing applications, the establishment of geoinformation systems and the management of geoinformation. To this end ITC concentrates on three main activities: education, research and advisory services.

- **IntelliGIS Inc.** (http://www.igis.com/) provides the Energy Industry with GIS Software, including a data navigator for geoscience databases and a system for the protection of underground utilities.
- **ISO/TC 211 Geographic Information/Geomatics** (http://www.isotc211.org/) This is the web-server of the ISO committee for geographic information/geomatics. Contains work programme, organization including working groups and experts, important information, calendar, list of documents, etc.

J

- **James W. Sewall Company** (http://www.jws.com/) Mapping, GIS, and resource management consultants to government, utilities, and private industry. Expertise in aerial photography, surveying/GPS, photogrammetric mapping, and orthophotography.
- The **Japan GIS/Mapping Sciences Resource Guide** (http://www.cast.uark.edu/jpgis/), which is a comprehensive reference for Japan's GIS, map and geospatial data products, activities and information sources.
- **GIS Jobs and Resources** (http://www.geojobsource.com/) An international listing of career opportunities in the spatial data professions and a growing directory of professional development resources.
- **GISLinx Job Department** (http://www.gislinx.com/) The GJD is a rapidly growing listing of GIS Job openings and GIS resumes from around the world. This listing is searchable, or may be viewed by categories.
- **JShape—Publish GIS to WWW** (http://shiuhlin.ml.org/) JShape is a Java applet to allow users to publish their GIS shapefile to WWW. JShape is a vector-based shapefile browser, which supports zoom/pan/query/select/label/... most GIS functions.

K

- **Land Info** (http://www.landinfo.com/) Provider of map data products. Land Info's Web site describes the company and its products, and includes a number of free downloadable sample map data product files.
- **Laser-Scan Ltd.** (http://www.lsl.co.uk/) are a U.K. GIS systems vendor. Details on products, case studies and technical papers.

M

- **MAPCOM Systems, Inc** (http://www.mapcom.com/) AM/FM-GIS Windows based software for utilities that Works! Descriptions of M4, Windows based AM/FM-GIS Software, applications and services.
- **Map*Factory—Raster Based GIS Modelling for the Macintosh** (http://www.keigansystems.com/) Keigan Systems specializes in the development of Macintosh spatial modelling, image processing, and GIS mapping products. Software tools include Map*Factory, Map*Factory Developer's Kit and MAP II.
- **MapInfo Corporation** (http://www.mapinfo.com/) MapInfo Corp website

- **Map Maker** (www.mapmaker.com) is a user friendly GIS for Windows.
- **MAPublisher** (http://www.avenza.com/) by Avenza Software Marketing Inc. MAPublisher is a software program that works with Adobe Illustrator and Macromedia Freehand to allow the import of GIS data into the graphics programs.
- **Massachusetts Electronic Atlas** (http://icg.harvard.edu/~maps/maatlas.htm) This is an interactive atlas showing data by minor civil division with over 200 thematic layers.
- **McGill University Department of Earth Sciences** (http://stoner.eps.mcgill.ca/) Some departmental information, together with an ftp server and a gopher server.
- **Meridian GIS—Consultants for Mineral Exploration** (http://www.meridian-gis.com.au/) Meridian GIS was formed in January 1997 to service the mapping and GIS community in Australia. We aim to supply MicroStation support in the mineral exploration and cartographic areas and more specifically the integration with GIS.
- **MicroImages, Inc.** (http://www.microimages.com/) Free TNTlite spatial data viewing and analysis software, plus the whole line of TNT professional GIS, Image Processing, and desktop cartography products.
- **Minnesota Legislative GIS Office** (http://www.commissions.leg.state.mn.us/). Information and mapping for Minnesota. Not much of great interest as yet.
- **Mississippi Automated Resource Information System** (http://www.maris.state.ms.us/) GIS clearinghouse for Mississippi. Online data, extensive metadata and links to other GIS sites around the country.
- **GIS/Census data for Missouri** (http://msdis.missouri.edu/) GIS datasets in ARC/INFO.e00 format for Missouri at the county level as well as the state level. Home pages for MGISAC and Missouri GIS user groups also available.
- **MIT Earth Resources Laboratory** (http://eaps.mit.edu/erl/) Environmental Engineering, Applied Geophysics, Centre for Advanced Geophysical Computing, and GPS information.
- **MultiGIS** (http://ourworld.compuserve.com/homepages/mgis) 3D photo realistic terrain visualization and DEM and planimetric digitisation. We offer FREE datum and coordinate software. Article preprints.
- **Municipal Software Corporation** (http://www.islandnet.com/municipal/) has been providing superior software solutions to local government for over 12 years.

N

- The U.S. **National Center for Geographic Information and Analysis (NCGIA)** (http://www.ncgia.ucsb.edu/)—Projects, References, Reports and Bibliographies relating to a range of GIS themes
- **State University of New York (SUNY) at Buffalo** (http://www.geog.buffalo.edu/ncgia/)
- **National Key Centre for Social Applications of GIS, Australia** (http://www.gisca.adelaide.edu.au/)
- **Natural History Museum** (http://www.nhm.ac.uk/) (U.K.)—Not much on GIS as yet, but we suspect they are working on it.
- The U.S. State of North Carolina's **Center for Geographic Information and Analysis** (http://cgia.cgia.state.nc.us/)—Spatial data and information about the state GIS in North Carolina.

- **Natural Resources Development Centre** (http://www2.tcd.ie/Natural_Resources/) The NRDC brings to environmental management and resource development an understanding of not only natural and human processes but also an expertise in information systems. By combining these abilities its develops new tools and provides new solutions for the optimal use of the environment.
- **NOAA National Geophysical Data Center** (http://www.ngdc.noaa.gov/) A major provider of environmental data (international data in addition to USA).

O

- Oak Ridge National Laboratory (U.S.) **Environmental Science Division,** (http://www.esd.ornl.gov/ern/embyr/embyr.html) including their **Fire Simulation Model (EMBYR)** (http://www.esd.ornl.gov/ern/embyr/embyr.html)
- **Object/FX Corp.—Embedded Tools and Interactive Internet Spatial Applications** (http://www.objectfx.com/) A software company offering spatial/mapping visualization tools and templates for viewing and interacting with enterprise data in either client/server or web environments.
- **Ohio State University—Center for Mapping** (http://www.cfm.ohio-state.edu/)
- Ohio State University, **Department of Geography** (http://thoth.sbs.ohio-state.edu/) provides a major GIS On-Line Bibliography.
- **On Target Mapping** (http://www.otmapping.com/) is a leading provider of desktop mapping solutions for service-intensive industries and is a leader in telecommunications mapping, routing, logistics and scheduling technologies.
- **Open GIS Consortium** (http://ogis.org/). Information on the OpenGIS Project, meetings, testbed and publications. OGIS also have pages at the **University of California at Berkeley** (http://www.regis.berkeley.edu/ogis/ogis.html) which include information on the public domain GIS.
- The **Ordnance Survey** (http://www.ordsvy.gov.uk/)—the U.K. National Mapping Agency. Including a query interface to the Spatial Information Enquiry Service (SINES).

P

- **PCI** (http://www.pci.on.ca/) produces software for Remote Sensing, GIS and Terrain Analysis, including EASI/PACE. Cartographic editing software in WYSIWYG environment reads 45+GIS file types. Produces high quality maps. Designed by Cartographers for cartographers.
- **PCI Pacific GeoSolutions, developers of PAMAP GIS** (http://www.pci-pacific.com/)
- **Pennsylvania Spatial Data Access** (http://www.pasda.psu.edu/) PASDA is PA's node on the NSDI. PASDA data includes Digital Elevation Models, Digital Raster Graphics, and Digital Ortho Photo Quadrangles as well as watersheds, streams, roads, etc.
- **Professional GEO Systems (PGS) Product Info (Java GIS Browser)** (http://www.pgs.nl/) PGS specializes in the development of Geographic Information Systems for Internet. We have developed a 100% Pure Java GIS browser.
- **PROGIS—Professional GIS software for MS Windows** (http://www.progis.com/) Professional GIS platform for Windows95/NT, add-on modules for GPS, Routing,

Multimedia, large Raster handling (TurboRaster) and Isoline generation. GIS downsizing with SDK, connection to any database.

R

- **GIS Conversion and Software Development** (http://www.ramtech-corp.com/) GIS Conversion and GIS Software Development Service Provider for Telecommunications, Electric, Gas, Water, Government, Mapping, Municipalities etc.
- **Retail Profit Management GIS Resources** (http://home.earthlink.net/~rpminfonet/gis.html) RPM, a provider of Atlas GIS solutions sponsors this comprehensive listing of GIS data sites, news and discussion groups, GIS and mapping related search engines, and more.
- **GIS data for Rhode Island** (http://www.edc.uri.edu/) Wetland, land use, images, roads, soils and utilities.
- The U.K. Regional Research Laboratory Initiative: **RRL Scotland** (http://www.geo.ed.ac.uk/home/research/rrl.html) Server includes information about RRL Scotland and links to the other RRLs.
- **RouteSmart Technologies (TM)** (http://www2.ari.net/distinct/) RouteSmart is a cutting edge vehicle routing and scheduling system for organizations that demand reductions in vehicle use, labour time and operating expenses.

S

- **SAFE Software Inc.** (http://www.safe.com/fme) Universal Spatial Data Translator—These pages contain a description, live demo, and downloadable version of the Feature Manipulation Engine (FME), which is a configurable, 'universal' Spatial and Attribute Data Translator. The FME supports most popular GIS formats, and more are being added all the time.
- **Sandia National Labs (USA) GIS Site** (http://www.sandia.gov/) Environmental, security, risk, transportation applications. ESRI ARC/INFO and ArcView technical information and code. MapObjects information.
- **Society of Cartographers** (http://www.shef.ac.uk/uni/projects/sc/index.html) Cartographic site organised by CARTOGRAPHERS for Cartographers and those interested in cartography and maps. Also has a listserver "CARTO-SOC" used by cartographers as an information and problem sharing resource.
- **South Carolina Dept. of Natural Resources GIS Data Clearinghouse** (http://www.dnr.state.sc.us/) Provides GIS data (soils, wetlands/land use, DLGs, DEMs, and DRGS) for public use in natural resource decision-making. All data is in Arc/Info EXPORT format and 1:24,000 scale.
- **State of Kansas GIS Data Access and Support Center** (http://gisdasc.kgs.ukans.edu/dasc.html)
- **The UK Street Map Web Site** (http://www.streetmap.co.uk/) Provides Street Maps of Greater London and Road Atlas coverage for the rest of the UK. Search by Post Code, Street Name, OS Grid & Landranger Grid (Lat/Long coming soon). Link to maps by Post Code or OS Grid.

- **Subway Navigator** (http://metro.jussieu.fr:10001/), An clever resource for navigating your way through the subway/underground systems of the world.
- Swiss Federal Institute of Technology—**GERMINAL Research Project** (http://dgrwww.epfl.ch/GERMINAL/Germinal.html), designing an Integrated Approach to Land Use Planning and Environmental Management.

T

- **TCI Software** (http://www.tcicorp.com/) An AutoDesk Registered Application Develope specializing in tools for creation, conversion and maintenance of AutoCAD basemaps since 1988.
- **Thomas Bros. Maps** (http://www.thomas.com/) of Irvine, California. Details on their PC GIS called GEOFINDER and their digital data.
- **TYDAC Research Inc.** (http://www.tydac.com/)

U

- **UNEP/GRID-Sioux Falls** (http://grid2.cr.usgs.gov/) Global Population Data Sets, Environmental Data-Sets Available On-Line, Digital Topographic Data, EDC's Landsat Archive and Landsat Pathfinder, Gap Analysis, Desertification, Land Cover Characterization.
- **The University of Arizona ARC/INFO User Group** (http://www.srnr.arizona.edu/UAAIUG/) The University of Arizona ARC/INFO User Group (UAAIUG) was established in 1994 to develop a forum for faculty, staff and students at the University of Arizona to share knowledge and ideas pertaining to Environmental Systems Research Institute (ESRI) software applications.
- United Nations Environment Programme (UNEP) **Global Resource Information Database (GRID) at Arendal, Norway** (http://www.grida.no/). Provide access to a number of environmental GIS databases.
- **The Center for Advanced Spatial Technologies (CAST)** (http://www.cast.uark.edu/) at the University of Arkansas.
- University of California at their **Research Program in Environmental Planning and Geographic Information Systems** (http://www.regis.berkeley.edu/) (REGIS). Information is available on the San Francisco Bay/Sacramento Area, also **GRASSLinks** (http://www.regis.berkeley.edu/grasslinks/index.html)—a WWW-GRASS interface.
- **University of California, Riverside Extension GIS Certificate** (http://www.unex.ucr.edu/GIS/GIS.html) Pages contain information regarding the curriculum of the innovative Certificate in Geographic Information Systems at UCR, the proposed Spring 1996 Institute and the current course offerings in GIS.
- **University College London, Dept. Photogrammetry and Surveying** (http://www.ps.ucl.ac.uk/) Information about research projects and master degrees in Photogrammetry, Surveying, GIS and Remote Sensing at UCL.
- **University of East Anglia, School of Environmental Sciences** (http://enva2.env.uea.ac.uk/)

- **University of Extremadura. Center of Teledetection and GIS.** (http://geot.unex.es/) Working group in GIS projects and digital image treatment, from satellite data, inside of the University of Extremadura in Spain.
- **University of Glasgow, Dept. of Geography and Topographic Science** (http://www.geog.gla.ac.uk/) Department Information, Undergraduate & Postgraduate courses, Staff Information, Specialist Department for cartography, digital mapping, GIS/LIS, remote sensing, photogrammetry, geodesy & surveying, Research Activities and Prospectus Information.
- **University of Leicester Department of Geography Research** (http://www.geog.le.ac.uk/argus/) Project Argus—Visualization for Spatial Scientists. These pages demonstrate the use of interactive computer visualization of spatially distributed data as a technique for helping teaching, learning and research in a variety of subjects, such as geography, geology, meteorology and the social sciences.
- **University of Maine, Spatial Information Science and Engineering** (http://www.spatial.maine.edu/) provides bibliographies to major GIS proceedings, full text of a GIS law proceedings, describes ongoing research and academic programs.
- **Department of Geography, University of Minnesota** (http://www.geog.umn.edu/) Complete description of degree programs in Geography at the University of Minnesota. Information for prospective and current students.
- **University of Minnesota Remote Sensing Laboratory** (http://www.gis.umn.edu/) for AML programs, general GIS/RS information, and GIS Jobs Clearinghouse.
- **University of Newcastle** (http://www.ncl.ac.uk/~ngraphic/) Information from the University Computing Service on Graphics and GIS at Newcastle University, including data, software, hardware and usage issues.
- **University of Salzburg, Institute for Geography** (http://www.sbg.ac.at/geo/home.htm) provide information which includes the UNIGIS German distance-learning diploma, and are an IDRISI Resource Center (http://www.sbg.ac.at/geo/idrisi/irchome.htm) (German language) which hosts an **IDRISI WWW Tutorial** (http://www.sbg.ac.at/geo/idrisi/wwwtutor/tuthome.htm) in English.
- University of South Carolina, teaches GIS at the **College of Liberal Arts** (http://www.cla.sc.edu/index.html#GIS)
- **University of Southampton, GeoData Institute** (http://www.geodata.soton.ac.uk/) is a research organisation and contract agency which provides tutorials and training aids.
- **University of Stuttgart, Institute for Photogrammetry** (http://www.ifp.uni-stuttgart.de/) Image processing, GIS projects and interests, and a GIS bibliography.
- **University of Virginia, Social Sciences Data Center** (**http://fisher.lib.virginia.edu/**) Virginia On-Line Atlas, Virginia Census Data.
- **State University of Utrecht, Faculty of Geographical Sciences** (http://www.frw.ruu.nl/) (The Netherlands) and their list of **"Nice" Geography/GIS/RS Servers** (http://www.frw.ruu.nl/nicegeo.html).
- **Urban Planning Job Links** (http://www.geocities.com/CollegePark/7442/jobs.html) This is the most complete collection of Urban Planning Job Links on the net. It includes a GIS section.
- **U.S. Army—Construction Engineering Research Laboratory (CERL)** (http://www.cecer.army.mil/) The home of the **GRASS** (http://www.cecer.army.mil/grass/

GRASS.main.html) public-domain GIS, the Open-GIS (OGIS) foundation, also spatial modelling information.

- **U.S. Bureau of the Census** (http://www.census.gov/)—General information and access TIGER census data and much other statistical information for the U.S.A.
- **U.S. Bureau Land Management** (http://www.blm.gov/)– includes a GIS page with data and information relevant to land management.
- **U.S. Environmental Protection Agency** (http://www.epa.gov/) includes information on projects using GIS such as their **National Estuary Program** (http://www.epa.gov/nep/nep.html). Also some information on software and data.
- The **U.S. Fish and Wildlife Service** (http://www.fws.gov/) supports the U.S. **National Wetlands Inventory** (http://www.nwi.fws.gov/) which provides access to GIS data relating to US Wetlands in a number of different formats.
- **U.S. Geological Survey (USGS)** (http://www.usgs.gov/), for general GIS information, application examples and lots more.
- **USGS Mid-Continent Mapping Center at Rolla, Missouri** (http://mcmcweb.er.usgs.gov/) USGS Map Products and Digital Data.
- **USGS EROS Data Center** (http://edcwww.cr.usgs.gov/)—Major Data Archive for world data, satellite imagery and etc.
- U.S. National Biological Service **Environmental Management Technical Center (EMTC)** (http://www.emtc.nbs.gov/http_data/image_maps/umrsys.html)—GIS Data Browsing and Retrieval System for the Upper Mississippi River System.
- **Universal Systems Ltd.** (http://www.universal.ca/) A software and systems integration company that develops, supports and markets, the CARIS GIS software.

V

- **Vermont Geographic Information System** (http://www.vcgi.org) The Vermont Center for Geographic Information offers GIS data for the state of Vermont.
- **Plug-in Visual WebMap for CAD/MAPS/GIS on WWW** (http://hem.passagen.se/project/npwebmap.htm) WWW Visual WebMap Plug-in for IT/CAD/GIS A Client/Server Concept for Intergraph or Bentley Systems vector and raster files or vector in database with API for user development.
- **Vysor Integration Inc.** (http://www.synapse.net/~vysor/)—V-Image Remote Sensing Software for Windows and downloadable Satellite Imagery.

W

- **Wageningen Agricultural University Department of Geographic Information Processing and Remote Sensing (GIRS)** (http://www.gis.wau.nl/girs/)
- Oliver Weatherbee's **GIS Software Listing** (http://triton.cms.udel.edu/~vinton/gis_gip/gis_gip_list.html) which contains contact information and short overviews of the majority of major GIS software packages currently available.
- **WWW Geospatial Metadata Server (GMS)** (http://www.geo.ed.ac.uk/~anp/gms/main.htm) On-line Demonstration software that includes a CSDGM metadata database,

metadata upload/parse/database entry, editing of database rows, data conversion utilities, a software robot for data mining and a coverage plotting utility.

- **WinCATS—GIS and Engineering Services** (http://www.appliedglobal.com/) WinCATS—Windows Cartographic Analysis Tool Set Powerful/Easy to use GIS for Windows.
- **Wireless Solutions through GIS** (http://www.geotranswireless.com/) Accelerated telecommunications PCS build-out solutions using GIS to site antenna facility identification, acquisition and construction management.
- **WorldSat International—satellite underlay for GIS** (http://www.worldsat.ca/) Complete cloud free images of the entire earth available as prints for display or in digital format for a GIS layer.

Additional Links

- **Geographic Engineering** (http://www.fleximage.fr/) FLEXIMAGE is a subsidiary of the French AEROSPATIALE MATRA Group. FLEXIMAGE is specialized in Image Processing software development and new technologies in GIS for Defence applications
- **MAPIX technologies Ltd** (http://www.mapix.com/) Hydrographic GIS applications using ESRI products, Oceanographic Databases for use in IS.
- **Update Description of Geodan** (http://www.geodan.nl/) Please skip the last part of the description: "Also on our site: Road planner, the planning tool for service engineers."
- **Versamap Digital Mapping—Cartography and Outline Maps** (http://www.versamap.com/index.htm) Home of the Versamap mapping program for drawing outline maps on many map projections, and of digital map databases designed for use with Versamap.
- **GIS Data for Australia and the World** (http://www.geomap.com.au/) Our extensive experience with GIS projects enables us to help clients design a GIS for their specific requirements and where necessary integrate GIS into existing operating procedures.
- **GIS Data Depot** (http://www.gisdatadepot.com/) Provides FREE GIS data downloads from over a terabyte of GIS data and creates custom spatial data CD-ROMs as an alternative.
- **Low-cost Desktop Digital Photogrammetry System (DDPS) for Windows** (http://www.3dmapper.com/) The DDPS is a fully featured, cost-effective, desktop soft photogrammetry solution that enables users to quickly and efficiently capture accurate 3D vector data and ortho-images...
- **Mapping Software Developer** (http://www.mapframe.com/) MapFrame Corporation develops smart simple solutions for field automation. The company's FieldSmart Solutions include map viewing, data collection, pole inspection, and mobile distribution design.
- **Innogistic Software plc** (http://www.innogistic.co.uk/) Developers and Suppliers of Geographical Information Systems, Land and Property Systems and AVL with two way data messaging.
- **The GeoCommunity** (http://www.geocomm.com/) Central location for GIS Data, News, Discussions, and much more.

- **Demographic and Spatial data products** (http://www.geolytics.com/) Data products on CD-Rom: CensusCD, CensusCD+Maps, CensusCD 1980, CensusCD Blocks, StreetCD (complete TIGER98)
- **Immense GIS software and GIS data site.** (http://www.manifold.net/)
- **Corporate GIS Consultants Web Site** (http://www.corp-gis.com.au/) An overview of Corporate GIS Consultants.
- **The only Interactive Webmap of Indonesia** (http://www.indomap.com/) If you want to find islands, mountains, straits, bays and street maps of Indonesian cities, find it here. And you can have your building/house (if it in Indonesia) to have a FREE page with the location map of course.
- **GIS Data Conversion** (http://www.kineticsolutions.net/)
- **Infrasoft Solutions Ltd—GIS and Engineering Solutions** (http://www.infrasoft-civil.com/) Company specialising in 3D data, OS products, 3D modelling and software for road, rail, air and GIS applications used in 62 countries.
- **ER Mapper Image Processing System** (http://www.ermapper.com/) Information on ER Mapper, ECW compression and Image Web Server Free GIS and other software plugins, Free software such as ER Viewer and Compressor and a free version of Image Web Server.
- **Image Web Server by Earth Resource Mapping** (http://www.earthetc.com/) Roam and zoom in realtime over terabytes of imagery using this revolutionary Image Web Server. Imagery ranges from scanned maps, aerial photo's and satellite data to traditional photography.
- **NYPIRG's Community Mapping Assistance Project** (http://www.cmap.nypirg.org/)
- **UNISIG Ecuador** (http://unisig.epn.edu.ec/) Unidad de Inteligencía Artificial y Sistemas de Información Geográfica (UNISIG) parte de la Facultad de Ingeniería de Sistemas de la Escuela Politécnica Nacional, Quito, Ecuador.
- **Discover the Benefits of a Proven GIS Solution.** (http://www.geosql.com/) Geo/SQL provides a cost effective way to create maps, integrate information, present powerful ideas, and develop effective solutions like never before.
- **GIS Software development recruitment consultants** (http://www.cityprojects.com/) Positions with the leading developers of GIS systems within the UK.
- **AM/FM/GIS for Telecommunications and Utility Companies** (http://www.enghouse.com/)

  Provide suites of specialist GIS software under the brand names TelcoWORKS, CATVWORKS, PowerWORKS & GasWORKS. For design and management of network infrastructures.
- **GIS applied to Marine Geology and Oceanography** (http://www.marinegis.com/) Resources for GIS systems applied to marine geology and oceanography (links to softwares, data, jobs, universities pages, etc.).
- **Remote Sensing Data and Geographic Information Systems** (http://lee1.en.a.u-tokyo.ac.jp/Rs&gis/Engl/Read_me.htm)
- **ENVI—The Environment for Visualizing Images** (http://www.rsinc.com/envi/index.cfm) Research Systems' remote sensing and image processing software specialising in hyperspectral analysis.

- **RiverTools—Surface Water and DEM modelling software** (http://www.rsinc.com/rivertools/index.cfm) Research Systems' hydrologic modelling software that can handle very large DEM's.
- **CgiMapScript** (http://imedia.king.ac.uk/) CGI script driven interactive web mapping using vector data to drive GIF & HTML generators or Java 'mapplets' derived from research into public transport information systems.
- **CAD/GIS/RemoteSensing/Digital Photogrammetry Solutions for 2000** (http://www.tricadinfo.com/) Total end to end Geomatics solutions.
- **The Engineers Resource** (http://engineersupply.com/) GIS Books, Engineering Reference, Products, Services, and Forums (discussion and chat) for Engineers of all types.
- **Devel-Tech Inc.** (http://www.develtech.ca/) Integrated Services in Geomatics & Information Management.
- **Smallworld** (http://www.gepower.com/dhtml/network_solutions/)
  Smallworld creates, markets and delivers spatial resource planning (SRP) software products to the Utility and Public Systems and Communications markets globally.
- **Planet Earth Home Page** (http://www.tidusa.com/PEHP2000/Planet_Earth/info.html) Connections to everywhere on every topic.
- **GIS Staff Consultants** (http://www.cityprojects.com/)
- **Mapping Services** (http://www.specksystems.com/)
  We offer services in mapping and GIS. We offer total solutions starting from data acquisition to the making of a final map
- **GIS and Mapping Services** (http://www.specksystems.com/)
  Facilities and capabilities of Speck Systems Limited, Hyderabad, India in the areas of GIS and mapping.
- **Florida Dept. of Environmental Protection GIS Section** (http://www.dep.state.fl.us/gis)
  Florida GIS data, satellite imagery, links to other GIS clearinghouses.
- **MapMart** (http://www.mapmart.com/) Online ordering of Digital Mapping and Aerial Photography. Specializing in USGS products as well as customized mapping. Provides aerial photography acquisition, photogrammetry services, digital mapping, ortho rectification, scanning and custom GIS solutions for public and private sector clients.
- **GIS Research Center Inasitute of Geograhy Russian AS** (http://geocnt.geonet.ru/)
  GIS—software, maps, data and projects, consultancy in Russia The most popular GIS software in Russia GeoDraw/GeoGraph/GeoConstructor ActiveX (2500 installations).
- **Developer of Mapping and Geographic Data Technologies** (http://www.chicagomap.com/) Developer of mapping and geographic data technologies. Key disciplines include vector map displays, geocoding, data compression, and other related software efforts.
- **Journalism and GIS** (http://groups.yahoo.com/group/JAGIS-L/)
  A resource site for journalists interested in using GIS for the analysis of a variety of phenomena and communicating of the results of that analysis.
- **GIS and GPS products and online support** (http://www.geoposition.com/)
  Products from ESRI, Trimble, Juniper, EDS. Support lists, data resources, forums, etc...

APPENDIX VIII

Recommended GIS and Related Books

Title	*Author*	*Description*
	1. Business GIS	
ArcView GIS Means Business: Geographic Information Systems Solutions for Business	Harder, Christian	Written for business professionals, this book presents success stories chronicling the application of desktop geographic information systems (GIS) to a variety of common business problems.
GIS for Business and Service Planning	Paul Longley (Editor),Graham Clarke (Editor)	Renowned contributors assess the links between technological change, analytical information and data customization which are now beginning to stimulate the wider adoption of GIS as a management and applied research tool.
GIS in Real Estate: Integrating, Analysing, and Presenting Locational Information	Gilbert H. Castle (Editor),Foreword by Joseph R. Stanfield	Advocates the use of Global Information. Systems (GIS) by real estate professionals, and demonstrates how this use can be achieved.
	2. GIS Developer Tutorials	
Focus on GIS Component Software: Featuring ESRI's Mapobjects	Hartman, Robert	Explaining what GIS component technology means for managers and developers, this text illustrates, through hands-on tutorials using Visual Basic and ESRI's MapObjects product, how to use GIS software components to create, display, and query geographic data. The CD-ROM includes an evaluation copy of Map Objects and tutorial code.
ArcView GIS/Avenue Developer's Guide, Third Edition	Razavi, Amir H.	ArcView GIS/Avenue Developer's Guide continues to offer readers one of the most complete introductions to Avenue, the programming language of ArcView GIS.

(contd.)

(contd.)

Title	Author	Description
3. Cartography		
How to Lie with Maps	Mark Monmonier	Originally published to wide acclaim, this lively, cleverly illustrated essay on the use and abuse of maps teaches us how to evaluate maps critically and promotes a healthy skepticism about these easy-to-manipulate models of reality.
AAA's Atlas of Population and Environment	Paul Harris, Fred Pearce	The AAAS Atlas of Population and Environment is an important analysis of the relationships between human population and the environment.
Longitude: The True Story of a Lone Genius Who Solved the Greatest Scientific Problem of his Time	Dava Sobel	Longitude is the dramatic human story of an epic scientific quest, and of John Harrison's forty-year obsession with building his perfect timekeeper, known today as the chronometer.
Cartographic Design Using ArcView GIS	Madej, Ed	Both an effective primer on digital map design, as well as a classic software tutorial, this book makes particular reference to cartographic methods available in one of today's most popular GIS applications software packages: ArcView GIS from ESRI, Inc. Structured around fundamental concepts of map design, each stand—a lone chapter introduces the reader to a general map design theory followed by opportunities to apply these theories.
Thematic Cartography and Visualization	Slocum, Terry A.	This comprehensive book bridges the evolution in cartography by presenting both traditional methods and recently developed computer-based cartographic techniques.
Mapping It Out: Expository Cartography for the Humanities and Social Sciences	Monmonier, Mark	Writers know only too well how long it can take—and how awkward it can be—to describe spatial relationships with words alone. And while a map might not always be worth a thousand words, a good one can help writers communicate an argument or explanation clearly, succinctly, and effectively.
Elements of Cartography	Robinson, Arthur H., Muehrcke, Phillip C. (Unk)	Recognized as the classic resource in cartography, this text continues to integrate the latest modern technology with traditional cartographic principles.

(contd.)

(contd.)

Title	*Author*	*Description*
Geographical Information Systems and Computer Cartography	Jones, Christopher B.	Presents the fundamental concepts in Geographical Information Systems (GIS), emphasizing the techniques in management, analysis, and graphic display of spatial information. (Paper)
	4. Data Conversion	
GIS Data Conversion: Strategies, Techniques, and Management	Hohl, Pat	This book provides an overview of the issues involved in GIS data conversion projects, including understanding and locating data; allocating personnel, budgets, and time; documenting the process; becoming familiar with the target system, and safeguarding data quality.
	5. Environment	
Air Apparent; How Meteorologists Learned to Map, Predict, and Dramatize Weather	Monmonier, Mark	Air Apparent is the singular history of the weather map, one that has developed around the twin poles of weather's many facets and the public's varied needs.
Cartographies of Danger: Mapping Hazards of America	Mark S. Monmonier	No place is perfectly safe, but some places are more dangerous than others. As Mark Monmonier shows in this entertaining and immensely informative book, maps can tell us a lot about where we can anticipate certain risks, but maps can also be dangerously misleading.
Environmentalism Unbound: Exploring New Pathways for Change	Robert Gottlieb	In Environmentalism Unbound, Robert Gottlieb proposes a new strategy for social and environmental change that involves reframing and linking the movements for environmental justice and pollution prevention.
Resource Wars: The New Landscape of Global Conflict	Michael T. Klare	This sobering look at the future of warfare predicts that conflicts will now be fought over diminishing supplies of our most precious natural resources.
	6. GIS/GPS Primers	
Understanding Gps: Principles and Applications {Artech	Kaplan, Elliott D.(Edt)	This practical design guide provides an in-depth description of GPS modern digital receiver, operation and explores future applications in air,

(contd.)

(contd.)

Title	*Author*	*Description*
House Telecommunications Library)		land, and marine navigation.
Fundamentals of Geographic Information Systems	DeMers, Michael N.	This comprehensive look at GIS doesn't overwhelm with details, equations, or computer codes.
The GIS Book 5th Edition	Korte, George	Discover what a geographic information system (GIS) is, how it applies to a variety of professions, and the benefits of using GIS for a variety of applications.
Principles of Geographical Information Systems	Burrough, P.A., Burrough, P.A.	Spatial Information Systems (Paper)
The GIS Book	Korte, George B.	Proven through three highly praised editions, The GIS Book, 4th Ed. is a completely revised and greatly expanded resource for anyone who needs to understand what a geographic system is, how it applies to their profession, and what it can do.
Extending ArcView GIS: With Network Analyst, Spatial Analyst, and 3D Analyst	Ormsby, Tim, Alvi, Jonell	Extending ArcView GIS is the companion workbook to ESRI's best-selling Getting to Know ArcView GIS. Taking up where that book leaves off, it explores the three most popular ESRI ArcView GIS extensions—ArcView Network Analyst, ArcView Spatial Analyst, and ArcView 3D Analyst.
Getting to Know ArcView GIS: The Geographic Information System (GIS) for Everyone	ESRI Press	Here's your opportunity to test drive ArcView GIS, the desktop geographic information system from ESRI, the leading global provider of GIS software. The CD packaged with this introductory book includes a working demonstration copy of ArcView software, a hands-on ArcView tutorial, and informative multimedia presentations.
GIS for Everyone—Exploring Your Neighbourhood and Your World with a Geographic Information System	Davis, David E.	Exploring your neighbourhood and your world with a Geographic Information System... This book with its illustrated exercises and CD-ROM support material, explain how GIS takes geographic information and makes it visually understandable to users.

(contd.)

(contd.)

Title	*Author*	*Description*
		7. GIS Tutorials
Map basic Developer's Guide	Whitener, Angela., Ryker, Breck	MapBasic Developer's Guide is a handbook for customizing MapInfo Professional. The book begins with a tutorial on MapBasic elements, the MapBasic development environment, and program building basics.
Inside MapInfo Professional: The Friendly User Guide to MapInfo Professional	Daniel, Larry/ Loree, Paula	Now fully updated to the software's latest features and functions, INSIDE MapInfo Professional, 2nd Edition continues to set the standard for desktop mapping tutorials.
Getting to Know ArcGIS Desktop: The Basics of ArcView, Arceditor, and Arcinfo	Tim Ormsby, Eileen Napoleon, Carolyn Groessl	ArcView is the world's most widely used Geographic Information Systems (GIS) software. Version 8 is the most significant upgrade to ArcView since its inception—it has been completely redesigned and engineered to be an easy-to-use, fast, modern, and powerful GIS, and requires a new guidebook for all users.
The ESRI Press Dictionary of GIS Terminology	ESRI Press	The ESRI Press Dictionary of GIS Terminology gives clear, concise definitions of over 1200 terms from GIS and related fields such as geography, GPS, cartography, remote sensing, photogrammetry, and geodesy.
Inside MapInfo Professional	Larry Daniel, Paula Loree, Angela Whitener	Inside MapInfo Professional is for desktop mapping neophytes who wish to quickly become proficient with MapInfo Professional software, MapInfo Corporation's desktop mapping market leader.
		8. GIS-Related Programming
Professional Java E-Commerce	Ronald Ashri, Robert Flenner, Tracie Karsjens, Ma Kerzner,Ronald Ashri,Chad Darby	This book has been designed for professional Java programmers who wish to leverage their in-depth knowledge of J2EE technologies to develop e-commerce solutions, either for consumer sites, or business-to-business scenarios.
Palm OS Programming for Dummies	Liz O'Hara, John Schettino	A one-stop reference for writing applications for the Palm platform used in the popular hand-held organizers.

(contd.)

(contd.)

Title	*Author*	*Description*
	9. Census	
Ecological Census Techniques: A Handbook	William J. Sutherland (Editor)	Virtually any exercise in ecology will require some knowledge of the techniques for carrying out a census of population numbers. This practical text outlines clearly, with worked examples, the main techniques used by field ecologists to enumerate plants and animals.
State Census Records	Ann Smith Lainhart	State censuses rank with federal censuses as a major genealogical resource, but, because they were taken randomly, remain a much under-utilized resource in American genealogy.
Counting on the Census?: Race, Group Identity and the Evasion of Politics	Peter Skerry	Skerry provides a skilful and very timely historicy of the various controversies involving the census and then lays out, and attemps to assess, the contemporary debate.
American Census: A Social History	Margo J. Anderson	This work examines the "history of the American census...how it was taken as well as an analysis of those... who made the decisions of how to gather, compile, and use the statistics.
	10. Adventure—With or Without GIS	
What the Ice Gets: Shackleton's Antarctic Expedition, 1914-1916	Melinda Mueller, Foreword by Gary Holthaus	Blending historical, scientific, and literary scholarship with an impressive range of poetic forms, Melinda Mueller masterfully brings us the legendary tale of Sir Ernest Shackleton's Trans-Antarctic Expedition.
Wilderness Navigation: Finding Your Way Using Map, Compass, Altimeter, and GPS	Bob Burns, Mike Burns,Paul Hughes	Wilderness Navigation answer your questions reading maps; buying a will all about compass; using a compass, a clinometer, an altimeter, and a GPS; orientation with map and compass; how to avoid getting lost, and what to do if that happens.
Power of Passion: Achieve Your Own Everests	Hobson, Alan, Clarke, Jamie	In this book, Jamie and Alan share the true story of their struggle to climb Everest. It is far more than a mountaineering tale, it's a real-life story—a story of hopes and dreams, of tragedy and tenacity to which everyone can relate. Mostly, it's a story about triumph—a triumph of the human spirit. It will help you climb your

(contd.)

(contd.)

Title	*Author*	*Description*
		own Everests—whatever mountains you are climbing in your own life.
	11. GPS	
Fundamentals of Global Positioning System Receivers: A Software Approach	James Bao-Yen Tsui,James B.Y. Tsui	Provides engineers who use and design GPS systems with a reference on the operational principles guiding this technology, with an emphasis on software-based signal processing. Covers all aspects of receiver technology as well as relevant navigation schemes, concentrating on civilian C/A code used by commercial GPS receivers, rather than military code. Describes GPS basics and satellites, examines signal structure and tracking, and presents mathematical formulas for calculating a user's position.
Using GPS: Finding Your Way with Global Positioning System	Bruce Grubbs	
Integrating GIS and the Global Positioning System with CD-ROM	Karen Steede-Terry, Jack Dangermond	The coupling of the Global Positioning System (GPS) with Geographic Information System (GIS) is an information revolution that will give unprecedented accuracy in measurement and completeness of coverage to anyone who works with geographic data of any kind.
Understanding GPS: Principles and Applications (Artech House Telecommunications Library)	Kaplan, Elliott D. (Edt)	This practical design guide provides an in-depth description of GPS modern digital receiver operation and explores future applications in air, land, and marine navigation.
	12. Government	
GIS in Public Policy	R. W. Greene	This case-study book shows how progressive policy makers have put GIS to use-the National Partnership for Reinventing Government being one example
Disaster Response: GIS for Public Safety	Gary Amdahl, Preface by Jack Dangermond	Local, state, and federal government units and the private sector need to learn how to best use the GIS technology now in the hands of those on the front lines fighting fires, managing

(contd.)

(contd.)

Title	*Author*	*Description*
		evacuations, and cleaning toxic spills. These case studies illustrate the best new strategies for mitigation, response, and recovery from both natural and human disasters. A wide variety of disasters and scenarios are represented, including a lethal mudslide and wildfires.
Opening Access: GIS in E-Government	R.W. Greene	As a result of the geographic aspect of most all government services, map-based information systems are critical to government performance. This guide shows how GIS is becoming an essential tool in the new world of e-government, demonstrating how GIS and the digital revolution are doing for government what they have done for business-making it more efficient in areas such as economic development and environmental protection.
Urban Planning and Development Applications of GIS	Said Easa, American Society of Civil Engineers Geographic Information Syste Yupo Chan (Editor)	Presents urban planning and development applications of Geographic Information Systems (GIS). The 13 chapters deal with GIS technology implementation, remote sensing, trends in spatial databases.
	13. Internet Mapping	
GIS Online: Information Retrieval, Mapping, and The Internet	Plewe, Brandon	Synopsis for people who want to build a Web site based around GIS and mapping technology or to include maps on their Web-sites, this text describes the concept of geographic information retrieval and data sharing, outlines motivations for sharing geographic data, and includes step-by-step instructions through the planning and development stages of the Web-site.
Serving Maps on the Internet: Geographic Information on the World Wide Web	Harder, Christian	Internet-based mapping applications are changing the way that people use and manage geographic information. Twelve fascinating case studies reveal how prominent organizations use Web-based geographic information systems (GIS) to improve customer service, increase employee efficiency, and inform the community at large.

(contd.)

(contd.)

Title	*Author*	*Description*
		14. Natural Resources
Managing Natural Resources with GIS	Laura Lang, Environmental Systems Research Institute	As global population continues to grow, mankind must learn to balance consumption and conservation of the planet's scarece commodities. Geographic information system (GIS) technology brings a new perspective to the challenges faced by natural resource managers.
Geodestinies: The Inevitable Influence of Earth Resources on Nations and Individuals	Walter Lewellyn Youngquist	Geologists: Where is the search for energy and minerals today? How is the world Earth resource base changing? What new roles will alternative energy sources play?
Handbook of Hydrology	Maidment, David R.	Introducing Hydrology's New Benchmark Reference Here's the first book in nearly 30 years to provide comprehensive coverage of the current state of hydrologic knowledge and practice—saving you hours of time tracking down the latest techniques in professional journals.
Land Administration	Peter F. Dale, John D. McLaughlin	Land Administration provides a high level overview of recent advances in building formal property systems throughout the world, and reviews the role of property in advancing a society's economic and social agenda.
Flattening the Earth; Two Thousand Years of Map Projections	John P. Snyder	As long as there have been maps, cartographers have grappled with the impossibility of portraying the earth in two dimensions. To solve this problem mapmakers have created hundreds of map projections, mathematical methods for drawing the round earth on a flat surface.
		15. Non-Traditional GIS
GIS for Landscape Architects	Hanna, Karen C.	Geographic information system (GIS) technology is an increasingly important software tool for organizing digital spatial data in the field of landscape architecture.
GIS for Health Organizations with CD-ROM	Laura Lang	Health care executives and managers will get a closer look at how to use geographic information systems (GIS) to improve patient care, better manage their facilities, and increase their market share with this book.

(contd.)

(contd.)

Title	*Author*	*Description*
GIS for Health Organizations with CD-ROM	Laura Lang	Health care executives and managers will get a closer look at how to use geographic information systems (GIS) to improve patient care, better manage their facilities, and increase their market share with this book. GIS is already used to monitor the spread of disease, assign patients to the nearest facility, and display injuries on a map of the human body.
Spatial Analysis, GIS, and Remote Sensing Applications in the Health Sciences	Donald Patrick Albert (Editor), Wilbert M. Gesler (Editor),Barbara Levergood (Editor)	Geographers and epidemiologists explain some of the ways geographical information systems can be applied in human and environmental health. Their topics include how spatial analysis can be used in medical geography, health services research, infectious disease, and a historical perspective on the development of remotely sensed data as applied to medical geography.
Remote Sensing and Geographical Information Systems in Epidemiology	Simon I. Hay (Editor),Sarah E. Randolph (Editor),Preface by J. R. Baker	With ever more sophisticated control methods being developed for infectious diseases, our ability to map spatial and temporal variation in risk is more important than ever. Only with global information may we plan control campaigns and deliver novel interventions and remedies where the need is greatest, and sustainable success is most likely.
Statistical Methods in Spatial Epidemiology	Andrew B. Lawson, A.B. Lawson	Epidemiology is the study of disease patterns. Spatial epidemiology considers the statistical aspects of the development of disease not only in relationship to time but in particular to geographical location.
Medical Geography, (Second Edition)	Melinda S. Meade,Robert J. Earickson	...this revised and expanded 2nd edition surveys the perspectives, methodologies and theories that geographers use to address the subject of human health and disease with clearly presented vignettes, tables, figures and maps
	16. Remote Sensing	
Remote Sensing of the Environment and Earth Resource Perspective	John R. R. Jensen	This book introduces the principles of remote sensing from an Earth resource perspective.

(contd.)

(contd.)

Title	*Author*	*Description*
An Introductory Digital Image Processing: A Remote Sensing Perspective	John R. Jensen	Explains how to apply digital image processing techniques to remotely sensed imagery, and how to obtain useful results and information during this process.
Computer Processing of Remotely Sensed Images: An Introduction	Paul M. Mather, Paul Mather	Remotely sensed images provide a valuable source of information concerning the nature and variability of the land and sea surfaces of the world.
Remote Sensing and Geographical Information Systems in Epidemiology	Simon I. Hay (Editor),Sarah E. Randolph (Editor),Preface by J. R. Baker	With ever more sophisticated control methods being developed for infectious diseases, our ability to map spatial and temporal variation in risk is more important than ever.
Laser Remote Sensing of the Ocean: Methods and Applications	Konstantin I. Voliak, Alexey F. Bunkin	Recent advances in lasers and measuring devices, as well as new capabilities in automated data acquisition, have made possible the design and practical implementation of self-sufficient lidar systems.
Remote Sensing and Image Interpretation	Thomas M. Lillesand, Ralph W. Kiefer	Extensively illustrated, this updated edition provides a balance between classical visual image interpretation and digital image processing techniques.
GIS: A Visual Approach	Bruce Ellsworth Davis, Bruce Davis	This fully updated edition of Geographic Information Systems: A Visual Approach offers a comprehensive introduction to the application of GIS concepts.
	17. Spatial Data Analysis	
Spatial Analysis, GIS, and Remote Sensing Applications in the Health Sciences	Donald Patrick Albert (Editor), Wilbert M. Gesler (Editor), Barbara Levergood (Editor)	Geographers and epidemiologists explain some of the ways geographical information systems can be applied in human and environmental health.
The ESRI Guide to GIS Analysis, Volume 1: Geographic Patterns and Relationships	Mitchell, Andy	By the author of the best-selling GIS classic Zeroing In: GIS at Work in the Community comes an important new book about how to do real analysis with a geographic information system. The ESRI Guide to GIS Analysis, Volume 1: Geographic Patterns and

(contd.)

(contd.)

Title	*Author*	*Description*
		Relationships focuses on six of the most common geographic analysis tasks.
Spatial Analysis: Modelling in a GIS Environment	Longley, Paul/ Batty, Mike	This book addresses spatial analysis with an emphasis on the integration of different spatial analysis functions within GIS.
Casebook for Spatial Statistical Data Analysis: A Compilation of Analyses of Different Thematic Data Sets	Griffith, Daniel A./ Sone, Akio	This volume compiles geostatistical and spatial autoregressive data analyses involving georeferenced socioeconomic, natural resources, agricultural, pollution, and epidemiological variables.
Interpolation of Spatial Data: Some Theory for Kriging	Stein, Michael Leonard	Kriging, which is the geostatistical term for optimal linear prediction of spatial processes, is widely used in geology, hydrology, environmental monitoring and other fields to interpolate spatial data.
Modelling Our World: The Esri Guide to Geodatabase Design	Zieler, Michael	This conceptual overview introduces and explains the concept of geodatabases-object-oriented data models—which are introduced in ArcInfo 8, the world's most advanced Geographic Information System (GIS) software package.
Beyond Mapping: Concepts, Algorithms, and Issues in GIS	Berry, Joseph K.	This book discusses maps as data with data structure implications; roving windows; spatial data modelling and management; assessing variability, shape, and pattern of map features; overlaying maps; cost-benefit analysis; algorithms involved in slope, distance, and connectivity; and cartographic and spatial modelling.
Cartography: Visualization of Spatial Data	Kraak, Menno-Jan/ Ormeling, Ferjan	Provides a comprehensive account of the acquisition of data, map design, topographic and statistical mapping and ends with examples of advanced mapping environments. (Paper)
Exploring Spatial Analysis in GIS	Chou, Yue-Hong	Written for Geographic Information Systems (GIS) professionals and students, this book provides an introduction to spatial analysis concepts and applications along with numerous examples and exercises.

(contd.)

(contd.)

Title	*Author*	*Description*
Geographic Information Systems and Science	John Wiley & Sons	There are many books integrating GIS technologies with disciplines such as remote sensing, cartography, geography, and computer science. This book, written by an authoritative and experienced author team, this book offers peerless coverage of issues related to geographical information science (GISc), GIS management, and real-world applications.
GIS in Telecommunications Management with CD-ROM	Lisa Winget Godin, Lisa Godin	Telecommunications companies use geographic information systems (GIS) to plan, build, and maintain their networks. Techniques for merging spatial data from different sources into an integrated system are included, and the case studies profiled present innovative techniques for such daunting tasks as managing outside plant engineering processes
	18. Transportation	
Transportation GIS	Laura Lang	This richly illustrated volume is an excellent introduction to GIS in the transportation industry. It's easy-to-read style and relevent case studies will appeal tq industry professionals, students, and lay people alike.

Bibliography

Aangeenbrug, R.T. (1992) "A Review of 5 GIS Textbooks" *IJGIS*, Vol. 6, Iss 4, pp. 341-350.

Aangeenbrug, R.T. (1992) "Geographic Information Systems and Management Perspectives, by S. Aronoff" *IJGIS*, Vol. 6, Iss 4, pp. 344-345.

Aangeenbrug, R.T. (1992) "Geographic Information Systems—A Guide to the Technology, by J.C. Antenucci, K. Brown, P.L. Crosswell, M.J. Kevany, H. Archer" *IJGIS*, Vol. 6, Iss 4, pp. 348-350.

Aangeenbrug, R.T. (1992) "GIS in Education" *IJGIS*, Vol. 6, Iss 4, pp. 263-265.

Aangeenbrug, R.T. (1992) "Introduction to Urban Geographic Information Systems, by W.E. Huxhold" *IJGIS*, Vol. 6, Iss 4, pp. 346-348.

Aangeenbrug, R.T. (1992) "Principles of Geographic Information Systems for Land Resources Assessment", by P.A. Burrough *IJGIS*, Vol. 6, Iss 4, pp. 342-343.

Abel, D.J. (1985) "Some Elemental Operations on Linear Quadtrees for Geographic Information Systems " *Computer Journal*, Vol. 28, Iss 1, pp. 73-77.

Abel, D.J., Yap, S.K., Ackland, R. Cameron, M.A., Smith, D.F., Walker, G. (1992) "Environmental Decision Support System Project—An Exploration of Alternative Architectures for Geographical Information Systems" *IJGIS*, Vol. 6, Iss 3, pp. 193-204.

Abler, R.F. (1993) "Everything in Its Place—GPS, GIS, and Geography in the 1990s" *PG*, Vol. 45, Iss 2, pp. 131-139.

Adamus, C.L., Bergman, M.J. (1995) "Estimating Nonpoint-Source Pollution Loads with a GIS Screening Model" *Water Resources Bulletin*, Vol. 31, Iss 4, pp. 647-655.

Adinarayana, J., Flach, J.D., Collins, W.G. (1994) "Mapping Land-Use Patterns in a River Catchment Using Geographical Information Systems", *Journal of Environemntal Management*, Vol. 42, Iss 1, pp. 55-61.

Aitken, S., Stutz, F., Prosser, R., Chandler, R., (1993) "Neighbourhood Integrity and Residents Familiarity—Using a Geographic Information System to Investigate Place Identity", *Tijdschrift voor Economische en Scciale Geografie*, Vol. 84, Iss 1, pp. 2-12.

Alankary, K.M. (1991) "An Incremental Approach for Establishing a Geographical Information System in a Developing-Country—Saudi-Arabia", *IJGIS*, Vol. 5, Iss 1, pp. 85-98.

Albertson, P.E., Bourne, S.G., Hennington, G.W. (1992) "An Integrated Groundwater and Land-Use GIS for Impact Assessment", *PERS*, 1992, Vol. 58, Iss 8, pp. 1203-1207.

Albrecht, J. (1997) "GIS—A Computing Perspective, by M. Worboys" *IJGISC*, Vol. 11, Iss 4, pp. 405-406.

Ali, C.Q., Ross, L.G., Beveridge, M.C.M. (1991) "Microcomputer Spreadsheets for the Implementation of Geographic Information Systems in Aquaculture—A Case-Study on Carp in Pakistan", *Aquaculture*, Vol. 92, Iss 2-3, pp. 199-205.

Almond, S., (1991) "Geographic Information Systems—The Microcomputer and Modern Cartography", by D.R.F. Taylor, *Cartographic Journal*, Vol. 28, Iss 2, pp. 269-270.

Almond, S. (1994) "Mountain Environments and Geographic Information Systems", by M. Price, I. Heywood, *Cartographic Journal*, Vol. 31, Iss 2, pp. 145-145.

Alshuwaikhat, H.M., Nassef, K. (1996) "A GIS-Based Spatial Decision-Support System for Suitability Assessment and Land-Use Allocation", *Arabian Journal for Science and Engineering*, Vol. 21, Iss 4A, pp. 525-543.

Andrianarivo, J. (1993) "Using GIS to Evaluate the Crown-Line Intersect Sampling Method in Forest Survey" *Forest Ecology and Management*, Vol. 59, Iss 1-2, pp. 87-103.

Andrle, R. (1991) "Atlas GIS" *PG,* Vol. 43, Iss 3, pp. 362-363.

Anselin, L., Getis, A. (1992) "Spatial Statistical-Analysis and Geographic Information Systems", *Annals of Regional Science*, Vol. 26, Iss 1, pp. 19-33.

Ansoult, M.M., Soille, P.J., Loodts, J.A. (1990) "Mathematical Morphology—A Tool for Automated GIS Data Acquisition from Scanned Thematic Maps", *PERS*, Vol. 56, Iss 9, pp. 1263-1271.

Anthony, S.J., Corr, D.G. (1988) "Data-Structures in an Integrated Geographical Information System", *ESA Journal-European Space Agency,* Vol. 12, Iss 1, pp. 69-72.

Any, J., Treitz, P.M., and Howarth, P.J. (1992), Road network detection from SPOT imagery for updating geographical information systems in the rural-urban fringe, *International Journal on Geographic Information Systems,* Vol. 6, No. 2, pp. 141-57.

Apan, A.A. (1996) "Tropical Landscape Characterization and Analysis for Forest Rehabilitation Planning Using Satellite Data and GIS", *Landscape and Urban Planning*, Vol. 34, Iss 1, pp. 45-54.

Arbeit, D. (1992) "Geographic Information Systems—A Guide to the Technology", by J.C. Antenucci, K. Brown, P.L. Croswell, M.J. Kevany, H.A.V. Reinhold, *Journal of the Americam Planning Association*, Vol. 58, Iss 4, pp. 531-532.

Arbia, G. (1993) "The Use of GIS in Spatial Statistical Surveys", *International Statistical Review*, Vol. 61, Iss 2, pp. 339-359.

Archer, H., Croswell, P.L. (1989) "Public-Access to Geographic Information Systems—An Emerging Legal Issue", *PERS*, Vol. 55, Iss 11, pp. 1575-1581.

Archer, J.C. (1992) "Atlas-Asterisk-GIS", *CGIS*, Vol. 19, Iss 2, pp. 111-116.

Ardito, A., Iorio, R., Santagostino, G., Porrino, A., Fujimoto, N. (1992), "Accurate Modelling of Capacitively Graded Bushings for Calculation of Fast Transient Overvoltages in GIS", *IEEE Transactions on Power Delivery*, 1992, Vol. 7, Iss 3, pp. 1316-1327.

Arend, R.B. (1990) "GIS—Useful Tool or Expensive Toy", *Journal of Surveying Engineering-ASCE*, Vol. 116, Iss 2, pp. 131-138.

Arentze, T.A., Borgers, A.W.J., Timmermans, H.J.P. (1994) "Multistop-Based Measurements of Accessibility in a GIS Environment", *IJGIS*, Vol. 8, Iss 4, pp. 343-356.

Arlinghaus, S.L. (1996) "Geographical Record—A Path Through Recent GIS Literature", *Geographical Review*, Vol. 86, Iss 1, pp. 101-107.

Armstrong, M.P. (1994) "Requirements for the Development of GIS-Based Group Decision-Support Systems" *Journal of the Americal Society for Information Science*, Vol. 45, Iss 9, pp. 669-677.

Arthur, J.L., Nalle, D.J. (1997) "Clarification on the Use of Linear-Programming and GIS for Land-Use Modelling" *IJGISC*, Vol. 11, Iss 4, pp. 397-402.

Aschbacher, J., Ofren, R., Delsol, J.P., Suselo, T.B., Vibulsresth, S., Charrupat, T. (1995) "An Integrated Comparative Approach to Mangrove Vegetation Mapping Using Advanced Remote-Sensing and GIS Technologies—Preliminary-Results", *Hydrobiologia,* Vol. 295, Iss 1-3, pp. 285-294.

Aspinall, R., Matthews, K. (1994) "Climate-Change Impact on Distribution and Abundance of Wildlife Species—An Analytical Approach Using GIS", *Environmental Pollution*, 1994, Vol. 86, Iss 2, pp. 217-223.

Aspinall, R., Veitch, N. (1993) "Habitat Mapping from Satellite Imagery and Wildlife Survey Data Using a Bayesian Modelling Procedure in a GIS", *PERS*, Vol. 59, Iss 4, pp. 537-543.

Aspinall, R.J., Miller, D.R., Birnie, R.V. (1993) "Geographical Information Systemsfor Rural Land-Use Planning", *Applied Geogrpahy,* Vol. 13, Iss 1, pp. 54-66.

Astroth, J.H., Trujillo, J., Johnson, G.E. (1990) "A Retrospective Analysis of GIS Performance—The Umatilla Basin Revisited", *PERS*, Vol. 56, Iss 3, pp. 359-363.

Aultmanhall, L., Roorda, M., Baetz, B.W. (1997) "Using GIS for Evaluation of Neighbourhood Pedestrian Accessibility", *Journal of Urban Planning and Development-ASCE*, Vol. 123, Iss 1, pp. 10-17.

Bailey, R.G. (1988) "Problems with Using Overlay Mapping for Planning and Their Implications for Geographic Information Systems", *Environemntal Management,* Vol. 12, Iss 1, pp. 11-17.

Bailey, T.C. (1990) "GIS and Simple Systems for Visual, Interactive, Spatial-Analysis", *Cartographic Journal,* Vol. 27, Iss 2, pp. 79-84.

Baker, C.P., Bradley, M.D., Bobiak, S.M.K. (1993) "Wellhead Protection Area Delineation—Linking Flow Model with GIS", *Journal of Water Resources Planning and Management-ASCE*, Vol. 119, Iss 2, pp. 275-287.

Baker, W.L., Honaker, J.J., Weisberg, P.J. (1995) "Using Aerial-Photography and GIS to Map the Forest-Tundra Ecotone in Rocky-Mountain National-Park, Colorado, for Global Change Research", *PERS*, Vol. 61, Iss 3, pp. 313-320.

Bakker, S.A., Vandenberg, N.J., Speleers, B.P. (1994) "Vegetation Transitions of Floating Wetlands in a Complex of Turbaries Between 1937 and 1989 as Determined from Aerial Photographs with GIS", *Vegetatio,* Vol. 114, Iss 2, pp. 161-167.

Ball, G.L. (1994) "Ecosystem Modelling with GIS", *Environmental Management*, 1994, Vol. 18, Iss 3, pp. 345-349.

Banai, R. (1993) "Fuzziness in Geographical Information Systems—Contributions from the Analytic Hierarchy Process", *IJGIS*, Vol. 7, Iss 4, pp. 315-329.

Barrs, H.D., Prathapar, S.A. (1994) "An Inexpensive and Effective Basis for Monitoring Rice Areas Using GIS and Remote-Sensing", *Austrian Journal of Experimental Agriculture*, Vol. 34, Iss 7, pp. 1079-1083.

Battaglin, W.A., Hay, L.E., Parker, R.S., Leavesley, G.H. (1994) "Applications of a GIS for Modelling the Sensitivity of Water-Resources to Alterations in Climate in the Gunnison River Basin, Colorado", *Water Resources Bulletin*, 1993, Vol. 29, Iss 6, pp. 1021-1028.

Batty, M. (1992) "Geographic Information Systems—Development and Applications", by L. Worrall, *Urban Studies*, Vol. 29, Iss 1, pp. 132-134.

Batty, M. (1992)"Urban Modelling in Computer-Graphic and Geographic Information System Environments", *Environment and Planning B-Planning and Design*, Vol. 19, Iss 6, pp. 663-688.

Batty, M., Xie, Y.C. (1994) "Modelling Inside GIS.1. Model Structures, Exploratory Spatial Data-Analysis and Aggregation", *IJGIS*, Vol. 8, Iss 3, pp. 291-307.

Batty, M., Xie, Y.C. (1994) "Modelling Inside GIS.2. Selecting and Calibrating Urban Models Using Arc-Info", *IJGIS*, Vol. 8, Iss 5, pp. 451-470.

Baumgartner, M.F., Apfl, G. (1994) "Towards an Integrated Geographic Analysis System with Remote-Sensing, GIS and Consecutive Modelling for Snow Cover Monitoring", *IJRS*, 1994, Vol. 15, Iss 7, pp. 1507-1517.

Beard, R. (1992) "Geographic Information Systems and Their Socio-economic Applications", by D. Martin, *Cartographic Journal,* Vol. 29, Iss 1, pp. 80-81.

Bedient, P.B. (1992) "GIS and the Environment" *Water Environment Research,* 1992, Vol. 64, Iss 3, pp. 195-195.

Belyaeva, T., Higgins, J.M., Kirpichnikova, N., Lanzova, I., Hagerman, J.R. (1993) "GIS Application to Water-Quality Management in the Upper Volga River Basin—Joint TVA Russia Project", *Water Science and Technology*, Vol. 28, Iss 3-5, pp. 119-127.

Berry, J.K. (1986) "Geographic Information Systems.3. Learning Computer-Assisted Map Analysis", *JF*, Vol. 84, Iss 10, pp. 39-43.

Beyers, N., Gie, R.P., Zietsman, H.L., Kunneke, M., Hauman, J., Tatley, M., Donald, P.R. (1996) "The Use of a Geographical Information System (GIS) to Evaluate the Distribution of Tuberculosis in a High-Incidence Community", *South African Medical Journal*, Vol. 86, Iss 1, pp. 40-44.

Bishop, I. (1991) "MacGIS—A Geographic Information System for the Macintosh", *Journal of Architectural and Planning Reserach*, Vol. 8, Iss 1, pp. 78-80.

Bishop, I.D., Hulse, D.W. (1994) "Prediction of Scenic Beauty Using Mapped Data and Geographic Information Systems", *Landscape and Urban Planning*, Vol. 30, Iss 1-2, pp. 59-70.

Bissonette, J.A., Sherburne, S.S., Ramsey, R.D. (1994) "Analysing Telemetry Data with a GIS-Based Vector Structure", *IJGIS*, Vol. 8, Iss 6, pp. 533-543.

Blaise, P., Gessler, C. (1991) "Geobase—A Simple Geographical Information System on a Personal Computer" *Computer Applications in the Biosciences*, Vol. 7, Iss 2, pp. 155-160.

Blakemore, M. (1984), Generalization and error in spatial data bases, *Cartographia,* vol. 21, pp. 131-9.

Blakemore, M. (1985) "Cartography and Geographic Information Systems", *Progress in Human Geography*, Vol. 9, Iss 4, pp. 566-574.

Blakemore, M. (1986) "Cartography and Geographic Information Systems", *Progress in Human Geograph*, Vol. 10, Iss 4, pp. 553-563.

Blakemore, M. (1987) "Cartography and Geographic Information Systems", *Progress in Human Geography*, Vol. 11, Iss 4, pp. 590-606.

Blakemore, M. (1988) "Cartography and Geographic Information Systems", *Progress in Human Geography*, Vol. 12, Iss 4, pp. 525-537.

Blat, J., Delgado, A., Ruiz, M., Segui, J.M. (1995) "Designing Multimedia GIS for Territorial Planning—The Parcbit Case", *Environment and Planning B-Planning & Design*, Vol. 22, Iss 6, pp. 665-678.

Boeck, W., Kruger, K., (1992) "Arc Motion and Burn Through in GIS", *IEEE Transcations on Power Delivery*, 1992, Vol. 7, Iss 1, pp. 254-261.

Boggs, S.A. (1990) "Issues in Defining a Type Test for GIS Disconnectors", *IEEE Transcations on Power Delivery*, Vol. 5, Iss 4, pp. 1839-1845.

Boggs, S.A., Petterson, K., Diessner, A., Fujimoto, N., Leibold, A.A. (1992) "Partial Discharge Testing of Gas Insulated Substations—A Discussion Paper from the IEEE Substations Committee Working Group K4, GIS Diagnostic Methods", *IEEE Transcations on Power Delivery,* 1992, Vol. 7, Iss 2, pp. 499-506.

Boggs, S.A., Wang, Y. (1995) "Trapped Charge-Induced Field Distortion on GIS Spacers", *IEEE Transcations on Power Delivery*, Vol. 10, Iss 3, pp. 1270-1275.

Bojorqueztapia, L.A., Azuara, I., Ezcurra, E., Floresvillela, O. (1995) "Identifying Conservation Priorities in Mexico Through Geographic Information Systems and Modelling", *Ecological Applications*, Vol. 5, Iss 1, pp. 215-231.

Bolstad, Paul and Lillesand, T.M. (1991), Automated GIS integration in land cover classification, in Technical papers of the 1991, ACSM/ASPRS Annual Convention, vol. 3, pp. 23-32.

Bolstad, P.V. (1992) "Geometric Errors in Natural-Resource GIS Data—Tilt and Terrain Effects in Aerial Photographs", *Forest Science,* 1992, Vol. 38, Iss 2, pp. 367-380.

Bowler, I. (1994) "Landscape Ecology and Geographic Information Systems", by R. Hainesyoung *Geographical Journal*, Vol. 160, Iss Nov., pp. 345-345.

Bracken, I. (1992) "Geographic Information Systems—Developments and Applications", by L. Worrall *Progress in Human Geography,* Vol. 16, Iss 4, pp. 663-664.

Bracken, I. (1993) "An Introduction to Urban Geographic Information Systems", by W.E. Huxhold *Urban Studies,* Vol. 30, Iss 1, pp. 203-206.

Bracken, I. (1993) "Geographic Information Systems and Their Socioeconomic Applications, by D. Martin" *IEEE Transcations on Power Delivery,* Vol. 30, Iss 1, pp. 203-206.

Braddock, M., Lapidus, G., Cromley, E., Cromley, R., Burke, G., Banco, L. (1994) "Using a Geographic Information System to Understand Child Pedestrian Injury" *American Journal of Public Health,* Vol. 84, Iss 7, pp. 1158-1161.

Brainard, J., Lovett, A., Parfitt, J. (1996) "Assessing Hazardous-Waste Transport Risks Using a GIS" *IJGIS*, Vol. 10, Iss 7, pp. 831-849.

Braman, S. (1992) "Geographic Information Systems and Their Socioeconomic Applications", by D. Martin" *Journal of Communication*, Vol. 42, Iss 3, pp. 203-203.

Brandt, R., Groenewoudt, B.J., Kvamme, K.L. (1992) "An Experiment in Archaeological Site Location—Modelling in the Netherlands Using GIS Techniques" *World Archaeology,* Vol. 24, Iss 2, pp. 268-282.

Brannon, D.P., Irish, G.J. (1985) "Satellite Remote-Sensing and Geographic Information Systems Applied to Biogeochemical Flux Estimates" *Estuaries*, Vol. 8, Iss 2B, pp. A7-A7.

Braun, J.M., Ford, G.L., Fujimoto, N., Rizzetto, S., Stone, G.C., Boggs, S. (1993) "Reliability of GIS EHV Epoxy Insulators—The Need and Prospects for More Stringen—Acceptance Criteria" *IEEE Transactions on Power Delivery,* Vol. 8, Iss 1, pp. 121-131.

Braun, J.M., Rizzetto, S., Fujimoto, N., Ford, G.L. (1991) "Modulation of Partial Discharge Activity in GIS Insulators by X-Ray-Irradiation" IEEE *Transactions on Electrical Insulation* 1991, Vol. 26, Iss 3, pp. 460-468.

Briggs, D., Mounsey, H. (1989) "Integrating Land Resource Data into a European Geographical Information System—Practicalities and Problems" *Applied Geography,* Vol. 9, Iss 1, pp. 5-20.

Briggs, D.J., Collins, S., Elliott, P., Fischer, P., Kingham, S., Lebret, E., Pryl, K., Vanreeuwijk, H. (1997) "Smallbone-K Vanderveen-AMapping Urban Air-Pollution Using GIS—A Regression-Based Approach" *IJGISC,* Vol. 11, Iss 7, pp. 699-718.

Brimcombe, A.J., (1993) "Time in Geographic Information Systems, by G. Lanegran" *AREA,* Vol. 25, Iss 4, pp. 408-408.

Brinker, R.W., Jackson, B.D. (1991) "Using a Geographic Information System to Study a Regional Wood Procurement Problem" *Forest Science,* Vol. 37, Iss 6, pp. 1614-1631.

Bromley, R.D.F., Selman, J.M. (1993) "Geographic Information Usage in a Local-Authority—The Human Environment for GIS" *Area,* Vol. 25, Iss 3, pp. 228-236.

Brooks, K.R., London, J.B., Henry, M.S., Singletary, L.A. (1993) "Analysis and Simulation of Employment and Income Impacts of Infrastructure Investments in a State-Wide GIS Framework" *Computers Environment and Urban Systems,* Vol. 17, Iss 2, pp. 129-151.

Browne, T.J. (1992) "The Great Divide—Central Versus Departmental Support for Teaching GIS.1." *Journal of Geography in Higher Education,* Vol. 16, Iss 2, pp. 225-231.

Brown, S., Schreier, H., Thompson, W.A., Vertinsky, I. (1994) "Linking Multiple Accounts with GIS as Decision-Support System to Resolve Forestry Wildlife Conflicts" *Journal of Environmental Management,* Vol. 42, Iss 4, pp. 349-364.

Brunsdon, C. (1994) "Spatial-Analysis and GIS", by S. Fotheringham, P. Rogerson *Cartographic Journal*, Vol. 31, Iss 1, pp. 85-85.

Brunsdon, C. (1997) "International GIS Dictionary", by R. Mcdonnell, K. Kemp, *Environment and Planning B-Planning and Design,* Vol. 24, Iss 6, pp. 937-938.

Buck, K.E., Kleinetebbe, A., Rees, V. (1993) "Verification of the Seismic Resistance of Gas-Insulated Switchgear (GIS)" *IEEE Transactions on Power Delivery,* Vol. 8, Iss 1, pp. 114-120.

Budic, Z.D. (1994) "Effectiveness of Geographic Information Systems in Local-Planning" *Journal of the American Planning Association,* Vol. 60, Iss 2, pp. 244-263.

Budic, Z.D., Godschalk, D.R. (1994) "Implementation and Management Effectiveness in Adoption of GIS Technology in Local Governments" *Computers Environment and Urban Systems,* Vol. 18, Iss 5, pp. 285-304.

Bullen, N., Moon, G., Jones, K. (1996) "Defining Localities for Health-Planning—A GIS Approach" *Social Science and Medicine,* Vol. 42, Iss 6, pp. 801-816.

Burrough, P.A. (1986), *Principals of Geographical Information System for Land Resource Assessment,* Oxford University Press.

Burrough, P.A. (1992), Development of intelligent geographical information system, *International Journal of Geographical Information System,* Vol. 6:1, pp. 1-12.

Burrough, P.A. (1992) "Development of Intelligent Geographic Information Systems" *IJGIS,* Vol. 6, Iss 1, pp. 1-11.

Burrough, P.A. (1992) "Geographic Information Systems and Cartographic Modelling", by C.D. Tomlin, *IJGIS,* Vol. 6, Iss 5, pp. 449-450.

Burrough, P.A., Frank, A.U. (1995) "Concepts and Paradigms in Spatial Information—Are Current Geographical Information SystemsTruly Generic" *IJGIS,* Vol. 9, Iss 2, pp. 101-116.

Butler, W.I., Stehn, R.A., Balogh, G.R. (1995) "GIS for Mapping Waterfowl Density and Distribution from Aerial Surveys" *Wildlife Society Bulletin,* Vol. 23, Iss 2, pp. 140-147.

Buttenfield, B.P. (1992) "Geographic Information Systems and Cartographic Modelling", by C.D. Tomlin, *CGIS,* Vol. 19, Iss 4, pp. 262-263.

Cadoux-Hudson, J. and Heywood, I. (1991) (eds), *Geographic Information 1991,* Taylor and Francis, London.

Camm, J.D., Chorman, T.E., Dill, F.A., Evans, J.R., Sweeney, D.J., Wegryn, G.W. (1997) "Blending or/MS, Judgment, and GIS—Restructuring P-and-GS Supply Chain" *Interfaces,* Vol. 27, Iss 1, pp. 128-142.

Campbell, H. (1992) "The Impact of Geographic Information Systems on British Local-Government" *Computers Environment and Urban Systems,* Vol. 16, Iss 6, pp. 531-541.

Campbell, H. (1994) "How Effective Are GIS in Practice—A Case-Study of British Local-Government" *IJGIS,* Vol. 8, Iss 3, pp. 309-325.

Campbell, H., Masser, I. (1992) "GIS in Local-Government—Some Findings from Great-Britain", *IJGIS,* Vol. 6, Iss 6, pp. 529-546.

Campbell, H. (1992) "Organizational Issues and the Implementation of GIS in Massachusetts and Vermont—Some Lessons for the United-Kingdom" *Environment and Planning B-Planning and Design,* Vol. 19, Iss 1, pp. 85-95.

Campbell, J.C., Radke, J., Gless, J.T., Wirtshafter, R.M. (1992) "An Application of Linear-Programming and Geographic Information Systems—Cropland Allocation in Antigua", *Environment and Planning A,* Vol. 24, Iss 4, pp. 535-549.

Can, A. (1992) "Residential Quality Assessment—Alternative Approaches Using GIS", *Annals of Regional Science,* Vol. 26, Iss 1, pp. 97-110.

Can, A. (1993) "Tiger Line Files in Teaching GIS" *IJGIS,* Vol. 7, Iss 6, pp. 561-572.

Can, A. (1994) "Tiger Line Files in Teaching GIS (Vol. 7, Pg 561, 1993)" *IJGIS,* Vol. 8, Iss 4, pp. 403-403.

Cannistra, J.R., Leadbeater, R., Humphries, R. (1992) "Washington Suburban Sanitary Commission Implements GIS" *Journal American Water Works Association,* Vol. 84, Iss 7, pp. 62-67.

Carpentier, L.C., Bosch, D.J. (1994) "Adaptive Management and Geographic Information System to Overcome Informational Barriers to Pollution Reduction Trading" *American Journal of Agricultural Economics,* Vol. 76, Iss 5, pp. 1272-1272.

Carrara, A., Cardinali, M., Detti, R., Guzzetti, F., Pasqui, V., Reichenbach, P. (1991) "GIS Techniques and Statistical-Models in Evaluating Landslide Hazard" *Earth Surface Processes and Landforms 1991,* Vol. 16, Iss 5, pp. 427-445.

Carstensen, Jr., Laurence W. and Campbell, James B. (1991), Desktop scanning for cartographic digitization and spatial analysis, *Photogrammertic Engineering and Remote Sensing,* Vol. 57, No.11, pp. 1437-46.

Carstensen, L.W., Shaffer, C.A., Morrill, R.W., Fox, E.A. (1993) "Geosim—A GIS-Based Simulation Laboratory for Introductory Geography" *Journal of Geography,* Vol. 92, Iss 5, pp. 217-222.

Carter, J.R. (1985) *Geographic Names Information-System,* by F.R.L. Payne PG, Vol. 37, Iss 2, pp. 241-242.

Carver, S. (1997) "Innovations in GIS-3—Selected Papers from the 3rd National Conference on GIS-Research-UK (GISruk)", by D. Parker *IJGISC,* Vol. 11, Iss 5, pp. 521-521.

Carver, S.J. (1991) "Integrating Multicriteria Evaluation with Geographical Information Systems" *IJGIS,* Vol. 5, Iss 3, pp. 321-339.

Cassettari, S. (1991) "Geographic Information Systemsat Kingston" *Cartographic Journal,* Vol. 28, Iss 1, pp. 71-75.

Cassettari, S. (1992) "Time in Geographical Information Systems, by G. Langran" *Cartographic Journal,* Vol. 29, Iss 2, pp. 193-194.

Champion, A.G. (1988) "Population Information-Systems—Report of a Population Geography Study-Group Conference Held at Mansfield-College, Oxford, 7-8 September 1987" *AREA,* Vol. 20, Iss 1, pp. 82-83.

Chang, N.B., Lu-HY Wei,Y.L. (1997) "GIS Technology for Vehicle-Routing and Scheduling in Solid-Waste Collection Systems" *Journal of Environmental Engineering, ASCE,* Vol. 123, Iss 9, pp. 901-910.

Chen, J. Newkirk, R.T. Davidson, G. Gong, P. (1994) "The Development of a Knowledge-Based Geographical Information System for the Zoning of Rural-Areas", *Environment and Planning B-Planning and Design,* Vol. 21, Iss 2, pp. 179-190.

Chou, Y.H. (1992), Management of wildfires with a geographical information system, *International Journal on Geographic Information Systems,* Vol. 6, No. 2, pp. 123-40.

Chou, Y.H. (1992) "Slope-Line Detection in a Vector-Based GIS" *PERS* 1992, Vol. 58, Iss 2, pp. 227-233.

Chou, Y.H. (1992) "Management of Wildfires with a Geographical Information System" *IJGIS,* Vol. 6, Iss 2, pp. 123-140.

Chrisman, N.R. (1987) "Design of Geographic Information Systems Based on Social and Cultural Goals" *PERS,* Vol. 53, Iss 10, pp. 1367-1370.

Chuvieco, E. (1993) "Integration of Linear-Programming and GIS for Land-Use Modelling" *IJGIS,* Vol. 7, Iss 1, pp. 71-83.

Chuvieco, E. (1997) "Clarification on the Use of Linear-Programming and GIS for Land-Use Modelling—Reply" *IJGISC,* Vol. 11, Iss 4, pp. 403-403.

Chuvieco, E., Salas, J. (1996) "Mapping the Spatial-Distribution of Forest-Fire Danger Using GIS" *IJGIS,* Vol. 10, Iss 3, pp. 333-345.

Civco, D. (1994) "Erdas Imagine Remote-Sensing, Image-Processing, and GIS Software" *PERS* 1994, Vol. 60, Iss 1, pp. 35-39.

Civco, D.L. (1992) "VGA-Erdas Image-Processing and GIS Software" *PERS* 1992, Vol. 58, Iss 9, pp. 1350-1354.

Clarke, G. (1996) "Bringing GIS into Business", by D. Grimshaw *IJGIS,* Vol. 10, Iss 1, pp. 138-139.

Clarke, G.P., Langley, R. (1996) "A Review of the Potential of GIS and Spatial Modelling for Planning in the New Education Market" *Environment and Planning C-Government and Policy,* Vol. 14, Iss 3, pp. 301-323.

Clark, G.L. (1992) "GIS—What Crisis" *Environment and Planning A,* Vol. 24, Iss 3, pp. 321-322.

Clematis, A., Falcidieno, B., Spagnuolo, M., (1996) "Parallel-Processing on Heterogeneous Networks for GIS Applications" *IJGIS,* Vol. 10, Iss 6, pp. 747-767.

Collier, P., Fontana, D., Pearson, A. (1995) "GIS Mapping of Langstone-Harbor for an Integrated Ecological and Archaeological Study" *Cartographic Journal,* Vol. 32, Iss 2, pp. 137-142.

Coombes, M. Openshaw, S., Wong, C., Raybould, S. (1993)"Community Boundary Definition—A GIS Design Specification" *Regional Studies,* Vol. 27, Iss 3, pp. 280-286.

Coombes, M., Saunders, C., Raybould, S. (1993) "Critical Problems in Developing Databases for Upland Areas—The Potential Value of GIS Techniques" *Journal of Environmental Management,* Vol. 39, Iss 4, pp. 271-280.

Coppock, J.T. (1992) "GIS Education in Europe" *IJGIS,* Vol. 6, Iss 4, pp. 333-335.

Coppock, T. (1992) "GIS in Education—Foreword" *IJGIS,* Vol. 6, Iss 4, pp. R1-R1.

Coppock, T. (1993) "International Journal of Geographical Information Systems" *IJGIS,* Vol. 7, Iss 2, pp. 101-101.

Coppock, T., Guptill, S. (1991) "GIS in Developing-Nations" *IJGIS,* Vol. 5, Iss 1, pp. 1-1.

Cornelius, S.C., Sear, D.A., Carver, S.J., Heywood, D.I. (1994) "GPS, GIS and Geomorphological Field Work" *Earth Surface Processes and Landforms,* Vol. 19, Iss 9, pp. 777-787.

Coroza, O., Evans, D., Bishop, I. (1997) "Enhancing Runoff Modelling with GIS" *Landscape and Urban Planning,* Vol. 38, Iss 1-2, pp. 13-23.

Cosetino, Brian L. and Lillesand, Thomas M. (1991), Towards automated statewise land cover mapping in Wisconsin using satellite remote sensing and GIS Techniques, 1991 ACSM/ ASPRS Technical Papers, 3, 93-102.

Coulson, M.R.C. (1991) "Tactile-Map Output from Geographical Information Systems—The Challenge and Its Importance" *IJGIS,* Vol. 5, Iss 3, pp. 353-360.

Cova, T.J., Church, R.L. (1997) "Modelling Community Evacuation Vulnerability Using GIS" *IJGISC,* Vol. 11, Iss 8, pp. 763-784.

Cracknell, A.P. (1986) "Editorial—Geographical Information Systems" *IJRS,* Vol. 7, Iss 6, pp. 721-722.

Cracknell, A.P. (1993) "Icorg-92—Remote-Sensing Applications and Geographic Information Systems—Recent Trends—Preface" *IJRS,* Vol. 14, Iss 17, pp. 3077-3077.

Craglia, M. (1994) "Geographical Information Systemsin Italian Municipalities" *Computers Environment and Urban Systems,* Vol. 18, Iss 6, pp. 381-&.

Craglia, M., Raper, J. (1995) "GIS and Multimedia" *Environment and Planning B-Planning and Design,* Vol. 22, Iss 6, pp. 634-636.

Craul, P.J., Herrington, L.P. (1992) "GIS Applications—A Wide Spectrum Not Without Problems" *PERS* 1992, Vol. 58, Iss 8, pp. 1092-1094.

Cress, J.J., Deister, R.R.P. (1990) "Development and Implementation of a Knowledge-Based GIS Geological Engineering Map Production System" *PERS* Vol. 56, Iss 11, pp. 1529-1535.

Croswell, P.L., Clark, S.R. (1988) "Trends in Automated Mapping and Geographic Information System Hardware" *PERS,* Vol. 54, Iss 11, pp. 1571-1576.

Curran, P.J. (1984) "Geographic Information Systems" *AREA,* Vol. 16, Iss 2, pp. 153-158.

Curry, M.R. (1994) "Image, Practice and the Hidden Impacts of Geographic Information Systems" *Progress in Human Geography,* Vol. 18, Iss 4, pp. 441-459.

Curtis, S.E. (1989) "The Development of Geographical Information Systems for Locality Planning in Health-Care" *AREA,* Vol. 21, Iss 4, pp. 391-399.

Dahlberg, R.E. (1991) "Development of an Integrated Cartography-Remote Sensing-Geographic Information Systems Program at Northern-Illinois-University" *CGIS,* Vol. 18, Iss 3, pp. 171-175.

Dai, Q., Fletcher, J., Phipps, T., Yuill, C. (1996) "Development of a GIS Decision-Support System for Water-Quality Policy Analysis" *American Journal of Agricultural Economics,* Vol. 77, Iss 5, pp. 1372-1372.

Dale, P.F., (1991) "Information Technology in Geography and Planning—Including Principles of GIS, by I. Bracken, C. Webster" *IJGIS,* Vol. 5, Iss 2, pp. 256-258.

Dangermond, J. (1989) "Trends in Geographic Information Systems" *EKISTICS-The Problems and Science of Human settlements,* Vol. 56, Iss 338-, pp. 318-330.

Dangermond, J., Smith, L.K., (1988) "Geographic Information Systems and the Revolution in Cartography—The Nature of the Role Played by a Commercial Organization" *American Cartographer,* Vol. 15, Iss 3, pp. 301-310.

Dasgupta, A.R. (2005), Standards and interoperability for a seamless world, *Geospatial Today,* Vol.4, issue 2, pp. 22-4.

Davidson, D.A., Theocharopoulos, S.P., Bloksma, R.J. (1994) "A Land Evaluation Project in Greece Using GIS and Based on Boolean and Fuzzy Set Methodologies" *IJGIS,* Vol. 8, Iss 4, pp. 369-384.

Davidson, D.A., Theocharopoulos, S.P., Bloksma, R.J. (1996) "A Land Evaluation Project in Greece Using GIS and Based on Boolean and Fuzzy Set Methodologies (Vol. 8, Pg 369, 1994)" *IJGIS,* Vol. 10, Iss 2, pp. 245-245.

Davies, C., Medyckyjscott, D. (1994) "GIS Usability—Recommendations Based on the Users View" *IJGIS,* Vol. 8, Iss 2, pp. 175-189.

Davies, C., Medyckyjscott, D. (1996) "GIS Users Observed" *IJGIS,* Vol. 10, Iss 4, pp. 363-384.

Davis, B.A., Carr, H.V., Schmidt, N., Hickerson, L. (1993) "Opportunities in the Commercial Uses of Remote-Sensing and GIS Technologies—An Overview of NASA Visiting Investigator Program at Stennis-Space-Center" *PERS* Vol. 59, Iss 6, pp. 935-939.

Davis, F.W., Quattrochi, D.A., Ridd, M.K., Lam, N.S.N., Walsh, S.J., Michaelsen, J.C., Franklin, J., Stow, D.A., Johannsen, C.J., Johnston, C.A. (1991) "Environmental-Analysis Using Integrated GIS and Remotely Sensed Data—Some Research Needs and Priorities" *PERS* 1991, Vol. 57, Iss 6, pp. 689-697.

Davis, J.L. (1993) "GIS Approaches to Regional Analysis—A Case Study of the Island of Hvar, by V. Gaffney, Z. Stancic" *American Journal of Archaeology,* Vol. 97, Iss 4, pp. 799-800.

Davis, J.L. (1993) "Interpreting Space—GIS and Archaeology", by K.M.S. Allen, S.W. Green, E.B.W. Zubrow, *American Journal of Archaeology,* Vol. 97, Iss 2, pp. 357-359.

Deane, G.C. (1994) "The Role of GIS in the Management of Natural Resources" *ASLIB Proceedings,* Vol. 46, Iss 6, pp. 157-161.

Defusco, R.P. (1993) "Modelling Bird Hazards to Aircraft—A GIS Application Study" *PERS* Vol. 59, Iss 10, pp. 1481+.

Degg, M. (1994) "International Workshop on Geographical Information Systems in Assessing Natural Hazards, University-for-Foreigners, Perugia, Italy, 20-22 September, 1993" *Disasters*, Vol. 18, Iss 2, pp. 177-178.

Demers, M.N. (1993) "Geographical Information Systems, Vol. 1—Principles, Vol. 2—Applications", by D.J. Maguire, M.F. Goodchild, D.W. Rhind, *Geographical Review,* Vol. 83, Iss 3, pp. 330-332.

Demers, M.N., Fisher, P.F. (1991) "Comparative Evolution of Statewide Geographic Information Systems in Ohio" *IJGIS,* Vol. 5, Iss 4, pp. 469-485.

Denning, J. (1993) "Small-Government GIS" *Civil Engineering,* Vol. 63, Iss 6, pp. 52-54.

Derenyi, E., Pollock, R. (1990) "Extending a GIS to Support Image-Based Map Revision", *PERS* Vol. 56, Iss 11, pp. 1493-1496.

Deroo, A.P.J., Burrough, P.A., Hazelhoff, L. (1989) "Soil-Erosion Modelling Using Answers and Geographical Information Systems" *Earth Surface Processes and Landforms,* Vol. 14, Iss 6-7, pp. 517-532.

Desmet, P.J.J., Govers, G. (1995) "GIS-Based Simulation of Erosion and Deposition Patterns in an Agricultural Landscape—A Comparison of Model Results with Soil Map Information" *CATENA* Vol. 25, Iss 1-4, pp. 389-401.

Desmet, P.J.J., Govers, G. (1997) "Modelling Topographic Potential for Erosion and Deposition Using GIS—Comment" *IJGISC,* Vol. 11, Iss 6, pp. 603-610.

Devantier, B.A. (1993) "Review of GIS Applications in Hydrologic Modelling (Vol. 119, Pg 248, 1993)" *Journal of Water Resources Planning and Management-ASCE* Vol. 119, Iss 4, pp. 504-504.

Devantier, B.A., Feldman, A.D. (1993) "Review of GIS Applications in Hydrologic Modelling" *Journal of Water Resources Planning and Management-ASCE* Vol. 119, Iss 2, pp. 246-261.

Dillworth, Mary E. (1991), Geographic windows in remote sensing: does window size matter? *ACSM/ASPRS Technical Papers,* Vol. 3, pp.122-28.

Dillworth, M.E., Stieg, R.T. (1992) "GIS in the Memphis, Tennessee Region—Working Together Saves Time" *PERS* 1992, Vol. 58, Iss 8, pp. 1213-1216.

Djokic, D., Maidment, D.R. (1993) "Application of GIS Network Routines for Water-Flow and Transport" *Journal of Water Resources Planning and Management-ASCE* Vol. 119, Iss 2, pp. 229-245.

Dobson, J.E. (1992) "Geographical Information Systems—Principles and Applications", by D.J. Maguire, M.F. Goodchild, D.W. Rhind, PG, Vol. 44, Iss 3, pp. 358-359.

Dobson, J.E. (1993) "Commentary—A Conceptual Framework for Integrating Remote Sensing, GIS, and Geography" *PERS* Vol. 59, Iss 10, pp. 1491-1496.

Dockrill, S.J. (1992) "GIS Approaches to Regional-Analysis—A Case Study of the Island of Hvar", by V. Gaffney, Z. Stancic, *Antiquity,* Vol. 66, Iss 250, pp. 266-268.

Dockrill, S.J. (1992) "Interpreting Space—GIS and Archaeology", by K.M.S. Allen, S.W. Green, E.B.W. Zubrow, *Antiquity,* Vol. 66, Iss 250, pp. 266-268.

Dockrill, S.J. (1992) "Introductory Readings in Geographic Information Systems", by D.J. Pequet, D.F. Marble *Antiquity,* Vol. 66, Iss 250, pp. 266-268.

Dorling, D.F.L. (1992) "Geographical Information Systems and Their Socio-economic Applications", by D. Martin, *IJGIS,* Vol. 6, Iss 5, pp. 447-448.

Duecker, K.J., Delacy, P.B. (1990) "GIS in the Land-Development Planning Process—Balancing the Needs of Land-Use Planners and Real-Estate Developers" *Journal of the American Planning Association,* Vol. 56, Iss 4, pp. 483-491.

Dueker, K.J. (1984) "Proceedings, United-States Australia Workshop on Design and Implementation of Computer-Based Geographic Information Systems", by D. Peuquet, J. Ocallaghan *PG,* Vol. 36, Iss 4, pp. 518-518.

Dueker, K.J. (1987) "Geographic Information Systems and Computer-Aided Mapping" *Journal of the American Planning Association,* Vol. 53, Iss 3, pp. 383-390.

Duguay, C.R., Beaudoin, M., Leroux, D. (1992) "Fundamentals of Geographic Information Systems—A Compendium", by W.J. Ripple, *Canadian Geographer-Geograhe Canadien,* Vol. 36, Iss 1, pp. 103-108.

Duguay, C.R., Beaudoin, M., Leroux, D. (1992) "Geographic Information Systems—An Introduction", by J. Star, J. Estes, *Canadian Geographer-Geograhe Canadien,* Vol. 36, Iss 1, pp. 103-108.

Duguay, C.R., Beaudoin, M., Leroux, D. (1992) "Geographic Information Systems for Urban and Regional Planning", by H.J. Scholten, J.C.H. Stillwell, *Canadian Geographer-Geograhe Canadien,* Vol. 36, Iss 1, pp. 103-108.

Duguay, C.R., Beaudoin, M., Leroux, D. (1992) "Geographic Information Systems—Developments and Applications", by L. Worrall, *Canadian Geographer-Geograhe Canadien,* Vol. 36, Iss 1, pp. 103-108.

Duguay, C.R., Beaudoin, M., Leroux, D. (1992) "Geographic Information Systems and Cartographic Modelling", by C.D. Tomlin, *Canadian Geographer-Geograhe Canadien,* Vol. 36, Iss 1, pp. 103-108.

Duncan, C.J., Epps, W.R. (1993) "GIS and the Role of the State Down Under—Comment" *Political Geography,* Vol. 12, Iss 1, pp. 3-7.

Dunn, C.E., Atkins, P.J. Townsend, J.G. (1997) "GIS for Development—A Contradiction in Terms" *AREA,* Vol. 29, Iss 2, pp. 151-159.

Dunn, C.E., Newton, D. (1992) "Optimal Routes in GIS and Emergency Planning Applications" *AREA,* Vol. 24, Iss 3, pp. 259-267.

Dunn, R. (1991) "Geographic Information Systems—Developments and Applications", by L. Worrall, *Local Government Studies,* Vol. 17, Iss 3, pp. 101-104.

Dunn, R. (1992) "Geographic Information Systems and Their Socioeconomic Applications", by D. Martin, *Environment and Planning A,* Vol. 24, Iss 7, pp. 1068-1069.

Eckhardt, D.W., Verdin, J.P., Lyford, G.R. (1990) "Automated Update of an Irrigated Lands GIS Using Spot HRV Imagery" *PERS* Vol. 56, Iss 11, pp. 1515-1522.

Edralin, J.S. (1991) "International-Conference on Geographic Information Systems Applications for Urban and Regional-Planning—Ciloto, Puncak, Indonesia, 3-6 October 1989" *IJGIS,* Vol. 5, Iss 1, pp. 147-154.

Egenhofer, M.J. (1992), Why not SQL!, *International Journal of Geographic Information Systems,* Vol. 6, No. 2, pp. 71-85.

Egenhofer, M.J., Frank, A.U. (1990) "Lobster—Combining AI and Database Techniques for GIS", *PERS* Vol. 56, Iss 6, pp. 919-926.

Ehlers, M., Greenlee, D., Smith, T., Star, J. (1991) "Integration of Remote-Sensing and GIS—Data and Data Access" *PERS* 1991, Vol. 57, Iss 6, pp. 669-675.

Ellis, J.R., Casavant, K.L., Jessup, E.L. (1996) "A GIS and Transportation Optimization Model Approach to Determining Highway and Rural Road Commodity Flows" *American Journal of Agricultural Economics,* Vol. 78, Iss 5, pp. 1394-1394.

Elston, D.A., Buckland, S.T. (1993) "Statistical Modelling of Regional GIS Data—An Overview" *Ecological Modelling,* Vol. 67, Iss 1, pp. 81-102.

Elston, D.A., Jayasinghe, G., Buckland, S.T., Macmillan, D.C., Aspinall, R.J. (1997) "Adapting Regression Equations to Minimize the Mean Squared Error of Predictions Made Using Covariate Data from a GIS" *IJGISC,* Vol. 11, Iss 3, pp. 265-280.

Elvis, M., Matsuoka, M., Siemiginowska, A., Fiore, F., Mihara, T., Brinkmann, W. (1994) "An Asca GIS Spectrum of S5-0014+813 at Z=3.384", *Altrophysical Journal,* 1994, Vol. 436, Iss 1, pp. L55-L58.

Engel, B.A., Srinivasan, R., Arnold, J., Rewerts, C., Brown, S.J. (1993) "Nonpoint-Source (Nps) Pollution Modelling Using Models Integrated with Geographic Information Systems (GIS)", *Water Science and Technology,* Vol. 28, Iss 3-5, pp. 685-690.

Esnard, A.M., Macdougall, E.B. (1997) "Common Ground for Integrating Planning-Theory and GIS Topics" *Journal of Planning Education and Research,* Vol. 17, Iss 1, pp. 55-62.

ESRI (2001), *Map Book,* Environmental System research Institute, Redlands.

Etzelmuller, B., Vatne, G., Odegard, R.S., Sollid, J.L. (1994) "Mass-Balance and Changes of Surface Slope, Crevasse and Flow Pattern of Erikbreen, Northern Spitsbergen—An Application of a Geographical Information System (GIS)" *Polar Research,* 1993, Vol. 12, Iss 2, pp. 131-146.

Evans, B.M., Myers, W.L. (1990) "A GIS-Based Approach to Evaluating Regional Groundwater Pollution Potential with Drastic" *Journal of Soil and Water Conservation,* Vol. 45, Iss 2, pp. 242-245.

Fairbairn, D. (1992) "Geographical Information Systems—Principles and Applications, Vol. 1, Vol. 2", by D. Maguire, M. Goodchild, D. Rhind, *Cartographic Journal,* Vol. 29, Iss 1, pp. 79-80.

Fairbairn, D., Openshaw, S. (1991)"Cartographic and GIS Activity at the University-of-Newcastle-upon-Tyne 1987-1991", *Cartographic Journal,* Vol. 28, Iss 1, pp. 79-82.

Farrell, T. (1995) "Closing with the Enemy—How GIS Fought the War in Europe, 1944-45", by M.D. Doubler, *Journal of Strategic Studies,* Vol. 18, Iss 4, pp. 139-140.

Farthing, K. (1996) "The GIS Book, 3rd Edition", by G.B. Korte, *IJGIS,* Vol. 10, Iss 1, pp. 139-141.

Farthing, S. (1991) "Information Technology in Geography and Planning—Including Principles of GIS", by I. Bracken, C. Webster, *Regional Studies,* Vol. 25, Iss 2, pp. 192-192.

Faust, N.L. (1987) "Automated Data Capture for Geographic Information- Systems—A Commentary" *PERS,* Vol. 53, Iss 10, pp. 1389-1390.

Faust, N.L., Anderson, W.H., Star, J.L. (1991) "Geographic Information Systems and Remote-Sensing Future Computing Environment" *PERS,* Vol. 57, Iss 6, pp. 655-668.

Felleman, J. (1990) "Theres a GIS in Your Future", *Government Information Quarterly,* Vol. 7, Iss 3, pp. 261-267.

Ferenth, M. (1991) "Gimms-(GIS)-LTD—Products and Platforms" *Cartographic Journal,* Vol. 28, Iss 1, pp. 47-48.

Fernandez, R.N., Lozanogarcia, D.F., Johannsen, C.J. (1989) "Development of Microcomputer Based Geo-Referenced Information-Systems (GIS)—Case Studies in Argentina and Mexico", *Interciencia,* Vol. 14, Iss 5, pp. 247-253.

Fernandez, R.N., Rusinkiewicz, M. (1993) "A Conceptual Design of a Soil Database for a Geographical Information System" *IJGIS,* Vol. 7, Iss 6, pp. 525-539.

Ferndndez, R. Norberto, D. Fabidn Lozano-Garcia, Greg Deeds and Cgria IJ. Johannsen (1991), Accuracy assessment of map coordinate retrieval, *Photogrammetric Engineering & Remote Sensing,* Vol. 37, No.11, pp. 1447-52.

Ferrier, G., Wadge, G. (1997) "An Integrated GIS and Knowledge-Based System as an Aid for the Geological Analysis of Sedimentary Basins" *IJGISC,* Vol. 11, Iss 3, pp. 281-297.

Feser, K., Sun, R., Eriksson, A., Frohlich, K. (1991) "On-Site Dielectric Testing of GIS—Theoretical and Practical Considerations" *IEEE Transactions on Power Delivery,* Vol. 6, Iss 2, pp. 615-625.

Fischer, M.M., Nijkamp, P. (1992) "Geographic Information Systems and Spatial-Analysis" *Annals of Regional Science,* Vol. 26, Iss 1, pp. 3-17.

Fisher, P. (1996) "Geographic Information Systems for Geoscientists—Modelling with GIS", by G.F. Bonhamcarter, *Environment and Planning B-Planning and Design,* Vol. 23, Iss 3, pp. 384-385.

Fisher, P. (1996) "Spatial-Analysis and GIS", by S. Fotheringham, P. Rogerson, *Environment and Planning B-Planning and Design,* Vol. 23, Iss 2, pp. 245-246 Fisher-P Strachan-A Stuart-N (1996) "UK Developments in Environmental GIS—In Honor of Coppock,T" *IJGIS,* Vol. 10, Iss 1, pp. 3+.

Fisher, P.F. (1989) "Geographical Information System Software for* Teaching", *Journal of Geography in Higher Education,* Vol. 13, Iss 1, pp. 69-80.

Fitzsimons, D. (1991), "Cartography Geographic Information Systems at Southwest-Texas-State-University" *CGIS,* Vol. 18, Iss 3, pp. 176-178.

Flowerdew, R. (1988) "Rrl Workshop on Geographical Information Systems—Report of a 2-Day Workshop Sponsored by Esrc for the Regional Research Laboratory Initiative at the University-of-Edinburgh, 18-19 September 1987", *AREA,* Vol. 20, Iss 1, pp. 89-90.

Flowerdew, R. (1990) "Population and Geographical Information Systems" *AREA,* Vol. 22, Iss 2, pp. 207-208.

Flowerdew, R., Green, M. (1992) "Developments in Areal Interpolation Methods and GIS" *Annals of Regional Science*, Vol. 26, Iss 1, pp. 67-78.

Fogarty, B.W. (1994) "Grand Challenges for GIS and Whats Really Relevant to Social-Science" *Social Science Computer Review,* Vol. 12, Iss 2, pp. 193-201.

Foody, G. (1991) "3-Dimensional Applications in Geographic Information Systems", by J. Raper, *AREA,* Vol. 23, Iss 3, pp. 278-279.

Forer, P. (1993) "An Introduction to Urban Geographic Information Systems", by W.E. Huxhold, *Australian Geographer,* Vol. 24, Iss 2, pp. 89-90.

Forrest, D. (1990) "3-Dimensional Applications in Geographic Information Systems", by J. Raper, *Cartographic Journal,* Vol. 27, Iss 1, pp. 70-72.

Fotheringham, A.S. (1991) "GIS and Spatial-Analysis—An Ncgia Research Initiative" *Environment and Planning A,* Vol. 23, Iss 10, pp. 1390-1391.

Fotheringham, A.S. (1992) "Exploratory Spatial Data-Analysis and GIS" *Environment and Planning A,* Vol. 24, Iss 12, pp. 1675-1678.

Fotheringham, A.S. (1993) "On the Future of Spatial-Analysis—The Role of GIS" *Environment and Planning A,* Iss NSI, pp. 30-34.

Fotheringham, A.S. (1994) "What Does Doing a PhD in GIS Mean", *Environment and Planning A,* Vol. 26, Iss 1, pp. 6-8.

Fotheringham, A.S. (1996) "1994 International GIS Sourcebook, by GIS-World", *Progress in Human Geography,* Vol. 20, Iss 3, pp. 426-426.

Fotheringham, A.S., Rogerson, P.A. (1993) "GIS and Spatial Analytical Problems", *IJGIS,* Vol. 7, Iss 1, pp. 3-19.

Fox, L., Carlson, S.A., Campbell, H. (1992) "Integrating Geographic Information Systems Education into the Curriculum at Humboldt-State-University" *PERS*, Vol. 58, Iss 8, pp. 1158-1158.

Frank, Andrew U., Egenhofer, Max J. and Kuhn, Werner (1992), A perspective on GIS Technology in the nineties, *Photogrammetric Engineering and Remote Sensing,* Vol. 57, No.11, pp. 1431-6.

Frank, A.U., Egenhofer, M.J., Kuhn, W. (1991) "A Perspective on GIS Technology in the Nineties" *PERS* Vol. 57, Iss 11, pp. 1431-1436.

Franklin, C. (1992) "An Introduction to Geographic Information Systems—Linking Maps to Databases", *Database-The Magazine of Database Reference and Review,* Vol. 15, Iss 2, pp. 12.

Franklin, C. (1993) "Using Geographic Information Systems (GIS) Software to Improve the Quality of Access to Us Census-Data on Cd-ROM" *Online aand CDRom* Review, Vol. 17, Iss 3, pp. 190-190.

French, S.P., Wiggins, L.L. (1990) "California Planning Agency Experiences with Automated Mapping and Geographic Information Systems" *Environment and Planning B-Planning and Design,* Vol. 17, Iss 4, pp. 441-450.

Furdada, G., Marti, G., Oller, P., Garcia, C., Mases, M., Vilaplana, J.M. (1995) "Avalanche Mapping and Related GIS Applications in the Catalan-Pyrenees", *Surveys in Geophysics,* Vol. 16, Iss 5-6, pp. 681-693.

Gaile, G.L. (1996) "Possibilities and Constants or GIS Applications in Developing-Countries", by P. Vanteeffelen, L. Vangrunsven, O. Verkoren, *Annals of the Association of American Geographers,* Vol. 86, Iss 4, pp. 802-803.

Gainey, Kevin and Tim Johnson (2000), Faster, flexible redistricting applications built with MapObjects, *ArcUser,* April-June, pp. 10-3.

Ganter, J.H. (1990) "3-Dimensional Applications in Geographical Information Systems, by J. Raper" *PG,* Vol. 42, Iss 4, pp. 508-509.

Garg, P.K., Harrison, A.R. (1992) "Land Degradation and Erosion Risk Analysis in S.E. Spain—A Geographic Information System Approach" *CATENA,* Vol. 19, Iss 5, pp. 411-425.

Gastelluetchegorry, J.P., Estreguil, C., Mougin, E., Laumonier, Y. (1993) "A GIS Based Methodology for Small-Scale Monitoring of Tropical Forests—A Case-Study in Sumatra" *IJRS* Vol. 14, Iss 12, pp. 2349-2368.

Gastelluetchegorry, J.P., Sinulingga, A.B. (1988) "Designing a GIS for the Study of Forest Evolution in Central Java" *TIJDSCHRIFT VOOR Economische en Sociale Geografie,* Vol. 79, Iss 2, pp. 93-103.

Gatrell, A. (1991) "Information Technology in Geography and Planning—Including Principles of GIS", by I. Bracken, C. Webster, *Urban Studies,* Vol. 28, Iss 6, pp. 1050-1052.

Gatrell, A. (1994) "Geographic Information Systems, Spatial Modelling and Policy Evaluation", by M.M. Fischer, P. Nijkamp, *Environment and Planning B-Planning and Design,* Vol. 21, Iss 5, pp. 647-648.

Gatrell, A.C. (1988) "A UK Geographic Information System for Environmental Monitoring, Resource Planning and Management Capable of Integrating and Using Satellite Remotely Sensed Data, by J.A.T. Young" *Regional Studies,* Vol. 22, Iss 2, pp. 172-173.

Gatrell, A.C., Dunn, C.E., Boyle, P.J. (1991) "The Relative Utility of the Central Postcode Directory and Pinpoint Address Code in Applications of Geographical Information Systems" *Environment and Planning A,* Vol. 23, Iss 10, pp. 1447-1458.

Gatrell, T. (1991) "Geographical Information Systems for Urban and Regional Planning", by H.J. Scholten, J.C.H. Stillwell, *Regional Studies,* Vol. 25, Iss 4, pp. 371-371.

Gatrell, T. (1997) "Spatial-Analysis—Modelling in a GIS Environment, by P. Longley, M. Batty" *Geography,* Vol. 82, Iss 356, pp. 288-288.

Gauthier, M.J. (1994) "Geographic Information System in Images—French", by C. Collet, *Canadian Geographer-Geograhe Canadien,* Vol. 38, Iss 2, pp. 191-192.

Geertman, S.C.M., Vaneck, J.R.R. (1995) "GIS and Models of Accessibility Potential—An Application in Planning" *IJGIS,* Vol. 9, Iss 1, pp. 67-80.

Gentile, J. (1993) "GIS Applications in Natural-Resources", by M. Heit, A. Shortreid, *PG,* Vol. 45, Iss 1, pp. 113-113.

Ghosh, R., Goel, R.K., Lole, B.S., Singh, T.P., Sastry, K.L.N., Patel, J.G., Vanikar, Y.V., Thakker, P.S., Navalgund, R.R. (1993) "District Level Planning—A Case-Study for the Panchmahals District Using Remote-Sensing and Geographic Information System Techniques," *International Journal of Remote Sensing,* Vol. 14, Iss 17, pp. 3163-3168.

Gifford, J.A. (1993) "Videography and Geographical Information Systems for Recording the Excavation of a Prehistoric Underwater Site," *International Journal of Nautical Archaeology,* Vol. 22, Iss 2, pp. 167-172.

Gilliland, M.W., Baxterpotter, W. (1987) "A Geographic Information System to Predict Non-Point Source Pollution Potential" *Water Resources Bulletin,* Vol. 23, Iss 2, pp. 281-291.

Gilmartin, P., Cowen, D. (1991) "Educational Essentials for Todays and Tomorrows Jobs in Cartography and Geographic Information Systems" *CGIS,* Vol. 18, Iss 4, pp. 262-267.

Gleave, D. (1992) "Spatial-Analysis and Spatial Policy Using Geographic Information Systems", by L. Worrall, *Regional Studies,* Vol. 26, Iss 6, pp. 600-601.

Gluck, M. (1992) "Prospecting for User-Based GIS Processes," *Proceedings, of the ASIS Annual Meeting,* Vol. 29, pp. 242-248.

Godschalk, D.R., Mcmahon, G., Kaplan, A., Qin, W. (1992) "Using GIS for Computer-Assisted Dispute Resolution" *PERS* 1992, Vol. 58, Iss 8, pp. 1209-1212.

Golledge, R.G., Loomis, J.M., Klatzky, R.L., Flury, A., Yang, X.L. (1991)"Designing a Personal Guidance-System to Aid Navigation Without Sight—Progress on the GIS Component" *IJGIS,* Vol. 5, Iss 4, pp. 373-395.

Gonzalez, C., Ortigosa, L., Marti, C., Garciaruiz, J.M. (1995) "The Study of the Spatial-Organization of Geomorphic Processes in Mountain Areas Using GIS," *Mountain Research and Development,* Vol. 15, Iss 3, pp. 241-249.

Goodchild, M.F. (1991), Geographical information science, *International Journal on Geographical Information System,* Vol. 6, No. 1, pp. 31-45.

Goodchild, M., Haining, R., Wise, S., Arbia, G., Anselin, L., Bossard, E., Brunsdon, C., Diggle, P., Flowerdew, R., Green, M., Griffith, D., Hepple, L., Krug, T., Martin, R., Openshaw, S. (1992) "Integrating GIS and Spatial Data-Analysis—Problems and Possibilities" *International Journal of Geographic Information Systems,* Vol. 6, Iss 5, pp. 407-423.

Goodchild, M.F. (1988) "Geographic Information Systems," *Progress in Human Geography,* Vol. 12, Iss 4, pp. 560-566.

Goodchild, M.F. (1990) "GIS and Basic Research—The National Center for Geographic Information and Analysis," *Government Information Quarterly,* Vol. 7, Iss 3,. pp. 343-355.

Goodchild, M.F., (1991) "Geographic Information Systems", *Journal of Retailing,* Vol. 67, Iss 1, pp. 3-15.

Goodchild, M.F. (1991) "Geographic Information Systems,"*Progress in Human Geography,* Vol. 15, Iss 2, pp. 194-200.

Goodchild, M.F., Mark, D.M. (1993) "GIS, Geography, and Ncgia—Response," *PG,* Vol. 45, Iss 2, pp. 216-220.

Goodchild, M.F., Yang, S.R. (1992) "A Hierarchical Spatial Data Structure for Global Geographic Information Systems," *CVGIP-Graphical Model and Image Processing,* Vol. 54, Iss 1, pp. 31-44.

Gordon, W.R. (1994) "A Role for Comprehensive Planning, Geographical Information System (GIS) Technologies and Program-Evaluation in Aquatic Habitat Development," *Bulletin of Marine Science,* Vol. 55, Iss 2-3, pp. 995-1013.

Gordon, W.R., Soubra, N.M. (1992) "Geographical Information Systems and Planning in the USA—Selected Municipal Adoption Trends and Educational Concerns" *IJGIS,* Vol. 6, Iss 4, pp. 267-278.

Gould, M.D. (1994) "GIS Design—A Hermeneutic View" *PERS* 1994, Vol. 60, Iss 9, pp. 1105-1116.

Gould, M.I. (1992) "The Use of GIS and Cac by Health Authorities—Results from a Postal Questionnaire" *AREA,* Vol. 24, Iss 4, pp. 391-401.

Goulter, I.C., Forrest, D. (1987) "Use of Geographic Information Systems (GIS) in River Basin Management," *Water Science and Technology,* Vol. 19, Iss 9, pp. 81-86.

Goulter, I.C., Forrest, D. (1987) "Use of Geographic Information Systems (GIS) in River Basin Management," *Water Science and Technology,* Vol. 19, Iss 9, pp. 81-86.

Graham, D.F., Bonhamcarter, G.F. (1993) "Airborne Radiometric Data—A Tool for Reconnaissance Geological Mapping Using a GIS" *PERS* Vol. 59, Iss 8, pp. 1243-1249.

Gratton, D.J., Howarth, P.J., Marceau, D.J. (1990) "Combining dem Parameters with LANDSAT MSS and TM Imagery in a GIS for Mountain Glacier Characterization" *IEEE Transactions on Geoscience and Remote Sensing,* Vol. 28, Iss 4, pp. 766-769.

Green, N.P., Finch, S. and Wiggins, J. (1986), The state of art in geographic information system, *Area,* Vol. 17.No. 4, pp. 295-301.

Green, K., Finney, M., Campbell, J., Weinstein, D., Landrum, V. (1995) "Fire—Using GIS to Predict Fire Behavior" *JF* Vol. 93, Iss 5, pp. 21-25.

Green, N.P., Finch, S., Wiggins, J. (1985) "The State of the Art in Geographical Information Systems" *AREA,* Vol. 17, Iss 4, pp. 295-301.

Green, R. (1990) "Geographic Information Systems in Europe" *Cartographic Journal,* Vol. 27, Iss 1, pp. 40-42.

Greer, J.D. (1994) "GIS and Remote-Sensing for Wildland Fire Suppression and Burned Area Restoration" *PERS* 1994, Vol. 60, Iss 9, pp. 1059-1064.

Grelot, J.P. (1994)"Cartography in the GIS Age" *Cartographic Journal,* Vol. 31, Iss 1, pp. 56-60.

Greve, C.W., Kelmelis, J.A., Fegeas, R., Guptill, S.C., Mouat, N, (1993) "Investigating United-States-Geological-Survey Needs for the Management of Temporal GIS Data" *PERS* Vol. 59, Iss 10, pp. 1503-1507.

Griffiths, A. (1989) "Sagis—A Proposal for a Sardinian Geographical Information System and an Assessment of Alternative Implementation Strategies," *Journal of Information Science,* Vol. 15, Iss 4-5, pp. 261-267.

Grimshaw, D.J. (1989) "Geographical Information Systems—A Tool for Business and Industry", *International Journal of Information Management,* Vol. 9, Iss 2, pp. 119-126.

Grimshaw, D.J. (1991) "Geographical Information Systems as Part of the Corporate Information Strategy," *International Journal of Information Management,* Vol. 11, Iss 4, pp. 292-297.

Griner, A.J. (1994) "Development of a Water-Supply Protection Model in a GIS," *Water Resources Bulletin* 1993, Vol. 29, Iss 6, pp. 965-971.

Grossmann, W.D., Eberhardt, S. (1992) "Geographical Information Systems and Dynamic Modelling—Potentials of a New Approach," *Annals of Regional Science,* Vol. 26, Iss 1, pp. 53-66.

Grove, M., Hohmann, M. (1992) "Social Forestry and GIS" *JF* 1992, Vol. 90, Iss 12, pp. 10-15.

Grunblatt, J., Ottichilo, W.K., Sinange, R.K., (1992) "A GIS Approach to Desertification Assessment and Mapping," *Journal of Arid Environments,* Vol. 23, Iss 1, pp. 81-102.

Gumbricht, T. (1996) "Application of GIS in-Training for Environmental-Management," *Journal of Environmental Management,* Vol. 46, Iss 1, pp. 17-30.

Guthe, W.G., Tucker, R.K., Murphy, E.A., England, R., Stevenson, E., Luckhardt, J.C. (1993) "Reassessment of Lead-Exposure in New-Jersey Using GIS Technology" *Environmental Research* 1992, Vol. 59, Iss 2, pp. 318-325.

Haining, R., Ma, J.S., Wise, S. (1996) "Design of a Software System for Interactive Spatial Statistical-Analysis Linked to a GIS," *Computation*al Statistics, Vol. 11, Iss 4, pp. 449-466.

Hall, G.B., Wang, F., Subaryono (1992) "Comparison of Boolean and Fuzzy Classification Methods in Land Suitability Analysis by Using Geographical Information Systems," *Environment and Planning A,* Vol. 24, Iss 4, pp. 497-516.

Healey, R.G. and Desa, G.B. (1989), Transputer based parallel processing for GIS analysis: problems and potentialities, *Auto-Carto, 9 Proceedings,* April, pp. 90-99.

Hendrix, W.G., Price, J.E., Fabos, J.G. (1988) "An Ecological Approach to Landscape Planning Using Geographic Information System Technology" *Landscape and Urban Planning,* Vol. 15, Iss 3-4, pp. 211-225.

Hernon, P. (1997) "GIS as a Service Option," *Journal of Academic Librarianship,* Vol. 23, Iss 3, pp. 235-237.

Herold, P. (1996) "Moving Geospatial Data to the Web—GIS at Mann-Library," *LIbrary Hi Tech,* Vol. 14, Iss 4, pp. 86-87.

Herr, A.M., Queen, L.P. (1993) "Crane Habitat Evaluation Using GIS and Remote-Sensing" *PERS* Vol. 59, Iss 10, pp. 1531-1538.

Herring, G.C. (1986) "And a Hard Rain Fell—A GIS True Story of the War in Vietnam", by J. Ketwig, *Journal of American History,* Vol. 73, Iss 1, pp. 152-164.

Hession, W.C., Shanholtz, V.O. (1988) "A Geographic Information System for Targeting Nonpoint Source Agricultural Pollution," *Journal of Soil and Water Conservation,* Vol. 43, Iss 3, pp. 264-266.

Heywood, I. (1990) "Geographic Information Systems in the Social-Sciences—Introduction" *Environment and Planning A,* Vol. 22, Iss 7, pp. 849-852.

Hickey, R. Jankowski, P. (1997) "GIS and Environmental Decision-Making to Aid Smelter Reclamation Planning," *Environment and Planning A,* Vol. 29, Iss 1, pp. 5-19.

Hicks, D.R. (1992) "The Role of Auburn-University Department of Geography in Remote-Sensing GIS for International Programs" *PERS* 1992, Vol. 58, Iss 8, pp. 1155-1157.

Higgs, G. (1994) "Landscape Ecology and GIS", by R.H. Hainesyoung, D.R. Green, S.H. Cousins, *Environment and Planning B-Planning and Design,* Vol. 21, Iss 3, pp. 383-385.

Higgs, G. (1996) "Environmental Modelling with GIS", by M.F. Goodchild, B.O. Parks, L.T. Steyaert, *Environment and Planning B-Planning and Design,* Vol. 23, Iss 4, pp. 501-503.

Hinton, J.C. (1996) "GIS and Remote-Sensing Integration for Environmental Applications" *IJGIS,* Vol. 10, Iss 7, pp. 877-890.

Hirschfield, A., Brown, P., Todd, P. (1995) "GIS and the Analysis of Spatially-Referenced Crime Data -Experiences in Merseyside, UK" *IJGIS,* Vol. 9, Iss 2, pp. 191-210.

Hirschfield, A., Brown, P.J.B., Bundred, P. (1995) "The Spatial-Analysis of Community-Health Services on Wirral Using Geographic Information Systems," *Journal of the Operational Research Society,* Vol. 46, Iss 2, pp. 147-159.

Hiscock, K.M., Lovett, A.A., Brainerd, J.S., Parfitt, J.P. (1995) "Groundwater Vulnerability Assessment—2 Case-Studies Using GIS Methodology," *Quarterly Journal of Engineering Geology* Vol. 28, Iss MAY, pp. 179-194.

Hodgson, M.E. (1992) "Geographic Information Systems—An Introduction", by J. Star, J. Estes *CGIS,* Vol. 19, Iss 3, pp. 181-182.

Hodgson, M.E., Jensen, J.R., Coulter, M.C., Mackey, H.E. (1988)"Monitoring Wood Stork Foraging Habitat Using Remote-Sensing and Geographic Information Systems" *PERS,* Vol. 54, Iss 11, pp. 1601-1607.

Hodgson, M.E., Price, K.P. (1993) "Special GIS Issue—Foreword" *PERS* Vol. 59, Iss 10, pp. 1489-1490.

Holden, R., Turner, T. (1997) "Western-Europe, Current City Expansion and the Use of GIS" *Landscape and Urban Planning,* Vol. 36, Iss 4, pp. 315-326.

Horowitz, A.J. (1997) "Integrating GIS Concepts into Transportation Network Data-Structures," *Transportation Planning and Technology,* Vol. 21, Iss 1-2, pp. 139-153.

Howard, D.C. (1996) "Landscape Ecology and GIS", by R.H. Young, D.R. Green, S.H. Cousins, *IJGIS,* Vol. 10, Iss 1, pp. 137-138.

Hrkal, Z., Trouillard, J.M. (1994) "Use of GIS for Optimalization of Human Activity in a Catchment-Area—An Example of the Beauce Region (France)," *Environmental Geology* 1994, Vol. 24, Iss 1, pp. 22-27.

Huber, T.P. (1993) "Integrated Remote-Sensing and GIS Techniques for Elk Habitat Management," *Journal of Environmental Systems,* Vol. 22, Iss 4, pp. 325-339.

Hultquist, N.B. (1993) "Geographic Information Systems—Developments and Applications", by L. Worrall, *PG,* Vol. 45, Iss 4, pp. 497-498.

Hultquist, N.B. (1994) "Understanding GIS Tutorial GIS Software" *PERS* 1994, Vol. 60, Iss 10, pp. 1203-1205.

Hunt, E.D. (1992) "Upgrading Site-Catchment Analyses with the Use of GIS—Investigating the Settlement-Patterns of Horticulturalists," *World Archaeology,* Vol. 24, Iss 2, pp. 283-309.

Hunter, A.A. (1996) "Modelling Prehistoric Agricultural Productivity in Southwestern Colorado—A GIS Approach", by C.R. Vanwest, *American Antiquity,* Vol. 61, Iss 2, pp. 441-442.

Hutchinson, C.F., Toledano, J. (1993) "Guidelines for Demonstrating Geographical Information Systems Based on Participatory Development," *IJGIS,* Vol. 7, Iss 5, pp. 453-461.

Hutchinson, M.F. (1988) "Principles of Geographical Information Systems for Land Resources Assessment", by P.A. Burrough, *Landscape and Urban Planning,* Vol. 15, Iss 3-4, pp. 357-358.

Huxhold, W.E. (1992) "Spatial-Analysis and Spatial Policy Using Geographic Information Systems", by L. Worrall, *Urban Studies,* Vol. 29, Iss 8, pp. 1352-1356.

Innes, J.E., Simpson, D.M. (1993) "Implementing GIS for Planning—Lessons from the History of Technological Innovation," *Journal of the American Planning Association,* Vol. 59, Iss 2, pp. 230-236.

Jackson, A. (1991) "Resource Assessment in Western Australia Using a Geographic Information System," *AAPG Bulletin-American Association of Petroleum Geologists,* Vol. 75, Iss 3, pp. 600-601.

Jackson, M.J. (1986) "Report of a Workshop on Geographical Information Systems Held in London on 13 September 1985," *IJRS,* Vol. 7, Iss 6, pp. 745-755.

Jackson, M.J. (1992) "Integrated Geographical Information Systems," *IJRS,* Vol. 13, Iss 6-7, pp. 1343-1351.

Jackson, M.J., James, W.J., Stevens, A. (1988) "The Design of Environmental Geographic Information Systems," *Philosophical Transactions of the Royal Society of London Series A-Mathematical and Physical Sciences,* Vol. 324, Iss 1579, pp. 373.

Jaga, R.M.N., Sundaram, A., Natarajan, T. (1993) "Wasteland Development Using Geographic Information System Techniques," *IJRS,* Vol. 14, Iss 17, pp. 3249-3257.

Jankowski, P., Richard, L. (1994) "Integration of GIS-Based Suitability Analysis and Multicriteria Evaluation in a Spatial Decision-Support System for Route Selection," *Environment and Planning B-Planning and Design,* Vol. 21, Iss 3, pp. 323-340.

Janssen, L.L.F., Molenaar, M. (1995) "Terrain Objects, Their Dynamics and Their Monitoring by the Integration of GIS and Remote-Sensing," *IEEE Transactions in Geoscience and Remote Sensing,* Vol. 33, Iss 3, pp. 749-758.

Jensen, J.R., Christensen, E.J. (1986) "Solid and Hazardous-Waste Disposal Site Selection Using Digital Geographic Information System Techniques," *Science of the Total Environment,* Vol. 56, Iss NOV, pp. 265-276.

Jensen, J.R., Cowen, D.J., Halls, J., Narumalani, S., Schmidt, N.J., Davis, B.A., Burgess, B. (1994) "Improved Urban Infrastructure Mapping and Forecasting for Bellsouth Using Remote-Sensing and GIS Technology," *PERS* 1994, Vol. 60, Iss 3, pp. 339-346.

Jensen, J.R., Narumalani, S., Weatherbee, O., Morris, K.S., Mackey, H.E. (1992) "Predictive Modelling of Cattail and Waterlily Distribution in a South-Carolina Reservoir Using GIS," *PERS* 1992, Vol. 58, Iss 11, pp. 1561-1568.

Jenson, S.K., Domingue, J.O. (1988) "Extracting Topographic Structure from Digital Elevation Data for Geographic Information System Analysis" *PERS*, Vol. 54, Iss 11, pp. 1593-1600.

Jing, T. (1995) "Surface-Charge Accumulation—An Inevitable Phenomenon in DC-GIS," *IEEE Transactions on Dielectrics and Electrical Insulation* Vol. 2, Iss 5, pp. 771-778.

John, S.A. (1993) "Data Integration in a GIS—The Question of Data Quality," *ASLIB Proceedings,* Vol. 45, Iss 4, pp. 109-119.

Johnson, D., Keys, L. (1991) "October 1990 GIS Observer," *PERS* Vol. 57, Iss 2, pp. 158-158.

Johnson, E., Klein, M., Mickus, K. (1993) "Assessment of the Feasibility of Utilizing Landsat for Detection and Monitoring of Landfills in a Statewide GIS," *Environmental Geology* Vol. 22, Iss 2, pp. 129-140.

Johnson, M.L. (1996) "GIS in Business—Issues to Consider in Curriculum Decision-Making," *Journal of Geography,* Vol. 95, Iss 3, pp. 98-105.

Johnson, R.M. (1989) "Scholars Guide to Washington, DC for Cartography and Remote-Sensing Imagery (Maps, Charts, Aerial Photographs, Satellite Images, Cartographic Literature and Geographic Information Systems)", by R.E. Ehrenberg, *Journal of American History,* Vol. 75, Iss 4, pp. 1393-1394.

Johnsson, K. (1995) "Fragmentation Index as a Region-Based GIS Operator," *IJGIS,* Vol. 9, Iss 2, pp. 211-220.

Johnston, C.A., Niemi, G.J., Detenbeck, N.E., Bonde, J.P. (1988)"Geographic Information Systems for Cumulative Impact Assessment," *PERS*, Vol. 54, Iss 11, pp. 1609-1615.

Johnston, K.M. (1987) "Natural-Resource Modelling in the Geographic Information System Environment," *PERS*, Vol. 53, Iss 10, pp. 1411-1415.

Jones, G.E. (1992) "Geographic Information Systems and Their Socioeconomic Applications", by D. Martin, *Geography,* Vol. 77, Iss 336, pp. 278-279.

Jones, G.E., Davidson, D.A. (1987) "A Computer-Based Land Resource Information-System—A Geographical Contribution to Land-Use Planning" *Town Planning Review,* Vol. 58, Iss 1, pp. 81-94.

Jones, K. (1996) "GIS for Business and Service Planning", by P. Longley, G. Clarke, *Cartographic Journal,* Vol. 33, Iss 2, pp. 157-157.

Jones, R.M., Copas, C.V., Edmonds, E.A. (1997) "GIS Support for Distributed Group-Work in Regional-Planning," *IJGISC,* Vol. 11, Iss 1, pp. 53-71.

Jordan, C., Mihalyfalvy, E., Garrett, M.K., Smith, R.V. (1994) "Modelling of Nitrate Leaching on a Regional-Scale Using a GIS," *Journal of Environmental Management,* Vol. 42, Iss 3, pp. 279-298.

Jordan, L.E., Rado, B.Q., Sperry, S.L. (1992) "Meeting the Needs of the GIS and Image-Processing Industry in the 1990s" *PERS* 1992, Vol. 58, Iss 8, pp. 1249-1251.

Jordan, L.E., Star, J. (1992) "A Call to Action—Standards for the GIS Community," *PERS* 1992, Vol. 58, Iss 6, pp. 863-864.

Junkin, B.G. (1982) "Development of 3-Dimensional Spatial Displays Using a Geographically Based Information-System," *PERS*, Vol. 48, Iss 4, pp. 577-586.

Kainz, W. (1989) "GIS Technology—Geographical Information Systems, Land Information-Systems and Their Fundamental Principles—German", by N. Bartelme, *Mitteilungen Der Osterreichischen Geographischen Gesellschaft,* Vol. 131, pp. 306-306.

Kainz, W. (1990) "Applications of Spatial Data-Structures—Computer Graphic, Image-Processing, and GIS", by H. Samet, *Mitteilungen Der Osterreichischen Geographischen Gesellschaft,* Vol. 132, pp. 325-325.

Kainz, W. (1991) "4th Vienna Symposium—Geographic Information Systems and Cartography—Theoretical Principles and Future Aspects—Vienna, July 8-9, 1991," *Mitteilungen Der Osterreichischen Geographischen Gesellschaft,* Vol. 133, pp. 293-294.

Kainz, W. (1991)"Applications of Spatial Data-Structures—Computer-Graphics, Image-Processing, and GIS", by H. Samet, *IJGIS,* Vol. 5, Iss 2, pp. 253-254.

Kapetsky, J.M., Worthy, L.D., Hill, J.M. (1988) "A Geographical Information System for Catfish Farming Development," *Aquaculture,* Vol. 68, Iss 4, pp. 311-320.

Karaki, S., Yamazaki, T., Nojima, K., Yokota, T., Murase, H., Takahashi, H., Kojima, S. (1995) "Transient Impedance of GIS Grounding Grid" *IEEE Transaction on Power Delivery,* Vol. 10, Iss 2, pp. 723-731.

Kay, Simon (1988), Geographic information systems: An·integrated database for Badiyah, *The Journal of Oman Studies,* Special Report No. 3 (The Scientific results of the Royal Geographical Society's Oman Wahiba Sands Project 1985-1987), pp. 47-54.

Keating, T., Ingram, K., Phillips, W. (1987) "An Integrated Topologic Database Design for Geographic Information Systems" *PERS,* Vol. 53, Iss 10, pp. 1399-1402.

Kelk, B. (1990) "3-Dimensional Applications in Geographical Information Systems", by J. Raper" *Geographical Journal,* Vol. 156, Iss JUL, pp. 226-227.

Keller, C.P., Davis, T.J., Canessa, R.R. (1996) "A Hypertext Tutor for Teaching Principles and Techniques of GIS," *Journal of Geography in Higher Education* 1996, Vol. 20, Iss 2, pp. 193-207.

Kemp, K.K., Frank, A.U. (1996) "Toward Consensus on a European GIS Curriculum—The International Postgraduate Course on GIS," *IJGIS,* Vol. 10, Iss 4, pp. 477-497.

Kemp, K.K., Goodchild, F.M. (1992) "Evaluating a Major Innovation in Higher-Education—The Ncgia Core Curriculum in GIS," *Journal of Geography in Higher Education,* Vol. 16, Iss 1, pp. 21-35.

Kemp, K.K., Goodchild, M.F. (1991) "Developing a Curriculum in Geographic Information Systems—The National Center for Geographic Information and Analysis Core Curriculum Project," *Journal of Geography in Higher Education*, Vol. 15, Iss 2, pp. 123-134.

Kemp, K.K., Goodchild, M.F., Dodson, R.F. (1992) "Teaching GIS in Geography" *PG,* Vol. 44, Iss 2, pp. 181-191.

Kennedy, G.A., Dai, M., Henning, S.A., Vandeveer, L.R. (1996) "A GIS-Based Approach for Including Topographic and Locational Attributes in the Hedonic Analysis of Rural Land Values," *American Journal of Agricultural Economics,* Vol. 78, Iss 5, pp. 1419-1419.

Kennedy, M. (1994) "Geographical Information Systems—Principles and Applications", by D.J. Maguire, M.F. Goodchild, D.W. Rhind, *Annals of the Association of American Geographers,* Vol. 84, Iss 1, pp. 172-173.

Kent, M., Jones, A., Weaver, R., (1993) "Geographical Information Systems and Remote-Sensing in Land-Use Planning—An Introduction," *Applied Geography,* Vol. 13, Iss 1, pp. 5-8.

Kessler, B, (1995) "PC-Based GIS Grows on Foresters," *JF* Vol. 93, Iss 5, pp. 28-29.

Kessler, B. (1995) "PC-Based GIS Grows on Foresters," *JF* Vol. 93, Iss 8, pp. 37-37.

Khorram, S., Devine, H., Brockhaus, J., Cheshire, H. (1992) "Remote-Sensing and GIS Research and Training at North-Carolina-State-University," *PERS* 1992, Vol. 58, Iss 8, pp. 1171-1172.

Kienegger, E.H. (1992) "Assessment of a Waste-Water Service Charge by Integrating Aerial-Photography and GIS," *PERS* 1992, Vol. 58, Iss 11, pp. 1601-1606.

Kim, K.Y., Ventura, S.J., Harris, P.M., Thum, P.G., Prey, J. (1993)"Urban Non-Point-Source Pollution Assessment Using a Geographical Information System" *Journal of Environmental Management,* Vol. 39, Iss 3, pp. 157-170.

King, G.Q. (1991) "Geography and GIS Technology," *Journal of Geography,* Vol. 90, Iss 2, pp. 66-72.

Kirkby, S.D. (1996) "Integrating a GIS with Arm Expert-System to Identify and Manage Dryland Salinization," *Applied Geography,* Vol. 16, Iss 4, pp. 289-303.

Kitron, U., Bouseman, J.K., Jones, C.J. (1992) "Use of the Arc Info GIS to Study the Distribution of Lyme-Disease Ticks in an Illinois County," *Preventive Veterinary Medicine* 1991, Vol. 11, Iss 3-4, pp. 243-248.

Kloos, H.W. (1992) "Graphics Data-Processing and Geographical Information Systems in Municipal Administration," *Wirtschaftsinformatik,* Vol. 34, Iss 6, pp. 577-583.

Klosterman, R.E. (1991) "An Introduction to Urban Geographic Information Systems", by W.E. Huxhold, *Journal of the American Planning Association,* Vol. 57, Iss 2, pp. 236-239.

Klosterman, R.E. (1991) "Geographic Information Systems and Cartographic Modelling", by C.D. Tomlin, *Journal of the American Planning Association,* Vol. 57, Iss 2, pp. 236-239.

Klosterman, R.E. (1991) "Geographic Information Systems—A Management Perspective", by S. Aronoff, *Journal of the American Planning Association,* Vol. 57, Iss 2, pp. 236-239.

Klosterman, R.E. (1991) "Geographic Information Systems—An Introduction", by J. Star, J. Estes, *Journal of the American Planning Association,* Vol. 57, Iss 2, pp. 236-239.

Klosterman, R.E. (1991) "Ncgia Core Curriculum in GIS", by M.F. Goodchild, K.K. Kemp, *Journal of the American Planning Association,* Vol. 57, Iss 2, pp. 236-239.

Klosterman, R.E. (1991) "Principles of Geographic Information Systems for Land Resource Assessment", by P.A. Burrough, *Journal of the American Planning Association,* Vol. 57, Iss 2, pp. 236-239.

Klosterman, R.E. (1991) "Understanding GIS—The Arc/Info Method, by Environm-Syst-Res, Inst," *Journal of the American Planning Association,* Vol. 57, Iss 2, pp. 236-239.

Knapp, W.G.M. van der (1992), The vector to raster conversion: (mis)use in geographical information systems, *International Journal on Geographic Information Systems,* Vol. 6, No. 2, pp.159-70.

Knapp, P.A., Bishop, K.A. Lancaster, J.M., Taylor, R.L. (1993) "Use of GIS in Optimizing Timber Thinning Strategies in the Eastern Sierra-Nevada," *PG,* Vol. 45, Iss 3, pp. 323-331.

Kollias, V.J., Voliotis, A. (1991) "Fuzzy-Reasoning in the Development of Geographical Information Systems—FRSIS—A Prototype Soil Information-System with Fuzzy Retrieval Capabilities," *IJGIS,* Vol. 5, Iss 2, pp. 209-223.

Kontoes, C., Wilkinson, G.G., Burrill, A., Goffredo, S., Megier, J. (1993)"An Experimental System for the Integration of GIS Data in Knowledge-Based Image-Analysis for Remote-Sensing of Agriculture," *IJGIS,* Vol. 7, I ss 3, pp. 247-262.

Kontoes, C., Wilkinson, G.G., Burrill, A., Goffredo, S., Megier, J. (1994) "An Experimental System for the Integration of GIS Data in Knowledge-Based Image-Analysis for Remote-Sensing of Agriculture (Vol. 7, Pg 247, 1993)," *International Journal of Geographic Information Systems,* Vol. 8, Iss 5, pp. 489-489.

Korp, P., Mckenzie, B. (1990) "Federal-Land and Geographic Information System Activities" *PERS*, Vol. 56, Iss 5, pp. 631-631.

Kosters, G., Pagel, B.U., Six, H.W. (1997) "GIS-Application Development with Geoooa," *IJGISC,* Vol. 11, Iss 4, pp. 307-335.

Koutnik, M.A., Padilla, D.K. (1994) "Predicting the Spatial-Distribution of Dreissena-Polymorpha (Zebra Mussel) Among Inland Lakes of Wisconsin—Modelling with a GIS" *Canadian Journal of Fisheries of Aquatic Sciences* 1994, Vol. 51, Iss 5, pp. 1189-1196.

Krautheim, K.R., Aldrich, T.E. (1997) "Geographic Information System (GIS) Studies of Cancer Around NPL-Sites (1997) Reprinted from Hazardous-Waste, Impacts on Human and Ecological Health, pp. 459-464) *Toxicology and Industrial Health,* 1997, Vol. 13, Iss 2-3, pp. 357-362.

Krist, F.J., Brown, D.G. (1994) "GIS Modelling of Paleo-Indian Period Caribou Migrations and Viewsheds in Northeastern Lower Michigan" *PERS* 1994, Vol. 60, Iss 9, pp. 1129-1137.

Kubo, S. (1984) "ALIS—A Geographical Information System for Urban Research" *IEEE Computer Graphics and Applications,* Vol. 4, Iss 5, pp. 68-76.

Kuijperslinde, M. (1992) "Geographic Information Systems—Developments and Applications", by L. Worrall, *Environment and Planning B-Planning and Design,* Vol. 19, Iss 3, pp. 359-360.

Kumar, L., Skidmore, A.K., Knowles, E. (1997) "Modelling Topographic Variation in Solar-Radiation in a GIS Environment," *IJGISC,* Vol. 11, Iss 5, pp. 475-497.

Kumar, N.K., Evans, I.S. (1997) "Map Output Format from GIS—Optimizing Visual Quality of Paper and Electronic Atlases," *Cartographic Journal,* Vol. 34, Iss 1, pp. 37-41.

Kupiszewski, M. (1993) "Transfer of Geographical Information Systems to Poland as an Example of the Transfer of Information Technology," *International Journal of Information Management,* Vol. 13, Iss 1, pp. 41-50.

Liengsakul, M., Mekpaiboonwatana, S., Pramojanee, P., Bronsveld, K., Huizing, H. (1994) "Use of GIS and Remote-Sensing for Soil Mapping and for Locating New Sites for Permanent Cropland—A Case-Study in the Highlands of Northern Thailand," *Geoderma* 1993, Vol. 60, Iss 1-4, pp. 293-307.

Lachowski, H., Maus, P., Platt, B. (1992) "Integrating Remote Sensing with GIS" *JF,* Vol. 90, Iss 12, pp. 16-21.

Lachowski, H.M., Wirth, T., Maus, P., Avers, P. (1994) "Remote-Sensing and GIS—Their Role in Ecosystem Management," *JF,* Vol. 92, Iss 8, pp. 39-40.

Lackaff, B.B., Hunt, B.J., Vonessen, I.E. (1994) "The Development and Implementation of a GIS-Based Contaminant Source Inventory over the Spokane Aquifer, Spokane County, Washington," *Water Resources Bulletin* 1993, Vol. 29, Iss 6, pp. 949-955.

Lai, P.C. (1990) "Feasibility of Geographic Information Systems Approach for Natural-Resource Management," *Environmental Management,* Vol. 14, Iss 1, pp. 73-80.

Lai, P.C. (1991) "Issues Concerning the Technology-Transfer of Geographic Information Systems," *Environmental Management,* Vol. 15, Iss 5, pp. 595-601.

Lai, P.C., Gillies, C.F. (1991) "The Impact of Geographical Information Systems on the Role of Spatial Data Libraries," *IJGIS,* Vol. 5, Iss 2, pp. 241-251.

Lake, R.W. (1993) "Planning and Applied Geography—Positivism, Ethics, and Geographic Information Systems," *Progress in Human Geography,* Vol. 17, Iss 3, pp. 404-413.

Lal, H., Hoogenboom, G., Calixte, J.P., Jones, J.W., Beinroth, F.H. (1993) "Using Crop Simulation-Models and GIS for Regional Productivity Analysis," *Transactions of the ASAE* Vol. 36, Iss 1, pp. 175-184.

Land Management Decisions, *Ecological Enomics,* Vol. 23, Iss 1, pp. 45-57.

Landis, J.D. (1990) "Microcomputer-Based GIS Systems," *Journal of the American Planning Association,* Vol. 56, Iss 3, pp. 373-376.

Lanfear, K.J. (1989) "Geographic Information Systems and Water-Resources Applications," *Water Resources Bulletin,* Vol. 25, Iss 3, pp. R5-R6.

Langford, M., Unwin, D.J. (1994) "Generating and Mapping Population Density Surfaces Within a Geographical Information System" *Cartographic Journal,* Vol. 31, Iss 1, pp. 21-26.

Lang, R., Muller, A., Lenz, R., Selige, T., Forster, E.M. (1997) "Landscape Modelling and GIS-Applications in the Munich-Research-Association-for-Agricultural-Ecosystems" *Landscape and Urban Planning,* Vol. 37, Iss 1-2, pp. 11-18.

Lanter, D.P. (1991) "Design of a Lineage-Based Meta-Data Base for GIS," *CGIS,* Vol. 18, Iss 4, pp. 255-261.

Lanter, D.P., Veregin, H. (1992) "A Research Paradigm for Propagating Error in Layer-Based GIS," *PERS* Vol. 58, Iss 6, pp. 825-833.

Lao, Y. (1996) "Innovations in GIS 1, by M.F. Worboys," *PG,* Vol. 48, Iss 2, pp. 235-236.

Lapalme, G., Rousseau, J.M., Chapleau, S., Cormier, M., Cossette, P., Roy, S. (1992) "Georoute—A Geographic Information System for Transportation Applications," *Communications of the ACM,* Vol. 35, Iss 1, pp. 80-88.

Lauer, D.T., Estes, J.E., Jensen, J.R., Greenlee, D.D. (1991) "Institutional Issues Affecting the Integration and Use of Remotely Sensed Data and Geographic Information Systems," *PERS,* Vol. 57, Iss 6, pp. 647-654.

Laxton, J.L. (1996) "Geographic Information Systems for Geoscientists—Modelling with GIS", by G.F. Bonhamcarter, *IJGIS,* Vol. 10, Iss 3, pp. 355-356.

Lee, C.T., Marsh, S.E. (1995) "The Use of Archival Landsat MSS and Ancillary Data in a GIS Environment to Map Historical Change in an Urban Riparian Habitat," *PERS* Vol. 61, Iss 8, pp. 999-1008.

Lee, J., Kretzschmar, W.A. (1993) "Spatial-Analysis of Linguistic Data with GIS Functions," *IJGIS,* Vol. 7, Iss 6, pp. 541-560.

Lees, B.G., Ritman, K., (1991) "Decision-Tree and Rule-Induction Approach to Integration of Remotely Sensed and GIS Data in Mapping Vegetation in Disturbed or Hilly Environments," *Environmental Management* Vol. 15, Iss 6, pp. 823-831.

Lee, Y.C. (1990) "Geographic Information Systems for Urban Applications Problems and Solutions," *Environment and Planning B-Planning and Design,* Vol. 17, Iss 4, pp. 463-473.

Lee, Y.C., Chin, F.L. (1995) "An Iconic Query Language for Topological Relationships in GIS," *IJGIS,* Vol. 9, Iss 1, pp. 25-46.

Leggett, D.J., Jones, A. (1996) "The Application of GIS for Flood Defense in the Anglian Region—Developing for the Future," *IJGIS,* Vol. 10, Iss 1, pp. 103-116.

Lehmann, A., Jaquet, J.M., Lachavanne, J.B. (1994) "Contribution of GIS to Submerged Macrophyte Biomass Estimation and Community Structure Modelling, Lake Geneva, Switzerland" *Aquati Botany* Vol. 47, Iss 2, pp. 99-117.

Leipnik, M.R., Kemp, K.K., Loaiciga, H.A. (1993) "Implementation of GIS for Water-Resources Planning and Management" *Journal of Water Resources Planning and Management-ASCE* Vol. 119, Iss 2, pp. 184-205.

Leung, Y., Leung, K.S. (1993) "An Intelligent Expert-System Shell for Knowledge-Based Geographical Information Systems.1. The Tools" *IJGIS,* Vol. 7, Iss 3, pp. 189-199.

Leung, Y., Leung, K.S. (1993) "An Intelligent Expert-System Shell for Knowledge-Based Geographical Information Systems.2. Some Applications," *IJGIS,* Vol. 7, Iss 3, pp. 201-213.

Levine, J., Landis, J.D. (1989) "Geographic Information Systems for Local-Planning," *Journal of the American Planning Association,* Vol. 55, Iss 2, pp. 209-220.

Levine, N. (1996) "Spatial Statistics and GIS—Software Tools to Quantify Spatial Patterns" *Journal of the American Planning Association,* Vol. 62, Iss 3, pp. 381-391.

Lew, A.A. (1992) "Geographic Information Systems and Their Socioeconomic Applications", by D. Martin, *PG,* Vol. 44, Iss 4, pp. 488-489.

Lewis, DM (1995) "Importance of GIS to Community-Based Management of Wildlife—Lessons from Zambia," *Ecological Applications* Vol. 5, Iss 4, pp. 861-871.

Lillesand, J.M. (1990), Satellite Remote Sensing: Its Evolution and Synergism with GIS technology, *Government Information Quarterly,* Vol. 7:3, pp. 307-27.

Lillesand, T. (1990) Satellite Remote-Sensing—Its Evolution and Synergism with GIS Technology," *Government Information Quarterly,* Vol. 7, Iss 3, pp. 307-327.

Limp, W.F. (1987) "Integrating Database and Geographic Information Systems for Archaeologists," *American Journal of Archaeology,* Vol. 91, Iss 2, pp. 323-323.

Linden, G. (1987) "An Integrated Study of Desertification—Applications of Remote-Sensing, GIS and Spatial Models in Semi-Arid Sudan", by L. Olsson, *Third World Planning Review,* Vol. 9, Iss 2, pp. 195-196.

Lindholm, M., Sarjakoski, T. (1992) "User Models and Information-Theory in the Design of a Query Interface for GIS," *Lecture Notes in Comptuer Science,* Vol. 639, pp. 328-347.

Lindhult, M.S., Price, N., Brown, P., Fabos, J. (1988) "Using Geographic Information Systems to Assess Conflicts Between Agriculture and Development," *Landscape and Urban Planning,* Vol. 16, Iss 4, pp. 333-343.

Lin, H., Wan, Q., Li, X., Chen, J., Kong,Y. (1997) "GIS-Based Multicriteria Evaluation for Investment Environment," *Environment and Planning B-Planning and Design,* Vol. 24, Iss 3, pp. 403-414.

Linsey, T.K., Raper, J.F. (1993) "Hyperarc—A Task-Oriented Hypertext GIS Interface," *IJGIS,* Vol. 7, Iss 5, pp. 435-452.

Lipschultz, M.S. Glaser, D.E. (1992) "Rainy Season Flooding Being Tracked by GIS" *Water-Engineering and Management,* Vol. 139, Iss 6, pp. 22-25.

Liu, J.G., Dunning, J.B., Pulliam, H.R. (1995) "Potential Effects of a Forest Management Plan on Bachman Sparrows (Aimophila-Aestivalis)—Linking a Spatially Explicit Model with GIS," *Conservation Biology* Vol. 9, Iss 1, pp. 62-75.

Llobera, M. (1996) "Exploring the Topography of Mind—GIS, Social Space and Archaeology" *Antiquity,* Vol. 70, Iss 269, pp. 612-622.

Lo, C.P. (1995) "Automated Population and Dwelling Unit Estimation from High-Resolution Satellite Images—A GIS Approach," *IJRS* Vol. 16, Iss 1, pp. 17-34.

Lo, C.P., Shipman, R.L. (1990) "A GIS Approach to Land-Use Change Dynamics Detection" *PERS* Vol. 56, Iss 11, pp. 1483-1491.

Loh, D.K., Rykiel, E.J. (1992) "Integrated Resource-Management Systems—Coupling Expert Systems with Database-Management and Geographic Information Systems," *Environmental Management,* Vol. 16, Iss 2, pp. 167-177.

Lombard, A.T. (1993) "Multispecies Conservation, Advanced Computer Architecture and GIS—Where Are We Today," *South African Journal of Science* Vol. 89, Iss 9, pp. 415-418.

Lombard, A.T. August, P.V., Siegfried, W.R. (1992) "A Proposed Geographic Information System for Assessing the Optimal Dispersion of Protected Areas in South-Africa," *South African Journal of Science,* Vol. 88, Iss 3, pp. 136-140.

Longley, P. (1990) "Geographical Information Systems and the Regional Research Laboratories," *AREA,* Vol. 22, Iss 2, pp. 199-200.

Longley, P. (1991) "Geographic Information Systems and Cartographic Modelling, by C.D. Tomlin," *Environment and Planning B-Planning and Design,* Vol. 18, Iss 2, pp. 256-257.

Longley, P. (1992) "Introductory Readings in Geographic Information Systems", by D.J. Peuquet, D.F. Marble, *Environment and Planning B-Planning and Design,* Vol. 19, Iss 1, pp. 116-117.

Longley, P. (1993) "Quantitative-Analysis, Technological Change, and the Dissemination of GIS Research—Some Reflections and Prospects," *Environment and Planning A,* Iss NSI, pp. 35-37.

Longley, P. (1995) "The 1995 European GIS Yearbook", by P.J. Shand, P.J. Ireland *Environment and Planning B-Planning and Design,* Vol. 22, Iss 6, pp. 772-773.

Longley, P., Higgs, G., Martin, D., (1994) "The Predictive Use of GIS to Model Property Valuations" *IJGIS,* Vol. 8, Iss 2, pp. 217-235.

Longley, P.A. (1989) "Applied Geographical Information Systems," *AREA,* Vol. 21, Iss 3, pp. 337-338.

Lopezblanco, J. Villersruiz, L. (1995) "Delineating Boundaries of Environmental Units for Land Management Using a Geomorphological Approach and GIS—A Study in Baja-California, Mexico," *Remote Sensing of Environment* Vol. 53, Iss 2, pp. 109-117.

Lopez, X.R. (1993) "Database Copyright Issues in the European GIS Community," *Government Information Quarterly,* Vol. 10, Iss 3, pp. 305-318.

Loukes, D.K., Mclaughlin, J. (1991) "GIS and Transportation—Canadian Perspective" *Journal of Surveying Engineering-ASCE,* Vol. 117, Iss 3, pp. 123-133.

Love, D., Lindquist, P. (1995) "The Geographical Accessibility of Hospitals to the Aged-A Geographic Information Systems Analysis Within Illinois," *Health Services Research,* Vol. 29, Iss 6, pp. 629-651.

Loveland, Thomas R., James W. Merchant, Donald O. Ohlen and Jesslyn F. Brown (1991), Development of a land-cover characteristics database for the conterminous U.S., *Photogrammertic Engineering and Remote Sensing,* Vol. 57, No. 11, pp. 1453-63.

Lovett, A.A., Parfitt, J.P., Brainard, J.S. (1997) "Using GIS in Risk Analysis—A Case-Study of Hazardous-Waste Transport" *Risk Analysis,* Vol. 17, Iss 5, pp. 625-633.

Lovett, J. (1993) "GIS Promise Applied," *SEARCH* Vol. 24, Iss 1, pp. 15-15.

Low, Cost Housing Development in Kuala-Lumpur, (Malaysia1990)" *Environment and Planning B-Planning and Design,* Vol. 17, Iss 4, pp. 385-393.

Lowell, K. (1991) "Utilizing Discriminant Function-Analysis with a Geographical Information System to Model Ecological Succession Spatially" *IJGIS,* Vol. 5, Iss 2, pp. 175-191.

Lowell, K.E. (1990) "Differences Between Ecological Land Type Maps Produced Using GIS or Manual Cartographic Methods," *PERS* Vol. 56, Iss 2, pp. 169-173.

Lowell, K.E. (1994), "Probabilistic Temporal GIS Modelling Involving More Than 2 Map Classes," *IJGIS,* Vol. 8, Iss 1, pp. 73-93.

Lowry, J.H., Miller, H.J., Hepner, G.F. (1995) "A GIS-Based Sensitivity Analysis of Community Vulnerability to Hazardous Contaminants on the Mexico/Us Border" *PERS* Vol. 61, Iss 11, pp. 1347-1359.

Ludeke, A.K., Maggio, R.C., Reid, L.M. (1990) "An Analysis of Anthropogenic Deforestation Using Logistic-Regression and GIS," *Journal of Environmental Management,* Vol. 31, Iss 3, pp. 247-259.

Lui, C.Y., Hiley, J. (1994) "Computational Study of Very Fast Transients in GIS with Special Reference to Effects of Trapped Charge and Risetime on Overvoltage Amplitude" *IEE Proceedings-Generation and Distribution,* Vol. 141, Iss 5, pp. 485-490.

Lull, K.J., Tindall, J.A., Potts, D.F. (1995) "Assessing Nonpoint-Source Pollution Risk—A GIS Application" JF Vol. 93, Iss 1, pp. 35-40.

Lundgaard, L.E., Tangen, G., Skyberg, B., Faugstad, K. (1992) "Acoustic Diagnoses of GIS—Field Experience and Development of Expert System" *IEEE Transactions of Power Delivery,* Vol. 7, Iss 1, pp. 287-294.

Lunetta, R.S., Congalton, R.G., Fenstermaker, L.K., Jensen, J.R., Mcgwire, K.C., Tinney, L.R. (1991) "Remote-Sensing and Geographic Information System Data Integration—Error Sources and Research Issues," *PERS,* Vol. 57, Iss 6, pp. 677 -687.

Lupien, A.E., Dangermond, J. Moreland, W.H. (1987) "Network Analysis in Geographic Information Systems," *PERS,* Vol. 53, Iss 10, pp. 1417-1421.

Lupton, M. Mather, C. (1997) "The Antipolitics Machine—GIS and the Reconstruction of the Johannesburg Local State," *Political Geography,* Vol. 16, Iss 7, pp. 565-580.

Macdougall, E.B. (1981) "Harvard Papers on Geographic Information Systems—1st International Advanced Study Symposium on Topological Data-Structures for Geographic Information Systems", by G. Dutton, *Environment and Planning A,* Vol. 13, Iss 5, pp. 650-651.

Macdougall, E.B. (1992), "Exploratory Analysis, Dynamic Statistical Visualization, and Geographic Information Systems," *CGIS,* Vol. 19, Iss 4, pp. 237-246.

Maceachren, A.M. (1988) "Principles of Geographical Information Systems for Land Resources Assessment", by P.A. Burrough, *PG,* Vol. 40, Iss 2, pp. 249-250.

Mace, T. (1991) "Special Issue—Integration of Remote-Sensing and GIS and Featuring the 1991 Asprs Directory of the Mapping Sciences—Foreword," *PERS* Vol. 57, Iss 6, pp. 641-641.

Macgill, S.M. (1990) "GIS in the 1990s" *Environment and Planning A,* Vol. 22, Iss 12, pp. 1559-1560.

Maclean, A.L. Davello, T.P. Shetron, S.G. (1993) "The Use of Variability Diagrams to Improve the Interpretation of Digital Soil Maps in a GIS," *PERS* Vol. 59, Iss 2, pp. 223-228.

Maclean, A.L., Reed, D.D., Mroz, G.D., Lyon, G.W. Edison, T. (1992) "Using GIS to Estimate Forest Resource Changes," *JF,* Vol. 90, Iss 12, pp. 22-25.

Madon, S., Sahay, S. (1997) "Managing Natural-Resources Using GIS—Experiences in India" *Information and Management,* Vol. 32, Iss 1, pp. 45-53.

Maguire, D.J., Goodchild, M.F. and Rhind. D, (1991), *Geographic Information System,* Longman, London.

Maguire, D. (1997) "GIS and Organizations—How Effective Are GIS in Practice", by H. Campbell, I. Masser, *AREA,* Vol. 29, Iss 2, pp. 189-191.

Maguire, D.J. (1991) "Information Technology in Geography and Planning—Including the Principles of GIS", by I. Bracken, C. Webster, *Transactions-Institute of British Geographers,* Vol. 16, Iss 2, pp. 245-246.

Maguire, D.J. (1992) "Geographic Information Systems—An Introduction", by L. Star, J.E. Estes, *Progress in Human Geography,* Vol. 16, Iss 3, pp. 486-487.

Malczewski, J. (1996) "A GIS-Based Approach to Multiple Criteria Group Decision-Making" *IJGIS,* Vol. 10, Iss 8, pp. 955-971.

Manning, E.W. (1990) "Geographic Information Systems and Sustainable Development" *Government Information Quarterly,* Vol. 7, Iss 3, pp. 329-342.

Marble, D.F. (1991) "Geographical Information Systems for Urban and Regional-Planning", by H.J. Scholten, J.C.H. Stillwell, *Geographical Analysis,* Vol. 23, Iss 4, pp. 371-373.

Marble, D.F. Amundson, S.E. (1988) "Microcomputer-Based Geographic Information Systems and Their Role in Urban and Regional-Planning," *Environment and Planning B-Planning and Design,* Vol. 15, Iss 3, pp. 305-324.

Mark, David M. and Gould, Michael D. (1991), Interacting with Geographic information: A commentary, *Photogrammertic Engineering and Remote Sensing,* Vol. 57, No.11, pp. 1427-30.

Mark, D.M. Calkins, H.W. (1991) "Cartography and Geographic Information Systems at the State-University-of-New-York at Buffalo," *CGIS,* Vol. 18, Iss 3, pp. 198-200.

Marsh, S.E., Walsh, J.L., Hutchinson, C.F. (1990) "Development of an Agricultural Land-Use GIS for Senegal Derived from Multispectral Video and Photographic Data" *PERS* Vol. 56, Iss 3, pp. 351-357.

Martin, D. (1988) "Principles of Geographical Information Systems for Land Resources Assessment", by P.A. Burrough, *Environment and Planning B-Planning and Design,* Vol. 15, Iss 2, pp. 231-232.

Martin, D (1991) "Geographic Information Systems—Developments and Applications", by L. Worrall" *Regional Studies,* Vol. 25, Iss 6, pp. 573-574.

Martin, D. (1994) "Recent Development in GIS" *AREA,* Vol. 26, Iss 2, pp. 199-200.

Martin, D. Longley, P. Higgs, G. (1994) "The Use of GIS in the Analysis of Diverse Urban Databases," *Computers Environment and Urban Systems,* Vol. 18, Iss 1, pp. 55-66.

Martin, F.C. (1985) "Using a Geographic Information System for Forest Land Mapping and Management" *PERS,* Vol. 51, Iss 11, pp. 1753-1759.

Masaki, K., Sakakibara, T., Murase, H., Akazaki, M., Uehara, K. Menju, S. (1994) "On-Site Measurement for the Development of Online Partial Discharge Monitoring-System in GIS" *IEEE Transactions of Power Delivery* Vol. 9, Iss 2, pp. 805-810.

Maslia, M.L., Aral, M.M., Williams, R.C. Susten, A.S., Heitgerd, J.L. (1994) "Exposure Assessment of Populations Using Environmental Modelling, Demographic-Analysis, and GIS," *Water Resources Bulletin,* Vol. 30, Iss 6, pp. 1025-1041.

Mason, D.C., Oconaill, M.A., Bell, S.B.M. (1994), "Handling 4-Dimensional Geo-Referenced Data in Environmental GIS," *IJGIS,* Vol. 8, Iss 2, pp. 191-215.

Masser, Ian (1998), *Governments and Geographic Information,* Taylor & Francis, London.

Masser, I. (1988) "The Development of Geographic Information Systems in Britain—The Chorley Report in Perspective," *Environment and Planning B-Planning and Design,* Vol. 15, Iss 4, pp. 489-494.

Masser, I. (1990) "Information Technology in Geography and Planning Including Principles of GIS", by I. Bracken, C. Webster, *Environment and Planning B-Planning and Design,* Vol. 17, Iss 3, pp. 367-368.

Masser, I. (1992) "Geographical Information Systems for Urban and Regional Planning", by H.J. Scholten, J.C.H. Stillwell, *IJGIS,* Vol. 6, Iss 1, pp. 59-60.

Mather, P.M. (1992) "Introductory Readings in Geographic Information Systems", by D.J. Peuquet, D.F. Marble, *Geographical Journal*, Vol. 158, Iss MAR, pp. 95-96.

Matsuyama, T., Hao, L.V., Nagao, M. (1984) "A File-Organization for Geographic Information Systems Based on Spatial Proximity," *Computer Vision Graphics and Image Processing,* Vol. 26, Iss 3, pp. 303-318.

Matthews, K.B., Macdonald, A., Aspinall, R.J., Hudson, G., Law, A.N.R., Paterson, E. (1994) "Climatic Soil-Moisture Deficit—Climate and Soil Data Integration in a GIS," *Climatic Change,* Vol. 28, Iss 3, pp. 273-287.

Matthews, S.A. (1992) "Geographic Information Systems and Their Socioeconomic Applications", by D. Martin, *Environment and Planning B-Planning and Design,* Vol. 19, Iss 2, pp. 238-239.

Mattson, M.D., Godfrey, P.J. (1994) "Identification of Road Salt Contamination Using Multiple-Regression and GIS," *Environmental Management,* Vol. 18, Iss 5, pp. 767-773.

Maurel, P. (1993) "Satellite Imagery and Geographical Information Systems," *RECHERCHE,* Vol. 24, Iss 258, pp. 1140-1142.

Mcclean, C. (1996) "Remote-Sensing and GIS in Ecosystem Management", by V.A. Sample, *Land Use Policy,* Vol. 13, Iss 3, pp. 237-238.

Mcdonald, G. (1991) "Map-II Map Processor—A Geographic Information System for the Macintosh," *Journal of Architectural and Planning Research*, Vol. 8, Iss 1, pp. 76-78.

Mcdonald, G.T. (1994) "Environmental Modelling with GIS", by M.F. Goodchild, B.O. Parks, L.T. Steyaert, *PG,* Vol. 46, Iss 3, pp. 393-394.

Mcglamery, P., Lamont, M. (1994) "New Opportunities and Challenges—Geographic Information Systems in Libraries," *Database-the Mgazine of Electronic Database Reviews,* Vol. 17, Iss 6, pp. 35-&.

Mclaren, R. (1997) "GIS and Organizations", by H. Campbell, I. Masser" *IJGISC,* Vol. 11, Iss 5, pp. 521-522.

Mcnaughton, R.B. (1992) "Geographic Information Systems—The Microcomputer and Modern Cartography", by D.R.F. Taylor, *Canadian Geographer-Geograhe Canadien,* Vol. 36, Iss 4, pp. 392-393.

Meehan, K.T. (1991) "Integrating Geologic and Economic Data Using a Geographic Information System (GIS)—Offshore Gulf of Mexico," *AAPG Bulletin-American Association of Petroleum Geologists,* Vol. 75, Iss 3, pp. 633-633.

Meijer, E. (1989) "From High-End to Low-End Geographic Information Systems—A Continuum," *EKISTICS-the Problems and Science of Human Settlements,* Vol. 56, Iss 338-, pp. 300-303.

Mellerowicz, K.T., Rees, H.W., Chow, T.L., Ghanem, I. (1994) "Soil Conservation Planning at the Watershed Level Using the Universal Soil Loss Equation with GIS and Microcomputer Technologies—A Case-Study," *Journal of Soil and Water Conservation* Vol. 49, Iss 2, pp. 194-200.

Merchant, J.W. (1988) "Special Issue—Geographic Information Systems (GIS)—Foreword" *PERS*, Vol. 54, Iss 11, pp. 1545-1545.

Merchant, J.W. (1994) "GIS-Based Groundwater Pollution Hazard Assessment—A Critical-Review of the Drastic Model," *PERS* Vol. 60, Iss 9, pp. 1117-1128.

Merchant, J.W. (1994) "Landscape Ecology and GIS", by R. Hainesyoung, D.R. Green, S. Cousins, *PG,* Vol. 46, Iss 4, pp. 516-517.

Michalak, W.Z. (1993) "GIS in Land-Use Change Analysis—Integration of Remotely Sensed Data into GIS," *Applied Geography,* Vol. 13, Iss 1, pp. 28-44.

Mikhova, D., Pickles, J. (1994) "GIS in Bulgaria—Development and Perspectives," *IJGIS,* Vol. 8, Iss 5, pp. 471-477.

Miller, H.J. (1991) "Modelling Accessibility Using Space-Time Prism Concepts Within Geographical Information Systems," *IJGIS,* Vol. 5, Iss 3, pp. 287-301.

Miller, H.J. (1996) "GIS and Geometric Representation in Facility Location Problems," *IJGIS,* Vol. 10, Iss 7, pp. 791-816.

Millette, T.L. (1994) "Geographic Information Systems, Spatial Modelling, and Policy Evaluation", by M.M. Fischer, P. Nijkamp, *Journal of Regional Science,* Vol. 34, Iss 3, pp. 465-468.

Milne, J.A., Sear, D.A. (1997) "Modelling River Channel Topography Using GIS," *IJGISC,* Vol. 11, Iss 5, pp. 499-519.

Mires, P.B. (1993) "Relationships of Louisiana Colonial Land Claims with Potential Natural Vegetation and Historic Standing Structures—A GIS Approach," *PG,* Vol. 45, Iss 3, pp. 342-350.

Missotten, R., Han, Q., Aureli, A. (1994) "Remote-Sensing and GIS Activities in UNESCO" *IJRS* Vol. 15, Iss 15, pp. 3111-3118.

Mitasova, H., Hofierka, J., Zlocha, M., Iverson, L. (1997) "Modelling Topographic Potential for Erosion and Deposition Using GIS—Reply," *IJGISC,* Vol. 11, Iss 6, pp. 611-618.

Mitasova, H., Hofierka, J., Zlocha, M., Iverson, L.R. (1996) "Modelling Topographic Potential for Erosion and Deposition Using GIS," *IJGIS,* Vol. 10, Iss 5, pp. 629-641.

Mitchell, D. (1991) "3-Dimensional Applications in Geographical Information Systems", by J. Raper" *IJGIS,* Vol. 5, Iss 2, pp. 255-256.

Monmonier, M. (1993) "What a Friend We Have in GIS" *PG,* Vol. 45, Iss 4, pp. 448-450.

Morad, M., Chalmers, A.I., Oregan, P.R. (1996) "The Role of Root-Mean-Square Error in the Geo-Transformation of Images in GIS," *IJGIS,* Vol. 10, Iss 3, pp. 347-353.

Morgan, J.M. (1987) "Academic Geographic Information Systems Education—A Commentary" *PERS,* Vol. 53, Iss 10, pp. 1443-1445.

Morgan, K.M., Busbey, A.B., Donovan, R.N., Newland, L.W. (1992) "Image-Processing and GIS at Tcus Center for Remote-Sensing and Energy Research" *PERS* Vol. 58, Iss 8, pp. 1173-1174.

Morgan, K.M., Newland, L.W. (1990) "Geographic Information Systems (GIS) for Environmental Data-Base Management and Analysis," *Toxicological and Environmental Chemistry,* Vol. 26, Iss 1-4, pp. 129-136.

Morse, A., Zarriello, T.J., Kramber, W.J. (1990) "Using Remote-Sensing and GIS Technology to Help Adjudicate Idaho Water Rights," *PERS,* Vol. 56, Iss 3, pp. 365-370.

Moskos, C.C. (1985) "Female GIS in the Field," *SOCIETY,* Vol. 22, Iss 6, pp. 28-33.

Mounsey, H.M. and Tomlinson,R.F., (1988), *Building Data Bases for Global Science,* Taylor and Francis, London.

Moyer, D.D. (1991) "Introductory Readings in Geographic Information Systems", by D.J. Peuquet, D.F. Marble", *Journal of Agricultural Economics Research,* Vol. 43, Iss 4, pp. 28-29.

Mozeto, A.A., Stone, T.A., Brown, I.F., Skole, D.L. (1990) "The Use of a Remote Sensing Geographic Information System in the Evaluation of Environmental-Impact in the Ecological Station of Uhe Samuel, Rondonia, Brazil," *Interciencia,* Vol. 15, Iss 5, pp. 265-271.

Muehrcke, P.C. (1990) "Cartography and Geographic Information Systems," *CGIS,* Vol. 17, Iss 1, pp. 7-15.

Mueller, M. (1988) "Geographic Information System—Small Area Applications," *Population Index,* Vol. 54, Iss 3, pp. 426-426.

Mukaiyama, Y., Takagi, I., Izumi, K., Sekiguchi, T., Kobayashi, A., Sumikawa, T. (1990) "Investigation on Abnormal Phenomena of Contacts Using Disconnecting Switch and Detachable Bus in 300Kv GIS" *IEEE Transactions of Power Delivery* Vol. 5, Iss 1, pp. 189-195.

Muller, J.C. (1993) "Latest Developments in GIS LIS," *IJGIS,* Vol. 7, Iss 4, pp. 293-303.

Muller, J.C. (1996) "Advances in GIS—Proceedings of the 6th International-Symposium on Spatial Data Handling", by R.G. Healey, T.C. Waugh, *IJGIS,* Vol. 10, Iss 2, pp. 238-241.

Muller, P.O. (1991) "An Introduction to Urban Geographic Information Systems", by W.E. Huxhold, *Urban Geography,* Vol. 12, Iss 4, p. 387.

Murty, C.V.S.S.B.R., Venkatachalam, P., (1992) "Delineation of Erosion Cells in a Watershed Using GIS," *Journal of Environmental Management,* Vol. 36, Iss 3, pp. 159-166.

Nag, P. and M. Kudrat (1998), *Digital Remote Sensing,* Concept Publishing Company, New Delhi.

Nag, P. (1986), A proposed base for geographical information system for India, *International Journal for Geographical Information System,* Vol. 1, No. 2, pp. 181-7.

Nag, P. (1987), A proposed geographical information system for India, *International Journal on Geographical Information System,* Vol. 1:2.

Namken, J.C. Stuth, J.W. (1997) "A Prototype Graphic Landscape Analysis System.1. Predicting Spatial Patterns of Grazing Pressure Using GIS" *IJGISC,* Vol. 11, Iss 8, pp. 785-798.

Neldner, V.J., Crossley, D.C., Cofinas, M. (1995) "Using Geographic Information Systems (GIS) to Determine the Adequacy of Sampling in Vegetation Surveys" *Biological Conservation,* Vol. 73, Iss 1, pp. 1-17.

Nellis, M.D. (1991) "Geographic Information Systems—The Microcomputer and Modern Cartography", by D.R.F. Taylor, *Journal of Rural Studies,* Vol. 7, Iss 3, pp. 352-353.

Nichol, J.E. (1994) "A GIS-Based Approach to Microclimate Monitoring in Singapore Highrise Housing Estates" *PERS* Vol. 60, Iss 10, pp. 1225-1232.

Nichol, J.E. (1994) "An Examination of Tropical Rain-Forest Microclimate Using GIS Modelling," *Global Ecology and Biogeography Letters,* Vol. 4, Iss 3, pp. 69-78.

Nielsen, G.A., Caprio, J.M., Mcdaniel, P.A., Snyder, R.D., Montagne, C. (1990) "Maps—A GIS for Land Resource-Management in Montana" *Journal of Soil and Water Conservation,* Vol. 45, Iss 4, pp. 450-453.

Nishiwaki, S., Nojima, K., Tatara, S., Kosakada, M., Tanabe, N., Yanabu, S. (1995) "Electromagnetic-Interference with Electronic Apparatus by Switching Surges in GIS—Cable System" *IEEE Transactions of Power Delivery* Vol. 10, Iss 2, pp. 739-746.

Nitta, T., Nakanishi, K. (1991) "Charge Accumulation on Insulating Spacers for HVDC GIS" *IEEE Transactions on Electrical Insulation,* Vol. 26, Iss 3, pp. 418-427.

Noronha, V.T. (1990) "Geographic Information Systems—A Management Approach", by S. Aronoff, *CGIS,* Vol. 17, Iss 2, pp. 178-179.

Novak, K. (1995) "Mobile Mapping Technology for GIS Data-Collection" *PERS* Vol. 61, Iss 5, pp. 493-501.

NSDI (197), *A Strategy for the National Spatial Data Infrastructure,* Federal Geographic Data Committee, US Geological Survey, Reston.

Nugent, J.L. (1995) "Quality-Control Techniques for a GIS Database Development Project" *PERS* Vol. 61, Iss 5, pp. 523-527.

Nyerges, T.L. (1991) "Geographical Information Systems for Urban and Regional-Planning", by H.J. Scholten, J.C.H. Stillwell, *Journal of Regional Science,* Vol. 31, Iss 2, pp. 231-233.

Nyerges, T.L., Chrisman, N.R. (1989) "A Framework for Model Curricula Development in Cartography and Geographic Information Systems," *PG,* Vol. 41, Iss 3, pp. 283-293.

Obermeyer, N.J. (1993) "Certification and Accreditation in GIS—Prospects and Pitfalls" *Computers Environment and Urban Systems,* Vol. 17, Iss 1, pp. 91-102.

Obermeyer, N.J. (1994) "Spatial-Analysis and Spatial Policy Using Geographic Information Systems", by L. Worral, *PG,* Vol. 46, Iss 2, pp. 276-276.

Obermeyer, N.J. (1994) "GIS—A New Profession" *PG,* Vol. 46, Iss 4, pp. 498-503.

Okabe, S., Kosakada, M., Toda, H., Suzuki, K., Ishikawa, M. (1993) "Investigations of Multiple Reignition Phenomena and Protection Scheme of Shunt Reactor Current Interruption in GIS Substations" *IEEE Transactions of Power Delivery* Vol. 8, Iss 1, pp. 197-202.

Oldak, A. (1992) "Geographical Information System Techniques in Land-Use Assessment Within Landscape Types in Eastern Poland," *Land Degradation and Rehabilitation,* Vol. 3, Iss 3, pp. 181-194.

Olivier, J.J., Greenwood, P.H., Cooper, A.K., Mcpherson, D.R., Engelbrecht, R. (1990) "Selecting a GIS for a National Water Management Authority" *PERS,* Vol. 56, Iss 11, pp. 1471-1475.

Ong, Y. (1994) "Mapping and Monitoring the Intertidal Zone of the East-Coast of England Using Remote-Sensing Techniques and a Coastal Monitoring GIS," *Marine Technology Society Journal* 1994, Vol. 28, Iss 2, pp. 19-29.

Openshaw, S. (1988) "An Evaluation of the BBC Domesday Interactive Videodisc As a GIS for Planners," *Environment and Planning B-Planning and Design,* Vol. 15, Iss 3, pp. 325-338.

Openshaw, S. (1991) "A View on the GIS Crisis in Geography, or, Using GIS to Put Humpty-Dumpty Back Together Again," *Environment and Planning A,* Vol. 23, Iss 5, pp. 621-628.

Openshaw, S. (1992) "Further Thoughts on Geography and GIS—A Reply" *Environment and Planning A,* Vol. 24, Iss 4, pp. 463-466.

Openshaw, S. (1992) "Some Suggestions Concerning the Development of Artificial-Intelligence Tools for Spatial Modelling and Analysis in GIS" *Annals of Regional Science,* Vol. 26, Iss 1, pp. 35-51.

Openshaw, S. (1993) "GIS Crime and GIS Criminality," *Environment and Planning A,* Vol. 25, Iss 4, pp. 451-458.

Openshaw, S., Charlton, M., Wymer, C. (1986) "A Geographical Information and Mapping System for the BBC Domesday Optical Disks," *Transactions-Institute of British Geographers,* Vol. 11, Iss 3, pp. 296-304.

Orland, B. (1994) "Visualization Techniques for Incorporation in Forest Planning Geographic Information Systems" *Landscape and Urban Planning,* Vol. 30, Iss 1-2, pp. 83-97.

Orr, R.A. (1991) "Bartholomew and the Giant GIS" *Cartographic Journal,* Vol. 28, Iss 1, pp. 30-33.

Osmokrovic, P., Krstic, S., Ljevak, M., Novakovic, D. (1992) "Influence of GIS Parameters on the Topler Constant" *IEEE Transactions on Electrical Insulation* Vol. 27, Iss 2, pp. 214-220.

Otoole, K. (1994) "GIS, Administrative Reform and Local-Government—The Issue of Privacy" *Australian Journal of Social Issues,* Vol. 29, Iss 4, pp. 345-354.

Ottens, H.F.L. (1991) "Geographical Information Systems in the Netherlands" *TIJDSCHRIFT Voor Economische en Sociale Geographie,* Vol. 82, Iss 4, pp. 306-309.

Oyama, M., Hanai, E., Aoyagi, H., Murase, H., Ohshima, I., Menju, S. (1994) "Development of Detection and Diagnostic-Techniques for Partial Discharges in GIS" *IEEE Transactions of Power Delivery* Vol. 9, Iss 2, pp. 811-818.

Ozernoy, V.M., Smith, D.R., Sicherman, A. (1981) "Evaluating Computerized Geographic Information Systems Using Decision Analysis" Interfaces, Vol. 11, Iss 5, pp. 92-100.

Padmavathy, A.S., Raj, K.G., Yogarajan, N., Thangavel, P., Chandrasekhar, M.G. (1993) "Checkdam Site Selection Using GIS Approach" *Advances in Scace Research,* Vol. 13, Iss 11, pp. 123-127.

Pal, Saroj K. and Dilip Kumar (2003), GIS and remote sensing in natural resources mapping in Ferozpur Jhirka Block, District Gurgaon, Harayana, Paper presented in the International Conference on National & Regional Atlases, Kolkata.

Parker, H. Dennison (1991), The role of private sector in GIS, *Photogrammetric Engineering and Remote Sensing,* vol. 37, No. 11, pp. 1465-6.

Parker, H. Dennison (1991), The role of the private sector in GIS, *Photogrammertic Engineering and Remote Sensing,* Vol. 57, No.11, pp. 1465-6.

Parker, H.D. (1988) "The Unique Qualities of a Geographic Information System-A Commentary," *PERS,* Vol. 54, Iss 11, pp. 1547-1549.

Parker, H.D. (1991) "The Role of the Private-Sector in GIS" *PERS* Vol. 57, Iss 11, pp. 1465-1466.

Parker, S., Cocklin, C. (1993) "The Use of Geographical Information Systems for Cumulative Environmental Effects Assessment," *Computers Environment and Urban Systems,* Vol. 17, Iss 5, pp. 393-407.

Pathan, S.K., Sastry, S.V.C., Dhinwa, P.S., Rao, M., Majumdar, K.L., Kumar, D.S., Patkar, V.N., Phatak, V.N. (1993) "Urban-Growth Trend Analysis Using GIS Techniques—A Case-Study of the Bombay Metropolitan Region," *IJRS,* Vol. 14, Iss 17, pp. 3169-3179.

Pearson, M., Wheaton, S. (1993) "GIS and Storm-Water Management,"*Civil Engineering,* Vol. 63, Iss 9, pp. 72-73.

Peled, A., Adler, R. (1993) "A Common Database for Digital Mapping and GIS," *IJGIS,* Vol. 7, Iss 5, pp. 425-434.

Pereira, J.M.C., Duckstein, L. (1993) "A Multiple Criteria Decision-Making Approach to GIS-Based Land Suitability Evaluation," *IJGIS,* Vol. 7, Iss 5, pp. 407-424.

Pereira, J.M.C., Itami, R.M. (1991) "GIS-Based Habitat Modelling Using Logistic Multiple-Regression—A Study of the Mt Graham Red Squirrel," *PERS* Vol. 57, Iss 11, pp. 1475-1486.

Perera, L.K., Tateishi, R. (1995) "Do Remote-Sensing and GIS Have a Practical Applicability in Developing-Countries—(Including Some Sri-Lankan Experiences)," *IJRS,* Vol. 16, Iss 1, pp. 35-51.

Perkins, C. (1994) "Quality in Map Librarianship and Documentation in the GIS Age," *Cartographic Journal,* Vol. 31, Iss 2, pp. 93-99.

Pesachowitz, A.M. (1992) "Using Geographic Information Systems (GIS) for Environmental Decision-Making," *Abstracts of Papers of the American Chemical Society* Vol. 204, Iss AUG, pp. 83-ENVR.

Petersen, J.F. (1987) "An Inexpensive Videotape About Geographic Information Systems" *Journal of Geography,* Vol. 86, Iss 1, pp. 34-34.

Petit, J. (1994) "Looking at Urban Forests with GIS Technology," *JF* Vol. 92, Iss 10, pp. 45-45.

Peuquet, D.J. (1994) "Its About Time—A Conceptual-Framework for the Representation of Temporal Dynamics in Geographic Information Systems," *Annals of the Association of American Geographers,* Vol. 84, Iss 3, pp. 441-461.

Peuquet, D.J., Bacastow, T. (1991) "Organizational Issues in the Development of Geographical Information Systems—A Case-Study of United States Army Topographic Information Automation," *IJGIS,* Vol. 5, Iss 3, pp. 303-319.

Pickles, J. (1992)"Geographic Information Systems and Their Socioeconomic Applications", by D. Martin, *Environment and Planning D-Society and Space,* Vol. 10, Iss 5, pp. 597-600.

Pickles, J. (1997) "Tool or Science—GIS, Technoscience, and the Theoretical Turn," *Annals of the Association of American Geographers,* Vol. 87, Iss 2, pp. 363-372.

Poiani, K.A., Bedford, B.L. (1995) "GIS Based Nonpoint-Source Pollution Modelling—Considerations for Wetlands," *Journal of Soil and Water Conservation,* Vol. 50, Iss 6, pp. 613-619.

Pollak, L.M., Corbett, J.D. (1994) "Using GIS Datasets to Classify Maize-Growing Regions in Mexico and Central-America," *Agronomy Journal,* 1993, Vol. 85, Iss 6, pp. 1133-1139.

Pollock, R.J., Mclaughlin, J.D. (1991) "Database Management System Technology and Geographic Information Systems," *Journal of Surveying Engineering-ASCE,* Vol. 117, Iss 1, pp. 9-26.

Polydorides, N.D. (1994) "GIS-Related Developments in Europe," *Social Science Computer Review,* Vol. 12, Iss 4, pp. 515-531.

Populus, J., Moreau, F., Coquelet, D., Xavier, J.P. (1995) "An Assessment of Environmental Sensitivity to Marine Pollutions—Solutions with Remote-Sensing and Geographical Information Systems (GIS)," *IJRS* Vol. 16, Iss 1, pp. 3-15.

Poulsen, M.F. (1994) "Human-Geography and Its Relationship to GIS—A Perspective from the Vantage Point of GIS," *Australian Geographer,* Vol. 25, Iss 2, pp. 170-177.

Pratt, Monica (2000), Works for labor information analysts, *ArcUser,* April-June, pp. 14-6.

Prevedel, D.A. (1995) "Project Sparkey—A Strategic Wildfire Monitoring Package Using AVHRR Satellites and GIS," *PERS* Vol. 61, Iss 3, pp. 271-278.

Priestnall, G. (1997) "GIS—A Computing Perspective", by M.F. Worboys, *Transactions of the Institute of British Geographers,* Vol. 22, Iss 2, pp. 271-272.

Probert, M. (1995) "Innovations in GIS-2—Selected Papers from the 2nd National Conference on GIS Research UK", by P. Fisher, *Cartographic Journal,* Vol. 32, Iss 2, pp. 187-188.

Prosperi, D.C. (1997) "GIS and Organizations", by H. Campbell, I. Masser, *Journal of the American Planning Association,* Vol. 63, Iss 3, pp. 414-414.

Puffer, R.L. (1994) "Closing with the Enemy—How GIS Fought the War in Europe, 1944-1945", by M.D. Doubler, *Library Journal,* Vol. 119, Iss 19, pp. 76-76.

Quattrochi, D.A. (1993) "The Need for a Lexicon of Scale Terms in Integrating Remote-Sensing Data with Geographic Information Systems," *Journal of Geography,* Vol. 92, Iss 5, pp. 206-212.

Qureshi, M.A., Khan, S. (1994) "Environmental Pollution Information System Using Remote-Sensing and GIS—Report on Training Program," *Journal of Scientific and Industrial Research,* Vol. 53, Iss 4, pp. 275-282 -R-.

Raafat, H.M., Xiao, Q.H., Gauthier, D.A. (1991) "An Extended Relational Database for Remotely Sensed Image Data Management Within GIS," *IEEE Transactions on Geoscience and Remote Sensing,* Vol. 29, Iss 4, pp. 651-655.

Ramasubramanian, L. (1995) "Building Communities—GIS and Participatory Decision-Making," *Journal of Urban Technology,* Vol. 3, Iss 1, pp. 67-79.

Rao, M.N. (1991) "Introductory Readings in Geographic Information Systems", by D.J. Peuquet, D.F. Marble, *PG,* Vol. 43, Iss 4, pp. 547-548.

Raper, J. (1993) "Geographical Information Systems," *Progress in Physical Geography,* Vol. 17, Iss 4, pp. 493-502.

Raper, J. (1993) "Time in Geographic Information Systems", by G. Langran, *Environment and Planning B-Planning and Design,* Vol. 20, Iss 6, pp. 734-735.

Raper, J., Green, N. (1989) "The Development of a Hypertext-Based Tutor for Geographical Information Systems," *British Journal of Educational Technology,* Vol. 20, Iss 3, pp. 164-172.

Raper, J., Green, N. (1992) "Teaching the Principles of GIS—Lessons from the GIStutor Project," *IJGIS,* Vol. 6, Iss 4, pp. 279-290.

Raper, J.F. (1992) "Map-II Map Processor—A Geographic Information System for the Macintosh, Vol. 1-2", by M. Pazner, K.C. Kirby, N. Thies, *Geographical Journal,* Vol. 158, Iss JUL, pp. 240-240.

Raubal, M., Gaupmann, B., Kuhn, W. (1997) "Teaching Raster GIS Operations with Spreadsheets," *Journal of Geography,* Vol. 96, Iss 5, pp. 258-263.

Rhind, D. (1992), Data Access, Charging and Copyright and Their Implications for Geographic Information Systems, *International Journal on Geographical Information System,* Vol. 6:1, pp. 13-30.

Rhind, D. (1991) "Geographical Information Systems and Environmental Problems," *International Social Science Journal,* Vol. 43, Iss 4, pp. 649-668.

Rhind, D. (1992) "Data Access, Charging and Copyright and Their Implications for Geographic Information Systems," *IJGIS,* Vol. 6, Iss 1, pp. 13-30.

Rhind, D., Openshaw, S., Armstrong, P. (1988) "The Domesday Machine—A Nationwide Geographical Information System," *Geographical Journal,* Vol. 154, Iss MAR, pp. 56-68.

Rhind, D.W. (1986) "Remote-Sensing, Digital Mapping, and Geographical Information Systems—The Creation of National Policy in the United Kingdom," *Environment and Planning C-Government and Policy,* Vol. 4, Iss 1, pp. 91-102.

Rhoads, A.F., Thompson, L. (1992) "Integrating Herbarium Data into a Geographic Information System—Requirements for Spatial-Analysis," *TAXON,* Vol. 41, Iss 1, pp. 43-49.

Rice, K.W. (1992) "Introductory Readings in Geographic Information Systems", by D.J. Pevquet, D.F. Marble, *CGIS,* Vol. 19, Iss 2, pp. 105-106.

Richards, C., Host, G. (1994) "Examining Land-Use Influences on Stream Habitats and Macroinvertebrates—A GIS Approach," *Water Resources Bulletin,* Vol. 30, Iss 4, pp. 729-738.

Rifai, H.S., Hendricks, L.A., Kilborn, K., Bedient, P.B. (1993) "A Geographic Information System (GIS) User Interface for Delineating Wellhead Protection Areas," *Ground Water,* Vol. 31, Iss 3, pp. 480-488.

Rifai, H.S., Newell, C.J., Bedient, P.B. (1993) "Getting to the Nonpoint-Source with GIS," *Civil Engineering,* Vol. 63, Iss 6, pp. 44-46.

Ringrose, S., Vanderpost, C., Matheson, W. (1996) "The Use of Integrated Remotely-Sensed and GIS Data to Determine Causes of Vegetation Cover Change in Southern Botswana," *Applied Geography,* Vol. 16, Iss 3, pp. 225-242.

Robinson, G.M. (1992) "Geographical Information Systems (GIS) and Planning in Scotland—The Rluis Project and Advances in GIS," *Scottish Geographical Magazine,* Vol. 108, Iss 1, pp. 22-28.

Robinson, G.M. (1992) "Geographical Information Systems (GIS) and Planning in Scotland—The Rluis Project and Advances in GIS," *Scottish Geographical Magazine,* Vol. 108, Iss 1, pp. 22-28.

Robinson, G.M., Furley, P.A., Gray, D.A., Healey, R.G. (1989) "Developing a Geographical Information System (GIS) for Agricultural-Development in Belize, Central-America," *Applied Geography,* Vol. 9, Iss 2, pp. 81-94.

Robinson, G.M., Furley, P.A., Gray, D.A., Healey, R.G. (1989) "Developing a Geographical Information System (GIS) for Agricultural-Development in Belize, Central-America," *Applied Geography,* Vol. 9, Iss 2, pp. 81-94.

Robinson, V.B., Frank, A.U. (1987) "Expert Systems for Geographic Information Systems," *PERS,* Vol. 53, Iss 10, pp. 1435-1441.

Robinson, V.B., Frank, A.U., Blaze, M.A. (1986) "Expert Systems and Geographic Information Systems—Review and Prospects," *Journal of Surveying Engineering-ASCE,* Vol. 112, Iss 2, pp. 119-130.

Rocheleau, B. (1992) "An Introduction to Urban Geographic Information Systems, by W.E. Huxhold," *Public Administration Review,* Vol. 52, Iss 4, pp. 398-400.

Roche, S.C., Gittings, B.M. (1996) "Parallel Polygon Line Shading—The Quest for More Computational Power from an Existing GIS Algorithm," *IJGIS,* Vol. 10, Iss 6, pp. 731-746.

Rogerson, R.J. (1992) "Teaching Generic GIS Using Commercial Software," *IJGIS,* Vol. 6, Iss 4, pp. 321-331.

Rogowski, A.S. (1995) "Quantifying Soil Variability in GIS Applications.1. Estimates of Position," *IJGIS,* Vol. 9, Iss 1, pp. 81-94.

Rogowski, A.S. (1996) "Quantifying Soil Variability in GIS Applications.2. Spatial-Distribution of Soil Properties," *IJGIS,* Vol. 10, Iss 4, pp. 455-475.

Roseberry, J.L., Richards, B.J., Hollenhorst, T.P. (1994) "Assessing the Potential Impact of Conservation Reserve Program Lands on Bobwhite Habitat Using Remote-Sensing, GIS, and Habitat Modelling," *PERS* Vol. 60, Iss 9, pp. 1139-1143.

Rosenthal, W.D., Srinivasan, R., Arnold, J.G. (1995) "Alternative River Management Using a Linked GIS-Hydrology Model," *Transactions of the ASAE* Vol. 38, Iss 3, pp. 783-790.

Roy, P.S. (2005),. Standards and Interoperability for Creating a Common Code, *Geospatial Today,* Vol.4, issue 2, pp. 40-3.

Rutten, B.J.C.M., Vonbeek, A., Vanderhaspel, B. (1988) "GISwa—Geographical Information System Wind Resource," *Journal of Wind Engineering and Industrial Aerodynamics,* Vol. 27, Iss 1-3, pp. 103-112.

Sader, S.A., Ahl, D., Liou, W.S. (1995) "Accuracy of Landsat-TM and GIS Rule-Based Methods for Forest Wetland Classification in Maine," *Remote Sensing of Environment* Vol. 53, Iss 3, pp. 133-144.

Sadler, J. (1992) "Interpreting Space—GIS and Archaeology", by K.M.S. Allen, S.W. Green, E.B.W. Zubrow, *IJGIS,* Vol. 6, Iss 1, pp. 60-61.

Sahay, S., Walsham, G. (1996) "Implementation of GIS in India—Organizational Issues and Implications," *IJGIS,* Vol. 10, Iss 4, pp. 385-404.

Sanders, P.F. Thornton, L.L., Louis, J.B., Ivahnenko, T., Evans, B.M. (1991)"Using a Geographical Information System to Assess the Susceptibility of Surface Drinking-Water Supplies to Pesticide Contamination," *Abstracts of Paper of the American Chemical Society,* Vol. 2 02, Iss AUG, pp. 53.

Savabi, M.R., Flanagan, D.C., Hebel, B., Engel, B.A. (1995) "Application of Wepp and GIS-Grass to a Small Watershed in Indiana," *Journal of Soil and Water Conservation* Vol. 50, Iss 5, pp. 477-483.

Scholten, Henk J. and Stillwell, John C.H. (1990)(eds), *Geographical Information System for Urban and Regional Planning,* Kluwer Academic Publishers, Dordrechht.

Scholten, H.J., Padding, P. (1990) "Working with Geographic Information Systems in a Policy Environment" *Environment and Planning B-Planning and Design,* Vol. 17, Iss 4, pp. 405-416.

Scholten, H.J., Vandervlugt, M. (1989) "Application of Geographic Information Systems in Europe," *EKISTICS, The Problems and Science of Human Settlements,* Vol. 56, Iss 338-, pp. 304-311.

Schreier, H., Brown, S., Kennedy, G., Shah, P.B. (1991) "Food, Feed, and Fuelwood Resources of Nepal—A GIS Evaluation," *Environmental Management,* Vol. 15, Iss 6, pp. 815-822.

Schreier, H., Brown, S., Schmidt, M., Shah, P., Shrestha, B., Nakarmi, G., Subba, K., Wymann, S. (1994) "Gaining Forests But Losing Ground—A GIS Evaluation in a Himalayan Watershed," *Environmental Management,* Vol. 18, Iss 1, pp. 139-150.

Schreier, H., Shah, P.B., Kennedy, G. (1990) "Evaluating Mountain Watersheds in Nepal Using Micro-GIS," *Mountain Research and Development,* Vol. 10, Iss 2, pp. 151-159.

Scott, L.M. (1994) "Identification of GIS Attribute Error Using Exploratory Data-Analysis," *PG,* Vol. 46, Iss 3, pp. 378-386.

See, R.B., Naftz, D.L., Qualls, C.L. (1992) "GIS-Assisted Regression-Analysis to Identify Sources of Selenium in Streams," *Water Resources Bulletin* Vol. 28, Iss 2, pp. 315-330.

Sellars, A.G., Farish, O., Hampton, B.F. (1994) "Assessing the Risk of Failure Due to Particle Contamination of GIS Using the UHF Technique," *IEEE Transactions on Dielectruics and Electrical Insulation* Vol. 1, Iss 2, pp. 323-331.

Sellars, A.G., Farish, O., Hampton, B.F. (1994) "Characterizing the Discharge Development Due to Surface Contamination in GIS Using the UHF Technique," *IEE Proceedings-Science Measurement and Technology,* Vol. 141, Iss 2, pp. 118-122.

Shaffer, C.A. (1992) "Data Representations for Geographic Information Systems," *Annual Review of Information Science and Technology,* Vol. 27, pp. 135-172.

Sharma, P. (1993) "Geographic Information Systems—Developments and Applications", by L. Worrall, *Australian Geographer,* Vol. 24, Iss 1, pp. 86-88.

Sharma, S.K., Anjaneyulu, D. (1994) "Application of Remote Sensing and GIS in Water-Resource Management," *IJRS* 1993, Vol. 14, Iss 17, pp. 3209-3220.

Shaw, S.L. (1993) "Geography—A Place for GIS," *Applied Geography,* Vol. 13, Iss 2, pp. 107-110.

Shaw, S.L. (1993) "GIS for Urban Travel Demand Analysis—Requirements and Alternatives," *Computers Environment and Urban Systems,* Vol. 17, Iss 1, pp. 15-29.

Shea, C., Grayman, W., Darden, D., Males, R.M., Sushinsky, P. (1993) "Integrated GIS and Hydrologic Modelling for Countywide Drainage Study," *Journal of Water Resources Planning and Management-ASCE* Vol. 119, Iss 2, pp. 112-128.

Shelton, R.L., Estes, J.E. (1981) "Remote-Sensing and Geographic Information Systems—An Unrealized Potential," *Geo-Processing,* Vol. 1, Iss 4, pp. 395-420.

Shennan, S.J. (1993) "Interpreting Space—GIS and Archaeology", by K.M.S. Allen, S.W. Green, E.B.W. Zubrow, *Journal of Archaeological Science,* Vol. 20, Iss 2, pp. 237-238.

Shih, S.F., Jordan, J.D. (1992) "Landsat Midinfrared Data and GIS in Regional Surface Soil-Moisture Assessment," *Water Resources Bulletin,* Vol. 28, Iss 4, pp. 713-719.

Shih, S.F., Jordan, J.D. (1993) "Landsat Midinfrared Data and GIS in Regional Surface Soil-Moisture Assessment—Discussion," *Water Resources Bulletin,* Vol. 29, Iss 2, pp. 313-314.

Shih, S.F., Jordan, J.D. (1993) "Use of Landsat Thermal-Ir Data and GIS in Soil-Moisture Assessment," *Journal of Irrigation and Drainage Engineering-ASCE* Vol. 119, Iss 5, pp. 868-879.

Short, N.M., Dickens, L. (1995) "Automatic-Generation of Products from Terabyte-Size Geographical Information Systems Using Planning and Scheduling" *IJGIS,* Vol. 9, Iss 1, pp. 47-65.

Siva Kumar, R. (2005), Standards and Interoperability for an Indian standard, *Geospatial Today,* Vol.4, issue 2, pp. 54-56.

Skidmore, A.K., Gauld, A., Walker, P. (1996) "Classification of Kangaroo Habitat Distribution Using 3 GIS Models," *IJGIS,* Vol. 10, Iss 4, pp. 441-454.

Skidmore, A.K., Ryan, P.J., Dawes, W., Short, D., Oloughlin, E. (1991) "Use of an Expert System to Map Forest Soils from a Geographical Information System," *IJGIS,* Vol. 5, Iss 4, pp. 431-445.

Sleight, P. (1996) "GIS for Business and Service Planning", by P. Longley, G. Clarke, *IJGIS,* Vol. 10, Iss 5, pp. 664-665.

Slocum, T.A. (1991) "Geographic Information Systems and Cartographic Modelling", by C.D. Tomlin, *PG,* Vol. 43, Iss 1, pp. 126-126.

Smith, C.L., Steel, B.S., List, P.C., Cordray, S. (1995) "Making Forest Policy—Integrating GIS with Social Processes," *JF* Vol. 93, Iss 5, pp. 31-36.

Smith, D.A., Tomlinson, R.F. (1992) "Assessing Costs and Benefits of Geographical Information Systems—Methodological and Implementation Issues," *IJGIS,* Vol. 6, Iss 3, pp. 247-256.

Smith, J.L., Logan, J.A., Gregoire, T.G. (1992) "Using Aerial-Photography and Geographic Information Systems to Develop Databases for Pesticide Evaluations," *PERS*, Vol. 58, Iss 10, pp. 1447-1452.

Smith, M.B., Brilly, M. (1992) "Automated Grid Element Ordering for GIS-Based Overland-Flow Modelling," *PERS* Vol. 58, Iss 5, pp. 579-585.

Smith, M.B., Vidmar, A. (1994) "Data Set Derivation for GIS-Based Urban Hydrological Modelling," *PERS* Vol. 60, Iss 1, pp. 67-76.

Sombroek, W.G., Antoine, J. (1994) "The Use of Geographic Information Systems (GIS) in Land Resources Appraisal," *Outlook on Agriculture,* Vol. 23, Iss 4, pp. 249-255.

Somers, R. (1987) "Geographic Information Systems in Local-Government—A Commentary," *PERS*, Vol. 53, Iss 10, pp. 1379-1382.

Sprunt, B. (1988) "A UK Geographic Information System for Environmental Monitoring, Resource Planning and Management Capable of Integrating and Using Satellite Remotely Sensed Data", by J.A.T. Young, *Cartographic Journal,* Vol. 25, Iss 1, pp. 69-70.

Srinivasan, A., Richards, J.A. (1993) "Analysis of GIS Spatial Data Using Knowledge-Based Methods," *IJGIS,* Vol. 7, Iss 6, pp. 479-500.

Srinivasan, R., Arnold, J.G. (1994) "Integration of a Basin-Scale Water-Quality Model with GIS," *Water Resources Bulletin,* Vol. 30, Iss 3, pp. 453-462.

Staebler, R.N. (1992) "Natural-Resource Management and GIS," *JF,* Vol. 90, Iss 12, pp. 2-2.

Star, J.L., Estes, J.E., Davis, F. (1991)"Improved Integration of Remote-Sensing and Geographic Information Systems—A Background to Ncgia Initiative 12," *PERS*, Vol. 57, Iss 6, pp. 643-645.

Steinke, T.R. (1990) "3-Dimensional Applications in Geographical Information Systems", by J. Raper, *CGIS,* Vol. 17, Iss 2, pp. 177-178.

Stillwell, J. (1992) "Geographic Information Systems—Developments and Applications", by L. Worrall, *IJGIS,* Vol. 6, Iss 1, pp. 56-59.

Stillwell, J. (1992) "Geographical Information Systems and Urban and Rural Planning", by T.W. Rideout, *IJGIS,* Vol. 6, Iss 5, pp. 445-447.

Stockwell, J.R., Sorensen, J.W., Eckert, J.W., Carreras, E.M. (1993) "The United-States EPA Geographic Information System for Mapping Environmental Releases of Toxic-Chemical Release Inventory (TRI) Chemicals," *Risk Analysis,* Vol. 13, Iss 2, pp. 155-164.

Stoms, D.M., Davis, F.W., Cogan, C.B. (1992) "Sensitivity of Wildlife Habitat Models to Uncertainties in GIS Data," *PERS* Vol. 58, Iss 6, pp. 843-850.

Stoorvogel, J.J. (1995) "Linking GIS and Models—Structure and Operationalization for a Costa-Rican Case Study," *Netherlands Journal of Agricultural Science* Vol. 43, Iss 1, pp. 19-29.

Strachan, A.J., Stuart, N. (1996) "UK Developments in Environmental GIS," *IJGIS,* Vol. 10, Iss 1, pp. 17-20.

Strand, E.J., Parker, H.D. (1992) "GIS Has a Bright Future in Business," *PERS* Vol. 58, Iss 8, pp. 1259-1262.

Strasser, T. (1997) "International GIS Dictionary", by R. Mcdonnell, K. Kemp, *Journal of Government Information,* Vol. 24, Iss 2, pp. 130-132.

Stuart, N. (1992) "Geographic Information Systems—Developments and Applications", by L. Worrall, *Scottish Geographical Magazine,* Vol. 108, Iss 2, pp. 134-135.

Stuart, N. (1992) "Time in Geographic Information Systems", by G. Langran, *IJGIS,* Vol. 6, Iss 6, pp. 547-547.

Stuart, N. (1995) "Visualization in Geographical Information Systems", by H. Hearnshaw, D.J. Unwin, *Applied Geography,* Vol. 15, Iss 1, pp. 91-92.

Stuebe, M.M., Johnston, D.M. (1990) "Runoff Volume Estimation Using GIS Techniques," *Water Resources Bulletin* Vol. 26, Iss 4, pp. 611-620.

Sui, D.Z. (1991) "Geographic Information System—A Management Perspective", by S. Aronoff, *PG,* Vol. 43, Iss 3, pp. 370-370.

Sui, D.Z. (1994) "GIS and Urban Studies—Positivism, Post-Positivism, and Beyond," *Urban Geography,* Vol. 15, Iss 3, pp. 258-278.

Sui, D.Z. (1997) "Reconstructing Urban Reality—From GIS to Electropolis," *Urban Geography,* Vol. 18, Iss 1, pp. 74-89.

Sussman, R. (1993) "Municipal GIS and the Enterprise Data Model," *IJGIS,* Vol. 7, Iss 4, pp. 367-377.

Takeyama, M., Couclelis, H. (1997) "Map Dynamics—Integrating Cellular-Automata and GIS Through Geo-Algebra," *IJGISC,* Vol. 11, Iss 1, pp. 73-91.

Tan, Y.R., Shih, S.F. (1990) "GIS in Monitoring Agricultural Land-Use Changes and Well Assessment," *Transactions of the ASAE* Vol. 33, Iss 4, pp. 1147-1152.

Tappan, G.G., Moore, D.G., Knausenberger, W.I. (1991) "Monitoring Grasshopper and Locust Habitats in Sahelian Africa Using GIS and Remote-Sensing Technology," *IJGIS,* Vol. 5, Iss 1, pp. 123-135.

Tarboton, K.C. (1992) "Interfacing GIS and Hydrological Modelling—Mgeni Case-Study," *Water SA* Vol. 18, Iss 4, pp. 273-278.

Taylor, D.R.F. (1991) "GIS in Developing-Nations," *IJGIS,* Vol. 5, Iss 1, pp. 3-3.

Taylor, P.J., Overton, M. (1991) "Further Thoughts on Geography and GIS—A Preemptive Strike," *Environment and Planning A,* Vol. 23, Iss 8, pp. 1087-1090.

Theocharopoulos, S.P., Davidson, D.A., Mcarthur, J.N., Tsouloucha, F. (1995) "GIS as an Aid to Soil Surveys and Land Evaluation in Greece," *Journal of Soil and Water Conservation* Vol. 50, Iss 2, pp. 118-124.

Thomas, A. (1992) "Agricultural Water-Balance of Yunnan Province, Pr China -Agroclimatic Zoning with a Geographical Information System," *Agricultural Water Management,* Vol. 21, Iss 4, pp. 249-263.

Thorpe, J.A., Gilbert, D.M. (1987) "AM/FM Requirements of Geographic Information Systems—A Commentary," *PERS,* Vol. 53, Iss 10, pp. 1403-1404.

Tim, U.S. (1995) "The Application of GIS in Environmental-Health Sciences—Opportunities and Limitations," *Environmental Research,* Vol. 71, Iss 2, pp. 75-88.

Tobin, R. (1989) "Introducing Geographic Information Systems to a Small Municipality," *EKISTICS-The Problems and Science of Human Settlements,* Vol. 56, Iss 338-, pp. 296-299.

Tobler, W. (1995) "The Resel-Based GIS," *IJGIS,* Vol. 9, Iss 1, pp. 95-100.

Tomlinson, R.F. (1989) "Geographic Information Systems and Geographers in the 1990s," *Canadian Geographer-Geograhe Canadien,* Vol. 33, Iss 4, pp. 290-298.

Towers, G. (1997) "GIS Versus the Community—Siting Power in Southern West-Virginia," *Applied Geography,* Vol. 17, Iss 2, pp. 111-125.

Treitz, P., Elliot, L., Howarth, P. (1993) "Differential Global Positioning System—Potential for Geographical Information System Database Management" *Environment and Planning A,* Vol. 25, Iss 6, pp. 883-898.

Treitz, P.M., Howarth, P.J., Gong, P. (1992) "Application of Satellite and GIS Technologies for Land-Cover and Land-Use Mapping at the Rural-Urban Fringe—A Case-Study," *PERS,* Vol. 58, Iss 4, pp. 439-448.

Trotter, C.M. (1991) "Remotely-Sensed Data as an Information Source for Geographical Information Systems in Natural-Resource Management—A Review," *IJGIS,* Vol. 5, Iss 2, pp. 225-239.

Tudor, G.S., Sugarbaker, L.J. (1993) "GIS Orthographic Digitizing of Aerial Photographs by Terrain Modelling," *PERS* Vol. 59, Iss 4, pp. 499-503.

Twigg, L. (1990) "Health Based Geographical Information Systems—Their Potential Examined in the Light of Existing Data Sources," *Social Science and Medicine,* Vol. 30, Iss 1, pp. 143-155.

Unwin, D. (1993) "Geographic Information Systems and Their Socioeconomic Applications", by D. Martin, *Progress in Human Geography,* Vol. 17, Iss 1, pp. 147-148.

Unwin, D. (1994) "Cartography, Visc and GIS," *Progress in Human Geography,* Vol. 18, Iss 4, pp. 516-522.

Unwin, D., Dale, P. (1990) "An Educationalists View of GIS—Some Educational and Sociological Concerns," *Journal of Geography in Higher Education*, Vol. 14, Iss 2, pp. 166-170.

Unwin, D.J. (1992) "Introductory Readings in Geographic Information Systems", by D.J. Peuquet, D.F. Marble, *AREA,* Vol. 24, Iss 2, pp. 189-190.

Unwin, D.J. (1996) "GIS, Spatial-Analysis and Spatial Statistics," *Progress in Human Geography,* Vol. 20, Iss 4, pp. 540-551.

Utsumi, T., Endo, F., Ishikawa, T., Iwaasa, S., Yamagiwa, T., Diessner, A., Gorablenkow, J., Hashoff, L. (1993) "Preventive Maintenance System with a Different Gas Injecting Facility for GIS," *IEEE Transactions of Power Delivery* Vol. 8, Iss 3, pp. 1107-1113.

Valenzuela, C.R. (1989) "The Digital Geographic Information System As a Management Tool," *Impact of Science on Society,* Vol. 39, Iss 4, pp. 315-324.

Vancleynenbreugel, J., Fierens, F., Suetens, P., Oosterlinck, A. (1990) "Delineating Road Structures on Satellite Imagery by a GIS-Guided Technique," *PERS* Vol. 56, Iss 6, pp. 893-898.

Vanderknaap, W.G.M. (1992) "The Vector to Raster Conversion—(MIS) Use in Geographical Information Systems," *IJGIS,* Vol. 6, Iss 2, pp. 159-170.

Vangroenigen, J.W., Stein, A., Zuurbier, R. (1997) "Optimization of Environmental Sampling Using Interactive GIS," *Soil Technology,* Vol. 10, Iss 2, pp. 83-97.

Vanlanen, H.A.J., Vandiepen, C.A., Reinds, G.J., Dekoning, G.H.J., Bulens, J.D., Bregt, A.K. (1992) "Physical Land Evaluation Methods and GIS to Explore the Crop Growth-Potential and Its Effects Within the European Communities," *Agricultural Systems,* Vol. 39, Iss 3, pp. 307-328.

Vaughan, P.J., Corwin, D.L. (1994) "A Method of Modelling Vertical Fluid-Flow and Solute Transport in a GIS Context," *GEODERMA,* Vol. 64, Iss 1-2, pp. 139-154.

Verbyla, D.L. (1993) "Landsat Midinfrared Data and GIS in Regional Surface Soil-Moisture Assessment—Discussion," *Water Resources Bulletin* Vol. 29, Iss 2, pp. 309-311.

Veregin, H. (1994) "Integration of Simulation Modelling and Error Propagation for the Buffer Operation in GIS," *PERS* Vol. 60, Iss 4, pp. 427-435.

Vertinsky, I., Brown, S., Schreier, H., Thompson, W.A., Vankooten, G.C. (1994) "A Hierarchical-GIS-Based Decision-Model for Forest Management—The Systems-Approach," *Interfaces,* Vol. 24, Iss 4, pp. 38-53.

Viet, C.P., Phuong, M.N. (1993) "Natural Resources Evaluation by the Use of Remote Sensing and GIS Technology for Agricultural-Development," *Advances in Sapce Research,* Vol. 13, Iss 11, pp. 117-121.

Villa, F., Ceroni, M., Mazza, A. (1996) "A GIS-Based Method for Multiobjective Evaluation of Park Vegetation," *Landscape and Urban Planning,* Vol. 35, Iss 4, pp. 203-212.

Vincent, P. (1993) "Geographical Information Systems, Principles and Applications", by D.J. Maguire, M.F. Goodchild, D.W. Rhind, *Progress in Human Geography,* Vol. 17, Iss 3, pp. 439-440.

Visvalingam, M. (1989) "Cartography, GIS and Maps in Perspective," *Cartographic Journal,* Vol. 26, Iss 1, pp. 26-32.

Visvalingam, M. (1992) "GIS in Malaysia," *Cartographic Journal,* Vol. 29, Iss 1, pp. 55-56.

Wadge, G. (1992) "Geological Applications of GIS," *Journal of the Geological Society,* Vol. 149, Iss Jul., pp. 672-672.

Wagner, D.E. (1995) "Reactive Data-Structures for Geographic Information Systems", by P.J.M. Vanoosterom, *PG,* Vol. 47, Iss 1, pp. 113-114.

Walford, N., Lane, M., Shearman, J. (1989) "The Rural-Areas Database—A Geographical Information and Mapping System for the Management and Integration of Data on Rural Britain," *Transactions-Institute of British Geographers,* Vol. 14, Iss 2, pp. 221-230.

Walker, D.J., Painter, K.M. (1994) "Measuring Rate and Onsite Damage Using GIS—A Policy Application," *American Journal of Agricultural Economics,* Vol. 76, Iss 5, pp. 1258-1258.

Walker, D.R.F., Newman, I.A., Medyckyjscott, D.J., Ruggles, C.L.N. (1992) "A System for Identifying Datasets for GIS Users," *IJGIS,* Vol. 6, Iss 6, pp. 511-527.

Walker, G. (1992) "GIS and Urban and Rural Planning," *AREA,* Vol. 24, Iss 2, pp. 230-231.

Walpole, S.C., Sinden, J.A. (1997) "BCA and GIS—Integration of Economic and Environmental Indicators to Aid."

Walsh, S.J. (1985) "Geographic Information Systems for Natural Resource Management," *Journal of Soil and Water Conservation,* Vol. 40, Iss 2, pp. 202-205.

Walsh, S.J. (1988) "Geographic Information Systems—An Instructional Tool for Earth Science Educators," *Journal of Geography,* Vol. 87, Iss 1, pp. 17-25.

Walsh, S.J. (1992) "Spatial Education and Integrated Hands-on Training—Essential Foundations of GIS Instruction," *Journal of Geography,* Vol. 91, Iss 2, pp. 54-61.

Walsh, S.J., Cooper, J.W., Vonessen, I.E., Gallager, K.R. (1990) "Image-Enhancement of LANDSAT Thematic Mapper Data and GIS Data Integration for Evaluation of Resource Characteristics," *PERS* Vol. 56, Iss 8, pp. 1135-1141.

Walsh, S.J., Lightfoot, D.R., Butler, D.R. (1987) "Recognition and Assessment of Error in Geographic Information Systems," *PERS,* Vol. 53, Iss 10, pp. 1423-1430.

Wang, F. (1994) "The Use of Artificial Neural Networks in a Geographical Information System for Agricultural Land Suitability Assessment," *Environment and Planning A,* Vol. 26, Iss 2, pp. 265-284.

Wang, F.J., Hall, G.B. (1996) "Fuzzy Representation of Geographical Boundaries in GIS," *IJGIS,* Vol. 10, Iss 5, pp. 573-590.

Wang, J.F., Treitz, P.M., Howarth, P.J. (1992) "Road Network Detection from Spot Imagery for Updating Geographical Information Systems in the Rural Urban Fringe," *IJGIS,* Vol. 6, Iss 2, pp. 141-157.

Wanless, B. (1992) "GPS Positioned Digital Video for Airborne GIS Data Acquisition," *Journal of Surveying Engineering—ASCE* Vol. 118, Iss 3, pp. 80-89.

Warwick, J.J., Haness, S.J. (1994) "Efficacy of Arc/Info GIS Application to Hydrologic Modelling," *Journal of Water Resources Planning and Management-ASCE* Vol. 120, Iss 3, pp. 366-381.

Webster, C.J. (1993) "GIS and the Scientific Inputs to Urban-Planning.1. Description," *Environment and Planning B-Planning and Design,* Vol. 20, Iss 6, pp. 709-728.

Webster, C.J. (1994) "GIS and the Scientific Inputs to Planning.2. Prediction and Prescription," *Environment and Planning B-Planning and Design,* Vol. 21, Iss 2, pp. 145-157.

Webster, C.J. (1994) "Structured Methods for GIS Design.1. A Relational System for Physical Plan Monitoring," *Computers Environment and Urban Systems,* Vol. 18, Iss 1, pp. 1-18.

Webster, C.J. (1997) "GIS and Organizations—How Effective Are GIS in Practice", by H. Campbell, I. Masser, *Environment and Planning B-Planning and Design,* Vol. 24, Iss 4, pp. 636-638.

Wehmeyer, M.G. (1994) "Integrating SCADA Systems with Am/FM-GIS Technology," *Water-Engineering and Management* Vol. 141, Iss 7, pp. 18+.

Weibel, R., Buttenfield, B.P. (1992) "Improvement of GIS Graphics for Analysis and Decision-Making," *IJGIS,* Vol. 6, Iss 3, pp. 223-245.

Welborn, H., Wagner, J., Weber, D. (1990) "Trends in Choosing a Geographic Information System," *Water-Engineering and Management,* Vol. 137, Iss 11, pp. 20-23.

Welch, R., Remillard, M., Alberts, J. (1992) "Integration of GPS, Remote-Sensing, and GIS Techniques for Coastal Resource-Management," *PERS* Vol. 58, Iss 11, pp. 1571-1578.

Welch, R., Remillard, M., Doren, R.F. (1995) "GIS Database Development for South Florida National-Parks and Preserves," *PERS* Vol. 61, Iss 11, pp. 1371-1381.

Wentz, E.A. (1996) "Innovations in GIS—Selected Papers from the First National Conference on GIS Research UK", by M.F. Worboys, *Annals of the Association of American Geographers,* Vol. 86, Iss 2, pp. 376-377.

Wheeler, J.O. (1994) "GIS and Critical Social-Theory—Whither Urban-Geography," *Urban Geography,* Vol. 15, Iss 2, pp. 103-105.

Wheeler, J.O. (1997) "Quantitative-Analysis and GIS in Urban-Geography—Then and Now," *Urban Geography,* Vol. 18, Iss 2, pp. 95-99.

White, K.L., Simms, M. (1993) "Geographic Information Systems as an Educational-Tool," *Journal of Geography,* Vol. 92, Iss 2, pp. 80-85.

Wiggins, L.L., Ferreira, J. (1992) "MITS Computer Resource Lab—A Research and Educational Facility for GIS in Urban-Planning," *IJGIS,* Vol. 6, Iss 4, pp. 299-307.

Wiggins, L.L., French, S.P. (1990) "Desk-Top Mapping and GIS," *Journal of the American Planning Association,* Vol. 56, Iss 3, pp. 370-373.

Wilkinson, G.G. (1996) "A Review of Current Issues in the Integration of GIS and Remote-Sensing Data," *IJGIS,* Vol. 10, Iss 1, pp. 85-101.

Williams, D.C., Lyon, J.G. (1991) "Use of a Geographic Information System Data Base to Measure and Evaluate Wetland Changes in the St-Marys River, Michigan," *Hydrobiologia,* Vol. 219, Iss JUL, pp. 83-95.

Williams, G.J. (1995) "Templates for Spatial Reasoning in Responsive Geographical Information Systems," *IJGIS,* Vol. 9, Iss 2, pp. 117-131.

Williamson, A.N. (1992) "Map Scanning for GIS Applications," *PERS* Vol. 58, Iss 8, pp. 1199-1202.

Wise, S. (1992) "Geographic Information Systems—Developments and Applications", by L. Worrall, *Applied Geography,* Vol. 12, Iss 1, pp. 85-85.

Wise, S. (1992) "Introductory Readings in Geographical Information Systems", by D.J. Peuquet, D.F. Marble, *Progress in Human Geography,* Vol. 16, Iss 3, pp. 474-476.

Wise, S. (1993) "Geographic Information Systems and Their Socioeconomic Applications", by D. Martin, *Geographical Journal,* Vol. 159, Iss JUL, pp. 229-231.

Wise, S. (1993) "Geographical Information Systems, Vol. 1, Principles, Vol. 2, Applications", by D.J. Maguire, M.F. Goodchild, D.W. Rhind, *Geography,* Vol. 78, Iss 338, pp. 89-90.

Wise, S. (1993) "Geographical Information Systems—Principles and Applications", by D. Maguire, M. Goodchild, D. Rhind, *Geographical Journal,* Vol. 159, Iss JUL, pp. 229-231.

Wise, S. (1996) "Innovations in GIS-2—Selected Papers from the 2nd National Conference on GIS Research UK", by P. Fisher, *IJGIS,* Vol. 10, Iss 2, pp. 241-242.

Wise, S.M. (1993) "Geographic Information Systems—A Guide to the Technology", by A.N. Antenucci, K. Brown, P.L. Croswell, M.J. Kevany, *Applied Geography,* Vol. 13, Iss 1, pp. 100-101.

Wong, D. (1997) "Enhancing Segregation Studies Using GIS," *Computers Environment and Urban Systems,* Vol. 20, Iss 2, pp. 99-109.

Wood, B., Washino, R., Beck, L., Hibbard, K., Pitcairn, M., Roberts, D., Rejmankova, E., Paris, J., Hacker, C., Salute, J., Sebesta, P., Legters, L. (1992) "Distinguishing High and Low Anopheline-Producing Rice Fields Using Remote-Sensing and GIS Technologies," *Preventive Veterinary Medicine* 1991, Vol. 11, Iss 3-4, pp. 277-288.

Woodcock, C.E., Sham, C.H., Shaw, B. (1990) "Comments on Selecting a Geographic Information System for Environmental Management," *Environmental Management,* Vol. 14, Iss 3, pp. 307-315.

Woodsford, P.A. (1991) "Cartographic and GIS Developments at Laser-Scan, 1987-91," *Cartographic Journal,* Vol. 28, Iss 1, pp. 43.

World, L. (1994) "GIS Feels the Earth Move in Los-Angeles," *IEEE Computer Graphics and Applications,* Vol. 14, Iss 3, pp. 16-16.

Worrall, Les (1990)(ed), *Geographic Information System: Development and Applications,* Taylor and Francis, London.

Worrall, L. (1994) "Justifying Investment in GIS—A Local-Government Perspective," *IJGIS,* Vol. 8, Iss 6, pp. 545-565.

Worrall, L. (1994) "The Role of GIS-Based Spatial-Analysis in Strategic Management in Local-Government," *Computers Environment and Urban Systems,* Vol. 18, Iss 5, pp. 323-332.

Wright, D.J., Goodchild, M.F. (1997) "Data from the Deep—Implications for the GIS Community," *IJGISC,* Vol. 11, Iss 6, pp. 523-528.

Wright, D.J., Goodchild, M.F., Proctor, J.D. (1997) "Demystifying the Persistent Ambiguity of GIS as Tool Versus Science," *Annals of the Association of American Geographers,* Vol. 87, Iss 2, pp. 346-362.

Wright, J. (1988) "The Plain Fellows Guide to Geographic Information Systems," *Geographical Journal,* Vol. 154, Iss JUL, pp. 161-168.

Wrigley, N. (1991) "Market-Based Systems of Health-Care Provision, the NHS Bill, and Geographical Information Systems," *Environment and Planning A,* Vol. 23, Iss 1, pp. 5-8.

Wyatt, P.J. (1997) "The Development of a GIS-Based Property Information-System for Real-Estate Valuation," *IJGISC,* Vol. 11, Iss 5, pp. 435-450.

Xiang, W.N. (1993) "Application of a GIS-Based Stream Buffer Generation Model to Environmental-Policy Evaluation," *Environmental Management,* Vol. 17, Iss 6, pp. 817-827.

Xiang, W.N. (1993) "A GIS Method for Riparian Water-Quality Buffer Generation," *IJGIS,* Vol. 7, Iss 1, pp. 57-70.

Xiang, W.N. (1993) "A GIS MMP-Based Coordination Model and Its Application to Distributed Environmental Planning," *Environment and Planning B-Planning and Design,* Vol. 20, Iss 2, pp. 195-220.

Xiang, W.N. (1996) "A GIS Based Method for Trail Alignment Planning," *Landscape and Urban Planning,* Vol. 35, Iss 1, pp. 11-23.

Xiang, W.N. (1996) "GIS-Based Riparian Buffer Analysis—Injecting Geographic Information into Landscape Planning," *Landscape and Urban Planning,* Vol. 34, Iss 1, pp. 1-10.

Xiang, W.N., Stratton, W.L. (1996) "The B-Function and Variable Stream Buffer Mapping—A Note on a GIS Method for Riparian Water-Quality Buffer Generation," *IJGIS,* Vol. 10, Iss 4, pp. 499-510.

Xu, F., Prato, T., Fulcher, C. (1993) "Broiler Litter Application to Land in an Agricultural Watershed—A GIS Approach," *Water Science and Technology,* Vol. 28, Iss 3-5, pp. 111-118.

Xu, H., Bailey, J.O., Barrett, E.C., Kelly, R.E.J. (1994) "Monitoring Snow Area and Depth with Integration of Remote-Sensing and GIS," *IJRS* 1993, Vol. 14, Iss 17, pp. 3259-3268.

Xu, J.P., Lathrop, R.G. (1995) "Improving Simulation Accuracy of Spread Phenomena in a Raster-Based Geographic Information System," *IJGIS,* Vol. 9, Iss 2, pp. 153-168.

Yaakup, A.B., Healey, R.G. (1994) "A GIS Approach to Spatial Modelling for Squatter Settlement Planning in Kuala-Lumpur, Malaysia," *Environment and Planning B-Planning and Design,* Vol. 21, Iss 1, pp. 21-34.

Yapa, L.S. (1991) "Is GIS Appropriate Technology," *IJGIS,* Vol. 5, Iss 1, pp. 41-58.

Yeh, A.G. (1991) "The Development and Applications of Geographic Information Systems for Urban and Regional-Planning in the Developing Countries," *IJGIS,* Vol. 5, Iss 1, pp. 5-27.

Yeh, A.G.O., Batty, M. (1990) "Applications of Geographic Information Systems in Urban and Regional-Planning," *Environment and Planning B-Planning and Design,* Vol. 17, Iss 4, pp. 369-374.

Yi, G.C., Risley, D., Koneff, M., Davis, C. (1994) "Development of Ohio GIS-Based Wetlands Inventory," *Journal of Soil and Water Conservation* Vol. 49, Iss 1, pp. 23-28.

Young, J.A.T. (1986) "Remote-Sensing and an Experimental Geographic Information System for Environmental Monitoring, Resource Planning and Management," *IJRS,* Vol. 7, Iss 6, pp. 741-744.

Zack, J.A., Minnich, R.A. (1991) "Integration of Geographic Information Systems with a Diagnostic Wind-Field Model for Fire Management," *Forest Science,* Vol. 37, Iss 2, pp. 560-573.

Zaima, E., Okabe, S., Nishiwaki, S., Ishikawa, M., Nakamoto, T., Murayama, Y., Kawaguchi, Y. (1990) "Reignition Surges At Reactor Current Interruption in Cable-System GIS," *IEEE Transactions of Power Delivery* Vol. 5, Iss 2, pp. 947-954.

Zech, Y., Sillen, X., Debources, C., Vanhauwaert, A. (1994) "Rainfall-Runoff Modelling of Partly Urbanized Watersheds—Comparison Between and Distributed Model Using GIS and Other Models Sensitivity Analysis," *Water Science and Technology,* Vol. 29, Iss 1-2, pp. 163-170.

Zhou, H.Z., Macdonald, K.B., Moore, A. (1991) "Some Cautions on the Use of Geographic Information System (GIS) Technology to Integrate Soil Site and Area Data," *Canadian Journal of Soil Science,* Vol. 71, Iss 3, pp. 389-394.

Zhou, J., Civco, D.L. (1996) "Using Genetic Learning Neural Networks for Spatial Decision-Making in GIS," *PERS* 1996, Vol. 62, Iss 11, pp. 1287-1295.

Zhou, Q.M. (1993) "Arcview for Windows GIS Software," *PERS* Vol. 59, Iss 7, pp. 1103-1106.

Zhu, Z.L., Evans, D.L. (1992) "Mapping Midsouth Forest Distributions—AVHRR Satellite Data and GIS Help Meet RPA Mandate," *JF* Vol. 90, Iss 12, pp. 27-30.

Zumbrunnen, C. (1991) "GIS tutor (GIST)—The Geographical Information Systems Tutor," *PG,* Vol. 43, Iss 2, pp. 235-236.

Zwarenstein, M., Krige, D., Wolff, B. (1991) "The Use of a Geographical Information System for Hospital Catchment-Area Research in Natal Kwazulu," *South African Medical Journal,* Vol. 80, Iss 10, pp. 497-500.

Zybach, B., Barrington, M., Downey, T. (1995) "Converting Historical Information to GIS—Political Boundaries of the Douglas-Fir Region, 1788 to 1995," *JF* Vol. 93, Iss 5, pp. 15-20

Abbreviations Used

IJGIS	:	International Journal of Geographical Information Systems
IJGISC	:	International Journal of Geographic Information Science
IJRS	:	International Journal of Remote Sensing
PERS	:	Phogrammmetric Engineering and Remote Sensing
Pg	:	PG
JF	:	JF

Index

(157) Concept